TECHNOLOGY AND SOCIETY

Inside Technology

edited by Wiebe E. Bijker, W. Bernard Carlson, and Trevor J. Pinch

A list of books in the series appears at the back of the book.

TECHNOLOGY AND SOCIETY

Building Our Sociotechnical Future

edited by Deborah G. Johnson and Jameson M. Wetmore

The MIT Press
Cambridge, Massachusetts
London, England

© 2009 Massachusetts Institute of Technology

All rights reserved. No part of this book may be reproduced in any form by any electronic or mechanical means (including photocopying, recording, or information storage and retrieval) without permission in writing from the publisher.

For information about special quantity discounts, please email special_sales@mitpress.mit.edu

This book was set in Stone Serif and Stone Sans by Asco Typesetters, Hong Kong.
Printed and bound in the United States of America.

Library of Congress Cataloging-in-Publication Data

Technology and society : building our sociotechnical future / [compiled and edited by] Deborah G. Johnson and Jameson M. Wetmore.
    p.   cm.
Includes bibliographical references and index.
ISBN 978-0-262-10124-0 (hardcover : alk. paper)—ISBN 978-0-262-60073-6 (pbk. : alk. paper)
1. Technology—Social aspects. 2. Technological innovations. 3. Technology and civilization. I. Johnson, Deborah G., 1945–. II. Wetmore, Jameson M.
T14.5.T44169 2008
303.48′3—dc22
                                                          2008002813

10 9 8 7 6 5 4 3

Contents

Acknowledgments   ix

Introduction   xi

I	**VISIONS OF A TECHNOLOGICAL FUTURE** 1	
1	"Technology and Social Justice" 5 Freeman J. Dyson	
2	"The Machine Stops" 13 E. M. Forster	
3	"The Prolongation of Life" 37 Francis Fukuyama	
4	"Reproductive Ectogenesis: The Third Era of Human Reproduction and Some Moral Consequences" 51 Stellan Welin	
5	"Nanotechnology: Shaping the World Atom by Atom" 63 Interagency Working Group on Nanoscience, Engineering, and Technology	
6	"Why the Future Doesn't Need Us" 69 Bill Joy	
II	**THE RELATIONSHIP BETWEEN TECHNOLOGY AND SOCIETY** 93	
7	"Do Machines Make History?" 97 Robert L. Heilbroner	
8	"The Social Construction of Facts and Artifacts" 107 Trevor J. Pinch and Wiebe Bijker	
9	"Technological Momentum" 141 Thomas P. Hughes	
10	"Where Are the Missing Masses? The Sociology of a Few Mundane Artifacts" 151 Bruno Latour	

11	"Code Is Law" 181 Lawrence Lessig	
12	"The Intersection of Culture, Gender, and Technology" 195 Patrick D. Hopkins	
III	**TECHNOLOGY AND VALUES 205**	
13	"Do Artifacts Have Politics?" 209 Langdon Winner	
14	"Control: Human and Nonhuman Robots" 227 George Ritzer	
15	*White* 257 Richard Dyer	
16	"Manufacturing Gender in Commercial and Military Cockpit Design" 265 Rachel N. Weber	
17	"Pas de Trois: Science, Technology, and the Marketplace" 275 Daniel Sarewitz	
18	"Amish Technology: Reinforcing Values and Building Community" 297 Jameson M. Wetmore	
IV	**THE COMPLEX NATURE OF SOCIOTECHNICAL SYSTEMS 319**	
19	"Will Small Be Beautiful? Making Policies for Our Nanotech Future" 323 W. Patrick McCray	
20	"Sociotechnical Complexity: Redesigning a Shielding Wall" 355 Dominique Vinck	
21	"The Naked Launch: Assigning Blame for the Challenger Explosion" 369 Harry Collins and Trevor Pinch	
22	"Bodies, Machines, and Male Power" 389 M. Carme Alemany Gomez	
23	"Crash!: Nuclear Fuel Flasks and Anti-Misting Kerosene on Trial" 407 Harry Collins and Trevor Pinch	
24	"When Is a Work Around? Conflict and Negotiation in Computer Systems Development" 423 Neil Pollock	

Contents

| V | TWENTY-FIRST-CENTURY CHALLENGES 441 |

25 "Shaping Technology for the 'Good Life': The Technological Imperative versus the Social Imperative" 445
Gary Chapman

26 "The Feminization of Work in the Information Age" 459
Judy Wajcman

27 "Nanotechnology and the Developing World" 475
Fabio Salamanca-Buentello, Deepa L. Persad, Erin B. Court, Douglas K. Martin, Abdallah S. Daar, and Peter A. Singer

28 "Nanotechnology and the Developing World: Will Nanotechnology Overcome Poverty or Widen Disparities?" 485
Noela Invernizzi and Guillermo Foladori

29 "People's Science in Action: The Politics of Protest and Knowledge Brokering in India" 499
Roopali Phadke

30 "Security Trade-Offs Are Subjective" and "Technology Creates Security Imbalances" 515
Bruce Schneier

31 "Questioning Surveillance and Security" 537
Torin Monahan

32 *Energy, Society, and Environment: Technology for a Sustainable Future* 565
David Elliott

33 Introduction to *Environmental Justice: Creating Equality, Reclaiming Democracy* 579
Kristin S. Shrader-Frechette

34 "Icarus 2.0: A Historian's Perspective on Human Biological Enhancement" 599
Michael Bess

Index 613

Acknowledgments

We thank students at Arizona State University and the University of Virginia for the valuable feedback they gave us as we tested versions of this volume in our courses.

Jane Carlson is a powerhouse of energy and competence, and we were fortunate to have her attention on this project. We are grateful to her for obtaining the permissions, creating the index, and giving the manuscript a last-minute proofread.

This book was developed with support from the National Science Foundation under Grant No. 0220748. Any opinions, findings, and conclusions or recommendations expressed in this material are those of the editors and authors and do not necessarily reflect the views of the National Science Foundation.

Introduction

Technology is a powerful component of the modern world. Without technology, many of the most significant feats of the twentieth century could not have been achieved such as sending people to the moon, erecting skyscrapers, damming enormous rivers, and destroying entire cities. But the most significant, though less often recognized, power of technology is its permeation of so many aspects of our everyday lives. We use technology to house, clothe, feed, entertain, and transport ourselves. It is a crucial component of our work, play, education, communication, child rearing, travel, and even reproduction. Technologies have even become a central part of our identity; when we differentiate our world from that of a hundred years ago, or even ten years ago, one of the most striking differences is the technologies we use and the way they shape our daily lives.

We have embraced technologies because of the wonders we have accomplished with them and the promise that they will continue to unlock new possibilities. Technologies have helped us to eradicate diseases, communicate with friends around the world, and even eliminate bad breath. New technologies are often equated with progress itself in large part because they can help us to do things that were not previously possible and solve problems that have plagued humanity for centuries.

Nevertheless, technological development has not been an unqualified blessing. Technologies frequently have negative social and environmental side effects; can enable a (desirable or undesirable) shift of power from one group to another; or can be "misused" and have destructive consequences. The effects and implications of some technologies were perceived as disturbing as soon as they were developed—like the atomic bombs dropped on Hiroshima and Nagasaki. Other technologies manifested problems more slowly—like the damage done to the environment that has accumulated over the past few hundred years from chemical pesticides and fertilizers.

The negative effects of technology present us with a formidable challenge. It seems obvious that technologies will be indispensable in solving current problems and ensuring a better future, and yet we have to acknowledge that deploying technology is likely to create new problems and constraints. The challenge of taking advantage of the benefits of technology while minimizing the negative consequences of doing so is one that faces all of us, because all of us make decisions that shape or direct the development and use of technologies. For instance, venture capitalists and government officials decide which areas of science and engineering to fund with their budgets. Corporate executives choose the types of products to bring to market. Engineers shape the specific design and applications of technologies. Marketers decide how technologies are presented to the public. Regulators develop systems to help ensure that industries develop products that are safe and efficient. Advocates working with Non-Government

Organizations (NGOs) encourage the development of alternative technologies corporations may have overlooked; they voice concerns about technologies they believe governments should regulate. And finally, every individual decides whether, where, when, and how they will use new technologies at home, in public places, and at work.

Often these decisions are difficult, must resolve competing values, and require some negotiation. In some cases, the negative effects of a particular technology have been reluctantly accepted as necessary in order to reap its benefits—like the yearly death toll on our highway transportation systems. In a few cases the potential disruptive effects of particular technologies have convinced some countries to ban or restrict their use—like prohibiting CFCs as propellants, not allowing thalidomide to be prescribed to women who are or could become pregnant, and using politics (and sometimes war) to prevent other countries from developing nuclear capabilities. In yet other cases, controversies persist as to how a particular technology should be integrated into society—like the unresolved debates about nuclear power, stem cell research, and file-sharing.

These debates are not simply about how to best limit injuries and deaths that could be caused by new technologies. Those are difficult and important questions, but they do not address all the possible implications. Debates about technology are discussions about how it should and should not shape the fabric of tomorrow's society. Technologies do not entirely determine what the world will be like in the future, but they structure our lives, our capabilities, and our relationships with other people. The implementation and use of new devices and techniques create new possibilities while simultaneously making other avenues more difficult. Thus, even if a technology is non-controversial, decisions about how to develop it, and when and how to use it, can make a significant difference for future economic, political, and social well-being. Because of the significant implications of technological decisions, they should not be taken lightly.

This book is designed to inform and enlighten decisions about technology. It is based on the idea that to direct technology in the most beneficial ways, we need to know more than science and engineering traditionally offer—we need an understanding of how devices, techniques, people, institutions, goals, and values are intertwined. This more complex understanding of the world can empower those who design, fund, market, distribute, regulate, use, and dispose of or recycle technology to build a better future. Making informed decisions about technology is not simply a process of maximizing benefits and minimizing problems; it requires careful reflection on the values that are at stake and thoughtful deliberation about the best strategies to realize those values.

The readings in this volume provide conceptual tools, theories, case studies, and a multitude of ideas for thinking about the relationships between technology, society, and values. They present the views of lawyers, engineers, philosophers, sociologists, anthropologists, economists, politicians, science fiction writers, journalists, and others who reflect on the ways in which technologies are conceived, created, and integral to society and how social, economic and cultural forces and conditions shape and are

Introduction xiii

shaped by technology. Together these readings present an interdisciplinary approach to technology studies while providing access to a wide array of specialized expertise.

The book continually poses three important questions. First, how does technology constitute the world? The volume explores the variety of ways by which machines and techniques are embedded in society and thereby shape institutions, relationships, and values. This exploration provides a basic understanding of how technology and society are intertwined, an understanding that is essential to any attempt to shape the future. Second, what kind of future is desirable? The authors of many of the articles argue that certain values are of key importance to a just society and challenge the reader to consider which values are most dear and which are most defensible. Careful deliberation and discussion of values and goals is vital if we are to make wise decisions about our future. And third, how can technology be steered to achieve a better future? Only when the role of technology in society is understood can strategies be developed to get from here to there. This book identifies a number of ways in which others have had an important effect on our future through technological decision making.

The starting point for answering these questions is, as already indicated, with an understanding of the ways in which technology and society are inextricably intertwined. Just as technology influences the kind of society we have, society influences the kind of technologies that are developed. This means that one cannot understand either as separate. Indeed, technology and society are complex interconnected systems.

The book argues that technology should not be understood to be material objects alone. Technology neither exists nor has meaning without the human activities of which it is a part and similarly, many social practices would be impossible or incomprehensible without material objects. Thus, to understand the ways in which technology permeates and constitutes our everyday lives, we have to examine material objects together with the social practices and social relationships that make the material objects possible and useful. The book refers to these combinations as "sociotechnical systems" to emphasize the idea that material objects and social practices are inseparable.

Thinking in terms of sociotechnical systems instead of artifacts is not just an academic exercise. This approach provides a window into the ways in which technologies are intertwined with values. Technologies are developed, promoted, and used because people perceive that they will fulfill a certain need, accomplish a certain task, or achieve a certain goal within a given set of circumstances. Technologies are not simply chosen at random. People are motivated to integrate technologies into the fabric of society because they have certain values and they want to promote those values. Once created, sociotechnical systems can sometimes seem to take on a power of their own. They facilitate and constrain certain actions and thereby facilitate and constrain certain values. In other words, the intertwining of society and technology is not neutral; it is value-laden. Values shape the technologies we get and technologies subsequently have a significant effect on the values that are realized in a society. For example, some technologies facilitate democratic decision making; others reinforce race and gender bias; yet others facilitate individual independence and autonomy and erode community cohesion. Whether these values are thought to be cultural, social, moral, political, or whatever, recognition of this aspect of technology has important implications for future technological development.

The processes by which technologies are created, imbued with values, and come to affect the world are far from simple. In order to effect change in such a system, one must understand the complexity of sociotechnical systems. Many actors and many kinds of uncertainty are at work in the development and functioning of sociotechnical systems. All of those who help shape technology—engineers, business people, policymakers, and even users—are not just cobbling together artifacts and giving them meanings and uses, they are building the framework and social structure for the society in which humans live. There are many different kinds of complexity here. The task of developing technology requires that individuals work on small components of enormously large projects and therefore must coordinate their work with many different actors and make sometimes-risky decisions in endless iterations of designs and decision making. Users may see something in a technology that the engineers who designed it or marketers who advertise it never even considered. Politicians and government agencies make decisions about technology in extremely complex economic and political environments. Cultural norms including those with regard to race, class, and gender are at work in sociotechnical systems. Once we acknowledge that technology is a combination of things and people, i.e., that it is sociotechnical systems, and that sociotechnical systems affect and are affected by values, then it follows that those who contribute to the production and use of a sociotechnical system are not just building a 'thing'; they are building society and making value decisions that will affect society.

With this understanding in hand, we are better equipped to think constructively about the future and explore the issues and challenges that face us in the twenty-first century. Many aspects of the current world now pose daunting challenges. How are we to continue to improve upon health care and assure equal access? How can privacy be protected while at the same time achieving adequate security? How will globalization affect the lives of individuals around the world? An understanding of how technology and society are intertwined and how values are woven into this relationship is essential to making informed choices to meet these challenges.

This book does not provide any simple algorithms for solving the challenges of the future. Rather, the readings are put together with the idea that individuals and institutions need a particular kind of understanding of technology, society, and values in order to make informed choices. Politicians, business people, engineers, technology users, and concerned citizens are not only affected by technology, they also play an important role in shaping modern technology. By reflecting on the relationship between society and technology and carefully considering how values may be affected, individuals are better situated to advance the values they cherish. This book has been designed to frame discussions about the future and how technology might be managed to get us there.

Many who are interested in technology are motivated by the possibility of making the world a better place—for themselves and for others. While this book provides an understanding of the relationship between technology and society, the ultimate intent is not simply to inform, but to challenge readers and equip them to be agents of change in our sociotechnical future. This book offers encouragement and guidance for that quest. It offers a way of thinking about technology and society that can lead to better technologies, and a society better able to make decisions about the future.

I VISIONS OF A TECHNOLOGICAL FUTURE

Visions of the future are an important part of Western culture. Stories about possible futures have served a variety of purposes. They reveal underlying trends and they caution, inspire, give meaning, and teach us about social change. Some visions are full of hope and promise; they explain how technologies will set us free, ease everyday drudgery, help us to live longer, and bring global justice. Those put forward by corporations and professional organizations are often aimed at encouraging and inspiring engineers, entrepreneurs, policymakers, and consumers. Yet others serve as cautionary tales. For instance, George Orwell's *1984*, Aldous Huxley's *Brave New World*, and many science fiction films present visions of the future that serve as ominous warnings about the ultimate effects of certain trends. The authors of such futures challenge readers with pictures of a dystopian world that could result from the implementation of certain technologies, often in conjunction with particular types of government or corporate practices.

For at least the last hundred years, technology has played an important role in visions of the future. Future societies are rarely ever conceived of or presented without future technologies and often the development of new technologies is used to explain how we could transition from present social structures to different social structures. Sometimes the force of technology is presented as unstoppable—as taking us to a specific future regardless of whether we want to go there. In other visions, technology enables new choices, but it is up to people to decide what type of society they want to live in. With either approach, technological change and social change are understood to be intimately linked so that they create the future together.

While we might not always be conscious of it, visions of the future involving technology are being presented to us each day. We commonly see them in the popular media, especially in advertising but also in television, film, and literature. These images may encourage us to adopt certain products or support certain policies with the idea that the new technologies will situate us more positively in the future to come. Commercials touting hair growth formulas do not simply claim that the product will prevent hairlines from receding, they claim that using the product will make the user more popular and more influential.

Ideas about the future are not simply used to legitimate policies and products to citizens and consumers. They also play an important role in the decision making processes of technology producers. Discussions in the boardrooms of major corporations, in government, and in the research labs of major universities are also focused on the future. They develop products, policies, and strategies in an effort to solve a current problem or push the boundaries of science and technology to get innovations that will achieve better security, safety, longevity, comfort, etc. A key part of this process is to think through how technologies might shape people, institutions, and the world in general.

These visions of the future are not simply predictions or educated guesses. As producers and consumers, we actively use visions of the future. They are useful tools that can help us to make decisions and thus they have an important effect on the future we ultimately create. Hence, it is important to consider them thoughtfully, to reflect on the motivations of those who created them, and try to understand their deeper meanings.

This section presents readings with both positive and negative images of technology and the future. Most are contemporary visions. The one exception—an E. M. Forster short story—was written at the beginning of the 20th century, but still strikes a chord today. The authors come from a variety of backgrounds including science fiction writing, science and engineering, philosophy, and social criticism. Each vision reflects the author's hopes and concerns as well as the state of society at the time of its writing. Whatever specific suggestions they make about the future, these stories all embody assumptions about technology's role in shaping society. They implicitly and explicitly assume a connection between the technologies we develop and use, the quality and character of the lives we live, and the accompanying social and political order.

Although some of these technological futures may seem far-fetched, it is difficult to deny that today's actions shape tomorrow's world and that those who envision, design, and develop technologies play a pivotal role in designing our future. We must think about the future because we are actively creating it. Thus, when the federal government allocates billions of dollars to nanotechnology research; when a drug company chooses to focus on fertility drugs rather than a cure for malaria; when engineers test a wing design in a wind tunnel or debug a program; or when a handful of electronics enthusiasts begin building affordable computers in their garage, they are making decisions that will affect the entire world. The future is being designed and built by all of those who develop, fund, regulate, and use technology.

While none of these groups are alone responsible for the future, it is important for all to understand their role in building the future so that they can take their contribution into account. While governments may not be held responsible for missed opportunities, they must ultimately answer to their constituents; corporations must answer to consumers for the safety and side-effects of their products; engineers are morally and legally responsible to ensure a reasonable level of safety in their inventions; and consumers are responsible for appropriate use of the products they consume. None of these groups can act responsibly unless they understand how their actions today affect the future. Creating visions of the future and thinking carefully through them can help us envision worlds we want to live in, worlds we don't want to live in, and how today's decisions can lead to different possibilities.

Questions to consider while reading the selections in this section:

1. What assumptions does each author make about the relationship between technology and society?

Introduction to Part I

2. What technological and social developments would have to occur to bring about the author's vision of the future?
3. Are the changes necessary to achieve the author's vision feasible or desirable?
4. What makes the author's vision utopian or dystopian?
5. Who will benefit and how will they benefit from realization of the author's vision? Who will lose and how will they lose if the author's vision were to be realized?

1 "Technology and Social Justice"

Freeman J. Dyson

The first chapter in this book is an idealistic vision of the future, one in which social justice is achieved globally through new technology. Freeman J. Dyson, a highly respected physicist, envisions a world of global equity brought about by the development and distribution of technologies that give each small community the resources it needs to compete in a global economy. It is, to be sure, a vision, not a blueprint or proposal. Dyson does not worry about the technical details of what he envisions, nor, for that matter, the social and political change necessary to produce and put into use the technologies he proposes. He adopts an approach that has been commonly advocated over the past century or more—the idea of the technological fix. He believes that society's many problems will be solved if we only choose the right technologies. While this might be naive, the idea of technology and ethics working together—with ethics guiding technological development—is an important ideal.

The new century will be a good time for new beginnings. Technology guided by ethics has the power to help the billions of poor people all over the earth. Too much of technology today is making toys for the rich. Ethics can push technology in a new direction, away from toys for the rich and towards necessities for the poor. The time is ripe for this to happen. The sun, the genome, and the internet are three revolutionary forces arriving with the new century. These forces are strong enough to reverse some of the worst evils of our time. One of the greatest evils is rural poverty. All over the earth, and especially in the poor countries to the south of us, millions of desperate people leave their villages and pour into overcrowded cities. There are now ten megacities in the world with populations twice as large as New York City. Soon there will be more. Mexico City is one of them. The increase of human population is one of the causes of the migration. The other cause is the poverty and lack of jobs in the villages. Both the population explosion and the poverty must be reversed if we are to have a decent future. Many experts on population say that if we can mitigate the poverty, the population will stabilize itself, as it has done in Europe and Japan.

I am not an expert on population, so I will say no more about that. I am saying that the poverty can be reduced by a combination of solar energy, genetic engineering, and the internet. And perhaps when the poverty stops increasing, the population will stop exploding.

I have seen with my own eyes what happens to a village when the economic basis of life collapses. My wife grew up in a village in East Germany that was under communist management. The communist regime took care of the village economy, selling the

---

From *The Sun, The Genome, & The Internet: Tools of Scientific Revolutions* (New York: Oxford University Press, 1999), pp. 61–74. © Oxford University Press, Inc. Reprinted with permission.

output of the farms to Russia at fixed prices which gave the farmers economic security. The village remained beautiful and on the whole pleasant to live in. Nothing much had changed in the village since 1910. One thing the communist regime did was to organize a zoo, with a collection of animals maintained by a few professionals with enthusiastic help from the local schoolchildren. The village was justly proud of its zoo. The zoo was subsidized by the regime so that it did not need to worry about being unprofitable. My wife and I visited the village under the old regime and found it very friendly.

Then came 1990 and the unification of Germany. Overnight the economy of the village was wrecked. The farmers could no longer farm because nobody would buy their product. Russia could not buy because the price had to be paid in West German marks. German consumers would not buy because the quality was not as good as the produce available in the supermarkets. The village farmers could not compete with the goods pouring in from France and Denmark. So the farmers were out of work. Most of the younger generation moved out of the village to compete for jobs in the cities. Most of the older generation remained. Many of them, both old and young, are still unemployed. The zoo, deprived of its subsidy, was closed. The sad exodus that we saw in the village of Westerhausen in 1991 is the same exodus that is happening in villages all over the world. Everywhere, the international market devalues the work of the village. The people of the village sink from poverty into destitution. Without work, the younger and more enterprising people move out.

In the nine years since the unification, Westerhausen has slowly been recovering. Recovery is possible because of the process of gentrification. The village is still a pleasant place to live if you have money. Wealthy people from the local towns move in and modernize the homes abandoned by the farmers. Cottages are demolished to make room for two-car garages. Ancient and narrow roads are widened to make room for two Mercedes cars to pass each other. The thousand-year-old church is being repaired and restored. The village will survive as a community of nature-lovers and commuters. Lying on the northern edge of the Harz mountains, it is close to the big cities of northern Germany and even closer to unspoiled mountain forests. Its permanent asset is natural beauty.

In 1997 my wife and I were back in the village. The change since we last visited in 1991 was startling. We stayed in the elegant new home of a friend who had been in my wife's class in the village elementary school fifty years earlier. The village now looks well maintained and prosperous. The recovery from the disaster of 1990 has been slow and difficult, but it has been steady. The government did two things to mitigate the harshness of the free market economy, to enable the native villagers who remained in the village to survive. Every homeowner can borrow money with low interest to modernize houses, and every farming cooperative can borrow money with low interest to modernize farms. As a result, many of the houses which were not bought by outsiders are being modernized, and the few farmers who remained as farmers are flourishing. The zoo has been revived. In addition, there are some new enterprises. A western immigrant has planted a large vineyard on a south-facing hillside and will soon be producing the first Westerhausen wines. My wife's family and many of her friends are remaining in the village. They gave us a warm and joyful welcome.

The probable future of Westerhausen can be seen in a thousand villages in England. The typical English village today is not primarily engaged in farming. The typical village remains beautiful and prosperous because of the process of gentrification. Wealthy homeowners pay large sums of money for the privilege of living under a thatched roof. The thatching of roofs is one of the few ancient village crafts that still survives. The thatchers are mostly young, highly skilled, and well paid. The farmers who remain are either gentleman amateurs who run small farms as a hobby, or well-educated professionals who run big farms as a business. The old population of peasant farmers, who used to live in the villages in poverty and squalor, disappeared long ago. Discreetly hidden in many of the villages are offices and factories engaged in high-tech industry. One of the head offices of IBM Europe is in the English village of Hursley, not far from where I was born. In the villages of France, at least in the area that I know around Paris, the picture is similar. Wealth came to the villages because they have what wealthy people seek: peace and security and beauty.

What would it take to reverse the flow of people from villages to megacities all over the world, to stop the transformation of beautiful cities into unmanageable slums? I believe that the flow can be reversed by the same process of gentrification that is happening in Westerhausen. To make gentrification possible, the villages themselves must become a source of wealth. How can a godforsaken Mexican village become a source of wealth? Three facts can make it possible. First, solar energy is distributed equitably over the earth. Second, genetic engineering can make solar energy usable everywhere for the local creation of wealth. Third, the internet can provide people in every village with the information and skills they need to develop their talents. The sun, the genome, and the internet can work together to bring wealth to the villages of Mexico, just as the older technology of electricity and automobiles brought wealth to the villages of England. Each of the three new technologies has essential gifts to offer.

Solar energy is most abundant where it is most needed, in the countryside rather than in cities, in the tropical countries where most of the population lives rather than in temperate latitudes. Recently I got to know a young man called Bob Freling who runs a venture called SELF, the Solar Electric Light Fund. Freling is not a scientist. He is a linguist with a passion for languages. He is fluent in Spanish, Russian, French, Chinese, Portuguese, and Indonesian. He has the know-how to operate small enterprises in many different countries with different cultures and different ways of doing business. The idea of SELF is to bring electricity generated by sunlight to remote places that have no other way to get electricity.

A working solar energy system can make an enormous difference to the quality of life in a tropical village. Thirty or fifty watts of direct current is enough to run a couple of fluorescent lights, a radio, or a small black-and-white television for several hours every night. Each hut in a village can have its own system. No central generator, no power lines, no transformers are needed. Sunlight distributes power equally to each rooftop. Children can read and study at night in their homes. The village is in touch with the outside world.

SELF is one of the organizations dedicated to making this happen. It is a charitable foundation, but it does not give solar-energy systems to the villagers for free. The villagers pay market prices for the hardware. SELF gives them credit so they can spread their payments over four years. SELF also pays the people who train the villagers to install and operate and maintain the hardware. SELF now has village projects working in eleven countries, all in remote places unlikely to be reached in the near future by electric power lines. The most recent start was on the island of Guadalcanal, not far from the battlefield where one of the bloodiest campaigns of World War II was fought. Technology now brings light to the island instead of death and destruction. The village that is now electrified is still accessible only by canoe. The solar hardware is light and rugged enough to travel by canoe.

One day there will be a global internet carried by a network of low-altitude satellites linked by radio and laser communications. Every point on earth will be within range of one or more of the satellites all the time. But not every point on earth will be able to communicate with the internet. For places without electricity to power transmitters and receivers, the satellites overhead will be useless. The villages where SELF has supplied solar-energy systems will be hooked up to the net. Their neighbors all around will still be isolated.

Why do we need a charitable foundation to do this work? Why cannot the villagers do it by themselves? Unfortunately, the technology of solar energy is still too expensive for an average third-world village to afford. Even with the help provided by SELF, only villages with a substantial cash economy can afford it. The present cost of a minimal solar-energy installation is about five hundred dollars per household.

About half of this cost is due to the photovoltaic collector panels that convert sunlight into electricity. The other half is spent on accessories such as storage batteries and control circuitry. The cost of collector panels is now about five dollars per watt. This is the cost of commercially available units that are properly packaged to work outdoors in rough weather. Experimental units that are under development will be substantially cheaper. The prevailing belief among economists is that solar energy will not supplant kerosene and other fossil fuels on a massive scale until the price comes down below one dollar per watt. Nobody can tell when, if ever, the existing photovoltaic technology will become cheap enough to supply the world's needs. All that we know for sure is that there is enough sunlight to supply our needs many times over, if we can find a way to use it.

Meanwhile, other ways of using sunlight are making slower progress. The traditional ways to use sunlight are to grow crops for food and trees for fuel. Even today, with the forests rapidly disappearing, firewood is still a major source of energy for the rural population of India. Less destructive ways of converting sunlight into fuel exist. Plant and animal wastes can be digested in tanks to make methane gas. Sugar cane can be fermented to make alcohol. These technologies are being used successfully in many places. Progress is slow because coal and oil and natural gas are still too cheap. It is possible that energy crops, either trees or other plants grown for conversion into fuel, will one day be a major source of energy for the world. But this will not happen until the cost of growing and processing energy crops falls far below its present level.

The SELF projects in remote regions of Asia and Africa and Australasia are cost effective because the usual alternative source of energy for the villagers is kerosene, and kerosene in remote areas is expensive. Kerosene has to be carried in cans on the backs of animals or people. Solar energy needs no carrying. But the SELF projects are limited to a small scale because that is all the villagers can afford.

Fifty watts per household is not enough to support a modern economy. For the people in a remote village to become a part of the modern world, they need electricity in far greater quantity. They need kilowatts per household rather than watts. They need refrigerators and power tools and electrical machinery. If solar energy is to satisfy their needs, it must be available on a massive scale.

The flux of solar energy incident upon the earth is enormous compared with all other energy resources. Each square mile in the tropics receives about a thousand megawatts, averaged over day and night. This quantity of energy would be ample to support a dense population with all modern conveniences. Solar energy has not yet been used on a large scale for one simple reason: it is too expensive. The country that has used solar energy on the largest scale is Brazil, where sugar was grown as an energy crop to make alcohol to be used as a substitute for gasoline in cars and trucks. Brazil protected and subsidized the local alcohol industry. The experiment was technically successful, but the cost was high. Brazil has now reverted to free market policies, and the experiment is at an end. What the world needs is not high-cost subsidized solar energy but solar energy cheap enough to compete with oil.

Solar energy is expensive today because it has to be collected from large areas and we do not yet have a technology that covers large areas cheaply. One of the virtues of solar energy is the fact that it can be collected in many ways. It is adaptable to local conditions.

The two main tools for collecting it are photoelectric panels producing electricity and energy crops producing fuel. With the exception of small units such as those used in the SELF projects, energy crops are the method of choice for farmland and forests, while photoelectric collection is the method of choice for deserts. Each method has its advantages and disadvantages. Photoelectric systems have high efficiency, typically between 10 and 15 percent, but are expensive to deploy and maintain. Energy crops have low efficiency, typically around 1 percent, and are expensive and messy to harvest. The electricity produced by photoelectric systems is intermittent and cannot be cheaply converted into storable form.

Fuels produced from energy crops are storable. To make solar energy cheap, we need a technology that combines the advantages of photovoltaic and biological systems. Two technical advances would make this possible. First, crop plants could be developed that convert sunlight to fuel with efficiency comparable to photovoltaic collectors, in the range of 10 percent. This would reduce the costs of land and harvesting by a large factor. Second, crop plants could be developed that do not need to be harvested at all.

An energy crop could be a permanent forest of trees that convert sunlight to liquid fuel and deliver the fuel directly through their roots to a network of underground pipelines. If these two advances could be combined, we would have a supply of solar energy that was cheap, abundant, and environmentally benign.

Our hopes for radical decrease of costs and increase of efficiency of energy crops must rest on the genome. Traditional farming has always been based on genetic engineering. Every major crop plant and farm animal has been genetically engineered by selective breeding until it barely resembles the wild species from which it originated. Genetic engineering as the basis of the world economy is nothing new. What is new is the speed of development. Traditional genetic engineering took centuries or millennia to produce the improved plants and animals that fed the world until a hundred years ago. Modern genetic engineering, based on detailed understanding of the genome, will be able to make radical improvements within a few years. That is why I look to the genome, together with the sun and the internet, as tools with which to build a brighter future for mankind.

The energy supply system of the future might be a large forest, with the species of trees varying from place to place to suit the local climate and topography. We may hope that substantial parts of the forest would be nature reserves, closed to human settlement and populated with wildlife so as to preserve the diversity of natural ecologies. But the greater part could be open to human settlement, teeming with towns and villages under the trees. Landowners outside the nature reserves could be given a free choice, whether to grow trees for energy or not. If the trees converted sunlight into fuel with 10 percent efficiency, landowners could sell the fuel for ten thousand dollars per year per acre and easily undercut the present price of gasoline. Owners of farmland and city lots would have a strong economic incentive to grow trees. Even without such an incentive, towns where wealthy people live are usually full of trees.

People like to live among trees. If the trees were also generating fuel from sunlight, the appearance of the towns would not be greatly altered. The future energy plantation need not be a monotonous expanse of trees of a single species planted in uniform rows. It could be as varied and as spontaneous as a natural woodland, interspersed with open spaces and houses and towns and factories and lakes.

To make this dream of a future landscape come true, the essential tool is genetic engineering. At present large sums of money are being spent on sequencing the human genome. The human genome project does not contribute directly to the engineering of trees. But alongside the human genome, many other genomes are being sequenced: bacteria and yeast and worms and fruit flies. For advancing the art of genetic engineering, the genomes of simpler organisms are more useful than the human genome. Before long we shall have sequenced the genomes of the major crop plants, wheat and maize and rice, and after that will come trees. Within a few decades we shall have achieved a deep understanding of the genome, an understanding that will allow us to breed trees that will turn sunlight into fuel and still preserve the diversity that makes natural forests beautiful.

While we are genetically engineering trees to use sunlight efficiently to make fuel, we shall also be breeding trees that use sunlight to make other useful products, such as silicon chips for computers and silicon film for photovoltaic collectors. Economic forces will then move industries from cities to the country. Mining and manufacturing could be economically based on locally available solar energy, with genetically engineered creatures consuming and recycling the waste products. It might even become possible to build roads and buildings biologically, breeding little polyps to lay down durable structures on land in the same way as their cousins build coral reefs in the ocean.

The third and most important of the triad of new technologies is the internet. The internet is essential, to enable businesses and farms in remote places to function as part of the modern global economy. The internet will allow people in remote places to make business deals, to buy and sell, to keep in touch with their friends, to continue their education, to follow their hobbies and avocations, with full knowledge of what is going on in the rest of the world. This will not be the internet of today, accessible only to computer-literate people in rich countries and to the wealthy elite in poor countries. It will be a truly global internet, using a network of satellites in space for communication with places that fiber optics cannot reach, and connected to local networks in every village. The new internet will end the cultural isolation of poor countries and poor people.

There are two technical problems that must be solved to make the internet accessible to almost everybody. There is the problem of large-scale architecture and the problem of the last mile. Large-scale architecture means choosing the most efficient combination of land-lines and satellite links to cover every corner of the globe. The Teledesic system of satellite communication now under development is intended to be a partial answer to this problem. The Teledesic system has 288 satellites in a dense network of low orbits, allowing any two points on the globe to be connected with minimum delay. If the Teledesic system fails or is not deployed, some other system will be designed to do the same job. The problem of the last mile is more difficult. This is the problem of connecting individual homes and families, wherever they happen to be, to the nearest internet terminal. The problem of the last mile has to be solved piecemeal, with methods depending on the local geography and the local culture.

An ingenious method of solving the last mile problem in urban American neighborhoods has been introduced recently by Paul Baran, the original inventor of the internet. Baran's system is called Ricochet and consists of a multitude of small wireless transmitters and receivers. Each user has a modem that communicates by radio with the local network. The feature that makes the system practical is the ability of the transmitters to switch their frequencies constantly so as not to interfere with one another. The system is flexible and cheap, avoiding the large expense of laying cables from the internet terminal to every apartment in every house. It works well in the environment of urban America. It remains to be seen whether it is flexible and cheap enough to work well in the environment of a Mexican village or a Peruvian barrio. An essential prerequisite for access to the internet is access to a reliable local supply of electricity. Villagers must have electricity first, before they can have modems.

Suppose that we can solve the technical problems of cheap solar energy, genetic engineering of industrial crop plants, and universal access to the internet. What then will follow? The solution of these three problems could bring about a world-wide social revolution, similar to the revolution that we have seen in the villages of England and Germany. Cheap solar energy and genetic engineering will provide the basis for primary industries in the countryside, modernized farming and mining and manufacturing. After that, the vast variety of secondary and tertiary economic activities that use the internet for their coordination, food processing and publishing and education and entertainment and health care, will follow the primary industries as they move from overgrown cities to country towns and villages. As soon as the villages become rich, they will attract people and wealth back from the cities.

I am not suggesting that in the brave new world of the future everyone will be compelled to live in villages. Many of us will always prefer to live in large cities or in towns of moderate size. I am suggesting only that people should be free to choose. When wealth has moved back to the villages, people who live there will no longer be compelled by economic necessity to move out, and people who live in megacities will no longer be compelled by economic necessity to stay there. Many of us who have the freedom to choose, like successful stockbrokers and business executives in England and Germany, will choose to live in villages.

This is my dream, that solar energy and genetic engineering and the internet will work together to create a socially just world, in which every Mexican village becomes as wealthy as Princeton. Of course that is only a dream. Inequalities will persist and poverty will not disappear. But I see a hope that the world will move far and fast along the road that I have been describing. We need to apply a strong ethical push to add force to the technological pull. Ethics must guide technology in the direction of social justice. Let us help to push the world in that direction as hard as we can. It does no harm to hope.

2 "The Machine Stops"

E. M. Forster

Despite being written a hundred years ago, this powerful story by E. M. Forster still seems fresh and pertinent today. On one level it can be interpreted as a cautionary tale about humans becoming too dependent on technology—a counterbalance to Dyson's utopian vision. But much of Forster's concern is not with artifacts themselves, but rather the deep cultural and psychological changes that often accompany technological change. "The Machine Stops" depicts a world in which machines transform not only things, but people—their capabilities, desires, inner lives, emotions, and family relationships. This future is disturbing in spite of the fact that it has achieved many of the ideals to which today's technologies seem to aspire—such as immediate communication without the need for physical presence, having everything we want at our fingertips, and being free from mundane tasks so that we can engage in what are considered "higher" activities, for example, art, music, poetry. Thus, the story implicitly raises profound questions that will resonate through the rest of the book: What is progress? Will our technical goals bring us the happiness we desire? What does it mean to worship technology? Technology mediates human experience, but how should this be done? How can we get a sociotechnical future we want and avoid getting one that is less desirable than what we have now? Forster challenges us to consider which values are most important to us and how technology might help or interfere with those values.

Part I: The Air-Ship

Imagine, if you can, a small room, hexagonal in shape, like the cell of a bee. It is lighted neither by window nor by lamp, yet it is filled with a soft radiance. There are no apertures for ventilation, yet the air is fresh. There are no musical instruments, and yet, at the moment that my meditation opens, this room is throbbing with melodious sounds. An arm-chair is in the centre, by its side a reading-desk—that is all the furniture. And in the arm-chair there sits a swaddled lump of flesh—a woman, about five feet high, with a face as white as a fungus. It is to her that the little room belongs.

An electric bell rang.

The woman touched a switch and the music was silent.

"I suppose I must see who it is," she thought, and set her chair in motion. The chair, like the music, was worked by machinery, and it rolled her to the other side of the room, where the bell still rang importunately.

From *Oxford and Cambridge Review*, November 1909, pp. 83–122. Reprinted with permission of Creative Education, Inc.

"Who is it?" she called. Her voice was irritable, for she had been interrupted often since the music began. She knew several thousand people; in certain directions, human intercourse had advanced enormously.

But when she listened into the receiver, her white face wrinkled into smiles, and she said:

"Very well. Let us talk, I will isolate myself. I do not expect anything important will happen for the next five minutes—for I can give you fully five minutes, Kuno. Then I must deliver my lecture on 'Music during the Australian Period.'"

She touched the isolation-knob, so that no one else could speak to her. Then she touched the lighting apparatus, and the little room was plunged into darkness.

"Be quick!" she called, her irritation returning. "Be quick, Kuno; here I am in the dark wasting my time."

But it was fully fifteen seconds before the round plate that she held in her hands began to glow. A faint blue light shot across it, darkening to purple, and presently she could see the image of her son, who lived on the other side of the earth, and he could see her.

"Kuno, how slow you are."

He smiled gravely.

"I really believe you enjoy dawdling."

"I have called you before, mother, but you were always busy or isolated. I have something particular to say."

"What is it, dearest boy? Be quick. Why could you not send it by pneumatic post?"

"Because I prefer saying such a thing. I want———"

"Well?"

"I want you to come and see me."

Vashti watched his face in the blue plate.

"But I can see you!" she exclaimed. "What more do you want?"

"I want to see you not through the Machine," said Kuno. "I want to speak to you not through the wearisome Machine."

"Oh, hush!" said his mother, vaguely shocked. "You mustn't say anything against the Machine."

"Why not?"

"One mustn't."

"You talk as if a god had made the Machine," cried the other. "I believe that you pray to it when you are unhappy. Men made it, do not forget that. Great men, but men. The Machine is much, but it is not everything. I see something like you in this plate, but I do not see you. I hear something like you through this telephone, but I do not hear you. That is why I want you to come. Come and stop with me. Pay me a visit, so that we can meet face to face, and talk about the hopes that are in my mind."

She replied that she could scarcely spare the time for a visit.

"The air-ship barely takes two days to fly between me and you."

"I dislike air-ships."

"Why?"

"I dislike seeing the horrible brown earth, and the sea, and the stars when it is dark. I get no ideas in an air-ship."

"I do not get them anywhere else."

"What kind of ideas can the air give you?"

He paused for an instant.

"Do you not know four big stars that form an oblong, and three stars close together in the middle of the oblong, and hanging from these stars, three other stars?"

"No, I do not. I dislike the stars. But did they give you an idea? How interesting; tell me."

"I had an idea that they were like a man."

"I do not understand."

"The four big stars are the man's shoulders and his knees. The three stars in the middle are like the belts that men wore once, and the three stars hanging are like a sword."

"A sword?"

"Men carried swords about with them, to kill animals and other men."

"It does not strike me as a very good idea, but it is certainly original. When did it come to you first?"

"In the air-ship———" He broke off, and she fancied that he looked sad. She could not be sure, for the Machine did not transmit *nuances* of expression. It only gave a general idea of people—an idea that was good enough for all practical purposes, Vashti thought. The imponderable bloom, declared by a discredited philosophy to be the actual essence of intercourse, was rightly ignored by the Machine, just as the imponderable bloom of the grape was ignored by the manufacturers of artificial fruit. Something 'good enough' had long since been accepted by our race.

"The truth is," he continued, "that I want to see these stars again. They are curious stars. I want to see them not from the air-ship, but from the surface of the earth, as our ancestors did, thousands of years ago. I want to visit the surface of the earth."

She was shocked again.

"Mother, you must come, if only to explain to me what is the harm of visiting the surface of the earth."

"No harm," she replied, controlling herself. "But no advantage. The surface of the earth is only dust and mud, no life remains on it, and you would need a respirator, or the cold of the outer air would kill you. One dies immediately in the outer air."

"I know; of course I shall take all precautions."

"And besides———"

"Well?"

She considered, and chose her words with care. Her son had a queer temper, and she wished to dissuade him from the expedition.

"It is contrary to the spirit of the age," she asserted.

"Do you mean by that, contrary to the Machine?"

"In a sense, but———"

His image in the blue plate faded.

"Kuno!"

He had isolated himself.

For a moment Vashti felt lonely.

Then she generated the light, and the sight of her room, flooded with radiance and studded with electric buttons, revived her. There were buttons and switches everywhere—buttons to call for food, for music, for clothing. There was the hot-bath button, by pressure of which a basin of (imitation) marble rose out of the floor, filled to the brim with a warm deodorised liquid. There was the cold-bath button. There was the button that produced literature. And there were of course the buttons by which she communicated with her friends. The room, though it contained nothing, was in touch with all that she cared for in the world.

Vashti's next move was to turn off the isolation-switch, and all the accumulations of the last three minutes burst upon her. The room was filled with the noise of bells, and speaking-tubes. What was the new food like? Could she recommend it? Had she had any ideas lately? Might one tell her one's own ideas? Would she make an engagement to visit the public nurseries at an early date?—say this day month.

To most of these questions she replied with irritation—a growing quality in that accelerated age. She said that the new food was horrible. That she could not visit the public nurseries through press of engagements. That she had no ideas of her own but had just been told one—that four stars and three in the middle were like a man: she doubted there was much in it. Then she switched off her correspondents, for it was time to deliver her lecture on Australian music.

The clumsy system of public gatherings had been long since abandoned; neither Vashti nor her audience stirred from their rooms. Seated in her arm-chair she spoke, while they in their arm-chairs heard her, fairly well, and saw her, fairly well. She opened with a humorous account of music in the pre-Mongolian epoch, and went on to describe the great outburst of song that followed the Chinese conquest. Remote and primæval as were the methods of I-San-So and the Brisbane school, she yet felt (she said) that study of them might repay the musician of to-day: they had freshness; they had, above all, ideas.

Her lecture, which lasted ten minutes, was well received, and at its conclusion she and many of her audience listened to a lecture on the sea; there were ideas to be got from the sea; the speaker had donned a respirator and visited it lately. Then she fed, talked to many friends, had a bath, talked again, and summoned her bed.

The bed was not to her liking. It was too large, and she had a feeling for a small bed. Complaint was useless, for beds were of the same dimension all over the world, and to have had an alternative size would have involved vast alterations in the Machine. Vashti isolated herself—it was necessary, for neither day nor night existed under the ground—and reviewed all that had happened since she had summoned the bed last. Ideas? Scarcely any. Events—was Kuno's invitation an event?

By her side, on the little reading-desk, was a survival from the ages of litter—one book. This was the Book of the Machine. In it were instructions against every possible contingency. If she was hot or cold or dyspeptic or at loss for a word, she went to the book, and it told her which button to press. The Central Committee published it. In accordance with a growing habit, it was richly bound.

Sitting up in the bed, she took it reverently in her hands. She glanced round the glowing room as if some one might be watching her. Then, half ashamed, half joyful, she murmured "O Machine! O Machine!" and raised the volume to her lips. Thrice she kissed it, thrice inclined her head, thrice she felt the delirium of acquiescence. Her ritual performed, she turned to page 1367, which gave the times of the departure of the air-ships from the island in the Southern Hemisphere, under whose soil she lived, to the island in the Northern Hemisphere, whereunder lived her son.

She thought, "I have not the time."

She made the room dark and slept; she awoke and made the room light; she ate and exchanged ideas with her friends, and listened to music and attended lectures; she made the room dark and slept. Above her, beneath her, and around her, the Machine hummed eternally; she did not notice the noise, for she had been born with it in her ears. The earth, carrying her, hummed as it sped through silence, turning her now to the invisible sun, now to the invisible stars. She awoke and made the room light.

"Kuno!"

"I will not talk to you," he answered, "until you come."

"Have you been on the surface of the earth since we spoke last?"

His image faded.

Again she consulted the book. She became very nervous and lay back in her chair palpitating. Think of her as without teeth or hair. Presently she directed the chair to the wall, and pressed an unfamiliar button. The wall swung apart slowly. Through the opening she saw a tunnel that curved slightly, so that its goal was not visible. Should she go to see her son, here was the beginning of the journey.

Of course she knew all about the communication-system. There was nothing mysterious in it. She would summon a car and it would fly with her down the tunnel until it reached the lift that communicated with the air-ship station: the system had been in use for thousands of years, long before the universal establishment of the Machine. And of course she had studied the civilisation that had immediately preceded her own—the civilisation that had mistaken the functions of the system, and had used it for bringing people to things, instead of for bringing things to people. Those funny old days, when men went for change of air instead of changing the air in their rooms! And yet—she was frightened of the tunnel: she had not seen it since her last child was born. It curved—but not quite as she remembered; it was brilliant—but not quite as brilliant as a lecturer had suggested. Vashti was seized with the terrors of direct experience. She shrank back into the room, and the wall closed up again.

"Kuno," she said, "I cannot come to see you. I am not well."

Immediately an enormous apparatus fell on to her out of the ceiling, a thermometer was automatically inserted between her lips, a stethoscope was automatically laid upon her heart. She lay powerless. Cool pads soothed her forehead. Kuno had telegraphed to her doctor.

So the human passions still blundered up and down in the Machine. Vashti drank the medicine that the doctor projected into her mouth, and the machinery retired into the ceiling. The voice of Kuno was heard asking how she felt.

"Better." Then with irritation: "But why do you not come to me instead?"

"Because I cannot leave this place."

"Why?"

"Because, any moment, something tremendous may happen."

"Have you been on the surface of the earth yet?"

"Not yet."

"Then what is it?"

"I will not tell you through the Machine."

She resumed her life.

But she thought of Kuno as a baby, his birth, his removal to the public nurseries, her one visit to him there, his visits to her—visits which stopped when the Machine had assigned him a room on the other side of the earth. "Parents, duties of," said the book of the Machine, "cease at the moment of birth. P. 422327483." True—but there was something special about Kuno—indeed there had been something special about all her children, and, after all, she must brave the journey if he desired it. And "something tremendous might happen." What did that mean? The nonsense of a youthful man, no doubt, but she must go. Again she pressed the unfamiliar button, again the wall swung back, and she saw the tunnel that curved out of sight. Clasping the Book, she rose, tottered on to the platform, and summoned the car. Her room closed behind her: the journey to the Northern Hemisphere had begun.

Of course it was perfectly easy. The car approached and in it she found arm-chairs exactly like her own. When she signalled, it stopped, and she tottered into the lift. One other passenger was in the lift, the first fellow creature she had seen face to face for months. Few travelled in these days, for, thanks to the advance of science, the earth was exactly alike all over. Rapid intercourse, from which the previous civilisation had hoped so much, had ended by defeating itself. What was the good of going to Pekin when it was just like Shrewsbury? Why return to Shrewsbury when it would be just like Pekin? Men seldom moved their bodies; all unrest was concentrated in the soul.

The air-ship service was a relic from the former age. It was kept up, because it was easier to keep it up than to stop it or to diminish it, but it now far exceeded the wants of the population. Vessel after vessel would rise from the vomitories of Rye or of Christchurch (I use the antique names), would sail into the crowded sky, and would draw up at the wharves of the South—empty. So nicely adjusted was the system, so independent of meteorology, that the sky, whether calm or cloudy, resembled a vast kaleidoscope whereon the same patterns periodically recurred. The ship on which Vashti sailed started now at sunset, now at dawn. But always, as it passed above Rheims, it would neighbour the ship that served between Helsingfors and the Brazils, and, every third time it surmounted the Alps, the fleet of Palermo would cross its track behind.

Night and day, wind and storm, tide and earthquake, impeded man no longer. He had harnessed Leviathan. All the old literature, with its praise of Nature, and its fear of Nature, rang false as the prattle of a child.

Yet as Vashti saw the vast flank of the ship, stained with exposure to the outer air, her horror of direct experience returned. It was not quite like the air-ship in the cinematophote. For one thing it smelt—not strongly or unpleasantly, but it did smell, and with her eyes shut she should have known that a new thing was close to her. Then she had to walk to it from the lift, had to submit to glances from the other passengers. The man in front dropped his Book—no great matter, but it disquieted them all. In the rooms, if the Book was dropped, the floor raised it mechanically, but the gangway to the air-ship was not so prepared, and the sacred volume lay motionless. They stopped—the thing was unforeseen—and the man, instead of picking up his property, felt the muscles of his arm to see how they had failed him. Then some one actually said with direct utterance: "We shall be late"—and they trooped on board, Vashti treading on the pages as she did so.

Inside, her anxiety increased. The arrangements were old-fashioned and rough. There was even a female attendant, to whom she would have to announce her wants during the voyage. Of course a revolving platform ran the length of the boat, but she was expected to walk from it to her cabin. Some cabins were better than others, and she did not get the best. She thought the attendant had been unfair, and spasms of rage shook her. The glass valves had closed, she could not go back. She saw, at the end of the vestibule, the lift in which she had ascended going quietly up and down, empty. Beneath those corridors of shining tiles were rooms, tier below tier, reaching far into the earth, and in each room there sat a human being, eating, or sleeping, or producing ideas. And buried deep in the hive was her own room. Vashti was afraid.

"O Machine! O Machine!" she murmured, and caressed her Book, and was comforted.

Then the sides of the vestibule seemed to melt together, as do the passages that we see in dreams, the lift vanished, the Book that had been dropped slid to the left and vanished, polished tiles rushed by like a stream of water, there was a slight jar, and the air-ship, issuing from its tunnel, soared above the waters of a tropical ocean.

It was night. For a moment she saw the coast of Sumatra edged by the phosphorescence of waves, and crowned by lighthouses, still sending forth their disregarded beams. These also vanished, and only the stars distracted her. They were not motionless, but swayed to and fro above her head, thronging out of one skylight into another, as if the universe and not the air-ship was careening. And, as often happens on clear nights, they seemed now to be in perspective, now on a plane; now piled tier beyond tier into the infinite heavens, now concealing infinity, a roof limiting for ever the visions of men. In either case they seemed intolerable. "Are we to travel in the dark?" called the passengers angrily, and the attendant, who had been careless, generated the light, and pulled down the blinds of pliable metal. When the air-ships had been built, the desire to look direct at things still lingered in the world. Hence the extraordinary number of skylights and windows, and the proportionate discomfort to those who were civilised and refined. Even in Vashti's cabin one star peeped through a flaw in the blind, and after a few hours' uneasy slumber, she was disturbed by an unfamiliar glow, which was the dawn.

Quick as the ship had sped westwards, the earth had rolled eastwards quicker still, and had dragged back Vashti and her companions towards the sun. Science could prolong the night, but only for a little, and those high hopes of neutralising the earth's diurnal revolution had passed, together with hopes that were possibly higher. To "keep pace with the sun," or even to outstrip it had been the aim of the civilisation preceding this. Racing aeroplanes had been built for the purpose, capable of enormous speed, and steered by the greatest intellects of the epoch. Round the globe they went, round and round, westward, westward, round and round, amidst humanity's applause. In vain. The globe went eastward quicker still, horrible accidents occurred, and the Committee of the Machine, at the time rising into prominence, declared the pursuit illegal, unmechanical, and punishable by Homelessness.

Of Homelessness more will be said later.

Doubtless the Committee was right. Yet the attempt to "defeat the sun" aroused the last common interest that our race experienced about the heavenly bodies, or indeed about anything. It was the last time that men were compacted by thinking of a power outside the world. The sun had conquered, yet it was the end of his spiritual dominion. Dawn, midday, twilight, the zodiacal path, touched neither men's lives nor their hearts, and science retreated into the ground, to concentrate herself upon problems that she was certain of solving.

So when Vashti found her cabin invaded by a rosy finger of light, she was annoyed, and tried to adjust the blind. But the blind flew up altogether, and she saw through the skylight small pink clouds, swaying against a background of blue, and as the sun crept higher, its radiance entered direct, brimming down the wall, like a golden sea. It rose and fell with the air-ship's motion, just as waves rise and fall, but it advanced steadily, as a tide advances. Unless she was careful, it would strike her face. A spasm of horror shook her and she rang for the attendant. The attendant too was horrified, but she could do nothing; it was not her place to mend the blind. She could only suggest that the lady should change her cabin, which she accordingly prepared to do.

People were almost exactly alike all over the world, but the attendant of the air-ship, perhaps owing to her exceptional duties, had grown a little out of the common. She had often to address passengers with direct speech, and this had given her a certain roughness and originality of manner. When Vashti swerved away from the sunbeams with a cry, she behaved barbarically—she put out her hand to steady her.

"How dare you!" exclaimed the passenger. "You forget yourself!"

The woman was confused, and apologised for not having let her fall. People never touched one another. The custom had become obsolete, owing to the Machine.

"Where are we now?" asked Vashti haughtily.

"We are over Asia," said the attendant, anxious to be polite.

"Asia?"

"You must excuse my common way of speaking. I have got into the habit of calling places over which I pass by their unmechanical names."

"Oh, I remember Asia. The Mongols came from it."

"Beneath us, in the open air, stood a city that was once called Simla."

"Have you ever heard of the Mongols and of the Brisbane school?"

"No."

"Brisbane also stood in the open air."

"Those mountains to the right—let me show you them." She pushed back a metal blind. The main chain of the Himalayas was revealed. "They were once called the Roof of the World, those mountains."

"What a foolish name!"

"You must remember that, before the dawn of civilisation, they seemed to be an impenetrable wall that touched the stars. It was supposed that no one but the gods could exist above their summits. How we have advanced, thanks to the Machine!"

"How we have advanced, thanks to the Machine!" said Vashti.

"How we have advanced, thanks to the Machine!" echoed the passenger who had dropped his Book the night before, and who was standing in the passage.

"And that white stuff in the cracks?—what is it?"

"I have forgotten its name."

"Cover the window, please. These mountains give me no ideas."

The northern aspect of the Himalayas was in deep shadow: on the Indian slope the sun had just prevailed. The forests had been destroyed during the literature epoch for the purpose of making newspaper-pulp, but the snows were awakening to their morning glory, and clouds still hung on the breasts of Kinchinjunga. In the plain were seen the ruins of cities, with diminished rivers creeping by their walls, and by the sides of these were sometimes the signs of vomitories, marking the cities of to-day. Over the whole prospect air-ships rushed, crossing and intercrossing with incredible *aplomb*, and rising nonchalantly when they desired to escape the perturbations of the lower atmosphere and to traverse the Roof of the World.

"We have indeed advanced, thanks to the Machine," repeated the attendant, and hid the Himalayas behind a metal blind.

The day dragged wearily forward. The passengers sat each in his cabin, avoiding one other with an almost physical repulsion and longing to be once more under the surface of the earth. There were eight or ten of them, mostly young males, sent out from the public nurseries to inhabit the rooms of those who had died in various parts of the earth. The man who had dropped his Book was on the homeward journey. He had been sent to Sumatra for the purpose of propagating the race. Vashti alone was travelling by her private will.

At midday she took a second glance at the earth. The air-ship was crossing another range of mountains, but she could see little, owing to clouds. Masses of black rock hovered below her, and merged indistinctly into grey. Their shapes were fantastic; one of them resembled a prostrate man.

"No ideas here," murmured Vashti, and hid the Caucasus behind a metal blind.

In the evening she looked again. They were crossing a golden sea, in which lay many small islands and one peninsula.

She repeated, "No ideas here," and hid Greece behind a metal blind.

Part II: The Mending Apparatus

By a vestibule, by a lift, by a tubular railway, by a platform, by a sliding door—by reversing all the steps of her departure did Vashti arrive at her son's room, which exactly resembled her own. She might well declare that the visit was superfluous. The buttons, the knobs, the reading desk with the Book, the temperature, the atmosphere, the illumination—all were exactly the same. And if Kuno himself, flesh of her flesh, stood close beside her at last, what profit was there in that? She was too well bred to shake him by the hand.

Averting her eyes, she spoke as follows:

"Here I am. I have had the most terrible journey and greatly retarded the development of my soul. It is not worth it, Kuno, it is not worth it. My time is too precious. The sunlight almost touched me, and I have met with the rudest people. I can only stop a few minutes. Say what you want to say, and then I must return."

"I have been threatened with Homelessness," said Kuno.

She looked at him now.

"I have been threatened with Homelessness, and I could not tell you such a thing through the Machine."

Homelessness means death. The victim is exposed to the air, which kills him.

"I have been outside since I spoke to you last. The tremendous thing has happened, and they have discovered me."

"But why shouldn't you go outside!" she exclaimed. "It is perfectly legal, perfectly mechanical, to visit the surface of the earth. I have lately been to a lecture on the sea; there is no objection to that; one simply summons a respirator and gets an Egression-permit. It is not the kind of thing that spiritually minded people do, and I begged you not to do it, but there is no legal objection to it."

"I did not get an Egression-permit."

"Then how did you get out?"

"I found out a way of my own."

The phrase conveyed no meaning to her, and he had to repeat it.

"A way of your own?" she whispered. "But that would be wrong."

"Why?"

The question shocked her beyond measure.

"You are beginning to worship the Machine," he said coldly. "You think it irreligious of me to have found out a way of my own. It was just what the Committee thought, when they threatened me with Homelessness."

At this she grew angry. "I worship nothing!" she cried. "I am most advanced. I don't think you irreligious, for there is no such thing as religion left. All the fear and the superstition that existed once have been destroyed by the Machine. I only meant that to find out a way of your own was——— Besides, there is no new way out."

"So it is always supposed."

"Except through the vomitories, for which one must have an Egression-permit, it is impossible to get out. The Book says so."

"Well, the Book's wrong, for I have been out on my feet."

For Kuno was possessed of a certain physical strength.

By these days it was a demerit to be muscular. Each infant was examined at birth, and all who promised undue strength were destroyed. Humanitarians may protest, but it would have been no true kindness to let an athlete live; he would never have been happy in that state of life to which the Machine had called him; he would have yearned for trees to climb, rivers to bathe in, meadows and hills against which he might measure his body. Man must be adapted to his surroundings, must he not? In the dawn of the world our weakly must be exposed on Mount Taygetus, in its twilight our strong will suffer euthanasia, that the Machine may progress, that the Machine may progress, that the Machine may progress eternally.

"You know that we have lost the sense of space. We say 'space is annihilated,' but we have annihilated not space, but the sense thereof. We have lost a part of ourselves. I determined to recover it, and I began by walking up and down the platform of the railway outside my room. Up and down, until I was tired, and so did recapture the meaning of 'Near' and 'Far.' 'Near' is a place to which I can get quickly *on my feet*, not a place to which the train or the air-ship will take me quickly. 'Far' is a place to which I cannot get quickly on my feet; the vomitory is 'far,' though I could be there in thirty-eight seconds by summoning the train. Man is the measure. That was my first lesson. Man's feet are the measure for distance, his hands are the measure for ownership, his body is the measure for all that is lovable and desirable and strong. Then I went further: it was then that I called to you for the first time, and you would not come.

"This city, as you know, is built deep beneath the surface of the earth, with only the vomitories protruding. Having paced the platform outside my own room, I took the lift to the next platform and paced that also, and so with each in turn, until I came to the topmost, above which begins the earth. All the platforms were exactly alike, and all that I gained by visiting them was to develop my sense of space and my muscles. I think I should have been content with this—it is not a little thing,—but as I walked and brooded, it occurred to me that our cities had been built in the days when men still breathed the outer air, and that there had been ventilation shafts for the workmen. I could think of nothing but these ventilation shafts. Had they been destroyed by all the food-tubes and medicine-tubes and music-tubes that the Machine has evolved lately? Or did traces of them remain? One thing was certain. If I came upon them anywhere, it would be in the railway-tunnels of the topmost story. Everywhere else, all space was accounted for.

"I am telling my story quickly, but don't think that I was not a coward or that your answers never depressed me. It is not the proper thing, it is not mechanical, it is not decent to walk along a railway-tunnel. I did not fear that I might tread upon a live rail and be killed. I feared something far more intangible—doing what was not contemplated by the Machine. Then I said to myself, 'Man is the measure' and I went, and after many visits I found an opening.

"The tunnels, of course, were lighted. Everything is light, artificial light; darkness is the exception. So when I saw a black gap in the tiles, I knew that it was an exception, and rejoiced. I put in my arm—I could put in no more at first—and waved it round and round in ecstasy. I loosened another tile, and put in my head, and shouted into the darkness: 'I am coming, I shall do it yet,' and my voice reverberated down endless passages. I seemed to hear the spirits of those dead workmen who had returned each evening to the starlight and to their wives, and all the generations who had lived in the open air called back to me, 'You will do it yet, you are coming.'"

He paused, and, absurd as he was, his last words moved her. For Kuno had lately asked to be a father, and his request had been refused by the Committee. His was not a type that the Machine desired to hand on.

"Then a train passed. It brushed by me, but I thrust my head and arms into the hole. I had done enough for one day, so I crawled back to the platform, went down in the lift, and summoned my bed. Ah what dreams! And again I called you, and again you refused."

She shook her head and said:

"Don't. Don't talk of these terrible things. You make me miserable. You are throwing civilisation away."

"But I had got back the sense of space and a man cannot rest then. I determined to get in at the hole and climb the shaft. And so I exercised my arms. Day after day I went through ridiculous movements, until my flesh ached, and I could hang by my hands and hold the pillow of my bed outstretched for many minutes. Then I summoned a respirator, and started.

"It was easy at first. The mortar had somehow rotted, and I soon pushed some more tiles in, and clambered after them into the darkness, and the spirits of the dead comforted me. I don't know what I mean by that. I just say what I felt. I felt, for the first time, that a protest had been lodged against corruption, and that even as the dead were comforting me, so I was comforting the unborn. I felt that humanity existed, and that it existed without clothes. How can I possibly explain this? It was naked, humanity seemed naked, and all these tubes and buttons and machineries neither came into the world with us, nor will they follow us out, nor do they matter supremely while we are here. Had I been strong, I would have torn off every garment I had, and gone out into the outer air unswaddled. But this is not for me, nor perhaps for my generation. I climbed with my respirator and my hygienic clothes and my dietetic tabloids! Better thus than not at all.

"There was a ladder, made of some primæval metal. The light from the railway fell upon its lowest rungs, and I saw that it led straight upwards out of the rubble at the bottom of the shaft. Perhaps our ancestors ran up and down it a dozen times daily, in their building. As I climbed, the rough edges cut through my gloves so that my hands bled. The light helped me for a little, and then came darkness and, worse still, silence which pierced my ears like a sword. The Machine hums! Did you know that? Its hum penetrates our blood, and may even guide our thoughts. Who knows! I was getting beyond its power. Then I thought: 'This silence means that I am doing wrong.' But I heard voices in the silence, and again they strengthened me." He laughed. "I had need of them. The next moment I cracked my head against something."

She sighed.

"I had reached one of those pneumatic stoppers that defend us from the outer air. You may have noticed them on the air-ship. Pitch dark, my feet on the rungs of an invisible ladder, my hands cut; I cannot explain how I lived through this part, but the voices still comforted me, and I felt for fastenings. The stopper, I suppose, was about eight feet across. I passed my hand over it as far as I could reach. It was perfectly smooth. I felt it almost to the centre. Not quite to the centre, for my arm was too short. Then the voice said: 'Jump. It is worth it. There may be a handle in the centre, and you may catch hold of it and so come to us your own way. And if there is no handle, so that you may fall and are dashed to pieces—it is still worth it: you will still come to us your own way.' So I jumped. There was a handle, and———"

He paused. Tears gathered in his mother's eyes. She knew that he was fated. If he did not die to-day he would die to-morrow. There was not room for such a person in the world. And with her pity disgust mingled. She was ashamed at having borne such a son, she who had always been so respectable and so full of ideas. Was he really the little boy to whom she had taught the use of his stops and buttons, and to whom she had given his first lessons in the Book? The very hair that disfigured his lip showed that he was reverting to some savage type. On atavism the Machine can have no mercy.

"There was a handle, and I did catch it. I hung tranced over the darkness and heard the hum of these workings as the last whisper in a dying dream. All the things I had cared about and all the people I had spoken to through tubes appeared infinitely little. Meanwhile the handle revolved. My weight had set something in motion and I span slowly, and then———

"I cannot describe it. I was lying with my face to the sunshine. Blood poured from my nose and ears and I heard a tremendous roaring. The stopper, with me clinging to it, had simply been blown out of the earth, and the air that we make down here was escaping through the vent into the air above. It burst up like a fountain. I crawled back to it—for the upper air hurts—and, as it were, I took great sips from the edge. My respirator had flown goodness knows where, my clothes were torn. I just lay with my lips close to the hole, and I sipped until the bleeding stopped. You can imagine nothing so curious. This hollow in the grass—I will speak of it in a minute,—the sun shining into it, not brilliantly but through marbled clouds,—the peace, the nonchalance, the sense of space, and, brushing my cheek, the roaring fountain of our artificial air! Soon I spied my respirator, bobbing up and down in the current high above my head, and higher still were many air-ships. But no one ever looks out of air-ships, and in my case they could not have picked me up. There I was, stranded. The sun shone a little way down the shaft, and revealed the topmost rung of the ladder, but it was hopeless trying to reach it. I should either have been tossed up again by the escape, or else have fallen in, and died. I could only lie on the grass, sipping and sipping, and from time to time glancing around me.

"I knew that I was in Wessex, for I had taken care to go to a lecture on the subject before starting. Wessex lies above the room in which we are talking now. It was once an important state. Its kings held all the southern coast from the Andredswald to Cornwall, while the Wansdyke protected them on the north, running over the high ground. The lecturer was only concerned with the rise of Wessex, so I do not know how long it remained an international power, nor would the knowledge have assisted me. To tell the truth I could do nothing but laugh, during this part. There was I, with a pneumatic stopper by my side and a respirator bobbing over my head, imprisoned, all three of us, in a grass-grown hollow that was edged with fern."

Then he grew grave again.

"Lucky for me that it was a hollow. For the air began to fall back into it and to fill it as water fills a bowl. I could crawl about. Presently I stood. I breathed a mixture, in which the air that hurts predominated whenever I tried to climb the sides. This was not so bad. I had not lost my tabloids and remained ridiculously cheerful, and as for the Machine, I forgot about it altogether. My one aim now was to get to the top, where the ferns were, and to view whatever objects lay beyond.

"I rushed the slope. The new air was still too bitter for me and I came rolling back, after a momentary vision of something grey. The sun grew very feeble, and I remembered that he was in Scorpio—I had been to a lecture on that too. If the sun is in Scorpio and you are in Wessex, it means that you must be as quick as you can, or it will get too dark. (This is the first bit of useful information I have ever got from a lecture, and I expect it will be the last.) It made me try frantically to breathe the new air, and to advance as far as I dared out of my pond. The hollow filled so slowly. At times I thought that the fountain played with less vigour. My respirator seemed to dance nearer the earth; the roar was decreasing."

He broke off.

"I don't think this is interesting you. The rest will interest you even less. There are no Ideas in it, and I wish that I had not troubled you to come. We are too different, mother."

She told him to continue.

"It was evening before I climbed the bank. The sun had very nearly slipped out of the sky by this time, and I could not get a good view. You, who have just crossed the Roof of the World, will not want to hear an account of the little hills that I saw— low colourless hills. But to me they were living and the turf that covered them was a skin, under which their muscles rippled, and I felt that those hills had called with incalculable force to men in the past, and that men had loved them. Now they sleep— perhaps for ever. They commune with humanity in dreams. Happy the man, happy the woman, who awakes the hills of Wessex. For though they sleep, they will never die."

His voice rose passionately.

"Cannot you see, cannot all your lecturers see, that it is we that are dying, and that down here the only thing that really lives is the Machine? We created the Machine, to do our will, but we cannot make it do our will now. It has robbed us of the sense of space and of the sense of touch, it has blurred every human relation and narrowed down love to a carnal act, it has paralysed our bodies and our wills, and now it compels us to worship it. The Machine develops—but not on our lines. The Machine proceeds—but not to our goal. We only exist as the blood corpuscles that course through its arteries, and if it could work without us, it would let us die. Oh, I have no remedy—or, at least, only one—to tell men again and again that I have seen the hills of Wessex as Ælfrid saw them when he overthrew the Danes.

"So the sun set. I forgot to mention that a belt of mist lay between my hill and other hills, and that it was the colour of pearl."

He broke off for the second time.

"Go on," said his mother wearily.

He shook his head.

"Go on. Nothing that you say can distress me now. I am hardened."

"I had meant to tell you the rest, but I cannot: I know that I cannot: goodbye."

Vashti stood irresolute. All her nerves were tingling with his blasphemies. But she was also inquisitive.

"This is unfair," she complained. "You have called me across the world to hear your story, and hear it I will. Tell me—as briefly as possible, for this is a disastrous waste of time—tell me how you returned to civilisation."

"Oh—that!" he said, starting. "You would like to hear about civilisation. Certainly. Had I got to where my respirator fell down?"

"No—but I understand everything now. You put on your respirator, and managed to walk along the surface of the earth to a vomitory, and there your conduct was reported to the Central Committee."

"By no means."

He passed his hand over his forehead, as if dispelling some strong impression. Then, resuming his narrative, he warmed to it again.

"My respirator fell about sunset. I had mentioned that the fountain seemed feebler, had I not?"

"Yes."

"About sunset, it let the respirator fall. As I said, I had entirely forgotten about the Machine, and I paid no great attention at the time, being occupied with other things. I had my pool of air, into which I could dip when the outer keenness became intolerable, and which would possibly remain for days, provided that no wind sprang up to disperse it. Not until it was too late, did I realise what the stoppage of the escape implied. You see—the gap in the tunnel had been mended; the Mending Apparatus; the Mending Apparatus, was after me.

"One other warning I had, but I neglected it. The sky at night was clearer than it had been in the day, and the moon, which was about half the sky behind the sun, shone into the dell at moments quite brightly. I was in my usual place—on the boundary between the two atmospheres—when I thought I saw something dark move across the bottom of the dell, and vanish into the shaft. In my folly, I ran down. I bent over and listened, and I thought I heard a faint scraping noise in the depths.

"At this—but it was too late—I took alarm. I determined to put on my respirator and to walk right out of the dell. But my respirator had gone. I knew exactly where it had fallen—between the stopper and the aperture—and I could even feel the mark that it had made in the turf. It had gone, and I realised that something evil was at work, and I had better escape to the other air, and, if I must die, die running towards the cloud that had been the colour of a pearl. I never started. Out of the shaft—it is too horrible. A worm, a long white worm, had crawled out of the shaft and was gliding over the moonlit grass.

"I screamed. I did everything that I should not have done, I stamped upon the creature instead of flying from it, and it at once curled round the ankle. Then we fought. The worm let me run all over the dell, but edged up my leg as I ran. 'Help!' I cried. (That part is too awful. It belongs to the part that you will never know.) 'Help!' I cried. (Why cannot we suffer in silence?) 'Help!' I cried. Then my feet were wound together, I fell, I was dragged away from the dear ferns and the living hills, and past the great metal stopper (I can tell you this part), and I thought it might save me again if I caught hold of the handle. It also was enwrapped, it also. Oh, the whole dell was full of the things. They were searching it in all directions, they were denuding it, and the white snouts of others peeped out of the hole, ready if needed. Everything that could be moved they brought—brushwood, bundles of fern, everything, and down we all went intertwined into hell. The last things that I saw, ere the stopper closed after us, were certain stars, and I felt that a man of my sort lived in the sky. For I did fight, I fought till the very end, and it was only my head hitting against the ladder that quieted me. I woke up in this room. The worms had vanished. I was surrounded by artificial air, artificial light, artificial peace, and my friends were calling to me down speaking-tubes to know whether I had come across any new ideas lately."

Here his story ended. Discussion of it was impossible, and Vashti turned to go.

"It will end in Homelessness," she said quietly.

"I wish it would," retorted Kuno.

"The Machine has been most merciful."

"I prefer the mercy of God."

"By that superstitious phrase, do you mean that you could live in the outer air?"

"Yes."

"Have you ever seen, round the vomitories, the bones of those who were extruded after the Great Rebellion?"

"Yes."

"They were left where they perished for our edification. A few crawled away, but they perished, too—who can doubt it? And so with the Homeless of our own day. The surface of the earth supports life no longer."

"Indeed."

"Ferns and a little grass may survive, but all higher forms have perished. Has any airship detected them?"

"No."

"Has any lecturer dealt with them?"

"No."

"Then why this obstinacy?"

"Because I have seen them," he exploded.

"Seen *what?*"

"Because I have seen her in the twilight—because she came to my help when I called—because she, too, was entangled by the worms, and, luckier than I, was killed by one of them piercing her throat."

He was mad. Vashti departed, nor, in the troubles that followed, did she ever see his face again.

Part III: The Homeless

During the years that followed Kuno's escapade, two important developments took place in the Machine. On the surface they were revolutionary, but in either case men's minds had been prepared beforehand, and they did but express tendencies that were latent already.

The first of these was the abolition of respirators.

Advanced thinkers, like Vashti, had always held it foolish to visit the surface of the earth. Air-ships might be necessary, but what was the good of going out for mere curiosity and crawling along for a mile or two in a terrestrial motor? The habit was vulgar and perhaps faintly improper: it was unproductive of ideas, and had no connection with the habits that really mattered. So respirators were abolished, and with them, of course, the terrestrial motors, and except for a few lecturers, who complained that they were debarred access to their subject-matter, the development was accepted quietly. Those who still wanted to know what the earth was like, had after all only to listen to some gramophone, or to look into some cinematophote. And even the lecturers acquiesced when they found that a lecture on the sea was none the less stimulating when compiled out of other lectures that had already been delivered on the same subject. "Beware of first-hand ideas!" exclaimed one of the most advanced of them. "First-hand ideas do not really exist. They are but the physical impressions produced by love and fear, and on this gross foundation who could erect a philosophy? Let your ideas be second-hand, and if possible tenth-hand, for then they will be far removed from that disturbing element—direct observation. Do not learn anything about this subject of mine—the French Revolution. Learn instead what I think that Enicharmon thought Urizen thought Gutch thought Ho-Yung thought Chi-Bo-Sing thought Lafcadio Hearn thought Carlyle thought Mirabeau said about the French Revolution. Through the medium of these ten great minds, the blood that was shed at Paris and the windows that were broken at Versailles will be clarified to an Idea which you may employ most profitably in your daily lives. But be sure that the intermediates are many and varied, for in history one authority exists to counteract another. Urizen must counteract the scepticism of Ho-Yung and Enicharmon, I must myself counteract the impetuosity of Gutch. You who listen to me are in a better position to judge about the French Revolution than I am. Your descendants will be even in a better position than you, for they will learn what you think I think, and yet another intermediate will be added to the chain. And in time"—his voice rose—"there will come a generation that has got beyond facts, beyond impressions, a generation absolutely colourless, a generation

"seraphically free
From taint of personality,"

which will see the French Revolution not as it happened, nor as they would like it to have happened, but as it would have happened, had it taken place in the days of the Machine."

Tremendous applause greeted this lecture, which did but voice a feeling already latent in the minds of men—a feeling that terrestrial facts must be ignored, and that the abolition of respirators was a positive gain. It was even suggested that air-ships should be abolished too. This was not done, because air-ships had somehow worked themselves into the Machine's system. But year by year they were used less, and mentioned less by thoughtful men.

The second great development was the re-establishment of religion.

This, too, had been voiced in the celebrated lecture. No one could mistake the reverent tone in which the peroration had concluded, and it awakened a responsive echo in the heart of each. Those who had long worshipped silently, now began to talk. They described the strange feeling of peace that came over them when they handled the Book of the Machine, the pleasure that it was to repeat certain numerals out of it, however little meaning those numerals conveyed to the outward ear, the ecstasy of touching a button, however unimportant, or of ringing an electric bell, however superfluously.

"The Machine," they exclaimed, "feeds us and clothes us and houses us; through it we speak to one another, through it we see one another, in it we have our being. The Machine is the friend of ideas and the enemy of superstition: the Machine is omnipotent, eternal; blessed is the Machine." And before long this allocution was printed on the first page of the Book, and in subsequent editions the ritual swelled into a complicated system of praise and prayer. The word "religion" was sedulously avoided, and in theory the Machine was still the creation and the implement of man. But in practice all, save a few retrogrades, worshipped it as divine. Nor was it worshipped in unity. One believer would be chiefly impressed by the blue optic plates, through which he saw other believers; another by the mending apparatus, which sinful Kuno had compared to worms; another by the lifts, another by the Book. And each would pray to this or to that, and ask it to intercede for him with the Machine as a whole. Persecution—that also was present. It did not break out, for reasons that will be set forward shortly. But it was latent, and all who did not accept the minimum known as "undenominational Mechanism" lived in danger of Homelessness, which means death, as we know.

To attribute these two great developments to the Central Committee, is to take a very narrow view of civilisation. The Central Committee announced the developments, it is true, but they were no more the cause of them than were the Kings of the Imperialistic period the cause of war. Rather did they yield to some invincible pressure, which came no one knew whither, and which, when gratified, was succeeded by some new pressure equally invincible. To such a state of affairs it is convenient to give the name of Progress. No one confessed the Machine was out of hand. Year by year it was served with increased efficiency and decreased intelligence. The better a man knew his own duties upon it, the less he understood the duties of his neighbour, and in all the world there was not one who understood the monster as a whole. Those master brains had perished. They had left full directions, it is true, and their successors had each of them mastered a portion of those directions. But Humanity, in its desire for comfort, had over-reached itself. It had exploited the riches of nature too far. Quietly and complacently, it was sinking into decadence, and Progress had come to mean the Progress of the Machine.

As for Vashti, her life went peacefully forward until the final disaster. She made her room dark and slept; she awoke and made the room light. She lectured and attended lectures. She exchanged ideas with her innumerable friends and believed she was growing more spiritual. At times a friend was granted Euthanasia, and left his or her room for the Homelessness that is beyond all human conception. Vashti did not much mind. After an unsuccessful lecture, she would sometimes ask for Euthanasia herself. But the death-rate was not permitted to exceed the birth-rate, and the Machine had hitherto refused it to her.

The troubles began quietly, long before she was conscious of them.

One day she was astonished at receiving a message from her son. They never communicated, having nothing in common, and she had only heard indirectly that he was still alive, and had been transferred from the northern hemisphere, where he had behaved so mischievously, to the southern—indeed, to a room not far from her own.

"Does he want me to visit him?" she thought. "Never again, never. And I have not the time."

No, it was madness of another kind.

He refused to visualise his face upon the blue plate, and speaking out of the darkness with solemnity said:

"The Machine stops."

"What do you say?"

"The Machine is stopping, I know it, I know the signs."

She burst into a peal of laughter. He heard her and was angry, and they spoke no more.

"Can you imagine anything more absurd?" she cried to a friend. "A man who was my son believes that the Machine is stopping. It would be impious if it was not mad."

"The Machine is stopping?" her friend replied. "What does that mean? The phrase conveys nothing to me."

"Nor to me."

"He does not refer, I suppose, to the trouble there has been lately with the music?"

"Oh no, of course not. Let us talk about music."

"Have you complained to the authorities?"

"Yes, and they say it wants mending, and referred me to the Committee of the Mending Apparatus. I complained of those curious gasping sighs that disfigure the symphonies of the Brisbane school. They sound like some one in pain. The Committee of the Mending Apparatus say that it shall be remedied shortly."

Obscurely worried, she resumed her life. For one thing, the defect in the music irritated her. For another thing, she could not forget Kuno's speech. If he had known that the music was out of repair—he could not know it, for he detested music—if he had known that it was wrong, "the Machine stops" was exactly the venomous sort of remark he would have made. Of course he had made it at a venture, but the coincidence annoyed her, and she spoke with some petulance to the Committee of the Mending Apparatus.

They replied, as before, that the defect would be set right shortly.

"Shortly! At once!" she retorted. "Why should I be worried by imperfect music? Things are always put right at once. If you do not mend it at once, I shall complain to the Central Committee."

"No personal complaints are received by the Central Committee," the Committee of the Mending Apparatus replied.

"Through whom am I to make my complaint, then?"

"Through us."

"I complain then."

"Your complaint shall be forwarded in its turn."

"Have others complained?"

This question was unmechanical, and the Committee of the Mending Apparatus refused to answer it.

"It is too bad!" she exclaimed to another of her friends. "There never was such an unfortunate woman as myself. I can never be sure of my music now. It gets worse and worse each time I summon it."

"I too have my troubles," the friend replied. "Sometimes my ideas are interrupted by a slight jarring noise."

"What is it?"

"I do not know whether it is inside my head, or inside the wall."

"Complain, in either case."

"I have complained, and my complaint will be forwarded in its turn to the Central Committee."

Time passed, and they resented the defects no longer. The defects had not been remedied, but the human tissues in that latter day had become so subservient, that they readily adapted themselves to every caprice of the Machine. The sigh at the crisis of the Brisbane symphony no longer irritated Vashti; she accepted it as part of the melody. The jarring noise, whether in the head or in the wall, was no longer resented by her friend. And so with the mouldy artificial fruit, so with the bath water that began to stink, so with the defective rhymes that the poetry machine had taken to emit. All were bitterly complained of at first, and then acquiesced in and forgotten. Things went from bad to worse unchallenged.

It was otherwise with the failure of the sleeping apparatus. That was a more serious stoppage. There came a day when over the whole world—in Sumatra, in Wessex, in the innumerable cities of Courland and Brazil—the beds, when summoned by their tired owners, failed to appear. It may seem a ludicrous matter, but from it we may date the collapse of humanity. The Committee responsible for the failure was assailed by complainants, whom it referred, as usual, to the Committee of the Mending Apparatus, who in its turn assured them that their complaints would be forwarded to the Central Committee. But the discontent grew, for mankind was not yet sufficiently adaptable to do without sleep.

"Some one is meddling with the Machine———" they began.

"Some one is trying to make himself king, to reintroduce the personal element."

"Punish that man with Homelessness."

"To the rescue! Avenge the Machine! Avenge the Machine!"

"War! Kill the man!"

But the Committee of the Mending Apparatus now came forward, and allayed the panic with well-chosen words. It confessed that the Mending Apparatus was itself in need of repair.

The effect of this frank confession was admirable.

"Of course," said a famous lecturer—he of the French Revolution, who gilded each new decay with splendour—"of course we shall not press our complaints now. The Mending Apparatus has treated us so well in the past that we all sympathise with it, and will wait patiently for its recovery. In its own good time it will resume its duties. Meanwhile let us do without our beds, our tabloids, our other little wants. Such, I feel sure, would be the wish of the Machine."

Thousands of miles away his audience applauded. The Machine still linked them. Under the seas, beneath the roots of the mountains, ran the wires through which they saw and heard, the enormous eyes and ears that were their heritage, and the hum of many workings clothed their thoughts in one garment of subserviency. Only the old and the sick remained ungrateful, for it was rumoured that Euthanasia, too, was out of order, and that pain had reappeared among men.

It became difficult to read. A blight entered the atmosphere and dulled its luminosity. At times Vashti could scarcely see across her room. The air, too, was foul. Loud were the complaints, impotent the remedies, heroic the tone of the lecturer as he cried: "Courage! courage! What matter so long as the Machine goes on? To it the darkness and the light are one." And though things improved again after a time, the old brilliancy was never recaptured, and humanity never recovered from its entrance into twilight. There was an hysterical talk of "measures," of "provisional dictatorship," and the inhabitants of Sumatra were asked to familiarise themselves with the workings of the central power station, the said power station being situated in France. But for the most part panic reigned, and men spent their strength praying to their Books, tangible proofs of the Machine's omnipotence. There were gradations of terror—at times came rumours of hope—the Mending Apparatus was almost mended—the enemies of the Machine had been got under—new "nerve-centres" were evolving which would do the work even more magnificently than before. But there came a day when, without the slightest warning, without any previous hint of feebleness, the entire of the communication-system broke down, all over the world, and the world, as they understood it, ended.

Vashti was lecturing at the time and her earlier remarks had been punctuated with applause. As she proceeded the audience became silent, and at the conclusion there was no sound. Somewhat displeased she called to a friend who was a specialist in sympathy. No sound: doubtless the friend was sleeping. And so with the next friend whom she tried to summon, and so with the next, until she remembered Kuno's cryptic remark, "The Machine stops."

The phrase still conveyed nothing. If Eternity was stopping it would of course be set going shortly.

For example, there was still a little light and air—the atmosphere had improved a few hours previously. There was still the Book, and while there was the Book there was security.

Then she broke down, for with the cessation of activity came an unexpected terror—silence.

She had never known silence, and the coming of it nearly killed her—it did kill many thousands of people outright. Ever since her birth she had been surrounded by the steady hum. It was to the ear what artificial air was to the lungs, and agonising pains shot across her head. And scarcely knowing what she did, she stumbled forward and pressed the unfamiliar button, the one that opened the door of her cell.

Now the door of the cell worked on a simple hinge of its own. It was not connected with the central power station, dying far away in France. It opened, rousing immoderate hopes in Vashti, for she thought that the Machine had been mended. It opened, and she saw the dim tunnel that curved far away towards freedom. One look, and then she shrank back. For the tunnel was full of people—she was almost the last in that city to have taken alarm.

People at any time repelled her, and these were nightmares from her worst dreams. People were crawling about, people were screaming, whimpering, gasping for breath, touching each other, vanishing in the dark, and ever and anon being pushed off the platform on to the live rail. Some were fighting round the electric bells, trying to summon trains which could not be summoned. Others were yelling for Euthanasia or for respirators, or blaspheming the Machine. Others stood at the doors of their cells fearing, like herself, either to stop in them or to leave them. And behind all the uproar was silence—the silence which is the voice of the earth and of the generations who have gone.

No—it was worse than solitude. She closed the door again and sat down to wait for the end. The disintegration went on, accompanied by horrible cracks and rumbling. The valves that restrained the Medical Apparatus must have been weakened, for it ruptured and hung hideously from the ceiling. The floor heaved and fell and flung her from her chair. A tube oozed towards her serpent fashion. And at last the final horror approached—light began to ebb, and she knew that civilisation's long day was closing.

She whirled round, praying to be saved from this, at any rate, kissing the Book, pressing button after button. The uproar outside was increasing, and even penetrated the wall. Slowly the brilliancy of her cell was dimmed, the reflections faded from her metal switches. Now she could not see the reading-stand, now not the Book, though she held it in her hand. Light followed the flight of sound, air was following light, and the original void returned to the cavern from which it had been so long excluded. Vashti continued to whirl, like the devotees of an earlier religion, screaming, praying, striking at the buttons with bleeding hands.

It was thus that she opened her prison and escaped—escaped in the spirit: at least so it seems to me, ere my meditation closes. That she escapes in the body—I cannot perceive that. She struck, by chance, the switch that released the door, and the rush of foul air on her skin, the loud throbbing whispers in her ears, told her that she was facing the tunnel again, and that tremendous platform on which she had seen men fighting. They were not fighting now. Only the whispers remained, and the little whimpering groans. They were dying by hundreds out in the dark.

She burst into tears.

Tears answered her.

They wept for humanity, those two, not for themselves. They could not bear that this should be the end. Ere silence was completed their hearts were opened, and they knew what had been important on the earth. Man, the flower of all flesh, the noblest of all creatures visible, man who had once made god in his image, and had mirrored his strength on the constellations, beautiful naked man was dying, strangled in the garments that he had woven. Century after century had he toiled, and here was his reward. Truly the garment had seemed heavenly at first, shot with the colours of culture, sewn with the threads of self-denial. And heavenly it had been so long as it was a garment and no more, so long as man could shed it at will and live by the essence that is his soul, and the essence, equally divine, that is his body. The sin against the body—it was for that they wept in chief; the centuries of wrong against the muscles and the nerves, and those five portals by which we can alone apprehend—glossing it over with talk of evolution, until the body was white pap, the home of ideas as colourless, last sloshy stirrings of a spirit that had grasped the stars.

"Where are you?" she sobbed.

His voice in the darkness said, "Here."

"Is there any hope, Kuno?"

"None for us."

"Where are you?"

She crawled towards him over the bodies of the dead. His blood spurted over her hands.

"Quicker," he gasped, "I am dying—but we touch, we talk, not through the Machine."

He kissed her.

"We have come back to our own. We die, but we have recaptured life, as it was in Wessex, when Ælfrid overthrew the Danes. We know what they know outside, they who dwelt in the cloud that is the colour of a pearl."

"But, Kuno, is it true? Are there still men on the surface of the earth? Is this—this tunnel, this poisoned darkness—really not the end?"

He replied:

"I have seen them, spoken to them, loved them. They are hiding in the mist and the ferns until our civilisation stops. To-day they are the Homeless—to-morrow———"

"Oh, to-morrow—some fool will start the Machine again, to-morrow."

"Never," said Kuno, "never. Humanity has learnt its lesson."

As he spoke, the whole city was broken like a honeycomb. An air-ship had sailed in through the vomitory into a ruined wharf. It crashed downwards, exploding as it went, rending gallery after gallery with its wings of steel. For a moment they saw the nations of the dead, and, ere they joined them, scraps of the untainted sky.

3 "The Prolongation of Life"

Francis Fukuyama

In this chapter of *Our Postmodern Future*, Francis Fukuyama explores the social and psychological implications of technological change. He supposes that we will continue to make progress in using technologies to extend human lifespan, and then speculates about the effects this will have on a range of social arrangements, practices, and ways of thinking. Like Forster, Fukuyama suggests that we may regret getting what we want or, at least, we will have to cope with profound changes to our world if we get what we seem to want. Forster and Fukuyama both imagine worlds in which important human values and aspirations are achieved—at least on a technical level—yet both authors seem to be warning us to "be careful what we wish for." Fukuyama suggests that longevity may radically transform how we think about our lives, one another, and the organization of our society. Increased longevity could wreak havoc on many of our social practices as those in power hold on to their power for much longer periods of time. In this way, Fukuyama tries to anticipate the social, ethical, and psychological changes that would accompany the achievement of one of the preeminent values in western culture—prolonging death. His exploration reminds us that even when those making technological decisions achieve their goal, the social changes required to create new technologies and the new abilities engendered by the technology can have broad reaching and pervasive implications.

Many die too late, and a few die too early. The doctrine sounds strange: "Die at the right time!"

Die at the right time—thus teaches Zarathustra. Of course, how could those who never live at the right time die at the right time? Would that they had never been born! Thus I counsel the superfluous. But even the superfluous still make a fuss about their dying; and even the hollowest nut wants to be cracked.

Friedrich Nietzsche, *Thus Spoke Zarathustra*, I.21

The third pathway by which contemporary biotechnology will affect politics is through the prolongation of life, and the demographic and social changes that will occur as a result. One of the greatest achievements of twentieth-century medicine in the United States was the raising of life expectancies at birth from 48.3 years for men and 46.3 for women in 1900 to 74.2 for men and 79.9 for women in 2000.[1] This shift, coupled with dramatically falling birthrates in much of the developed world, has

---

From *Our Posthuman Future: Consequences of the Biotechnology Revolution* (New York: Farrar, Straus and Giroux, 2002), pp. 57–71. Copyright © 2002 by Francis Fukuyama. Reprinted with permission of Farrar, Straus and Giroux, LLC.

already produced a very different global demographic backdrop for world politics, whose effects are arguably being felt already. Based on birth and mortality patterns already in place, the world will look substantially different in the year 2050 than it does today, even if biomedicine fails to raise life expectancies by a single year over that period. The likelihood that there will not be significant advances in the prolongation of life in this period is small, however, and there is some possibility that biotechnology will lead to very dramatic changes.

One of the areas most affected by advances in molecular biology has been gerontology, the study of aging. There are at present a number of competing theories as to why people grow old and eventually die, with no firm consensus as to the ultimate reasons or mechanisms by which this occurs.[2] One stream of theory comes out of evolutionary biology and holds, broadly, that organisms age and die because there are few forces of natural selection that favor the survival of individuals past the age at which they are able to reproduce.[3] Certain genes may favor an individual's ability to reproduce but become dysfunctional at later periods of life. For evolutionary biologists, the big mystery is not why individuals die but why, for example, human females have a long postmenopausal life span. Whatever the explanation, they tend to believe that aging is the result of the interaction of a large number of genes, and that therefore there are no genetic shortcuts to the postponement of death.[4]

Another stream of theory on aging comes out of molecular biology and concerns the specific cellular mechanisms by which the body loses its functionality and dies. There are two types of human cells: germ cells, which are contained in the female ovum and male sperm, and somatic cells, which include the other hundred trillion or so cells that constitute the rest of the body. All cells replicate by cell division. In 1961, Leonard Hayflick discovered that somatic cells had an upper limit in the total number of divisions they could undergo. The number of possible cell divisions decreased with the age of the cell.

There are a number of theories as to why the so-called Hayflick limit exists. The leading one has to do with the accumulation of random genetic damage as cells replicate.[5] With each cellular division, environmental factors like smoke and radiation, as well as chemicals known as free hydroxyl radicals and cellular waste products, can prevent the accurate copying of the DNA from one cell generation to the next. The body has a number of DNA repair enzymes that oversee the copying process and fix transcription problems as they arise, but these fail to catch all mistakes. With continued cell replication, the DNA damage builds up in the cells, leading to faulty protein synthesis and impaired functioning. These impairments are in turn the basis for diseases characteristic of aging, such as arteriosclerosis, heart disease, and cancer.

Another theory that seeks to explain the Hayflick limit is related to telomeres, the noncoding bits of DNA attached to the end of each chromosome.[6] Telomeres act like the leaders in a filmstrip and ensure that the DNA is accurately replicated. Cell division involves the splitting apart of the two strands of the DNA molecule and their reconstitution into complete new copies of the molecule in the daughter cells. But with each cell division, the telomeres get a bit shorter, until they are unable to protect the ends of the DNA strand and the cell, recognizing the short telomeres as damaged DNA, ceases growth. Dolly the sheep, cloned from somatic cells of an adult animal, had the shortened telomeres of an adult rather than the longer ones of a newborn lamb, and presumably will not live as long as a naturally born sibling.

There are three major types of cells that are not subject to the Hayflick limit: germ cells, cancer cells, and certain types of stem cells. The reason these cells can reproduce indefinitely has to do with the presence of an enzyme called telomerase, first isolated in 1989, which prevents the shortening of telomeres. This is what permits the germ line to continue through the generations without end, and is also what lies behind the explosive growth of cancer tumors.

Leonard Guarente of the Massachusetts Institute of Technology reported findings that calorie restriction in yeast increased longevity, through the action of a single gene known as SIR2 (silent information regulator No. 2). The SIR2 gene represses genes that generate ribosomal wastes that build up in yeast cells and lead to their eventual death; low-calorie diets restrict reproduction but are helpful to the functioning of the SIR2 gene. This may provide a molecular explanation for why laboratory rats fed a low-calorie diet live up to 40 percent longer than other rats.[7]

Biologists such as Guarente have suggested that there might someday be a relatively simple genetic route to life extension in humans: while it is not practical to feed people such restricted diets, there may be other ways of enhancing the functioning of the SIR genes. Other gerontologists, such as Tom Kirkwood, assert flatly that aging is the result of a complex series of processes at the level of cells, organs, and the body as a whole, and that there is therefore no single, simple mechanism that controls aging and death.[8]

If a genetic shortcut to immortality exists, the race is already on within the biotech industry to find it. The Geron Corporation has already cloned and patented the human gene for telomerase and, along with Advanced Cell Technology, has an active research program into embryonic stem cells. The latter are cells that make up an embryo at the earliest stages of development, before there has been any differentiation into different types of tissue and organs. Stem cells have the potential to become any cell or tissue in the body, and hence hold the promise of generating entirely new body parts to replace ones worn out through the aging process. Unlike organs transplanted from donors, such cloned body parts will be almost genetically identical to cells in the body into which they are placed, and so presumably free from the kinds of immune reactions that lead to transplant rejection.

Stem cell research represents one of the great frontiers of contemporary biomedical research. It is also hugely controversial as a result of its use of embryos as sources of stem cells—embryos which must be destroyed in the process.[9] The embryos usually come from the extra embryos "banked" by in vitro fertilization clinics. (Once created, stem cell "lines" can be replicated almost indefinitely.) Out of concern that stem cell research would encourage abortion or lead to the deliberate destruction of human embryos, the U.S. Congress imposed a ban on funding from the National Institutes of Health for research that could harm embryos,[10] pushing U.S. stem cell research into the private sector. In 2001 a bitter policy debate exploded in the United States as the Bush administration considered lifting the ban. In the end, the administration decided to permit federally funded research, but only on the sixty or so existing stem cell lines that had already been created.

It is impossible to know at this point whether the biotech industry will eventually be able to come up with a shortcut to the prolongation of life, such as a simple pill that will add another decade or two to people's life spans.[11] Even if this never happens, however, it seems fairly safe to say that the *cumulative* impact of all the biomedical research going on at present will be to further increase life expectancies over time and therefore to continue the trend that has been under way for the last century. So it is not at all premature to think through some of the political scenarios and social consequences that might emerge from demographic trends that are already well under way.

In Europe at the beginning of the eighteenth century, half of all children died before they reached the age of 15. The French demographer Jean Fourastié has pointed out that reaching the age of 52 was then an accomplishment, since only a small minority of the population did so, and that such a person might legitimately consider himself or herself a "survivor."[12] Since most people reached the peak of their productive lives during their 40s and 50s, a huge amount of human potential was wasted. In the 1990s, by contrast, over 83 percent of the population could expect to live to the age of 65, and more than 28 percent would still be alive at age 85.[13]

Increasing life expectancies are only part of the story of what has happened to populations in the developed world by the end of the twentieth century. The other major development has been the dramatic fall in fertility rates. Countries such as Italy, Spain, and Japan have total fertility rates (that is, the average number of children born to a woman in her lifetime) of between 1.1 and 1.5, far below the replacement rate of about 2.2. The combination of falling birthrates and increasing life expectancies has dramatically shifted the age distribution in developed countries. While the median age of the U.S. population was about 19 years in 1850, it had risen to 34 years by the 1990s.[14] This is nothing compared to what will happen in the first half of the twenty-first century. While the median age in the United States will climb to almost 40 by the year 2050, the change will be even more dramatic in Europe and Japan, where rates of immigration and fertility are lower. In the absence of an unanticipated increase in fertility, the demographer Nicholas Eberstadt estimates, based on UN data, that the median age in Germany will be 54, in Japan 56, and in Italy 58.[15] These estimates, it should be noted, do *not* assume any dramatic increases in life expectancies. If only some of the promises of biotechnology for gerontology pan out, it could well be the case that *half* of the populations of developed countries will be retirement age or older by this point.

Up to now, the "graying" of the populations of developed countries has been discussed primarily in the context of the social security liability that it will create. This looming crisis is real enough: Japan, for instance, will go from a situation in which there were four active workers for every retired person at the end of the twentieth century, to one in which there are only two workers per retired person a generation or so down the road. But there are other political implications as well.

Take international relations.[16] While some developing countries have succeeded in approaching or even crossing the demographic transition to subreplacement fertility and declining population growth, as the developed world has, many of the poorer parts of the world, including the Middle East and sub-Saharan Africa, continue to experience high rates of growth. This means that the dividing line between the First and Third Worlds in two generations will be a matter not simply of income and culture but of age as well, with Europe, Japan, and parts of North America having a median age of nearly 60 and their less developed neighbors having median ages somewhere in the early 20s.

In addition, voting age populations in the developed world will be more heavily feminized, in part because more women in the growing elderly cohort will live to advanced ages than men, and in part because of a long-term sociological shift toward greater female political participation. Indeed, elderly women will emerge as one of the most important blocs of voters courted by twenty-first-century politicians.

What this will mean for international politics is of course far from clear, but we do know on the basis of past experience that there are important differences in attitudes toward foreign policy and national security between women and men, and between older and younger voters. American women, for example, have always been less supportive than American men of U.S. involvement in war, by an average margin of seven to nine percentage points. They are also consistently less supportive of defense spending and the use of force abroad. In a 1995 Roper survey conducted for the Chicago Council on Foreign Relations, men favored U.S. intervention in Korea in the event of a North Korean attack by a margin of 49 to 40 percent, while women were opposed by a margin of 30 to 54. Fifty-four percent of men felt that it was important to maintain superior worldwide military power, compared with only 45 percent of women. Women, moreover, are less likely than men to see force as a legitimate tool for resolving conflicts.[17]

Developed countries will face other obstacles to the use of force. Elderly people, and particularly elderly women, are not the first to be called to serve in military organizations, so the pool of available military manpower will shrink. The willingness of people in such societies to tolerate battle casualties among their young may fall as well.[18] Nicholas Eberstadt estimates that given current fertility trends, Italy in 2050 will be a society in which only 5 percent of all children have any collateral relatives (that is, brothers, sisters, aunts, uncles, cousins, and so forth) at all. People will be primarily related to their parents, grandparents, great-grandparents, and to their own offspring. Such a tenuous generational line is likely to increase the reluctance to go to war and accept death in battle.

The world may well be divided, then, between a North whose political tone is set by elderly women, and a South driven by what Thomas Friedman labels super-empowered angry young men. It was a group of such men that carried out the September 11 attacks on the World Trade Center. This does not, of course, mean that the North will fail to rise to challenges posed by the South, or that conflict between the two regions is inevitable. Biology is not destiny. But politicians will have to work within frameworks established by basic demographic facts, and one of those facts may be that many countries in the North will be both shrinking and aging.

There is another, perhaps more likely, scenario that will bring these worlds into direct contact: immigration. The estimates of falling populations in Europe and Japan given above assume no large increases in net immigration. This is unlikely, however, simply because developed countries will want economic growth and the population necessary to sustain it. This means that the North-South divide will be replicated within each country, with an increasingly elderly native-born population living alongside a culturally different and substantially younger immigrant population. The United States and other English-speaking countries have traditionally been good at assimilating culturally diverse groups of immigrants, but other countries, such as Germany and Japan, have not. Europe has already seen the rise of anti-immigrant backlash movements, such as the National Front in France, the Vlaams Blok in Belgium, the Lega Lombarda in Italy, and Jörg Haider's Freedom Party in Austria. For these countries, changes in the age structure of their populations, abetted by increasing longevity, are likely to lay the ground for growing social conflict.

The prolongation of life through biotechnology will have dramatic effects on the internal structures of societies as well. The most important of these has to do with the management of social hierarchies.

Human beings are by nature status-conscious animals who, like their primate cousins, tend from an early age to arrange themselves in a bewildering variety of dominance hierarchies.[19] This hierarchical behavior is innate and has easily survived the arrival of modern ideologies like democracy and socialism that purport to be based on universal equality. (One has only to look at pictures of the politburos of the former Soviet Union and China, where the top leadership is arrayed in careful order of dominance.) The nature of these hierarchies has changed as a result of cultural evolution, from traditional ones based on physical prowess or inherited social status, to modern ones based on cognitive ability or education. But their hierarchical nature remains.

If one looks around at a society, one quickly discovers that many of these hierarchies are age-graded. Sixth graders feel themselves superior to fifth graders and dominate the playground if both have recess together; tenured professors lord it over untenured ones and carefully control entry into their august circle. Age-graded hierarchies make functional sense insofar as age is correlated in many societies with physical prowess, learning, experience, judgment, achievement, and the like. But past a certain age, the correlation between age and ability begins to go in the opposite direction. With life expectancies only in the 40s or 50s for most of human history, societies could rely on normal generation succession to take care of this problem. Mandatory retirement ages came into vogue only in the late nineteenth century, when increasing numbers of people began to survive into old age.*

---

*Bismarck, who established Europe's first social security system, set retirement at 65, an age to which virtually no one at that time lived.

Life extension will wreak havoc with most existing age-graded hierarchies. Such hierarchies traditionally assume a pyramidal structure because death winnows the pool of competitors for the top ranks, abetted by artificial constraints such as the widely held belief that everyone has the "right" to retire at age 65. With people routinely living and working into their 60s, 70s, 80s, and even 90s, however, these pyramids will increasingly resemble squat trapezoids or even rectangles. The natural tendency of one generation to get out of the way of the up-and-coming one will be replaced by the simultaneous existence of three, four, even five generations.

We have already seen the deleterious consequences of prolonged generational succession in authoritarian regimes that have no constitutional requirements limiting tenure in office. As long as dictators like Francisco Franco, Kim Il Sung, and Fidel Castro physically survive, their societies have no way of replacing them, and all political and social change is effectively on hold until they die.[20] In the future, with technologically enhanced life spans, such societies may find themselves locked in a ludicrous deathwatch not for years but for decades.

In societies that are more democratic and/or meritocratic, there are institutional mechanisms for removing leaders, bosses, or CEOs who are past their prime. But the problem does not go away by any stretch of the imagination.

The root problem lies, of course, in the fact that people at the top of social hierarchies generally do not want to lose status or power and will often use their considerable influence to protect their positions. Age-related declines in capabilities have to be fairly pronounced before other people will go to the trouble of removing a leader, boss, ballplayer, professor, or board member. Impersonal formal rules like mandatory retirement ages are useful precisely because they don't require institutions to make nuanced personal judgments about an individual older person's capability. But impersonal rules often discriminate against older people who are perfectly capable of continuing to work and for that reason have been abolished in many American workplaces.

There is at present a tremendous amount of political correctness regarding age: *ageism* has entered the pantheon of proscribed prejudices, next to racism, sexism, and homophobia. There is of course discrimination against older people, particularly in a youth-obsessed society like that of the United States. But there are also a number of reasons why generational succession is a good thing. Chief among them is that it is a major stimulant of progress and change.

Many observers have noted that political change often occurs at generational intervals—from the Progressive Era to the New Deal, from the Kennedy years to Reaganism.[21] There is no mystery as to why this is so: people born in the same age cohort experience major life events—the Great Depression, World War II, or the sexual revolution—together. Once people's life views and preferences have been formed by these experiences, they may adapt to new circumstances in small ways, but it is very difficult to get them to change broad outlooks. A black person who grew up in the old South has a hard time seeing a white cop as anything but an untrustworthy agent of an oppressive system of racial segregation, regardless of whether this makes sense given the realities of life in a northern city. Those who lived through the Great Depression cannot help feeling uneasy at the lavish spending habits of their grandchildren.

This is true not just in political but in intellectual life as well. There is a saying that the discipline of economics makes progress one funeral at a time, which is unfortunately truer than most people are willing to admit. The survival of a basic "paradigm" (for example, Keynesianism or Friedmanism) that shapes the way most scientists and intellectuals think about things at a particular time depends not just on empirical evidence, as some would like to think, but on the physical survival of the people who created that paradigm. As long as they sit on top of age-graded hierarchies like peer review boards, tenure committees, and foundation boards of trustees, the basic paradigm will often remain virtually unshakable.

It stands to reason, then, that political, social, and intellectual change will occur much more slowly in societies with substantially longer average life spans. With three or more generations active and working at the same time, the younger age cohorts will never constitute more than a small minority of voices clamoring to be heard, and generational change will never be fully decisive. To adjust more rapidly, such societies will have to establish rules mandating constant retraining and downward social mobility at later stages in life. The idea that one can acquire skills and education during one's 20s that will remain useful for the next forty years is implausible enough at present, given the pace of technological change. The idea that these skills would remain relevant over working lives of fifty, sixty, or seventy years becomes even more preposterous. Older people will have to move down the social hierarchy not just to retrain but to make room for new entrants coming up from the bottom. If they don't, generational warfare will join class and ethnic conflict as a major dividing line in society. Getting older people out of the way of younger ones will become a significant struggle, and societies may have to resort to impersonal, institutionalized forms of ageism in a future world of expanded life expectancies.

Other social effects of life extension will depend heavily on the exact way that the geriatric revolution plays itself out—that is, whether people will remain physically and mentally vigorous throughout these lengthening life spans, or whether society will increasingly come to resemble a giant nursing home.

The medical profession is dedicated to the proposition that anything that can defeat disease and prolong life is unequivocally a good thing. The fear of death is one of the deepest and most abiding human passions, so it is understandable that we should celebrate any advance in medical technology that appears to put death off. But people worry about the quality of their lives as well—not just the quantity. Ideally, one would like not merely to live longer but also to have one's different faculties fail as close as possible to when death finally comes, so that one does not have to pass through a period of debility at the end of life.

While many medical advances have increased the quality of life for older people, many have had the opposite effect by prolonging only one aspect of life and increasing dependency. Alzheimer's disease—in which certain parts of the brain waste away, leading to loss of memory and eventually dementia—is a good example of this, because the likelihood of getting it rises proportionately with age. At age 65, only one person in a hundred is likely to come down with Alzheimer's; at 85, it is one in six.[22] The rapid growth in the population suffering from Alzheimer's in developed countries is thus a direct result of increased life expectancies, which have prolonged the health of the body without prolonging resistance to this terrible neurological disease.

There are in fact two periods of old age that medical technology has opened up, at least for people in the developed world.[23] Category I extends from age 65 until sometime in one's 80s, when people can increasingly expect to live healthy and active lives, with enough resources to take advantage of them. Much of the happy talk about increased longevity concerns this period, and indeed the emergence of this new phase of life as a realistic expectation for most people is an achievement of which modern medicine can be proud. The chief problem for people in this category will be the encroachment of working life on their domain: for simple economic reasons, there will be powerful pressures to raise retirement ages and keep the over-65 cohort in the workforce for as long as possible. This does not imply any kind of social disaster: older workers may have to retrain and accept some degree of downward social mobility, but many of them will welcome the opportunity to contribute their labor to society.

The second phase of old age, Category II, is much more problematic. It is the period that most people currently reach by their 80s, when their capabilities decline and they return increasingly to a childlike state of dependency. This is the period that society doesn't like to think about, much less experience, since it flies in the face of ideals of personal autonomy that most people hold dear. Increases in the number of people in both Category I and Category II have created a novel situation in which individuals approaching retirement age today find their own choices constrained by the fact that they still have an elderly parent alive and dependent on them for care.

The social impact of ever-increasing life expectancies will depend on the relative sizes of these two groups, which in turn will depend on the "evenness" of future life-prolonging advances. The best scenario would be one in which technology simultaneously pushes back parallel aging processes—for instance, by the discovery of a common molecular source of aging in all somatic cells, and the delaying of this process throughout the body. Failure of the different parts would come at the same time, just later; people in Category I would be more numerous and those in Category II less so. The worst scenario would be one of highly uneven advance, in which, for example, we found ways to preserve bodily health but could not put off age-related mental deterioration. Stem cell research might yield ways to grow new body parts, as William Haseltine...suggest[ed].... But without a parallel cure for Alzheimer's disease, this wonderful new technology would do no more than allow more people to persist in vegetative states for years longer than is currently possible.

An explosion in the number of people in Category II might be labeled the national nursing home scenario, in which people routinely live to be 150 but spend the last fifty years in a state of childlike dependence on caretakers. There is of course no way of predicting whether this or the happier extension of the Category I period will play itself out. If there is no molecular shortcut to postponing death because aging is the result of the gradual accumulation of damage to a wide range of different biological systems, then there is no reason to think that future medical advances will proceed with a neat simultaneity, any more than they have in the past. That existing medical technology is capable only of keeping people's bodies alive at a much reduced quality of life is the reason assisted suicide and euthanasia, as well as figures like Jack Kevorkian, have come to the fore as public issues in the United States and elsewhere in recent years.

In the future, biotechnology is likely to offer us bargains that trade off length of life span for quality of life. If they are accepted, the social consequences could be dramatic. But assessing them will be very difficult: slight changes in mental capabilities such as loss of short-term memory or growing rigidity in one's beliefs are inherently difficult to measure and evaluate. The political correctness about aging noted earlier will make a truly frank assessment nearly impossible, both for individuals dealing with elderly relatives and for societies trying to formulate public policies. To avoid any hint of discrimination against older people, or the suggestion that their lives are somehow worth less than those of the young, anyone who writes on the future of aging feels compelled to be relentlessly sunny in predicting that medical advances will increase both the quantity and quality of life.

This is most evident with regard to sexuality. According to one writer on aging, "One of the factors inhibiting sexuality with ageing is undoubtedly the brain-washing that all of us experience which says that the older person is less sexually attractive."[24] Would that sexuality were only a matter of brainwashing! Unfortunately, there are good Darwinian reasons that sexual attractiveness is linked to youth, particularly in women. Evolution has created sexual desire for the purpose of fostering reproduction, and there are few selective pressures for humans to develop sexual attraction to partners past their prime reproductive years.[25] The consequence is that in another fifty years, most developed societies may have become "postsexual," in the sense that the vast majority of their members will no longer put sex at the top of their "to do" lists.

There are a number of unanswerable questions about what life in this kind of future would be like, since there have never in human history been societies with median ages of 60, 70, or higher. What would such a society's self-image be? If you go to a typical airport newsstand and look at the people pictured on magazine covers, their median age is likely to be in the low 20s, the vast majority good-looking and in perfect health. For most historical human societies, these covers would have reflected the actual median age, though not the looks or health, of the society as a whole. What will magazine covers look like in another couple of generations, when people in their early 20s constitute only a tiny minority of the population? Will society still want to think of itself as young, dynamic, sexy, and healthy, even though the image departs from the reality that people see around them to an even more extreme degree than today? Or will tastes and habits shift, with the youth culture going into terminal decline?

A shift in the demographic balance toward societies with a majority of people in Categories I and II will have much more profound implications for the meaning of life and death as well. For virtually all of human history up to the present, people's lives and identities were bound up either with reproduction—that is, having families and raising children—or with earning the resources to support themselves and their families. Family and work both enmesh individuals in a web of social obligations over which they frequently have little control and which are a source of struggle and anxiety but also of tremendous satisfaction. Learning to meet those social obligations is a source of both morality and character. People in Categories I and II, by contrast, will have a much more attenuated relationship to both family and work. They will be beyond reproductive years, with links primarily to ancestors and descendants. Some in Category I may choose to work, but the obligation to work and the kinds of mandatory social ties that work engenders will be replaced largely by a host of elective occupations. Those in Category II will not reproduce, not work, and indeed will see a flow of resources and obligation moving one way: toward them.

This does not mean that people in either category will suddenly become irresponsible or footloose. It does mean, however, that they may find their lives both emptier and lonelier, since it is precisely those obligatory ties that make life worth living for many people. When retirement is seen as a brief period of leisure following a life of hard work and struggle, it may seem like a well-deserved reward; if it stretches on for twenty or thirty years with no apparent end, it may seem simply pointless. And it is hard to see how a prolonged period of dependency or debility for people in Category II will be experienced as joyful or fulfilling.

People's relationship to death will change as well. Death may come to be seen not as a natural and inevitable aspect of life, but a preventable evil like polio or the measles. If so, then accepting death will appear to be a foolish choice, not something to be faced with dignity or nobility. Will people still be willing to sacrifice their lives for others, when their lives could potentially stretch out ahead of them indefinitely, or condone the sacrifice of the lives of others? Will they cling desperately to the life that biotechnology offers? Or might the prospect of an unendingly empty life appear simply unbearable?

Notes

1. See http://www.demog.berkeley.edu/~andrew/1918/figure2.html for the 1900 figures, and https://www.cia.gov/library/publications/the-world-factbook/index.html.

2. For an overview of these theories, see Michael R. Rose, *Evolutionary Biology of Aging* (New York: Oxford University Press, 1991). p. 160 ff; Caleb E. Finch and Rudolph E. Tanzi, "Genetics of Aging," *Science* 278 (1997): 407–411; S. Michal Jazwinski, "Longevity, Genes, and Aging," *Science* 273 (1996): 54–59; and David M. A. Mann, "Molecular Biology's Impact on Our Understanding of Aging," *British Medical Journal* 315 (1997): 1078–1082.

3. Michael R. Rose, "Finding the Fountain of Youth," *Technology Review* 95, no. 7 (October 1992): 64–69.

4. Nicholas Wade, "A Pill to Extend Life? Don't Dismiss the Notion Too Quickly," *The New York Times*, September 22, 2000, p. A20.

5. Tom Kirkwood, *Time of Our Lives: Why Ageing Is Neither Inevitable nor Necessary* (London: Phoenix, 1999), pp. 100–117.

6. Dwayne A. Banks and Michael Fossel, "Telomeres, Cancer, and Aging: Altering the Human Life Span," *Journal of the American Medical Association* 278 (1997): 1345–1348.

7. Nicholas Wade, "Searching for Genes to Slow the Hands of Biological Time," *The New York Times*, September 26, 2000, p. D1; Cheol-Koo Lee and Roger G. Klopp et al., "Gene Expression Profile of Aging and Its Retardation by Caloric Restriction," *Science* 285 (1999): 1390–1393.

8. Kirkwood (1999), p. 166.

9. For a sample of the discussion on stem cells, see Eric Juengst and Michael Fossel, "The Ethics of Embryonic Stem Cells—Now and Forever, Cells without End," *Journal of the American Medical Association* 284 (2000): 3180–3184; Juan de Dios Vial Correa and S. E. Mons. Elio Sgreccia, *Declaration on the Production and the Scientific and Therapeutic Use of Human Embryonic Stem Cells* (Rome: Pontifical Academy for Life, 2000); and M. J. Friedrich, "Debating Pros and Cons of Stem Cell Research," *Journal of the American Medical Association* 284, no. 6 (2000): 681–684.

10. Gabriel S. Gross, "Federally Funding Human Embryonic Stem Cell Research: An Administrative Analysis," *Wisconsin Law Review* 2000 (2000): 855–884.

11. For some research strategies into therapies for aging, see Michael R. Rose, "Aging as a Target for Genetic Engineering," in Gregory Stock and John Campbell, eds., *Engineering the Human Germline: An Exploration of the Science and Ethics of Altering the Genes We Pass to Our Children* (New York: Oxford University Press, 2000), pp. 53–56.

12. Jean Fourastié, "De la vie traditionelle à la vie tertiaire," *Population* 14 (1963): 417–432.

13. Kirkwood (1999), p. 6.

14. "Resident Population Characteristics—Percent Distribution and Median Age, 1850–1996, and Projections, 2000–2050," www.doi.gov/nrl/statAbst/Aidemo.pdt.

15. Nicholas Eberstadt, "World Population Implosion?," *Public Interest*, no. 129 (February 1997): 3–22.

16. On this issue, see Francis Fukuyama, "Women and the Evolution of World Politics," *Foreign Affairs* 77 (1998): 24–40.

17. Pamela J. Conover and Virginia Sapiro, "Gender, Feminist Consciousness, and War," *American Journal of Political Science* 37 (1993): 1079–1099.

18. Edward N. Luttwak, "Toward Post-Heroic Warfare," *Foreign Affairs* 74 (1995): 109–122.

19. For a longer discussion of this, see Francis Fukuyama, *The Great Disruption: Human Nature and the Reconstitution of Social Order* (New York: Free Press, 1999), pp. 212–230.

20. This point is made by Fred Charles Iklé, "The Deconstruction of Death," *The National Interest*, no. 62 (Winter 2000/01): 87–96.

21. Generational change is the theme, inter alia, of Arthur M. Schlesinger, Jr.'s, *Cycles of American History* (Boston: Houghton Mifflin, 1986); see also William Strauss and Neil Howe, *The Fourth Turning: An American Prophecy* (New York: Broadway Books, 1997).

22. Kirkwood (1999), pp. 131–132.

23. Michael Norman, "Living Too Long," *The New York Times Magazine*, January 14, 1996, pp. 36–38.

24. Kirkwood (1999), p. 238.

25. On the evolution of human sexuality, see Donald Symons, *The Evolution of Human Sexuality* (Oxford: Oxford University Press, 1979).

4 "Reproductive Ectogenesis: The Third Era of Human Reproduction and Some Moral Consequences"

Stellan Welin

In this piece by Stellan Welin, we again have a writer following a trend and imagining how an area of current research and development might simultaneously expand in a particular direction because of social and cultural pressure, lead to profound social changes, and raise issues that are difficult to grasp. Welin analyzes the possibility that success in human xenotransplantation—the transplantation of an organ from an animal to a human—will lead to technology that enables ectogenesis for human embryos—the ability to develop embryos outside the womb. He argues that while such a technology might help to resolve certain ethical issues, the ability to bring embryos to maturation in such a way would give rise to ethical questions that could never have existed before. Such a technology would void many of the assumptions that underlie the ethical decisions we've made about the rights of fetuses, the rights of expectant mothers, and the role of governments. Whether one agrees or disagrees with Welin on the likelihood of such technologies, his account is important for illustrating the profound social, political, and ethical changes that can accompany technological change.

"**ectogenesis** n. Biol. the production of structures outside the organism"
The Concise Oxford Dictionary

It is very hard to seriously discuss future issues in science and their moral implications. Usually, one is trapped in one of two embarrassing positions. The first is *the science fiction trap*. To discuss ectogenesis today will very easily elicit the simple response: "this is not something that is going to happen for many years—if ever. It is at present pure science fiction." The conclusion drawn from this is simply that the discussion is pointless. It should wait until this is a serious issue.

However, following the cloning of Dolly it is difficult to know what is science fiction and what is just around the corner. In a (then) seminal paper, Davor Solter and Jim McGrath declared that cloning of mammals was biologically impossible.[1] Then came Dolly in 1997 and Solter has obviously changed his mind.[2] Science fiction seems to be moving quickly towards science.

The other trap is that *everything has already happened*. If some new biomedical technology is already extensively used in the clinic, saving life, it is very hard to radically question that. The most one can do is to modify practices and avoid, hopefully, the worst abuses. At present, the technology of In Vitro Fertilisation (IVF) is in this state.

From *Science and Engineering Ethics* 10, no. 4 (October 2004): 615–626. Reprinted with permission from Springer Verlag.

In the discussion on human embryonic stem cells and their possibilities, there has been much discussion about the destruction of embryos. President Bush's decision on 9 August 2001 is an example.[3] While he bans destruction of human embryos as a means of deriving stem cells in the federal sector—but allows it in the private sector—he does not question the practice of IVF. In most countries IVF is a closed discussion, but human embryonic stem cells are still open for discussion.

In this paper I briefly comment on the three eras of human reproduction—and primarily on the relation between the new individual and the woman—and then spend some time on a fictional story illustrating some moral consequences of the third era. I will also comment on some of the possible consequences—moral, social and psychological—and try to answer the question on the desirability of the third era.

Using Fictional Stories to Explore Conceptual and Moral Problems

The Parfit Teletransporter

Being a philosopher by training I have become used to fantastic stories which illustrate some philosophical points and give arguments for various positions. One famous example in this semi-fictional and partly science-fiction style is Derek Parfit's story about teletransportation as a background to a discussion of personal identity.[4] In first person narrative Parfit tells how he would use a teletransporter to go from earth to Mars, a seemingly standard procedure. The teletransporter would separate the atoms in his body, beam the information to Mars, and set the atoms together. In Parfit's story something goes wrong. He enters the teletransporter, presses the button but does not as expected lose consciousness. To cut the story short, the teletransporter had worked in a new way. The information had been beamed to Mars where a replica appeared of the traveller still on earth. There is now one traveller on earth and one very similar traveller on Mars.

Parfit goes on to discuss personal identity. Who is who? There seems to be two very similar persons. Had the traveller's body disappeared on earth there would have been only one person. No one would have questioned that the person on Mars really was the traveller. In Parfit's story, the traveller on earth is told a little later that his heart had been damaged by the malfunction of the teletransporter and that he will shortly die. He talks to the traveller on Mars who assures the traveller on earth that he truly loves his wife (the same for both) and will carry on the same work as was important for the traveller on earth. As a reader I feel rather scared. The traveller on earth is a completely redundant person. No one will miss him when he dies. Strangely, somehow Parfit seems to regard the story as an antidote to being afraid of death.

What Parfit gives us is a disconnection between two entities that normally (in real life) travel together through space and time, namely your personal identity consisting of both mind and body. Realising the possibility of separation, (or in Parfit's case of doubling) even if it might never happen in real life, new questions will arise and cast doubt on old solutions.

Separating the Woman and the Fetus

In human reproduction, if we focus on the two "entities" that exist in space and time, namely the fetus/future child and the woman (leaving the man out of the picture), we can envisage different relations between them, corresponding to three different eras of human reproduction.

Historically, the first era is the normal conception inside the woman, the growth of the fetus in the womb, after nine months—birth, and the appearance of a new individual. The second era is In Vitro Fertilisation (IVF). The fetus starts outside the woman as a fertilised egg, moves to the body of the woman and spends nine months there. The body of the woman and the fetus travel together in space-time to separate at birth. In the third era of reproductive ectogenesis, the two never travel together. The fetus spends its gestational time entirely outside the woman's body. We have two entities separated in space-time continuously. The intimate connection consisting of the fetus being inside the woman's body is gone.

Obviously, new questions can be asked in the third era. Old answers depending on the intimate bodily connection between fetus and the woman will no longer be valid. This might, for example, have consequences for the morality of abortion, the rights of the fetus and might radically change parenthood. For example, Rosemarie Tong writes "Most feminists believe that the abortion debate should centre not on the question of whether fetuses are the moral equivalent of adult persons but, rather, on the fact that fertilized eggs develop into infants inside wombs of women."[5] In the third era (ectogenesis), the fertilised egg does not any longer develop inside the womb.

There are many well-known philosophical stories, not just Parfit's. We have Robert Nozick's story of the experience machine as some kind of argument that it is not just our experiences that matter,[6] Jonathan Glover on the moral (un)importance of the "dreamworld,"[7] Daniel Dennet on the (un)importance of the body for the location of the self,[8] and Judith Jarvis Thomson's abortion parable of the case when some one has been connected to and gives essential life support to another person for nine months.[9] There also have been stories of the future written by scientists and social scientists. My favourites are J. B. S. Haldane from 1923, where he discusses (the future) ectogenes,[10] and the sociological fiction of Michael Young on the coming of meritocracy.[11] In all humility I will give you mine.

The Future Scenario of Pig Pharmaceuticals Limited

The Reproductive Engineer's View

It is the year 2050 and the conference "Frontiers of medicine" is having its annual meeting. There is a special session celebrating the 20th anniversary of successful clinical applications of xenotransplantation. Gone are the shortage of organs for transplantation and the industrial-medical farm business is flourishing. After having listened to all the tales of heroes—actually it was never the surgeons who were heroic; all the suffering and the failures fell on the patients—the audience finally listens to another talk.

There is a special lecture on xenotransplantation and human reproduction. The chief scientific director of Pig Pharmaceuticals mounts the rostrum. Expectations are high; the firm has advertised that a new development will be announced at this meeting. Everyone attending the session is also invited to the evening barbecue at the poolside as an extra temptation to achieve maximum attendance.

Ladies and gentlemen, Mr. Chairman, It is a great honour for me as chief director of scientific research at Pig Pharmaceuticals to appear at such a distinguished session as this. I wish to present to you a new project, and I hope we will have a fruitful discussion. The aim of Pig Pharmaceuticals is to serve humanity by developing science-based new medical technologies related to the medical use of pigs.

As you all know xenotransplantation made its breakthrough at the beginning of this century. It was a combination of genetically modified pigs, new immunosupressive drugs and more effective methods of handling rejection episodes which laid the groundwork. Gone are the days when recipients had to wait for someone to die, when relatives had to be approached in their mourning and asked about donations. No longer must anyone make the agonising choice of who should have the organ. There are organs for everyone. Fortunately, the once widely-discussed risk of infectious disease caused by the retroviruses of the pig never materialised.

A picture showing the increasing curves for post-operative survival for xenotransplantation is projected on the screen. In the background happy pigs can be seen.

It is most appropriate that this session has paid tribute to the many pioneers, both physicians and patients, who made all this happen. However, I want to spend my time presenting our latest work in reproductive medicine.

Pregnancy is a difficult period for many women. Our present project started when some women who had had many consecutive miscarriages contacted our researchers. In some cases there had been In Vitro Fertilisation, but the pregnancies were never carried to term. These women wondered if it would be possible to save the fetuses by letting the human embryo develop in the uterus of our pigs. When the question first was raised, we were not sure if this was technically feasible, nor were we at Pig Pharmaceuticals certain of the ethical acceptability.

Our scientists were able to answer the technical question rather quickly. Such a pig-related pregnancy, as we now call it, is possible. In principle, it is the same kind of problem as in the xeno case.

The pregnancies of pigs are shorter than human pregnancies. And pigs carry many fetuses. But as you all know, today we are very good at regulating the length of pregnancy. As the pig-related pregnancy would be handled by our competent personnel there would obviously be no problem with the length of the pregnancy, and naturally no porcine fetuses need to coexist with the human fetus in the porcine uterus. However, even if the technical medical problems could be overcome, what about ethics and social acceptability?

Pig Pharmaceuticals arranged many meetings with distinguished ethicists. They soon convinced us that it would not be possible to get ethical permission anywhere in the world to carry out the research necessary. The fundamental problem was that it would not be in the interests of a human embryo conceived by In Vitro Fertilisation to be implanted into a pig; it would have a much better chance in a human uterus. It would be even more difficult to obtain human embryos for the research into the early phases of the process, that is, research that would involve destruction of embryos, especially if we could not reveal our long-term goals. Therefore, any carefully planned research with human embryos in the porcine uterus was out of the question.

As you all know, the traditional regulation of medical research does not cover the use of new innovative techniques to save the life or restore the health of the patient. This is clinical care, not research, as is clear from the Helsinki Declaration of 1964 and onwards. We were given an opening for the project when a physician contacted us with a pregnant patient who had a history of serious early problems with pregnancies. It was the third month, complications were developing, and the risk was great that the fetus might be seriously injured by certain reactions developed by the woman. She objected to abortion on religious grounds. The doctor thought it was his duty to help the woman and the developing fetus. He could, obviously, not overrule her objections to abortion. He therefore contacted us.

One of our pigs was treated by gene therapy and the necessary operation was performed. The fetus was transferred from the human uterus to a porcine one. Of course, it was of great help that the similar technique for transferring fetuses to surrogate mothers is such a well-established technique. Naturally it was all done in the greatest privacy. Six months later the child was delivered by Caesarean section.

A picture of a baby girl is projected onto the screen. Next a round-bellied pig is seen. A storm of applause. The audience is standing up and cheering.

Yes, success is a wonderful thing. Today more than 40 pig-related pregnancies are under way. With your help Pig Pharmaceuticals hopes to be able to develop this new therapeutic tool. We also believe that pig-related pregnancies will be a big gain for society, a help for women and—as this will be explained by our ethicist—a matter of real moral progress. I thank you for your attention and wish you all welcome to our barbecue. And now I give the word to the staff ethicist of Pig Pharmaceuticals.

More applause. The audience stands up again. Eventually, calm returns as the ethicist enters the rostrum.

The Ethicist's View

Ladies and gentlemen. As you have just heard, great advances have been made in human reproduction. As always, when science and technology advance, new possibilities appear and new moral problems. But there are also moral gains. I will argue in my short talk that pig-related pregnancies constitute a very great moral gain for women, for fetuses and for society—and also for pigs.

Let me make a parallel with xenotransplantation. The serious shortage of human organs for transplantation in the beginning of this century caused the public debate to move to a slightly distributive standpoint. Is it my duty to donate one of my kidneys, a part of my liver while still alive? Should the state intervene and force us, as it forces us to pay taxes? Progress in xenotransplantation ended this moral and political discussion. There was no longer any need. Xenotransplantation had done away with the conflict between the interest of the bearer of the organ and the duty of helping a fellow human being in desperate need of the organ or part of it. There are now enough organs for everyone. Unfortunately, no one has come up with a similar solution that would make the tax system redundant.

The ethicist looks up. The audience smiles obediently, although the joke was not very funny. But what can you expect of an ethicist?

I will discuss the serious conflict between the interests of the pregnant woman and the fetus. I will also consider the interests of society and of the pigs. All these interests are legitimate and important. We have a moral obligation to give equal considerations to interests.

Let me start with the child. It has an interest in having a good life. We have an educational system, a social care system, etc. to ensure that the child has a reasonable chance of prospering. And we oblige parents to care for their children both physically and psychologically. If the parents disregard this duty or fail to execute it, society will intervene in order to safeguard the interests of the child. The situation is very different before birth.

The fetus is completely dependent on the good will of the pregnant woman. Most pregnant women do their best to ensure that the fetus will have a good start. They abstain from dangerous behaviour, heavy drinking, smoking, etc. Unfortunately, not everyone follows this pattern. And society has been helpless. There has been no possibility, apart from advising, to ensure that these women live without impairing the future of their expected child.

We live in a society where we firmly believe that there should be no discrimination on a sex or gender basis. Being pregnant is exclusively for women; any special legislation to safeguard fetuses would at the same time impinge upon the rights and autonomy of women. Legally speaking, during the first period when the pregnant woman can have an abortion on demand, the fetus is truly regarded as just a part of the woman's body. Then it is a little unclear, but if you murder a pregnant woman you are prosecuted for one murder, not two.

I think many of you who work in the reproductive area have witnessed the disturbing case when a delivery is protracted, the situation is getting serious for the expected child and something has to be done. The autonomy of the woman reigns sovereign and she may legally say no to a Caesarean section, the child may die or be seriously damaged. There is nothing to be done, if the woman does not consent. She has all the legal rights.

At birth, everything changes. A human being with a full set of human rights enters the world. The child-in-the-family is protected in a radically different way than the child-in-the-uterus. When a born child is abused, society will do something. When an unborn child is similarly abused, society will normally do nothing. There has previously been no possibility of safeguarding the interests of the fetus without infringement of the rights of women. In my opinion, the feminists have been absolutely right in putting the interests of the pregnant woman before the interest of the fetus.

All this changes with the prospect of pig-related pregnancies. In a pig-related pregnancy the human fetus is no longer a part of the body of the woman. In the porcine uterus, the human fetus can have the same legal status as the newborn child and have the same protection. And Ladies and Gentlemen, this is consistent with all the rights and liberties of women that the feminist movements successfully fought for. This is a moral gain.

I want to state the following moral principle: If a new technology makes it possible to avoid a conflict between legitimate interests, it is our duty to use the new technology. Hence, it is a duty to use the new technology of pig-related pregnancies. Should society therefore forbid "natural" pregnancies and force women into having pig-related pregnancies?

I think this is going too far. After all, most of us regard infidelity between spouses as a morally bad thing. Still we do not legally punish it. And we might ask when conditions for pregnant women are not just another kind of discrimination? I think a reasonable compromise is the following. Given the possibility of a pig-related pregnancy, society may indeed demand that women should first and foremost safeguard the well-being of the fetus. This can be done in two ways.

One is a pig-related pregnancy with no restrictions on the lifestyle of the woman. The other is a "natural" pregnancy with legally enforced restrictions on the lifestyle. On the other hand, women undergoing a "natural" pregnancy should have the option to switch to a pig-related one at any time.

What about the pigs? Pregnancy is a positive experience for the pig. Our pigs have a good life before, during and after the pregnancy. Last, but not least, pig-related pregnancies will give more pigs the opportunity to exist, and this is a good thing.

Ladies and Gentlemen, thank you for your attention. Welcome to participate in this new exciting development. And welcome to our party.

Comments on the Possibility, the Consequences and Desirability of Reproductive Ectogenesis

What are we to conclude from the story of Pig Pharmaceuticals Limited? Below I offer some comments on some of the issues, in particular: will it happen, will this change our view of the fetus, and what about the rights of pregnant women?

Will Reproductive Ectogenesis Ever Happen?
Personally I think so. It was already prophesied by J. B. S. Haldane in 1923. The chief scientific director of Pig Pharmaceuticals has a real point on how it could become a reality. Many things that are ruled out as unethical under a research regime by the rather restrictive demands of the Helsinki Declaration may be acceptable in the clinic. One example is the development of heart transplantation. This was not done as a research project. The driving force will be to save the fetuses and allow women to have children.

Pigs as surrogate mothers may not be very probable. After Novartis closed down Imutran in Cambridge in Spring 2001, the field of xenotransplantation did not look so promising.[12] The particular transgenic pig was not good enough and we will have to wait for new transgenic pigs—which seem to be on their way—or some other breakthrough. It is perhaps more feasible to believe that reproductive ectogenesis will start from what we learn from growing stem cells and, hopefully, learn to develop them into various cell types and later to grow organs for transplantation. Growing organs outside the human body is a very desirable form of tissue engineering with regard to the desperate need for organ transplants. It will at the same time give us knowledge helpful for sustaining a full embryonic development. Once this is given, it is a short step to use the therapeutical argument that we need to apply ectogenes to save the fetus for some woman. I do believe that a *therapeutical imperative* exists—at least with regard to introducing advanced medical technologies—in our societies. Whatever techniques that can be used to save life will be developed. Application may not become very widespread but it will be introduced.

Will This Change Our View of the Fetus?

Speaking generally it seems that fetuses are not considered as having rights or, alternatively, that the legal rights recognised are derivative on the child being born.[13] In my own country (Sweden) the legal status of the fetus changes dramatically at birth. It seems that viability is the criteria used. This is of course a criteria that is technology dependent and in the era of ectogenesis the embryo will be viable (capable of living outside the womb) from the very start. Using another more plausible view, the moral status of the fetus is related to the fetus being capable of sentience. This does not (at present) give a precise time but we can be rather certain that the moment does not occur at birth. It is either before or after.[14]

Ectogenesis will give new possibilities of safeguarding early embryos without interfering with women's rights to control over their bodies, a topic already discussed in the debate over human embryonic stem cells. Many might think that there is a direct coupling between holding (early) abortions admissible and thinking it is admissible to destroy and discard IVF embryos. However, it is possible to argue for special safeguards for the IVF embryo while allowing (early) abortion on women's demand. In a well-known discussion on abortion and other issues Ronald Dworkin makes the point that there can be no rights without interests and these cannot exist without some rudimentary form of consciousness. Therefore the fetus in its early stage does not have interests and therefore no rights.[15] (p.18) He introduces the distinction between two types of objections to abortion. There is *the derivative objection* which assumes that the fetus has interests and rights. The *detached objection* does not assume that the early fetus has rights but argues from the need to protect the intrinsic value of life.[15] (p.11) He then rules out the derivative objection as invalid and relegates the detached objection to the private sphere of religious and ethical conviction. Essentially, Dworkin agrees that there is a legitimate interest for the community to safeguard the intrinsic value of human life but this should not conflict with the rights of women to self-determination and control of their bodies. In the first phase of the pregnancy, the right of the woman, according to Dworkin, overrules the interest to protect the value of life.

Dworkin does not discuss the IVF embryo but his discussion of abortion can be applied to ectogenesis. Let us suppose that Dworkin is right about the interest of safeguarding the intrinsic value of life. This can obviously be done in the third era of reproductive ectogenesis without interfering with the woman's body. The case of the child-in-the-incubator/pig uterus is in that respect similar to the child-in-the-family. Also, the child-in-the-female-uterus will, in the third era, be in a similar position to the child-in-the-family. Most of us agree that the society should interfere into the family to safeguard the well-being of the child. The child can be removed. Now the child (in the maybe fictious) third era can be removed from the uterus and placed in some artificial womb. Such removal was proposed in 1984 as an alternative to abortion by Peter Singer and Deane Wells.[16]

The Rights of Pregnant Women
In the third era of reproductive ectogenesis the legal protections of the embryo no longer interfere with the woman's body as the pregnancy is outside the body. And furthermore, if we concede that society should not protect the early embryo in the incubator, it is very hard to find an argument why the woman *alone* should decide whether or not the embryonic development should be interrupted. The man should obviously share the decision. The third era of human reproduction will mean reproductive empowerment of men and end the historically short monopoly of women in regard to deciding the fate of the embryo and fetus. In the second ongoing era of IVF we already see the first signs. When the embryo is outside the woman, the man and the woman share decision-making.

Another possible change was pointed out by the ethicist in the future talk. In the era of ectogenesis, women who choose to have a natural pregnancy will have to face restrictions on lifestyles. At least, I believe it will be very hard to argue against such restrictions in order to protect the fetus. This will be a dramatic change and put the pregnant woman in a special situation. Maybe, for this kind of restriction to be ethically acceptable, every woman should have a choice between ectogenesis and a natural pregnancy. Ideally, it should be as depicted in the 2050 case in that it should be possible at any time to change from a natural pregnancy to ectogenesis.

If not everyone should be able to afford ectogenesis—which may be the case in the USA—then restrictions imposed on those who cannot pay does not seem fair. However, one might still make an argument for restrictions on those who can afford it. Furthermore, "not being able to afford" is a very vague concept.

It is perhaps possible to give arguments for prohibiting ectogenesis altogether and preserve the present situation that pregnant women can (more or less) live without any legal sanctions against hurting their fetuses and keep the sole decision to abort or not. I am not able to think of any good arguments for that, however. I think this is the way it must be today when the fetus is inside the woman's body, but I am doubtful if there are any good arguments for this position in the third era.

The third era will be good news for pro-lifers. Embryo or fetus adoption will be an alternative to abortion. The introduction of embryo adoption has already been proposed for IVF in the second era (that is now) by a committee of the European Parliament.[17] That particular proposal was not adopted.

Is Reproductive Ectogenesis Desirable?
The ethicist of Pig Pharmaceuticals Limited has given his (future) arguments for reproductive ectogenesis. I find ectogenesis in many ways repugnant but I must confess that I lack good arguments against its introduction, at least as an option for therapeutical reasons. But given an option, I suspect we are on the slippery slope.

Acknowledgment

Work on this paper has been supported (partly) by the Swedish Research Council grant K2002-31x-14012-02B. A preliminary version of the paper was presented at "The End of Natural Motherhood," Tulsa, Oklahoma 22–23 February 2002. I am grateful for comments at that meeting.

Personal Note

Together with my co-worker Anders Persson I have been studying the ethical and social issues in xenotransplantation. I got the idea for the story some years ago when I was told that the Xenotransplantation Society was contemplating new ethical guidelines. One proposal was, as I remembered, to forbid the insertion of a human fetus into an animal. This was presented at a workshop in Tübingen, Germany in 1998 by Claus Hammer of Munich University and is reprinted in *Leben als Labor-Material. Zur Problematik der Embryoforschung*. Ed. Wuerling, Hans-Bernard. Düsseldorf: Patmos Verlag 2000, p. 35. (I am grateful to Claus Hammer for confirming my memory and for the reference.)

References

1. McGrath, J. & Solter, D. (1984) Inability of mouse blastomere nuclei transferred to enucleated zygotes to support development in vitro. *Science* 226: 1317–183.

2. Solter, D. (2001) "Mammalian Cloning: Advances and Limitations". *Nature Review Genetics* 1: 199–206.

3. Bush, G. W. (2001) Remarks by the President on Stem Cell Research. http://www.nih.gov/news/stemcell/index.htm (1 Nov. 2001).

4. Parfit, D. (1987) *Reasons and Persons*, Clarendon Press, Oxford, pp. 199–200.

5. Tong, R. (1997) *Feminist Approaches to Bioethics*, Westview Press, Boulder, p. 129.

6. Nozick, R. (1980) *Anarchy, State, and Utopia*, Basil Blackwell, Oxford, pp. 42–45.

7. Glover, J. (1984) *What sort of people should there be?* Penguin Books, Harmondsworth, pp. 102–113.

8. Dennet, D. (1982) Where am I?, in: Hofstadter, D. & Dennet, D. (eds) *The Mind's I. Fantasies and Reflections on Self and Soul*, Penguin Books, Harmondsworth, pp. 217–229.

9. Thomson, J. J. (1985) A Defence of Abortion, in: Wasserstrom, R. (ed.) *Today's Moral Problems*, Macmillan, New York.

10. Haldane, J. B. S. (1923) *Daedalus or Science and the Future*, in: Krishna R. Dronamraju (ed.) (1995) *Haldane's Daedalus Revisited*, Oxford University Press, Oxford, New York and Tokyo.

11. Young, M. (2002) *The Rise of the Meriticracy*, Transaction Publishers, New Brunswick (US) and London.

12. Person, A. (2001) "Optimismen falnar om xenotranplantation". Läkartidningen 98: 3200–02 ("Optimism fades for xenotransplantation").

13. Wellman, C. (2002) The Concept of fetal rights, *Law and Philosophy* 21: 65–93.

14. Welin, S. (2002) Ethical Issues in human embryonic stem cell research, *Acta Obstet Gynecol Scand* 81: 277–382.

15. Dworkin, R. (1993) *Life's Dominion*, HarperCollins, London.

16. Singer, P. & Wells, D. (1984) *The reproduction revolution: new ways of making babies*, Oxford University Press, Oxford.

17. European Parliament, Temporary Committee on Human Genetics and Other New Technologies in Modern Medicine (2001) *Report on the ethical, legal, economic and social implication of human genetics*. (A5-0391/2001).

5   "Nanotechnology: Shaping the World Atom by Atom"

Interagency Working Group on Nanoscience, Engineering, and Technology

Futures are not only created by scholars, fiction writers, and independent visionaries. Sometimes those who predict the future have a great deal of power to push us in that direction. In 1999 policymakers from powerful government agencies including the Department of Defense, the Office of Management and Budget, the National Air and Space Administration, the Department of Energy, the National Institutes of Health, and the National Science Foundation (NSF) formed an Interagency Working Group on Nanoscience, Engineering and Technology (IWGN). Under the leadership of Mihail Roco of the NSF, they created this vision of the future of nanotechnology for the National Science and Technology Council in order to convince the U.S. Congress to increase funding of nanoscale science and technology. In the year this document was published (1999), the federal government spent 260 million dollars on nanotechnology research. By 2005, federal spending on nanoscience and technology was over a billion dollars a year. Whether the technologies and social changes predicted by this report will be realized is yet to be seen. But the report, along with the concerted effort of the IWGN members and others, was able to convince Congress and the President that this was a future worth spending lots of money to achieve.

If you were to deconstruct a human body into its most basic ingredients, you'd get a little tank each of oxygen, hydrogen, and nitrogen. There would be piddling piles of carbon, calcium, and salt. You'd squint at pinches of sulfur, phosphorus, iron, and magnesium, and tiny dots of 20 or so other chemical elements. Total street value: not much.

With its own version of what scientists call nanoengineering, nature transforms these inexpensive, abundant, and inanimate ingredients into self-generating, self-perpetuating, self-repairing, self-aware creatures that walk, wiggle, swim, sniff, see, think, and even dream. Total value: immeasurable.

Now, a human brand of nanoengineering is emerging. The field's driving question is this: What could we humans do if we could assemble the basic ingredients of the material world with even a glint of nature's virtuosity? What if we could build things the way nature does—atom by atom and molecule by molecule?

Scientists already are finding answers to these questions. The more they learn, the more they suspect nanoscience and nanoengineering will become as socially transforming as the development of running water, electricity, antibiotics, and microelectronics.

Interagency Working Group on Nanoscience, Engineering, and Technology "Nanotechnology: Shaping the World Atom by Atom," September 1999, Washington, D.C., pp. 1–2, 8.

The field is roughly where the basic science and technology behind transistors was in the late 1940s and 1950s.

In April 1998, Neal Lane, Assistant to the President for Science and Technology and former Director of the National Science Foundation (NSF), stated at a Congressional hearing, "If I were asked for an area of science and engineering that will most likely produce the breakthroughs of tomorrow, I would point to nanoscale science and engineering."

Lane is not alone in this view. Many scientists, including physicist and Nobel laureate Horst Stormer of Lucent Technologies and Columbia University, are themselves amazed that the emerging nanotechnology may provide humanity with unprecedented control over the material world. Says Stormer: "Nanotechnology has given us the tools... to play with the ultimate toy box of nature—atoms and molecules. Everything is made from it.... The possibilities to create new things appear limitless."

So what do scientists like Lane and Stormer mean by nanotechnology? In the language of science, the prefix nano means one-billionth of something like a second or a meter. Nanoscience and nanotechnology generally refer to the world as it works on the nanometer scale, say, from one nanometer to several hundred nanometers. That's the natural spatial context for molecules and their interactions, just as a 100 yard gridiron is the relevant spatial context for football games. Naturally-occurring molecular players on the nanoscale field range from tiny three-atom water molecules to much larger protein molecules like oxygen-carrying hemoglobin with thousands of atoms to gigantic DNA molecules with millions of atoms. Whenever scientists and engineers push their understanding and control over matter to finer scales, as they now are doing on the nanoscale, they invariably discover qualitatively new phenomena and invent qualitatively new technologies. "Nanotechnology is the builder's final frontier," remarks Nobel laureate Richard Smalley, Rice University.

For years now, scientists have been developing synthetic nanostructures that could become the basis for countless improved and completely new technologies. The way molecules of various shapes and surface features organize into patterns on nanoscales determines important material properties, including electrical conductivity, optical properties, and mechanical strength. So by controlling how that nanoscale patterning unfolds, researchers are learning to design new materials with new sets of properties.

Some of these nanostructures may turn out to be useful as discrete nanostructures. New types of vaccines and medicines come to mind here. The value of others may emerge only as they are assembled into larger structures like particles or fibers, which then would be processed into yet larger structures like textiles, films, coatings, bricks, and beams.

Forward looking researchers believe they could end up with synthetic creations with life-like behaviors. Cover an airplane with paint containing nanoscale pigment particles that instantly reconfigure, chameleon-like, to mimic the aircraft's surroundings. You would end up with an airplane indistinguishable from the sky, that is, an invisible plane. How about bricks and other building materials that can sense weather conditions and then respond by altering their inner structures to be more or less permeable to air and humidity? That would go a long way toward improving the comfort and energy efficiency of buildings. And how about synthetic antibody-like nanoscale drugs or devices that might seek out and destroy malignant cells wherever they might be in the body?

For many years futurists steeped in the culture of science fiction and prone to thinking in time frames that reach decades ahead have been dreaming up a fantastic future built using nanotechnologies. More recently, more cautious, established researchers, who are developing the tools and methods for a nanotechnological future, have been making projections of their own based on their expanding base of knowledge and experience. "As we enter the 21st century, nanotechnology will have a major impact on the health, wealth and security of the world's people that will be at least as significant in this century as antibiotics, the integrated circuit, and manmade polymers," according to a committee of leading scientists that convened in January 1999 at the National Science Foundation to assess the potential roles of nanotechnology in the coming years.

Not that it will be easy. Nanoscience and nanoengineering remain in an exploratory phase. Scientists have yet to understand all of the scientific and engineering issues that define what can happen and what can be done in the nanoscale regime. Still, laboratory accomplishments so far are making scientists in this country and elsewhere bullish about the future. So much so that the quest to master the nanoscale is becoming a global competition. New lightweight materials for future generations of more fuel efficient cars, military aircraft that can go farther and carry more payload, new classes of pharmaceuticals, materials that last longer and thereby reduce pollution from manufacturing, are just a few of the goals. Companies and countries are experimenting with new organizational, industrial and budgetary models they hope will give them the competitive edge toward these ends.

The U.S. Government, for one, invested approximately $116 million in fiscal year 1997 in nanotechnology research and development. For FY 1999, that figure has risen to an estimated $260 million. Japan and Europe are making similar investments. Whoever becomes most knowledgeable and skilled on these nanoscopic scales probably will find themselves well positioned in the ever more technologically-based and globalized economy of the 21st century.

That helps explain why the White House National Science and Technology Council (NSTC) created the Interagency Working Group on Nanoscience, Engineering and Technology (IWGN) in 1998. With members from eight Federal agencies interacting closely with the academic and industrial community, the IWGN's charge has been to assess the potential of nanotechnology and to formulate a national research and development plan. The Office of Science and Technology Policy (OSTP) and the Office of Management and Budget (OMB) have since issued a joint memorandum to Federal agency heads that recommends nanotechnology as a research priority area for Federal investment in FY 2001. The memorandum calls for a broad-based coalition in which academe, the private sector, and local, state, and Federal governments work together to push the envelope of nanoscience and nanoengineering to reap nanotechnology's potential social and economic benefits. The working group has recommended a doubling of the annual investment for research in these areas to about a half billion dollars.

. . .

Nanotechnologists Project That Their Work Will Leave No Stone Unturned

The list of nanotechnologies in various stages of conception, development and even commercialization already is vast and growing. If present trends in nanoscience and nanotechnology continue, most aspects of everyday life are subject to change. Consider these:

- Electronics Central   By patterning recording media in nanoscale layers and dots, the information on a thousand CDs could be packed into the space of a wristwatch. Besides the thousandfold to millionfold increase in storage capacity, computer processing speeds will make today's Pentium IIIs seem slow. Devices to transmit electromagnetic signals—including radio and laser signals—will shrink in size while becoming inexpensive and more powerful. Everyone and everything conceivably could be linked all the time and everywhere to a future World Wide Web that feels more like an all-encompassing information environment than just a computer network.
- Nanodoc   Nanotechnology will lead to new generations of prosthetic and medical implants whose surfaces are molecularly designed to interact with the body. Some of these even will help attract and assemble raw materials in bodily fluids to regenerate bone, skin or other missing or damaged tissues. New nanostructured vaccines could eliminate hazards of conventional vaccine development and use, which rely on viruses and bacteria. Nanotubules that act like tiny straws could conceivably take up drug molecules and release them slowly over time. A slew of chip-sized home diagnostic devices with nanoscale detection and processing components could fundamentally alter patient-doctor relationships, the management of illnesses, and medical culture in general.
- Smokeless Industry   More and more materials and products will be made from the bottom-up, that is, by building them up from atoms, molecules, and the nanoscale powders, fibers and other small structural components made from them. This differs from all previous manufacturing, in which raw materials like sheet metal, polymer, fabric and concrete get pressed, cut, molded and otherwise coerced into parts and products. Bottom-up manufacturing should require less material and pollute less. What's more, engineers expect to be able to embed sophisticated, life-like functions into materials. Even concrete will get smart enough to internally detect signs of weakness and life-like enough to respond by, say, releasing chemicals that combat corrosive conditions. In effect, the constructed world itself could become sensitive to damaging conditions and automatically take corrective or evasive action like a hand recoiling from a flame.

- Planes, Trains and Automobiles   Materials with an unprecedented combination of strength, toughness and lightness will make all kinds of land, sea, air and space vehicles lighter and more fuel efficient. Fighter aircraft designed with lighter and stronger nanostructured materials will be able to fly longer missions and carry more payload. Plastics that wear less because their molecular chains are trapped by ceramic nanoparticles will lead to materials that last a lifetime. Some long-view researchers are taking steps toward self-repairing metallic alloys that automatically fill in and reinforce tiny cracks that can grow and merge into larger ones, including catastrophic ones that have caused plane crashes.
- But, Wait, There's More!   Nanotechnology advocates say their field will leave no stone unturned. Their lengthy lists include artificial photosynthesis systems for clean energy; molecular layer-by-layer crystal growth to make new generations of more efficient solar cells; tiny robotic systems for space exploration; selective membranes that can fish out specific toxic or valuable particles from industrial waste or that can inexpensively desalinate sea water; chameleon-like camouflage that changes shape and color to blend in anywhere, anytime; and blood substitutes.

Ready or Not

No one knows how much of nanotechnology's promise will prove out. Technology prediction has never been too reliable. In the March 1949 edition of *Popular Mechanics*, hardly a year after the invention of the transistor, experts predicted computers of the future would add as many as 5000 numbers per second, weigh only 3000 pounds, and consume only 10 kilowatts of power. Today's five-pound laptops add several million numbers per second using only a watt or so of power. And thumbnail-sized microprocessors run washing machines and kids' toys as well as hundreds of millions of computers. What's more, computer technology spawned a new social epoch that some dub the Information Age or the Silicon Age.

And yet, many believe nanotechnology may do even more. In the collective opinion of the committee of scientists, engineers and technology professionals convened in January 1999 by the IWGN, "The total societal impact of nanotechnology is expected to be much greater than that of the silicon integrated circuit because it is applicable in many more fields than just electronics."

Despite the advances researchers have made, it is hard to work on the nanoscale. And even assuming something like [Richard] Feynman's vision of total nanoscale control comes about, the consequences are bound to be mixed. Like any extremely powerful new technology, nanotechnology will bring with it social and ethical issues.

Just consider quantum computers. Theorists expect them to be so good at factoring huge numbers that the toughest encryption schemes in use today—which are enabling revolutionary things like e-commerce—will become easy to crack. Or consider the claim that nanobiology will enable people to live longer, healthier lives. Longer average lifetimes will mean more people on Earth. But how many more people can the Earth sustain?

For the moment, it's nanotechnology's promise that's on most peoples' minds. "Never has such a comprehensive technology promised to change so much so fast.... Inevitably nanotech will give people more time, more value for less cost and provide for a higher quality of existence," predicts James Canton, president of the Institute for Global Futures. But maybe not for everyone. Says Canton: "Those nations, governments, organizations and citizens who are unaware of this impending power shift must be informed and enabled so that they may adequately adapt."

It no longer seems a question of whether nanotechnology will become a reality. The big questions are how important and transformative nanotechnology will become, will it become affordable, who will be the leaders, and how can it be used to make the world a better place?—questions that will, in time, be answered.

6 "Why the Future Doesn't Need Us"

Bill Joy

When this article by Bill Joy was first published in 2000, it caused some rumblings in scientific, engineering, and public policy circles. Joy was an extremely well respected and accomplished engineer who had been a major player in the IT revolution. Yet in this piece he raises questions and expresses doubts about the direction of current research. Joy does something parallel to the authors of the previous readings in that he takes certain research and development trends and extrapolates out to where they might take us. He is struck in particular by the possibilities of convergence of genetics, nanotechnology, and robotics and especially the potential of this convergence to lead to self-replicating beings that might effectively take over the world because of their superior-to-human capacities. He insists that he does not hate technology and is not advocating that we smash modern technology to bits. But he is concerned about its direction. Joy struggles with the negative vision to which his extrapolations lead and in so doing raises many of the questions that will be pursued in this book. His goal is to have us reflect on the technologies we are developing today to make sure they will help us create a future we want to live in.

From the moment I became involved in the creation of new technologies, their ethical dimensions have concerned me, but it was only in the autumn of 1998 that I became anxiously aware of how great are the dangers facing us in the 21st century. I can date the onset of my unease to the day I met Ray Kurzweil, the deservedly famous inventor of the first reading machine for the blind and many other amazing things.

Ray and I were both speakers at George Gilder's Telecosm conference, and I encountered him by chance in the bar of the hotel after both our sessions were over. I was sitting with John Searle, a Berkeley philosopher who studies consciousness. While we were talking, Ray approached and a conversation began, the subject of which haunts me to this day.

I had missed Ray's talk and the subsequent panel that Ray and John had been on, and they now picked right up where they'd left off, with Ray saying that the rate of improvement of technology was going to accelerate and that we were going to become robots or fuse with robots or something like that, and John countering that this couldn't happen, because the robots couldn't be conscious.

While I had heard such talk before, I had always felt sentient robots were in the realm of science fiction. But now, from someone I respected, I was hearing a strong argument that they were a near-term possibility. I was taken aback, especially given Ray's proven ability to imagine and create the future. I already knew that new technologies

Originally published in *WIRED* 8, no. 4 (April 2000): 238–262. © August 4, 2000, by Bill Joy. Reprinted by permission of the author.

like genetic engineering and nanotechnology were giving us the power to remake the world, but a realistic and imminent scenario for intelligent robots surprised me.

It's easy to get jaded about such breakthroughs. We hear in the news almost every day of some kind of technological or scientific advance. Yet this was no ordinary prediction. In the hotel bar, Ray gave me a partial preprint of his then-forthcoming book *The Age of Spiritual Machines*, which outlined a utopia he foresaw—one in which humans gained near immortality by becoming one with robotic technology. On reading it, my sense of unease only intensified; I felt sure he had to be understating the dangers, understating the probability of a bad outcome along this path.

I found myself most troubled by a passage detailing a *dys*topian scenario:

The New Luddite Challenge

First let us postulate that the computer scientists succeed in developing intelligent machines that can do all things better than human beings can do them. In that case presumably all work will be done by vast, highly organized systems of machines and no human effort will be necessary. Either of two cases might occur. The machines might be permitted to make all of their own decisions without human oversight, or else human control over the machines might be retained.

If the machines are permitted to make all their own decisions, we can't make any conjectures as to the results, because it is impossible to guess how such machines might behave. We only point out that the fate of the human race would be at the mercy of the machines. It might be argued that the human race would never be foolish enough to hand over all the power to the machines. But we are suggesting neither that the human race would voluntarily turn power over to the machines nor that the machines would wilfully seize power. What we do suggest is that the human race might easily permit itself to drift into a position of such dependence on the machines that it would have no practical choice but to accept all of the machines' decisions. As society and the problems that face it become more and more complex and machines become more and more intelligent, people will let machines make more of their decisions for them, simply because machine-made decisions will bring better results than man-made ones. Eventually a stage may be reached at which the decisions necessary to keep the system running will be so complex that human beings will be incapable of making them intelligently. At that stage the machines will be in effective control. People won't be able to just turn the machines off, because they will be so dependent on them that turning them off would amount to suicide.

On the other hand it is possible that human control over the machines may be retained. In that case the average man may have control over certain private machines of his own, such as his car or his personal computer, but control over large systems of machines will be in the hands of a tiny elite—just as it is today, but with two differences. Due to improved techniques the elite will have greater control over the masses; and because human work will no longer be necessary the masses will be superfluous, a useless burden on the system. If the elite is ruthless they may simply decide to exterminate the mass of humanity. If they are humane they may use propaganda or other psychological or biological techniques to reduce the birth rate until the mass of humanity becomes extinct, leaving the world to the elite. Or, if the elite consists of soft-hearted liberals, they may decide to play the role of good shepherds to the rest of the human race. They will see to it that everyone's physical needs are satisfied, that all children are raised under psychologically hygienic conditions, that everyone has a wholesome hobby to keep him busy, and that anyone who may become dissatisfied undergoes "treatment" to cure his "problem." Of course, life will be so purposeless that people will have to be biologically or psychologically engineered either to remove their need for the power process or make them "sublimate" their drive for power into some harmless hobby. These engineered human beings may be happy in such a society, but they will most certainly not be free. They will have been reduced to the status of domestic animals.[1]

In the book, you don't discover until you turn the page that the author of this passage is Theodore Kaczynski—the Unabomber. I am no apologist for Kaczynski. His bombs killed three people during a 17-year terror campaign and wounded many others. One of his bombs gravely injured my friend David Gelernter, one of the most brilliant and visionary computer scientists of our time. Like many of my colleagues, I felt that I could easily have been the Unabomber's next target.

Kaczynski's actions were murderous and, in my view, criminally insane. He is clearly a Luddite, but simply saying this does not dismiss his argument; as difficult as it is for me to acknowledge, I saw some merit in the reasoning in this single passage. I felt compelled to confront it.

Kaczynski's dystopian vision describes unintended consequences, a well-known problem with the design and use of technology, and one that is clearly related to Murphy's law—"Anything that can go wrong, will." (Actually, this is Finagle's law, which in itself shows that Finagle was right.) Our overuse of antibiotics has led to what may be the biggest such problem so far: the emergence of antibiotic-resistant and much more dangerous bacteria. Similar things happened when attempts to eliminate malarial mosquitoes using DDT caused them to acquire DDT resistance; malarial parasites likewise acquired multi-drug-resistant genes.[2]

The cause of many such surprises seems clear: The systems involved are complex, involving interaction among and feedback between many parts. Any changes to such a system will cascade in ways that are difficult to predict; this is especially true when human actions are involved.

I started showing friends the Kaczynski quote from *The Age of Spiritual Machines*; I would hand them Kurzweil's book, let them read the quote, and then watch their reaction as they discovered who had written it. At around the same time, I found Hans Moravec's book *Robot: Mere Machine to Transcendent Mind*. Moravec is one of the leaders in robotics research, and was a founder of the world's largest robotics research program, at Carnegie Mellon University. *Robot* gave me more material to try out on my friends—material surprisingly supportive of Kaczynski's argument. For example:

The Short Run (Early 2000s)
Biological species almost never survive encounters with superior competitors. Ten million years ago, South and North America were separated by a sunken Panama isthmus. South America, like Australia today, was populated by marsupial mammals, including pouched equivalents of rats, deers, and tigers. When the isthmus connecting North and South America rose, it took only a few thousand years for the northern placental species, with slightly

more effective metabolisms and reproductive and nervous systems, to displace and eliminate almost all the southern marsupials.

In a completely free marketplace, superior robots would surely affect humans as North American placentals affected South American marsupials (and as humans have affected countless species). Robotic industries would compete vigorously among themselves for matter, energy, and space, incidentally driving their price beyond human reach. Unable to afford the necessities of life, biological humans would be squeezed out of existence.

There is probably some breathing room, because we do not live in a completely free marketplace. Government coerces nonmarket behavior, especially by collecting taxes. Judiciously applied, governmental coercion could support human populations in high style on the fruits of robot labor, perhaps for a long while.

A textbook dystopia—and Moravec is just getting wound up. He goes on to discuss how our main job in the 21st century will be "ensuring continued cooperation from the robot industries" by passing laws decreeing that they be "nice,"[3] and to describe how seriously dangerous a human can be "once transformed into an unbounded superintelligent robot." Moravec's view is that the robots will eventually succeed us—that humans clearly face extinction.

I decided it was time to talk to my friend Danny Hillis. Danny became famous as the cofounder of Thinking Machines Corporation, which built a very powerful parallel supercomputer. Despite my current job title of Chief Scientist at Sun Microsystems, I am more a computer architect than a scientist, and I respect Danny's knowledge of the information and physical sciences more than that of any other single person I know. Danny is also a highly regarded futurist who thinks long-term—four years ago he started the Long Now Foundation, which is building a clock designed to last 10,000 years, in an attempt to draw attention to the pitifully short attention span of our society. (See "Test of Time," *Wired* 8.03, page 78.)

So I flew to Los Angeles for the express purpose of having dinner with Danny and his wife, Pati. I went through my now-familiar routine, trotting out the ideas and passages that I found so disturbing. Danny's answer—directed specifically at Kurzweil's scenario of humans merging with robots—came swiftly, and quite surprised me. He said, simply, that the changes would come gradually, and that we would get used to them.

But I guess I wasn't totally surprised. I had seen a quote from Danny in Kurzweil's book in which he said, "I'm as fond of my body as anyone, but if I can be 200 with a body of silicon, I'll take it." It seemed that he was at peace with this process and its attendant risks, while I was not.

While talking and thinking about Kurzweil, Kaczynski, and Moravec, I suddenly remembered a novel I had read almost 20 years ago—*The White Plague*, by Frank Herbert—in which a molecular biologist is driven insane by the senseless murder of his family. To seek revenge he constructs and disseminates a new and highly contagious plague that kills widely but selectively. (We're lucky Kaczynski was a mathematician, not a molecular biologist.) I was also reminded of the Borg of *Star Trek*, a hive of partly biological, partly robotic creatures with a strong destructive streak. Borg-like disasters are a staple of science fiction, so why hadn't I been more concerned about such robotic dystopias earlier? Why weren't other people more concerned about these nightmarish scenarios?

Part of the answer certainly lies in our attitude toward the new—in our bias toward instant familiarity and unquestioning acceptance. Accustomed to living with almost routine scientific breakthroughs, we have yet to come to terms with the fact that the most compelling 21st-century technologies—robotics, genetic engineering, and nanotechnology—pose a different threat than the technologies that have come before. Specifically, robots, engineered organisms, and nanobots share a dangerous amplifying factor: They can self-replicate. A bomb is blown up only once—but one bot can become many, and quickly get out of control.

Much of my work over the past 25 years has been on computer networking, where the sending and receiving of messages creates the opportunity for out-of-control replication. But while replication in a computer or a computer network can be a nuisance, at worst it disables a machine or takes down a network or network service. Uncontrolled self-replication in these newer technologies runs a much greater risk: a risk of substantial damage in the physical world.

Each of these technologies also offers untold promise: The vision of near immortality that Kurzweil sees in his robot dreams drives us forward; genetic engineering may soon provide treatments, if not outright cures, for most diseases; and nanotechnology and nanomedicine can address yet more ills. Together they could significantly extend our average life span and improve the quality of our lives. Yet, with each of these technologies, a sequence of small, individually sensible advances leads to an accumulation of great power and, concomitantly, great danger.

What was different in the 20th century? Certainly, the technologies underlying the weapons of mass destruction (WMD)—nuclear, biological, and chemical (NBC)—were powerful, and the weapons an enormous threat. But building nuclear weapons required, at least for a time, access to both rare—indeed, effectively unavailable—raw materials and highly protected information; biological and chemical weapons programs also tended to require large-scale activities.

The 21st-century technologies—genetics, nanotechnology, and robotics (GNR)—are so powerful that they can spawn whole new classes of accidents and abuses. Most dangerously, for the first time, these accidents and abuses are widely within the reach of individuals or small groups. They will not require large facilities or rare raw materials. Knowledge alone will enable the use of them.

Thus we have the possibility not just of weapons of mass destruction but of knowledge-enabled mass destruction (KMD), this destructiveness hugely amplified by the power of self-replication.

I think it is no exaggeration to say we are on the cusp of the further perfection of extreme evil, an evil whose possibility spreads well beyond that which weapons of mass destruction bequeathed to the nation-states, on to a surprising and terrible empowerment of extreme individuals.

Nothing about the way I got involved with computers suggested to me that I was going to be facing these kinds of issues.

My life has been driven by a deep need to ask questions and find answers. When I was 3, I was already reading, so my father took me to the elementary school, where I sat on the principal's lap and read him a story. I started school early, later skipped a grade, and escaped into books—I was incredibly motivated to learn. I asked lots of questions, often driving adults to distraction.

As a teenager I was very interested in science and technology. I wanted to be a ham radio operator but didn't have the money to buy the equipment. Ham radio was the Internet of its time: very addictive, and quite solitary. Money issues aside, my mother put her foot down—I was not to be a ham; I was antisocial enough already.

I may not have had many close friends, but I was awash in ideas. By high school, I had discovered the great science fiction writers. I remember especially Heinlein's *Have Spacesuit Will Travel* and Asimov's *I, Robot*, with its Three Laws of Robotics. I was enchanted by the descriptions of space travel, and wanted to have a telescope to look at the stars; since I had no money to buy or make one, I checked books on telescope-making out of the library and read about making them instead. I soared in my imagination.

Thursday nights my parents went bowling, and we kids stayed home alone. It was the night of Gene Roddenberry's original *Star Trek*, and the program made a big impression on me. I came to accept its notion that humans had a future in space, Western-style, with big heroes and adventures. Roddenberry's vision of the centuries to come was one with strong moral values, embodied in codes like the Prime Directive: to not interfere in the development of less technologically advanced civilizations. This had an incredible appeal to me; ethical humans, not robots, dominated this future, and I took Roddenberry's dream as part of my own.

I excelled in mathematics in high school, and when I went to the University of Michigan as an undergraduate engineering student I took the advanced curriculum of the mathematics majors. Solving math problems was an exciting challenge, but when I discovered computers I found something much more interesting: a machine into which you could put a program that attempted to solve a problem, after which the machine quickly checked the solution. The computer had a clear notion of correct and incorrect, true and false. Were my ideas correct? The machine could tell me. This was very seductive.

I was lucky enough to get a job programming early supercomputers and discovered the amazing power of large machines to numerically simulate advanced designs. When I went to graduate school at UC Berkeley in the mid-1970s, I started staying up late, often all night, inventing new worlds inside the machines. Solving problems. Writing the code that argued so strongly to be written.

In *The Agony and the Ecstasy*, Irving Stone's biographical novel of Michelangelo, Stone described vividly how Michelangelo released the statues from the stone, "breaking the marble spell," carving from the images in his mind.[4] In my most ecstatic moments, the software in the computer emerged in the same way. Once I had imagined it in my mind I felt that it was already there in the machine, waiting to be released. Staying up all night seemed a small price to pay to free it—to give the ideas concrete form.

After a few years at Berkeley I started to send out some of the software I had written—an instructional Pascal system, Unix utilities, and a text editor called vi (which is still, to my surprise, widely used more than 20 years later)—to others who had similar small PDP-11 and VAX minicomputers. These adventures in software eventually turned into the Berkeley version of the Unix operating system, which became a personal "success disaster"—so many people wanted it that I never finished my PhD. Instead I got a job working for Darpa putting Berkeley Unix on the Internet and fixing it to be reliable and to run large research applications well. This was all great fun and very rewarding. And, frankly, I saw no robots here, or anywhere near.

Still, by the early 1980s, I was drowning. The Unix releases were very successful, and my little project of one soon had money and some staff, but the problem at Berkeley was always office space rather than money—there wasn't room for the help the project needed, so when the other founders of Sun Microsystems showed up I jumped at the chance to join them. At Sun, the long hours continued into the early days of workstations and personal computers, and I have enjoyed participating in the creation of advanced microprocessor technologies and Internet technologies such as Java and Jini.

From all this, I trust it is clear that I am not a Luddite. I have always, rather, had a strong belief in the value of the scientific search for truth and in the ability of great engineering to bring material progress. The Industrial Revolution has immeasurably improved everyone's life over the last couple hundred years, and I always expected my career to involve the building of worthwhile solutions to real problems, one problem at a time.

I have not been disappointed. My work has had more impact than I had ever hoped for and has been more widely used than I could have reasonably expected. I have spent the last 20 years still trying to figure out how to make computers as reliable as I want them to be (they are not nearly there yet) and how to make them simple to use (a goal that has met with even less relative success). Despite some progress, the problems that remain seem even more daunting.

But while I was aware of the moral dilemmas surrounding technology's consequences in fields like weapons research, I did not expect that I would confront such issues in my own field, or at least not so soon.

Perhaps it is always hard to see the bigger impact while you are in the vortex of a change. Failing to understand the consequences of our inventions while we are in the rapture of discovery and innovation seems to be a common fault of scientists and technologists; we have long been driven by the overarching desire to know that is the nature of science's quest, not stopping to notice that the progress to newer and more powerful technologies can take on a life of its own.

I have long realized that the big advances in information technology come not from the work of computer scientists, computer architects, or electrical engineers, but from that of physical scientists. The physicists Stephen Wolfram and Brosl Hasslacher introduced me, in the early 1980s, to chaos theory and nonlinear systems. In the 1990s, I learned about complex systems from conversations with Danny Hillis, the biologist Stuart Kauffman, the Nobel-laureate physicist Murray Gell-Mann, and others. Most recently, Hasslacher and the electrical engineer and device physicist Mark Reed have been giving me insight into the incredible possibilities of molecular electronics.

In my own work, as codesigner of three microprocessor architectures—SPARC, picoJava, and MAJC—and as the designer of several implementations thereof, I've been afforded a deep and firsthand acquaintance with Moore's law. For decades, Moore's law has correctly predicted the exponential rate of improvement of semiconductor technology. Until last year I believed that the rate of advances predicted by Moore's law might continue only until roughly 2010, when some physical limits would begin to be reached. It was not obvious to me that a new technology would arrive in time to keep performance advancing smoothly.

But because of the recent rapid and radical progress in molecular electronics—where individual atoms and molecules replace lithographically drawn transistors—and related nanoscale technologies, we should be able to meet or exceed the Moore's law rate of progress for another 30 years. By 2030, we are likely to be able to build machines, in quantity, a million times as powerful as the personal computers of today—sufficient to implement the dreams of Kurzweil and Moravec.

As this enormous computing power is combined with the manipulative advances of the physical sciences and the new, deep understandings in genetics, enormous transformative power is being unleashed. These combinations open up the opportunity to completely redesign the world, for better or worse: The replicating and evolving processes that have been confined to the natural world are about to become realms of human endeavor.

In designing software and microprocessors, I have never had the feeling that I was designing an intelligent machine. The software and hardware is so fragile and the capabilities of the machine to "think" so clearly absent that, even as a possibility, this has always seemed very far in the future.

But now, with the prospect of human-level computing power in about 30 years, a new idea suggests itself: that I may be working to create tools which will enable the construction of the technology that may replace our species. How do I feel about this? Very uncomfortable. Having struggled my entire career to build reliable software systems, it seems to me more than likely that this future will not work out as well as some people may imagine. My personal experience suggests we tend to overestimate our design abilities.

Given the incredible power of these new technologies, shouldn't we be asking how we can best coexist with them? And if our own extinction is a likely, or even possible, outcome of our technological development, shouldn't we proceed with great caution?

The dream of robotics is, first, that intelligent machines can do our work for us, allowing us lives of leisure, restoring us to Eden. Yet in his history of such ideas, *Darwin Among the Machines*, George Dyson warns: "In the game of life and evolution there are three players at the table: human beings, nature, and machines. I am firmly on the side of nature. But nature, I suspect, is on the side of the machines." As we have seen, Moravec agrees, believing we may well not survive the encounter with the superior robot species.

How soon could such an intelligent robot be built? The coming advances in computing power seem to make it possible by 2030. And once an intelligent robot exists, it is only a small step to a robot species—to an intelligent robot that can make evolved copies of itself.

A second dream of robotics is that we will gradually replace ourselves with our robotic technology, achieving near immortality by downloading our consciousnesses; it is this process that Danny Hillis thinks we will gradually get used to and that Ray Kurzweil elegantly details in *The Age of Spiritual Machines*. (We are beginning to see intimations of this in the implantation of computer devices into the human body, as illustrated on the cover of *Wired* 8.02.)

But if we are downloaded into our technology, what are the chances that we will thereafter be ourselves or even human? It seems to me far more likely that a robotic existence would not be like a human one in any sense that we understand, that the robots would in no sense be our children, that on this path our humanity may well be lost.

Genetic engineering promises to revolutionize agriculture by increasing crop yields while reducing the use of pesticides; to create tens of thousands of novel species of bacteria, plants, viruses, and animals; to replace reproduction, or supplement it, with cloning; to create cures for many diseases, increasing our life span and our quality of life; and much, much more. We now know with certainty that these profound changes in the biological sciences are imminent and will challenge all our notions of what life is.

Technologies such as human cloning have in particular raised our awareness of the profound ethical and moral issues we face. If, for example, we were to reengineer ourselves into several separate and unequal species using the power of genetic engineering, then we would threaten the notion of equality that is the very cornerstone of our democracy.

Given the incredible power of genetic engineering, it's no surprise that there are significant safety issues in its use. My friend Amory Lovins recently cowrote, along with Hunter Lovins, an editorial that provides an ecological view of some of these dangers. Among their concerns: that "the new botany aligns the development of plants with their economic, not evolutionary, success." (See "A Tale of Two Botanies," *Wired* 8.04, page 247.) Amory's long career has been focused on energy and resource efficiency by taking a whole-system view of human-made systems; such a whole-system view often finds simple, smart solutions to otherwise seemingly difficult problems, and is usefully applied here as well.

After reading the Lovins' editorial, I saw an op-ed by Gregg Easterbrook in the *New York Times* (November 19, 1999) about genetically engineered crops, under the headline: "Food for the Future: Someday, rice will have built-in vitamin A. Unless the Luddites win."

Are Amory and Hunter Lovins Luddites? Certainly not. I believe we all would agree that golden rice, with its built-in vitamin A, is probably a good thing, if developed with proper care and respect for the likely dangers in moving genes across species boundaries.

Awareness of the dangers inherent in genetic engineering is beginning to grow, as reflected in the Lovins' editorial. The general public is aware of, and uneasy about, genetically modified foods, and seems to be rejecting the notion that such foods should be permitted to be unlabeled.

But genetic engineering technology is already very far along. As the Lovins note, the USDA has already approved about 50 genetically engineered crops for unlimited release; more than half of the world's soybeans and a third of its corn now contain genes spliced in from other forms of life.

While there are many important issues here, my own major concern with genetic engineering is narrower: that it gives the power—whether militarily, accidentally, or in a deliberate terrorist act—to create a White Plague.

The many wonders of nanotechnology were first imagined by the Nobel-laureate physicist Richard Feynman in a speech he gave in 1959, subsequently published under the title "There's Plenty of Room at the Bottom." The book that made a big impression on me, in the mid-'80s, was Eric Drexler's *Engines of Creation*, in which he described beautifully how manipulation of matter at the atomic level could create a utopian future of abundance, where just about everything could be made cheaply, and almost any imaginable disease or physical problem could be solved using nanotechnology and artificial intelligences.

A subsequent book, *Unbounding the Future: The Nanotechnology Revolution*, which Drexler cowrote, imagines some of the changes that might take place in a world where we had molecular-level "assemblers." Assemblers could make possible incredibly low-cost solar power, cures for cancer and the common cold by augmentation of the human immune system, essentially complete cleanup of the environment, incredibly inexpensive pocket supercomputers—in fact, any product would be manufacturable by assemblers at a cost no greater than that of wood—spaceflight more accessible than transoceanic travel today, and restoration of extinct species.

I remember feeling good about nanotechnology after reading *Engines of Creation*. As a technologist, it gave me a sense of calm—that is, nanotechnology showed us that incredible progress was possible, and indeed perhaps inevitable. If nanotechnology was our future, then I didn't feel pressed to solve so many problems in the present. I would get to Drexler's utopian future in due time; I might as well enjoy life more in the here and now. It didn't make sense, given his vision, to stay up all night, all the time.

Drexler's vision also led to a lot of good fun. I would occasionally get to describe the wonders of nanotechnology to others who had not heard of it. After teasing them with all the things Drexler described I would give a homework assignment of my own: "Use nanotechnology to create a vampire; for extra credit create an antidote."

With these wonders came clear dangers, of which I was acutely aware. As I said at a nanotechnology conference in 1989, "We can't simply do our science and not worry about these ethical issues."⁵ But my subsequent conversations with physicists convinced me that nanotechnology might not even work—or, at least, it wouldn't work anytime soon. Shortly thereafter I moved to Colorado, to a skunk works I had set up, and the focus of my work shifted to software for the Internet, specifically on ideas that became Java and Jini.

Then, last summer, Brosl Hasslacher told me that nanoscale molecular electronics was now practical. This was *new* news, at least to me, and I think to many people—and it radically changed my opinion about nanotechnology. It sent me back to *Engines of Creation*. Rereading Drexler's work after more than 10 years, I was dismayed to realize how little I had remembered of its lengthy section called "Dangers and Hopes," including a discussion of how nanotechnologies can become "engines of destruction." Indeed, in my rereading of this cautionary material today, I am struck by how naive some of Drexler's safeguard proposals seem, and how much greater I judge the dangers to be now than even he seemed to then. (Having anticipated and described many technical and political problems with nanotechnology, Drexler started the Foresight Institute in the late 1980s "to help prepare society for anticipated advanced technologies"—most important, nanotechnology.)

The enabling breakthrough to assemblers seems quite likely within the next 20 years. Molecular electronics—the new subfield of nanotechnology where individual molecules are circuit elements—should mature quickly and become enormously lucrative within this decade, causing a large incremental investment in all nanotechnologies.

Unfortunately, as with nuclear technology, it is far easier to create destructive uses for nanotechnology than constructive ones. Nanotechnology has clear military and terrorist uses, and you need not be suicidal to release a massively destructive nanotechnological device—such devices can be built to be selectively destructive, affecting, for example, only a certain geographical area or a group of people who are genetically distinct.

An immediate consequence of the Faustian bargain in obtaining the great power of nanotechnology is that we run a grave risk—the risk that we might destroy the biosphere on which all life depends.

As Drexler explained:

"Plants" with "leaves" no more efficient than today's solar cells could out-compete real plants, crowding the biosphere with an inedible foliage. Tough omnivorous "bacteria" could out-compete real bacteria: They could spread like blowing pollen, replicate swiftly, and reduce the biosphere to dust in a matter of days. Dangerous replicators could easily be too tough, small, and rapidly spreading to stop—at least if we make no preparation. We have trouble enough controlling viruses and fruit flies.

Among the cognoscenti of nanotechnology, this threat has become known as the "gray goo problem." Though masses of uncontrolled replicators need not be gray or gooey, the term "gray goo" emphasizes that replicators able to obliterate life might be less inspiring than a single species of crabgrass. They might be superior in an evolutionary sense, but this need not make them valuable.

The gray goo threat makes one thing perfectly clear: We cannot afford certain kinds of accidents with replicating assemblers.

Gray goo would surely be a depressing ending to our human adventure on Earth, far worse than mere fire or ice, and one that could stem from a simple laboratory accident.[6]

Oops.

It is most of all the power of destructive self-replication in genetics, nanotechnology, and robotics (GNR) that should give us pause. Self-replication is the modus operandi of genetic engineering, which uses the machinery of the cell to replicate its designs, and the prime danger underlying gray goo in nanotechnology. Stories of runamok robots like the Borg, replicating or mutating to escape from the ethical constraints imposed on them by their creators, are well established in our science fiction books and movies. It is even possible that self-replication may be more fundamental than we thought, and hence harder—or even impossible—to control. A recent article by Stuart Kauffman in *Nature* titled "Self-Replication: Even Peptides Do It" discusses the discovery that a 32-amino-acid peptide can "autocatalyse its own synthesis." We don't know how widespread this ability is, but Kauffman notes that it may hint at "a route to self-reproducing molecular systems on a basis far wider than Watson-Crick base-pairing."[7]

In truth, we have had in hand for years clear warnings of the dangers inherent in widespread knowledge of GNR technologies—of the possibility of knowledge alone enabling mass destruction. But these warnings haven't been widely publicized; the public discussions have been clearly inadequate. There is no profit in publicizing the dangers.

The nuclear, biological, and chemical (NBC) technologies used in 20th-century weapons of mass destruction were and are largely military, developed in government laboratories. In sharp contrast, the 21st-century GNR technologies have clear commercial uses and are being developed almost exclusively by corporate enterprises. In this age of triumphant commercialism, technology—with science as its handmaiden—is delivering a series of almost magical inventions that are the most phenomenally lucrative ever seen. We are aggressively pursuing the promises of these new technologies within the now-unchallenged system of global capitalism and its manifold financial incentives and competitive pressures.

This is the first moment in the history of our planet when any species, by its own voluntary actions, has become a danger to itself—as well as to vast numbers of others.

It might be a familiar progression, transpiring on many worlds—a planet, newly formed, placidly revolves around its star; life slowly forms; a kaleidoscopic procession of creatures evolves; intelligence emerges which, at least up to a point, confers enormous survival value; and then technology is invented. It dawns on them that there are such things as laws of Nature, that these laws can be revealed by experiment, and that knowledge of these laws can be made both to save and to take lives, both on unprecedented scales. Science, they recognize, grants immense powers. In a flash, they create world-altering contrivances. Some planetary civilizations see their way through, place limits on what may and what must not be done, and safely pass through the time of perils. Others, not so lucky or so prudent, perish.

That is Carl Sagan, writing in 1994, in *Pale Blue Dot*, a book describing his vision of the human future in space. I am only now realizing how deep his insight was, and how sorely I miss, and will miss, his voice. For all its eloquence, Sagan's contribution was not least that of simple common sense—an attribute that, along with humility, many of the leading advocates of the 21st-century technologies seem to lack.

I remember from my childhood that my grandmother was strongly against the overuse of antibiotics. She had worked since before the first World War as a nurse and had a commonsense attitude that taking antibiotics, unless they were absolutely necessary, was bad for you.

It is not that she was an enemy of progress. She saw much progress in an almost 70-year nursing career; my grandfather, a diabetic, benefited greatly from the improved treatments that became available in his lifetime. But she, like many levelheaded people, would probably think it greatly arrogant for us, now, to be designing a robotic "replacement species," when we obviously have so much trouble making relatively simple things work, and so much trouble managing—or even understanding—ourselves.

I realize now that she had an awareness of the nature of the order of life, and of the necessity of living with and respecting that order. With this respect comes a necessary humility that we, with our early-21st-century chutzpah, lack at our peril. The commonsense view, grounded in this respect, is often right, in advance of the scientific evidence. The clear fragility and inefficiencies of the human-made systems we have built should give us all pause; the fragility of the systems I have worked on certainly humbles me.

We should have learned a lesson from the making of the first atomic bomb and the resulting arms race. We didn't do well then, and the parallels to our current situation are troubling.

The effort to build the first atomic bomb was led by the brilliant physicist J. Robert Oppenheimer. Oppenheimer was not naturally interested in politics but became painfully aware of what he perceived as the grave threat to Western civilization from the Third Reich, a threat surely grave because of the possibility that Hitler might obtain nuclear weapons. Energized by this concern, he brought his strong intellect, passion for physics, and charismatic leadership skills to Los Alamos and led a rapid and successful effort by an incredible collection of great minds to quickly invent the bomb.

What is striking is how this effort continued so naturally after the initial impetus was removed. In a meeting shortly after V-E Day with some physicists who felt that perhaps the effort should stop, Oppenheimer argued to continue. His stated reason seems a bit strange: not because of the fear of large casualties from an invasion of Japan, but because the United Nations, which was soon to be formed, should have foreknowledge of atomic weapons. A more likely reason the project continued is the momentum that had built up—the first atomic test, Trinity, was nearly at hand.

We know that in preparing this first atomic test the physicists proceeded despite a large number of possible dangers. They were initially worried, based on a calculation by Edward Teller, that an atomic explosion might set fire to the atmosphere. A revised calculation reduced the danger of destroying the world to a three-in-a-million chance. (Teller says he was later able to dismiss the prospect of atmospheric ignition entirely.) Oppenheimer, though, was sufficiently concerned about the result of Trinity that he arranged for a possible evacuation of the southwest part of the state of New Mexico. And, of course, there was the clear danger of starting a nuclear arms race.

Within a month of that first, successful test, two atomic bombs destroyed Hiroshima and Nagasaki. Some scientists had suggested that the bomb simply be demonstrated, rather than dropped on Japanese cities—saying that this would greatly improve the chances for arms control after the war—but to no avail. With the tragedy of Pearl Harbor still fresh in Americans' minds, it would have been very difficult for President Truman to order a demonstration of the weapons rather than use them as he did—the desire to quickly end the war and save the lives that would have been lost in any invasion of Japan was very strong. Yet the overriding truth was probably very simple: As the physicist Freeman Dyson later said, "The reason that it was dropped was just that nobody had the courage or the foresight to say no."

It's important to realize how shocked the physicists were in the aftermath of the bombing of Hiroshima, on August 6, 1945. They describe a series of waves of emotion: first, a sense of fulfillment that the bomb worked, then horror at all the people that had been killed, and then a convincing feeling that on no account should another bomb be dropped. Yet of course another bomb was dropped, on Nagasaki, only three days after the bombing of Hiroshima.

In November 1945, three months after the atomic bombings, Oppenheimer stood firmly behind the scientific attitude, saying, "It is not possible to be a scientist unless you believe that the knowledge of the world, and the power which this gives, is a thing which is of intrinsic value to humanity, and that you are using it to help in the spread of knowledge and are willing to take the consequences."

Oppenheimer went on to work, with others, on the Acheson-Lilienthal report, which, as Richard Rhodes says in his recent book *Visions of Technology*, "found a way to prevent a clandestine nuclear arms race without resorting to armed world government"; their suggestion was a form of relinquishment of nuclear weapons work by nation-states to an international agency.

This proposal led to the Baruch Plan, which was submitted to the United Nations in June 1946 but never adopted (perhaps because, as Rhodes suggests, Bernard Baruch had "insisted on burdening the plan with conventional sanctions," thereby inevitably dooming it, even though it would "almost certainly have been rejected by Stalinist Russia anyway"). Other efforts to promote sensible steps toward internationalizing nuclear power to prevent an arms race ran afoul either of US politics and internal distrust, or distrust by the Soviets. The opportunity to avoid the arms race was lost, and very quickly.

Two years later, in 1948, Oppenheimer seemed to have reached another stage in his thinking, saying, "In some sort of crude sense which no vulgarity, no humor, no overstatement can quite extinguish, the physicists have known sin; and this is a knowledge they cannot lose."

In 1949, the Soviets exploded an atom bomb. By 1955, both the US and the Soviet Union had tested hydrogen bombs suitable for delivery by aircraft. And so the nuclear arms race began.

Nearly 20 years ago, in the documentary *The Day After Trinity*, Freeman Dyson summarized the scientific attitudes that brought us to the nuclear precipice:

I have felt it myself. The glitter of nuclear weapons. It is irresistible if you come to them as a scientist. To feel it's there in your hands, to release this energy that fuels the stars, to let it do your bidding. To perform these miracles, to lift a million tons of rock into the sky. It is something that gives people an illusion of illimitable power, and it is, in some ways, responsible for all our troubles—this, what you might call technical arrogance, that overcomes people when they see what they can do with their minds.[8]

Now, as then, we are creators of new technologies and stars of the imagined future, driven—this time by great financial rewards and global competition—despite the clear dangers, hardly evaluating what it may be like to try to live in a world that is the realistic outcome of what we are creating and imagining.

In 1947, *The Bulletin of the Atomic Scientists* began putting a Doomsday Clock on its cover. For more than 50 years, it has shown an estimate of the relative nuclear danger we have faced, reflecting the changing international conditions. The hands on the clock have moved 15 times and today, standing at nine minutes to midnight, reflect continuing and real danger from nuclear weapons. The recent addition of India and Pakistan to the list of nuclear powers has increased the threat of failure of the nonproliferation goal, and this danger was reflected by moving the hands closer to midnight in 1998.

In our time, how much danger do we face, not just from nuclear weapons, but from all of these technologies? How high are the extinction risks?

The philosopher John Leslie has studied this question and concluded that the risk of human extinction is at least 30 percent,[9] while Ray Kurzweil believes we have "a better than even chance of making it through," with the caveat that he has "always been accused of being an optimist." Not only are these estimates not encouraging, but they do not include the probability of many horrid outcomes that lie short of extinction.

Faced with such assessments, some serious people are already suggesting that we simply move beyond Earth as quickly as possible. We would colonize the galaxy using von Neumann probes, which hop from star system to star system, replicating as they go. This step will almost certainly be necessary 5 billion years from now (or sooner if our solar system is disastrously impacted by the impending collision of our galaxy with the Andromeda galaxy within the next 3 billion years), but if we take Kurzweil and Moravec at their word it might be necessary by the middle of this century.

What are the moral implications here? If we must move beyond Earth this quickly in order for the species to survive, who accepts the responsibility for the fate of those (most of us, after all) who are left behind? And even if we scatter to the stars, isn't it likely that we may take our problems with us or find, later, that they have followed us? The fate of our species on Earth and our fate in the galaxy seem inextricably linked.

Another idea is to erect a series of shields to defend against each of the dangerous technologies. The Strategic Defense Initiative, proposed by the Reagan administration, was an attempt to design such a shield against the threat of a nuclear attack from the Soviet Union. But as Arthur C. Clarke, who was privy to discussions about the project, observed: "Though it might be possible, at vast expense, to construct local defense systems that would 'only' let through a few percent of ballistic missiles, the much touted idea of a national umbrella was nonsense. Luis Alvarez, perhaps the greatest experimental physicist of this century, remarked to me that the advocates of such schemes were 'very bright guys with no common sense.'"

Clarke continued: "Looking into my often cloudy crystal ball, I suspect that a total defense might indeed be possible in a century or so. But the technology involved would produce, as a by-product, weapons so terrible that no one would bother with anything as primitive as ballistic missiles."[10]

In *Engines of Creation*, Eric Drexler proposed that we build an active nanotechnological shield—a form of immune system for the biosphere—to defend against dangerous replicators of all kinds that might escape from laboratories or otherwise be maliciously created. But the shield he proposed would itself be extremely dangerous—nothing could prevent it from developing autoimmune problems and attacking the biosphere itself.[11]

Similar difficulties apply to the construction of shields against robotics and genetic engineering. These technologies are too powerful to be shielded against in the time frame of interest; even if it were possible to implement defensive shields, the side effects of their development would be at least as dangerous as the technologies we are trying to protect against.

These possibilities are all thus either undesirable or unachievable or both. The only realistic alternative I see is relinquishment: to limit development of the technologies that are too dangerous, by limiting our pursuit of certain kinds of knowledge.

Yes, I know, knowledge is good, as is the search for new truths. We have been seeking knowledge since ancient times. Aristotle opened his *Metaphysics* with the simple statement: "All men by nature desire to know." We have, as a bedrock value in our society, long agreed on the value of open access to information, and recognize the problems that arise with attempts to restrict access to and development of knowledge. In recent times, we have come to revere scientific knowledge.

But despite the strong historical precedents, if open access to and unlimited development of knowledge henceforth puts us all in clear danger of extinction, then common sense demands that we reexamine even these basic, long-held beliefs.

It was Nietzsche who warned us, at the end of the 19th century, not only that God is dead but that "faith in science, which after all exists undeniably, cannot owe its origin to a calculus of utility; it must have originated *in spite of* the fact that the disutility and dangerousness of the 'will to truth,' of 'truth at any price' is proved to it constantly." It is this further danger that we now fully face—the consequences of our truth-seeking. The truth that science seeks can certainly be considered a dangerous substitute for God if it is likely to lead to our extinction.

If we could agree, as a species, what we wanted, where we were headed, and why, then we would make our future much less dangerous—then we might understand what we can and should relinquish. Otherwise, we can easily imagine an arms race developing over GNR technologies, as it did with the NBC technologies in the 20th century. This is perhaps the greatest risk, for once such a race begins, it's very hard to end it. This time—unlike during the Manhattan Project—we aren't in a war, facing an implacable enemy that is threatening our civilization; we are driven, instead, by our habits, our desires, our economic system, and our competitive need to know.

I believe that we all wish our course could be determined by our collective values, ethics, and morals. If we had gained more collective wisdom over the past few thousand years, then a dialogue to this end would be more practical, and the incredible powers we are about to unleash would not be nearly so troubling.

One would think we might be driven to such a dialogue by our instinct for self-preservation. Individuals clearly have this desire, yet as a species our behavior seems to be not in our favor. In dealing with the nuclear threat, we often spoke dishonestly to ourselves and to each other, thereby greatly increasing the risks. Whether this was politically motivated, or because we chose not to think ahead, or because when faced with such grave threats we acted irrationally out of fear, I do not know, but it does not bode well.

The new Pandora's boxes of genetics, nanotechnology, and robotics are almost open, yet we seem hardly to have noticed. Ideas can't be put back in a box; unlike uranium or plutonium, they don't need to be mined and refined, and they can be freely copied. Once they are out, they are out. Churchill remarked, in a famous left-handed compliment, that the American people and their leaders "invariably do the right thing, after they have examined every other alternative." In this case, however, we must act more presciently, as to do the right thing only at last may be to lose the chance to do it at all.

As Thoreau said, "We do not ride on the railroad; it rides upon us"; and this is what we must fight, in our time. The question is, indeed, Which is to be master? Will we survive our technologies?

We are being propelled into this new century with no plan, no control, no brakes. Have we already gone too far down the path to alter course? I don't believe so, but we aren't trying yet, and the last chance to assert control—the fail-safe point—is rapidly approaching. We have our first pet robots, as well as commercially available genetic engineering techniques, and our nanoscale techniques are advancing rapidly. While the development of these technologies proceeds through a number of steps, it isn't necessarily the case—as happened in the Manhattan Project and the Trinity test—that the last step in proving a technology is large and hard. The breakthrough to wild self-replication in robotics, genetic engineering, or nanotechnology could come suddenly, reprising the surprise we felt when we learned of the cloning of a mammal.

And yet I believe we do have a strong and solid basis for hope. Our attempts to deal with weapons of mass destruction in the last century provide a shining example of relinquishment for us to consider: the unilateral US abandonment, without preconditions, of the development of biological weapons. This relinquishment stemmed from the realization that while it would take an enormous effort to create these terrible weapons, they could from then on easily be duplicated and fall into the hands of rogue nations or terrorist groups.

The clear conclusion was that we would create additional threats to ourselves by pursuing these weapons, and that we would be more secure if we did not pursue them. We have embodied our relinquishment of biological and chemical weapons in the 1972 Biological Weapons Convention (BWC) and the 1993 Chemical Weapons Convention (CWC).[12]

As for the continuing sizable threat from nuclear weapons, which we have lived with now for more than 50 years, the US Senate's recent rejection of the Comprehensive Test Ban Treaty makes it clear relinquishing nuclear weapons will not be politically easy. But we have a unique opportunity, with the end of the Cold War, to avert a multipolar arms race. Building on the BWC and CWC relinquishments, successful abolition of nuclear weapons could help us build toward a habit of relinquishing dangerous technologies. (Actually, by getting rid of all but 100 nuclear weapons worldwide—roughly the total destructive power of World War II and a considerably easier task—we could eliminate this extinction threat.[13])

Verifying relinquishment will be a difficult problem, but not an unsolvable one. We are fortunate to have already done a lot of relevant work in the context of the BWC and other treaties. Our major task will be to apply this to technologies that are naturally much more commercial than military. The substantial need here is for transparency, as difficulty of verification is directly proportional to the difficulty of distinguishing relinquished from legitimate activities.

I frankly believe that the situation in 1945 was simpler than the one we now face: The nuclear technologies were reasonably separable into commercial and military uses, and monitoring was aided by the nature of atomic tests and the ease with which radioactivity could be measured. Research on military applications could be performed at national laboratories such as Los Alamos, with the results kept secret as long as possible.

The GNR technologies do not divide clearly into commercial and military uses; given their potential in the market, it's hard to imagine pursuing them only in national laboratories. With their widespread commercial pursuit, enforcing relinquishment will require a verification regime similar to that for biological weapons, but on an unprecedented scale. This, inevitably, will raise tensions between our individual privacy and desire for proprietary information, and the need for verification to protect us all. We will undoubtedly encounter strong resistance to this loss of privacy and freedom of action.

Verifying the relinquishment of certain GNR technologies will have to occur in cyberspace as well as at physical facilities. The critical issue will be to make the necessary transparency acceptable in a world of proprietary information, presumably by providing new forms of protection for intellectual property.

Verifying compliance will also require that scientists and engineers adopt a strong code of ethical conduct, resembling the Hippocratic oath, and that they have the courage to whistleblow as necessary, even at high personal cost. This would answer the call—50 years after Hiroshima—by the Nobel laureate Hans Bethe, one of the most senior of the surviving members of the Manhattan Project, that all scientists "cease and desist from work creating, developing, improving, and manufacturing nuclear weapons and other weapons of potential mass destruction."[14] In the 21st century, this requires vigilance and personal responsibility by those who would work on both NBC and GNR technologies to avoid implementing weapons of mass destruction and knowledge-enabled mass destruction.

Thoreau also said that we will be "rich in proportion to the number of things which we can afford to let alone." We each seek to be happy, but it would seem worthwhile to question whether we need to take such a high risk of total destruction to gain yet more knowledge and yet more things; common sense says that there is a limit to our material needs—and that certain knowledge is too dangerous and is best forgone.

Neither should we pursue near immortality without considering the costs, without considering the commensurate increase in the risk of extinction. Immortality, while perhaps the original, is certainly not the only possible utopian dream.

I recently had the good fortune to meet the distinguished author and scholar Jacques Attali, whose book *Lignes d'horizons* (*Millennium*, in the English translation) helped inspire the Java and Jini approach to the coming age of pervasive computing, as previously described in this magazine. In his new book *Fraternités*, Attali describes how our dreams of utopia have changed over time:

"At the dawn of societies, men saw their passage on Earth as nothing more than a labyrinth of pain, at the end of which stood a door leading, via their death, to the company of gods and to *Eternity*. With the Hebrews and then the Greeks, some men dared free themselves from theological demands and dream of an ideal City where *Liberty* would flourish. Others, noting the evolution of the market society, understood that the liberty of some would entail the alienation of others, and they sought *Equality*."

Jacques helped me understand how these three different utopian goals exist in tension in our society today. He goes on to describe a fourth utopia, *Fraternity*, whose foundation is altruism. Fraternity alone associates individual happiness with the happiness of others, affording the promise of self-sustainment.

This crystallized for me my problem with Kurzweil's dream. A technological approach to Eternity—near immortality through robotics—may not be the most desirable utopia, and its pursuit brings clear dangers. Maybe we should rethink our utopian choices.

Where can we look for a new ethical basis to set our course? I have found the ideas in the book *Ethics for the New Millennium*, by the Dalai Lama, to be very helpful. As is perhaps well known but little heeded, the Dalai Lama argues that the most important thing is for us to conduct our lives with love and compassion for others, and that our societies need to develop a stronger notion of universal responsibility and of our interdependency; he proposes a standard of positive ethical conduct for individuals and societies that seems consonant with Attali's Fraternity utopia.

The Dalai Lama further argues that we must understand what it is that makes people happy, and acknowledge the strong evidence that neither material progress nor the pursuit of the power of knowledge is the key—that there are limits to what science and the scientific pursuit alone can do.

Our Western notion of happiness seems to come from the Greeks, who defined it as "the exercise of vital powers along lines of excellence in a life affording them scope."[15]

Clearly, we need to find meaningful challenges and sufficient scope in our lives if we are to be happy in whatever is to come. But I believe we must find alternative outlets for our creative forces, beyond the culture of perpetual economic growth; this growth has largely been a blessing for several hundred years, but it has not brought us unalloyed happiness, and we must now choose between the pursuit of unrestricted and undirected growth through science and technology and the clear accompanying dangers.

It is now more than a year since my first encounter with Ray Kurzweil and John Searle. I see around me cause for hope in the voices for caution and relinquishment and in those people I have discovered who are as concerned as I am about our current predicament. I feel, too, a deepened sense of personal responsibility—not for the work I have already done, but for the work that I might yet do, at the confluence of the sciences.

But many other people who know about the dangers still seem strangely silent. When pressed, they trot out the "this is nothing new" riposte—as if awareness of what could happen is response enough. They tell me, There are universities filled with bioethicists who study this stuff all day long. They say, All this has been written about before, and by experts. They complain, Your worries and your arguments are already old hat.

I don't know where these people hide their fear. As an architect of complex systems I enter this arena as a generalist. But should this diminish my concerns? I am aware of how much has been written about, talked about, and lectured about so authoritatively. But does this mean it has reached people? Does this mean we can discount the dangers before us?

Knowing is not a rationale for not acting. Can we doubt that knowledge has become a weapon we wield against ourselves?

The experiences of the atomic scientists clearly show the need to take personal responsibility, the danger that things will move too fast, and the way in which a process can take on a life of its own. We can, as they did, create insurmountable problems in almost no time flat. We must do more thinking up front if we are not to be similarly surprised and shocked by the consequences of our inventions.

My continuing professional work is on improving the reliability of software. Software is a tool, and as a toolbuilder I must struggle with the uses to which the tools I make are put. I have always believed that making software more reliable, given its many uses, will make the world a safer and better place; if I were to come to believe the opposite, then I would be morally obligated to stop this work. I can now imagine such a day may come.

This all leaves me not angry but at least a bit melancholic. Henceforth, for me, progress will be somewhat bittersweet.

Do you remember the beautiful penultimate scene in *Manhattan* where Woody Allen is lying on his couch and talking into a tape recorder? He is writing a short story about people who are creating unnecessary, neurotic problems for themselves, because it keeps them from dealing with more unsolvable, terrifying problems about the universe.

He leads himself to the question, "Why is life worth living?" and to consider what makes it worthwhile for him: Groucho Marx, Willie Mays, the second movement of the Jupiter Symphony, Louis Armstrong's recording of "Potato Head Blues," Swedish movies, Flaubert's Sentimental Education, Marlon Brando, Frank Sinatra, the apples and pears by Cézanne, the crabs at Sam Wo's, and, finally, the showstopper: his love Tracy's face.

Each of us has our precious things, and as we care for them we locate the essence of our humanity. In the end, it is because of our great capacity for caring that I remain optimistic we will confront the dangerous issues now before us.

My immediate hope is to participate in a much larger discussion of the issues raised here, with people from many different backgrounds, in settings not predisposed to fear or favor technology for its own sake.

As a start, I have twice raised many of these issues at events sponsored by the Aspen Institute and have separately proposed that the American Academy of Arts and Sciences take them up as an extension of its work with the Pugwash Conferences. (These have been held since 1957 to discuss arms control, especially of nuclear weapons, and to formulate workable policies.)

It's unfortunate that the Pugwash meetings started only well after the nuclear genie was out of the bottle—roughly 15 years too late. We are also getting a belated start on seriously addressing the issues around 21st-century technologies—the prevention of knowledge-enabled mass destruction—and further delay seems unacceptable.

So I'm still searching; there are many more things to learn. Whether we are to succeed or fail, to survive or fall victim to these technologies, is not yet decided. I'm up late again—it's almost 6 am. I'm trying to imagine some better answers, to break the spell and free them from the stone.

Notes

1. The passage Kurzweil quotes is from Kaczynski's Unabomber Manifesto, which was published jointly, under duress, by the *New York Times* and *The Washington Post* to attempt to

bring his campaign of terror to an end. I agree with David Gelernter, who said about their decision:

"It was a tough call for the newspapers. To say yes would be giving in to terrorism, and for all they knew he was lying anyway. On the other hand, to say yes might stop the killing. There was also a chance that someone would read the tract and get a hunch about the author; and that is exactly what happened. The suspect's brother read it, and it rang a bell.

"I would have told them not to publish. I'm glad they didn't ask me. I guess." (*Drawing Life: Surviving the Unabomber*. Free Press, 1997: 120)

2. Garrett, Laurie. *The Coming Plague: Newly Emerging Diseases in a World Out of Balance*. Penguin, 1994: 47–52, 414, 419, 452.

3. Isaac Asimov described what became the most famous view of ethical rules for robot behavior in his book *I, Robot* in 1950, in his Three Laws of Robotics: 1. A robot may not injure a human being, or, through inaction, allow a human being to come to harm. 2. A robot must obey the orders given it by human beings, except where such orders would conflict with the First Law. 3. A robot must protect its own existence, as long as such protection does not conflict with the First or Second Law.

4. Michelangelo wrote a sonnet that begins:

Non ha l' ottimo artista alcun concetto
Ch' un marmo solo in sè non circonscriva
Col suo soverchio; e solo a quello arriva
La man che ubbidisce all' intelletto.

Stone translates this as:

The best of artists hath no thought to show
which the rough stone in its superfluous shell
doth not include; to break the marble spell
is all the hand that serves the brain can do.

Stone describes the process: "He was not working from his drawings or clay models; they had all been put away. He was carving from the images in his mind. His eyes and hands knew where every line, curve, mass must emerge, and at what depth in the heart of the stone to create the low relief" (*The Agony and the Ecstasy*. Doubleday, 1961: 6, 144).

5. First Foresight Conference on Nanotechnology in October 1989, a talk titled "The Future of Computation." Published in Crandall, B. C. and James Lewis, editors. *Nanotechnology: Research and Perspectives*. MIT Press, 1992: 269. See also www.foresight.org/Conferences/MNT01/Nano1.html.

6. In his 1963 novel *Cat's Cradle*, Kurt Vonnegut imagined a gray-goo-like accident where a form of ice called ice-nine, which becomes solid at a much higher temperature, freezes the oceans.

7. Kauffman, Stuart. "Self-replication: Even Peptides Do It." *Nature*, 382, August 8, 1996: 496.

8. Else, Jon. *The Day After Trinity: J. Robert Oppenheimer and The Atomic Bomb* (available at www.pyramiddirect.com).

9. This estimate is in Leslie's book *The End of the World: The Science and Ethics of Human Extinction*, where he notes that the probability of extinction is substantially higher if we accept Brandon Carter's Doomsday Argument, which is, briefly, that "we ought to have some reluctance to believe that we are very exceptionally early, for instance in the earliest 0.001 percent, among all humans who will ever have lived. This would be some reason for thinking that humankind will not survive for many more centuries, let alone colonize the galaxy. Carter's doomsday argument doesn't generate any risk estimates just by itself. It is an argument for *revising* the estimates which we generate when we consider various possible dangers" (Routledge, 1996: 1, 3, 145).

10. Clarke, Arthur C. "Presidents, Experts, and Asteroids." *Science*, June 5, 1998. Reprinted as "Science and Society" in *Greetings, Carbon-Based Bipeds! Collected Essays, 1934–1998*. St. Martin's Press, 1999: 526.

11. And, as David Forrest suggests in his paper "Regulating Nanotechnology Development," available at www.foresight.org/nano/forrest1989.html, "If we used strict liability as an alternative to regulation it would be impossible for any developer to internalize the cost of the risk (destruction of the biosphere), so theoretically the activity of developing nanotechnology should never be undertaken." Forrest's analysis leaves us with only government regulation to protect us—not a comforting thought.

12. Meselson, Matthew. "The Problem of Biological Weapons." Presentation to the 1,818th Stated Meeting of the American Academy of Arts and Sciences, January 13, 1999.

13. Doty, Paul. "The Forgotten Menace: Nuclear Weapons Stockpiles Still Represent the Biggest Threat to Civilization." *Nature*, 402, December 9, 1999: 583.

14. See also Hans Bethe's 1997 letter to President Clinton, at www.fas.org/bethecr.htm.

15. Hamilton, Edith. *The Greek Way*. W. W. Norton & Co., 1942: 35.

II THE RELATIONSHIP BETWEEN TECHNOLOGY AND SOCIETY

Technology plays a pivotal role in the futures presented in section I. Implicit in all of them is the idea that technology shapes, if not determines, society. This idea has generated considerable controversy, analysis, and research among those who study the social implications of technology. Technology plays an important role in shaping society, but what exactly is its role? The readings in this section delve more deeply into the relationship between technology and society; they articulate and make explicit ideas that the authors of the futures were implicitly thinking about (and perhaps what they weren't thinking about) as they imagined what might come to be.

Recent scholarship aimed at understanding the relationship between technology and society has focused on two sets of arguments that can be labeled *technological determinism* and *social construction*. The technological determinist argument comes in many forms, but the articulation most useful for this section is the claim that the introduction of new technologies produces direct and unalterable social changes. For example, a technological determinist might argue that the Internet will lead to greater democracy because it creates the capacity for people around the world to communicate with one another with ease and without the intervention of policies. Advertisers love to play on the idea of technological determinism by suggesting that their products will inevitably lead to certain kinds of outcomes, e.g., the sexy car that assures that the owner will get the girl, or the handy kitchen appliance that will produce tasty, gourmet meals with little effort.

This claim that technology determines society is often accompanied by or conflated with the claim that people have little, if any, control over technology and the effect it has on the world. This is often referred to as 'the autonomy of technology' or it is said that technology is autonomous. Technological development, it is claimed, follows a logical or natural path of progression; one invention builds on another; and humans do not and cannot influence the order or direction. Technological determinism presumes that the development of technology is unaffected (or affected very little) by social and political forces and that the technologies that we have today are simply the latest step in the linear progression of science and engineering. Technologically deterministic arguments present technology as a powerful force that requires people and institutions to behave in certain ways.

The social constructivist argument was developed as a direct challenge to the idea of technological determinism. Social Constructivists contend that technological development has no natural or logical order of progression and is not out of human control. Instead, they maintain that society (through interest groups, laws, the economy, political decisions, etc.) shapes and directs technology in every phase of its development—from conception to production to use (or even non-use). Many advocates of this approach believe that the determinist argument is not just wrong, but inherently dangerous because it pushes out of sight the social forces and interests at work in directing the development of technology and conceals the infinite array of alternative possibilities. Technological determinism leads people to believe that a given future is inescapable and allows individuals to abdicate responsibility for controlling the direction of technology. To counter this idea Social Constructivists demonstrate how individuals and groups—engineers, corporations, regulatory agencies, lawyers, politicians etc.—contribute to the direction of technological development. But their argument does not end with production. Social Constructivists maintain that once a technology is developed, society still has the ability to decide if, where, when, and how it will be used. Even users play an active role in this process by interpreting and reinterpreting technologies and using them for purposes for which they weren't designed.

The goal of this section is not to strongly advocate either technological determinism or social constructivism as the better way to describe the world. Crucial aspects of the relationships between technology and society are concealed if either is taken to the extreme. A firm technological determinist stance ignores the fact that while there may be painful consequences, we can turn a machine off and throw it away if we decide that is necessary. A hard core social constructivist argument might ignore the fact that while technologies can be interpreted in multiple ways, their physical limits prevent them from being interpreted in any way imaginable. This section will draw on the insights of both approaches and ultimately argue that we must consider the ways in which technology and society simultaneously influence and even constitute each other.

The first important step in doing this is to develop a better understanding of what technology is. Technology is not simply an assemblage of mechanical and electrical pieces; rather it is complex systems of people, relationships, and artifacts. Viewing technologies merely as physical objects is like trying to understand a chess piece separately from the game, the thirty-one other pieces, the board, the players, and the rules. Certainly one can learn some things about a "rook" by focusing on its design and studying how plastics are created or marble is carved, but such knowledge will give few clues as to why the piece was made the way it is or the role it will play in the world. To understand how technologies are developed, what they mean, and how they are used, it is best to focus on what we will call, 'sociotechnical systems'—assemblages of things, people, practices, and meanings. Few of the authors in this section (and throughout the book) explicitly use the term 'sociotechnical system' but nearly all of them have, at the heart of their argument, the idea that technology is a cluster of material objects, social practices, social relationships, and social organization.

In order to grasp what a sociotechnical system is, it helps to look at an example. Consider the safety air bag in an automobile. It can be seen as a system of sensors, an inflator, and a bag, but this tells one very little about why it was created, what it does, or even how it works. The sociotechnical system of an air bag includes the numerous relationships the device has with people and other devices. For instance, the modern air bag is meaningless if not viewed in the context of modern transportation, automobiles, and car collisions. Its purpose cannot be understood without looking at how insurance companies encouraged its development, how government regulations shaped its design, and how engineers had certain users in mind when they built it. The air bag can work only when there are automobile manufacturers and distributors, a road system, and drivers with certain habits. The device itself is not without meaning, but its purpose, value, and implications are best understood in its broader sociotechnical context.

The readings in this section discuss a variety of different ways in which technology and society influence one another. If one wants to influence the direction of technology and society, one must first understand their relationship. This book will continue to explore the different ways in which society and technology are intertwined. For instance, it will examine the ways in which technology can be used by employers to subvert the autonomy of employees, can reinforce or break down racial classification, and can be associated with lofty goals like equity, security, and progress. The temptation to draw a line and argue that everything on one side of the line should be categorized as "technical" and everything on the other side of the line should be labeled "social" is powerful but the readings in this section suggest that attempts to draw such a line are misguided and doomed to failure. Technology and society are intimately interwoven.

Questions to consider while reading the selections in this section:

1. Are technological determinism and social constructivism incompatible? How might they work together?
2. Which of the pieces in section I can be characterized as technologically deterministic? Why?
3. Which technologies are still "in the making" today, that is, which are still in the stage of interpretive flexibility?
4. Which current technologies have the momentum that Hughes describes? Are there technologies that would be difficult to change or replace (except around the edges) because so many individuals and groups have invested in them and/or rely upon them?
5. What are the major factors influencing the development of new technologies? What are the major factors influencing the development of society today?

7 "Do Machines Make History?"

Robert L. Heilbroner

In this classic article, Robert Heilbroner introduces two aspects of technological determinism. The first is the claim that technology develops in a fixed, naturally determined sequence. The second is the claim that adoption of a given technology imposes certain social and political characteristics upon the society in which it is used. This view of the technology-society relationship is a common one although it is not often articulated so explicitly. Many of the visions of the future that were presented in the last part were written as though specific technologies necessarily have specific social ramifications. Neither Heilbroner nor the authors of selections in section I are, however, "hard-core" technological determinists since they seem to acknowledge at least some social influence on technology. While Heilbroner wrote this piece in 1967, it remains one of the best articulations of technological determinism, and some of the technologies he mentions are still being debated today.

The hand-mill gives you society with the feudal lord; the steam-mill, society with the industrial capitalist.

Marx, *The Poverty of Philosophy*

That machines make history in some sense—that the level of technology has a direct bearing on the human drama—is of course obvious. That they do not make all of history, however that word be defined, is equally clear. The challenge, then, is to see if one can say something systematic about the matter, to see whether one can order the problem so that it becomes intellectually manageable.

To do so calls at the very beginning for a careful specification of our task. There are a number of important ways in which machines make history that will not concern us here. For example, one can study the impact of technology on the *political* course of history, evidenced most strikingly by the central role played by the technology of war. Or one can study the effect of machines on the *social* attitudes that underlie historical evolution: one thinks of the effect of radio or television on political behavior. Or one can study technology as one of the factors shaping the changeful content of life from one epoch to another: when we speak of "life" in the Middle Ages or today we define an existence much of whose texture and substance is intimately connected with the prevailing technological order.

None of these problems will form the focus of this essay. Instead, I propose to examine the impact of technology on history in another area—an area defined by the famous quotation from Marx that stands beneath our title. The question we are interested in, then, concerns the effect of technology in determining the nature of the *socioeconomic order*. In its simplest terms the question is: did medieval technology bring about feudalism? Is industrial technology the necessary and sufficient condition for capitalism? Or, by extension, will the technology of the computer and the atom constitute the ineluctable cause of a new social order?

Even in this restricted sense, our inquiry promises to be broad and sprawling. Hence, I shall not try to attack it head-on, but to examine it in two stages:

1. If we make the assumption that the hand-mill does "give" us feudalism and the steam-mill capitalism, this places technological change in the position of a prime mover of social history. Can we then explain the "laws of motion" of technology itself? Or to put the question less grandly, can we explain why technology evolves in the sequence it does?
2. Again, taking the Marxian paradigm at face value, exactly what do we mean when we assert that the hand-mill "gives us" society with the feudal lord? Precisely how does the mode of production affect the superstructure of social relationships?

These questions will enable us to test the empirical content—or at least to see if there *is* an empirical content—in the idea of technological determinism. I do not think it will come as a surprise if I announce now that we will find *some* content, and a great deal of missing evidence, in our investigation. What will remain then will be to see if we can place the salvageable elements of the theory in historical perspective—to see, in a word, if we can explain technological determinism historically as well as explain history by technological determinism.

I

We begin with a very difficult question hardly rendered easier by the fact that there exist, to the best of my knowledge, no empirical studies on which to base our speculations. It is the question of whether there is a fixed sequence to technological development and therefore a necessitous path over which technologically developing societies must travel.

I believe there is such a sequence—that the steam-mill follows the hand-mill not by chance but because it is the next "stage" in a technical conquest of nature that follows one and only one grand avenue of advance. To put it differently, I believe that it is impossible to proceed to the age of the steam-mill until one has passed through the age of the hand-mill, and that in turn one cannot move to the age of the hydroelectric plant before one has mastered the steam-mill, nor to the nuclear power age until one has lived through that of electricity.

Before I attempt to justify so sweeping an assertion, let me make a few reservations. To begin with, I am fully conscious that not all societies are interested in developing a technology of production or in channeling to it the same quota of social energy. I am very much aware of the different pressures that different societies exert on the direction in which technology unfolds. Lastly, I am not unmindful of the difference between the discovery of a given machine and its application as a technology—for example, the invention of a steam engine (the aeolipile) by Hero of Alexandria long before its incorporation into a steam-mill. All these problems, to which we will return in our last section, refer however to the way in which technology makes its peace with the social, political, and economic institutions of the society in which it appears. They do not directly affect the contention that there exists a determinate sequence of productive technology for those societies that are interested in originating and applying such a technology.

What evidence do we have for such a view? I would put forward three suggestive pieces of evidence: -

1. The Simultaneity of Invention   The phenomenon of simultaneous discovery is well known.[1] From our view, it argues that the process of discovery takes place along a well-defined frontier of knowledge rather than in grab-bag fashion. Admittedly, the concept of "simultaneity" is impressionistic,[2] but the related phenomenon of technological "clustering" again suggests that technical evolution follows a sequential and determinate rather than random course.[3]

2. The Absence of Technological Leaps   All inventions and innovations, by definition, represent an advance of the art beyond existing base lines. Yet, most advances, particularly in retrospect, appear essentially incremental, evolutionary. If nature makes no sudden leaps, neither, it would appear, does technology. To make my point by exaggeration, we do not find experiments in electricity in the year 1500, or attempts to extract power from the atom in the year 1700. On the whole, the development of the technology of production presents a fairly smooth and continuous profile rather than one of jagged peaks and discontinuities.

3. The Predictability of Technology   There is a long history of technological prediction, some of it ludicrous and some not.[4] What is interesting is that the development of technical progress has always seemed *intrinsically* predictable. This does not mean that we can lay down future timetables of technical discovery, nor does it rule out the possibility of surprises. Yet I venture to state that many scientists would be willing to make *general* predictions as to the nature of technological capability twenty-five or even fifty years ahead. This too suggests that technology follows a developmental sequence rather than arriving in a more chancy fashion.

I am aware, needless to say, that these bits of evidence do not constitute anything like a "proof" of my hypothesis. At best they establish the grounds on which a prima facie case of plausibility may be rested. But I should like now to strengthen these grounds by suggesting two deeper-seated reasons why technology *should* display a "structured" history.

The first of these is that a major constraint always operates on the technological capacity of an age, the constraint of its accumulated stock of available knowledge. The application of this knowledge may lag behind its reach; the technology of the hand-mill, for example, was by no means at the frontier of medieval technical knowledge, but technical realization can hardly precede what men generally know (although experiment may incrementally advance both technology and knowledge concurrently). Particularly from the mid-nineteenth century to the present do we sense the loosening constraints on technology stemming from successively yielding barriers of scientific knowledge—loosening constraints that result in the successive arrival of the electrical, chemical, aeronautical, electronic, nuclear, and space stages of technology.[5]

The gradual expansion of knowledge is not, however, the only order-bestowing constraint on the development of technology. A second controlling factor is the material competence of the age, its level of technical expertise. To make a steam engine, for example, requires not only some knowledge of the elastic properties of steam but the ability to cast iron cylinders of considerable dimensions with tolerable accuracy. It is one thing to produce a single steam-machine as an expensive toy, such as the machine depicted by Hero, and another to produce a machine that will produce power economically and effectively. The difficulties experienced by Watt and Boulton in achieving a fit of piston to cylinder illustrate the problems of creating a technology, in contrast with a single machine.

Yet until a metal-working technology was established—indeed, until an embryonic machine-tool industry had taken root—an industrial technology was impossible to create. Furthermore, the competence required to create such a technology does not reside alone in the ability or inability to make a particular machine (one thinks of Babbage's ill-fated calculator as an example of a machine born too soon), but in the ability of many industries to change their products or processes to "fit" a change in one key product or process.

This necessary requirement of technological congruence[6] gives us an additional cause of sequencing. For the ability of many industries to co-operate in producing the equipment needed for a "higher" stage of technology depends not alone on knowledge or sheer skill but on the division of labor and the specialization of industry. And this in turn hinges to a considerable degree on the sheer size of the stock of capital itself. Thus the slow and painful accumulation of capital, from which springs the gradual diversification of industrial function, becomes an independent regulator of the reach of technical capability.

In making this general case for a determinate pattern of technological evolution—at least insofar as that technology is concerned with production—I do not want to claim too much. I am well aware that reasoning about technical sequences is easily faulted as *post hoc ergo propter hoc*. Hence, let me leave this phase of my inquiry by suggesting no more than that the idea of a roughly ordered progression of productive technology seems logical enough to warrant further empirical investigation. To put it as concretely as possible, I do not think it is just by happenstance that the steam-mill follows, and does not precede, the hand-mill, nor is it mere fantasy in our own day when we speak of the coming of the automatic factory. In the future as in the past, the development of the technology of production seems bounded by the constraints of knowledge and capability and thus, in principle at least, open to prediction as a determinable force of the historic process.

II

The second proposition to be investigated is no less difficult than the first. It relates, we will recall, to the explicit statement that a given technology imposes certain social and political characteristics upon the society in which it is found. Is it true that, as Marx wrote in *The German Ideology*, "A certain mode of production, or industrial stage, is always combined with a certain mode of cooperation, or social stage,"[7] or as he put it in the sentence immediately preceding our hand-mill, steam-mill paradigm, "In acquiring new productive forces men change their mode of production, and in changing their mode of production they change their way of living—they change all their social relations"?

As before, we must set aside for the moment certain "cultural" aspects of the question. But if we restrict ourselves to the functional relationships directly connected with the process of production itself, I think we can indeed state that the technology of a society imposes a determinate pattern of social relations on that society.

We can, as a matter of fact, distinguish at least two such modes of influence:

1. The Composition of the Labor Force   In order to function, a given technology must be attended by a labor force of a particular kind. Thus, the hand-mill (if we may take this as referring to late medieval technology in general) required a work force composed of skilled or semiskilled craftsmen, who were free to practice their occupations at home or in a small atelier, at times and seasons that varied considerably. By way of contrast, the steam-mill—that is, the technology of the nineteenth century—required a work force composed of semiskilled or unskilled operatives who could work only at the factory site and only at the strict time schedule enforced by turning the machinery on or off. Again, the technology of the electronic age has steadily required a higher proportion of skilled attendants; and the coming technology of automation will still further change the needed mix of skills and the locale of work, and may as well drastically lessen the requirements of labor time itself.

2. The Hierarchical Organization of Work   Different technological apparatuses not only require different labor forces but different orders of supervision and co-ordination. The internal organization of the eighteenth-century handicraft unit, with its typical man-master relationship, presents a social configuration of a wholly different kind from that of the nineteenth-century factory with its men-manager confrontation, and this in turn differs from the internal social structure of the continuous-flow, semi-automated plant of the present. As the intricacy of the production process increases, a much more complex system of internal controls is required to maintain the system in working order.

Does this add up to the proposition that the steam-mill gives us society with the industrial capitalist? Certainly the class characteristics of a particular society are strongly implied in its functional organization. Yet it would seem wise to be very cautious before relating political effects exclusively to functional economic causes. The Soviet Union, for example, proclaims itself to be a socialist society although its technical base resembles that of old-fashioned capitalism. Had Marx written that the steam-mill gives you society with the industrial *manager*, he would have been closer to the truth.

What is less easy to decide is the degree to which the technological infrastructure is responsible for some of the sociological features of society. Is anomie, for instance, a disease of capitalism or of all industrial societies? Is the organization man a creature of monopoly capital or of all bureaucratic industry wherever found? These questions tempt us to look into the problem of the impact of technology on the existential quality of life, an area we have ruled out of bounds for this paper. Suffice it to say that superficial evidence seems to imply that the similar technologies of Russia and America are indeed giving rise to similar social phenomena of this sort.

As with the first portion of our inquiry, it seems advisable to end this section on a note of caution. There is a danger, in discussing the structure of the labor force or the nature of intrafirm organization, of assigning the sole causal efficacy to the visible presence of machinery and of overlooking the invisible influence of other factors at work. Gilfillan, for instance, writes, "engineers have committed such blunders as saying the typewriter brought women to work in offices, and with the typesetting machine made possible the great modern newspaper, forgetting that in Japan there are women office workers and great modern newspapers getting practically no help from typewriters and typesetting machines."[8] In addition, even where technology seems unquestionably to play the critical role, an independent "social" element unavoidably enters the scene in the *design* of technology, which must take into account such facts as the level of education of the work force or its relative price. In this way the machine will reflect, as much as mould, the social relationships of work.

These caveats urge us to practice what William James called a "soft determinism" with regard to the influence of the machine on social relations. Nevertheless, I would say that our cautions qualify rather than invalidate the thesis that the prevailing level of technology imposes itself powerfully on the structural organization of the productive side of society. A foreknowledge of the shape of the technical core of society fifty years hence may not allow us to describe the political attributes of that society, and may perhaps only hint at its sociological character, but assuredly it presents us with a profile of requirements, both in labor skills and in supervisory needs, that differ considerably from those of today. We cannot say whether the society of the computer will give us the latter-day capitalist or the commissar, but it seems beyond question that it will give us the technician and the bureaucrat.

III

Frequently, during our efforts thus far to demonstrate what is valid and useful in the concept of technological determinism, we have been forced to defer certain aspects of the problem until later. It is time now to turn up the rug and to examine what has been swept under it. Let us try to systematize our qualifications and objections to the basic Marxian paradigm:

1. Technological Progress Is Itself a Social Activity   A theory of technological determinism must contend with the fact that the very activity of invention and innovation is an attribute of some societies and not of others. The Kalahari bushmen or the tribesmen of New Guinea, for instance, have persisted in a neolithic technology to the present day; the Arabs reached a high degree of technical proficiency in the past and have since suffered a decline; the classical Chinese developed technical expertise in some fields while unaccountably neglecting it in the area of production. What factors serve to encourage or discourage this technical thrust is a problem about which we know extremely little at the present moment.[9]

2. The Course of Technological Advance Is Responsive to Social Direction   Whether technology advances in the area of war, the arts, agriculture, or industry depends in part on the rewards, inducements, and incentives offered by society. In this way the direction of technological advance is partially the result of social policy. For example, the system of interchangeable parts, first introduced into France and then independently into England failed to take root in either country for lack of government interest or market stimulus. Its success in America is attributable mainly to government support and to its appeal in a society without guild traditions and with high labor costs.[10] The general *level* of technology may follow an independently determined sequential path, but its areas of application certainly reflect social influences.

3. Technological Change Must Be Compatible with Existing Social Conditions   An advance in technology not only must be congruent with the surrounding technology but must also be compatible with the existing economic and other institutions of society. For example, labor-saving machinery will not find ready acceptance in a society where labor is abundant and cheap as a factor of production. Nor would a mass production technique recommend itself to a society that did not have a mass market. Indeed, the presence of slave labor seems generally to inhibit the use of machinery and the presence of expensive labor to accelerate it.[11]

These reflections on the social forces bearing on technical progress tempt us to throw aside the whole notion of technological determinism as false or misleading.[12] Yet, to relegate technology from an undeserved position of *primum mobile* in history to that of a mediating factor, both acted upon by and acting on the body of society, is not to write off its influence but only to specify its mode of operation with greater precision. Similarly, to admit we understand very little of the cultural factors that give rise to technology does not depreciate its role but focuses our attention on that period of history when technology is clearly a major historic force, namely Western society since 1700.

IV

What is the mediating role played by technology within modern Western society? When we ask this much more modest question, the interaction of society and technology begins to clarify itself for us:

1. The Rise of Capitalism Provided a Major Stimulus for the Development of a Technology of Production   Not until the emergence of a market system organized around the principle of private property did there also emerge an institution capable of systematically guiding the inventive and innovative abilities of society to the problem of facilitating production. Hence the environment of the eighteenth and nineteenth centuries provided both a novel and an extremely effective encouragement for the development of an *industrial* technology. In addition, the slowly opening political and social framework of late mercantilist society gave rise to social aspirations for which the new technology offered the best chance of realization. It was not only the steam-mill that gave us the industrial capitalist but the rising inventor-manufacturer who gave us the steam-mill.

2. The Expansion of Technology within the Market System Took on a New "Automatic" Aspect   Under the burgeoning market system not alone the initiation of technical improvement but its subsequent adoption and repercussion through the economy was largely governed by market considerations. As a result, both the rise and the proliferation of technology assumed the attributes of an impersonal diffuse "force" bearing on social and economic life. This was all the more pronounced because the political control needed to buffer its disruptive consequences was seriously inhibited by the prevailing laissez-faire ideology.

3. The Rise of Science Gave a New Impetus to Technology   The period of early capitalism roughly coincided with and provided a congenial setting for the development of an independent source of technological encouragement—the rise of the self-conscious activity of science. The steady expansion of scientific research, dedicated to the exploration of nature's secrets and to their harnessing for social use, provided an increasingly important stimulus for technological advance from the middle of the nineteenth century. Indeed, as the twentieth century has progressed, science has become a major historical force in its own right and is now the indispensable precondition for an effective technology.

It is for these reasons that technology takes on a special significance in the context of capitalism—or, for that matter, of a socialism based on maximizing production or minimizing costs. For in these societies, both the continuous appearance of technical advance and its diffusion throughout the society assume the attributes of autonomous process, "mysteriously" generated by society and thrust upon its members in a manner as indifferent as it is imperious. This is why, I think, the problem of technological determinism—of how machines make history—comes to us with such insistence despite the ease with which we can disprove its more extreme contentions.

Technological determinism is thus peculiarly a problem of a certain historic epoch— specifically that of high capitalism and low socialism—*in which the forces of technical change have been unleashed, but when the agencies for the control or guidance of technology are still rudimentary.*

The point has relevance for the future. The surrender of society to the free play of market forces is now on the wane, but its subservience to the impetus of the scientific ethos is on the rise. The prospect before us is assuredly that of an undiminished and very likely accelerated pace of technical change. From what we can foretell about the direction of this technological advance and the structural alterations it implies, the pressures in the future will be toward a society marked by a much greater degree of organization and deliberate control. What other political, social, and existential changes the age of the computer will also bring we do not know. What seems certain, however, is that the problem of technological determinism—that is, of the impact of machines on history—will remain germane until there is forged a degree of public control over technology far greater than anything that now exists.

Notes

1. See Robert K. Merton, "Singletons and Multiples in Scientific Discovery: A Chapter in the Sociology of Science," *Proceedings* of the American Philosophical Society, CV (October 1961), 470–86.

2. See John Jewkes, David Sawers, and Richard Stillerman, *The Sources of Invention* (New York, 1960 [paperback edition]), p. 227, for a skeptical view.

3. "One can count 21 basically different means of flying, at least eight basic methods of geophysical prospecting; four ways to make uranium explosive;...20 or 30 ways to control birth....If each of these separate inventions were autonomous, i.e., without cause, how could one account for their arriving in these functional groups?" S. C. Gilfillan, "Social Implications of Technological Advance," *Current Sociology*, I (1952), 197. See also Jacob Schmookler, "Economic Sources of Inventive Activity," *Journal of Economic History* (March 1962), pp. 1–20; and Richard Nelson, "The Economics of Invention: A Survey of the Literature," *Journal of Business*, XXXII (April 1959), 101–19.

4. Jewkes *et al.* (see n. 2) present a catalogue of chastening mistakes (p. 230 f.). On the other hand, for a sober predictive effort, see Francis Bello, "The 1960s: A Forecast of Technology," *Fortune*, LIX (January 1959), 74–78; and Daniel Bell, "The Study of the Future," *Public Interest*, I (Fall 1965), 119–30. Modern attempts at prediction project likely avenues of scientific advance or technological function rather than the feasibility of specific machines.

5. To be sure, the inquiry now regresses one step and forces us to ask whether there are inherent stages for the expansion of knowledge, at least insofar as it applies to nature. This is a very uncertain question. But having already risked so much, I will hazard the suggestion that the roughly parallel sequential development of scientific understanding in those few cultures that have cultivated it (mainly classical Greece, China, the high Arabian culture, and the West since the Renaissance) makes such a hypothesis possible, provided that one looks to broad outlines and not to inner detail.

6. The phrase is Richard LaPiere's in *Social Change* (New York, 1965), p. 263 f.

7. Karl Marx and Friedrich Engels, *The German Ideology* (London, 1942), p. 18.

8. Gilfillan (see n. 3), p. 202.

9. An interesting attempt to find a line of social causation is found in E. Hagen, *The Theory of Social Change* (Homewood, Ill., 1962).

10. See K. R. Gilbert, "Machine-Tools," in Charles Singer, E. J. Holmyard, A. R. Hall, and Trevor I. Williams (eds.), *A History of Technology* (Oxford, 1958), IV. chap. xiv.

11. See LaPiere (see n. 6), p. 284; also H. J. Habbakuk, *British and American Technology in the 19th Century* (Cambridge, 1962), *passim*.

12. As, for example, in A. Hansen, "The Technological Determination of History," *Quarterly Journal of Economics* (1921), pp. 76–83.

8 "The Social Construction of Facts and Artifacts"

Trevor J. Pinch and Wiebe Bijker

This article formed the basis of what is now known as the Social Construction of Technology (SCOT) approach. The piece can be understood as a concerted effort to disprove the idea that technology is deterministic. Advocates of the SCOT approach argue that social groups direct nearly every aspect of technology. It is people, not machines, that design, build, and give meaning to technologies and ultimately decide which ones to adopt and which ones to reject. For Pinch and Bijker, the capacity of human beings to define and change the technology around them is not limited to just a handful of powerful groups like CEOs, industrialists, or even engineers. Even after an artifact has been built and sold by a corporation, individuals still have the power to redefine what the technology means and come up with unanticipated uses for it. Pinch and Bijker argue that technological determinism is a myth that results when one looks backwards and believes that the path taken to the present was the only possible path. They encourage us instead to consider all the possibilities that were available in the past and realize that the technologies, designs, and uses that were chosen had as much to do with the social circumstances at the time as with the nature and the state of technical knowledge. Pinch and Bijker provide a theoretical framework for explaining technological development as a social process. They argue that in the early stages of a technology's development there are a broad array of possibilities—in their terms, the technology has "interpretive flexibility." Different social groups decide the uses, meaning, and specific design of a technology based on their needs, values, etc. As different interest groups coalesce around a particular design and meaning for a technology the technical design begins to stabilize and becomes much more difficult to reinterpret. Through this process, a technology is given uses and meanings that later appear to be an essential and somehow "natural" part of the technology rather than something arrived at through social negotiations.

Technology Studies

There is a large amount of writing that falls under the rubric of "technology studies." It is convenient to divide the literature into three parts: innovation studies, history of technology, and sociology of technology. We discuss each in turn.

Most innovation studies have been carried out by economists looking for the conditions for success in innovation. Factors researched include various aspects of the innovating firm (for example, size of R&D effort, management strength, and marketing

From W. Bijker, T. P. Hughes, and T. J. Pinch, eds., *The Social Construction of Technological Systems* (Cambridge, Mass., MIT Press, 1987), pp. 17–50. Reprinted with permission.

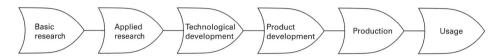

Figure 8.1
A six-stage model of the innovation process.

capability) along with macroeconomic factors pertaining to the economy as a whole.¹ This literature is in some ways reminiscent of the early days in the sociology of science, when scientific knowledge was treated like a "black box" (Whitley 1972) and, for the purpose of such studies, scientists might as well have produced meat pies. Similarly, in the economic analysis of technological innovation everything is included that might be expected to influence innovation, except any discussion of the technology itself. As Layton notes:

What is needed is an understanding of technology from inside, both as a body of knowledge and as a social system. Instead, technology is often treated as a "black box" whose contents and behaviour may be assumed to be common knowledge. (Layton 1977, p. 198)

Only recently have economists started to look into this black box.²

The failure to take into account the content of technological innovations results in the widespread use of simple linear models to describe the process of innovation. The number of developmental steps assumed in these models seems to be rather arbitrary (for an example of a six-stage process see figure 8.1).³ Although such studies have undoubtedly contributed much to our understanding of the conditions for economic success in technological innovation, because they ignore the technological content they cannot be used as the basis for a social constructivist view of technology.⁴

This criticism cannot be leveled at the history of technology, where there are many finely crafted studies of the development of particular technologies. However, for the purposes of a sociology of technology, this work presents two kinds of problem. The first is that descriptive historiography is endemic in this field. Few scholars (but there are some notable exceptions) seem concerned with generalizing beyond historical instances, and it is difficult to discern any overall patterns on which to build a theory of technology (Staudenmaier 1983, 1985). This is not to say that such studies might not be useful building blocks for a social constructivist view of technology—merely that these historians have not yet demonstrated that they are doing sociology of knowledge in a different guise.⁵

The second problem concerns the asymmetric focus of the analysis. For example, it has been claimed that in twenty-five volumes of *Technology and Culture* only nine articles were devoted to the study of failed technological innovations (Staudenmaier 1985). This contributes to the implicit adoption of a linear structure of technological development, which suggests that the whole history of technological development had followed an orderly or rational path, as though today's world was the precise goal toward which all decisions, made since the beginning of history, were consciously directed. (Ferguson 1974, p. 19)

This preference for successful innovations seems to lead scholars to assume that the success of an artifact is an explanation of its subsequent development. Historians of technology often seem content to rely on the manifest success of the artifact as evidence that there is no further explanatory work to be done. For example, many histories of synthetic plastics start by describing the "technically sweet" characteristics of Bakelite; these features are then used implicitly to position Bakelite at the starting point of the glorious development of the field:

God said: "let Baekeland be" and all was plastics! (Kaufman 1963, p. 61)

However, a more detailed study of the developments of plastic and varnish chemistry, following the publication of the Bakelite process in 1909 (Baekeland 1909a, b), shows that Bakelite was at first hardly recognized as the marvelous synthetic resin that it later proved to be.[6] And this situation did not change much for some ten years. During the First World War the market prospects for synthetic plastics actually grew worse. However, the dumping of war supplies of phenol (used in the manufacture of Bakelite) in 1918 changed all this (Haynes 1954, pp. 137–138) and made it possible to keep the price sufficiently low to compete with (semi-)natural resins, such as celluloid.[7] One can speculate over whether Bakelite would have acquired its prominence if it had not profited from that phenol dumping. In any case it is clear that a historical account founded on the retrospective success of the artifact leaves much untold.

Given our intention of building a sociology of technology that treats technological knowledge in the same symmetric, impartial manner that scientific facts are treated within the sociology of scientific knowledge, it would seem that much of the historical material does not go far enough. The success of an artifact is precisely what needs to be explained. For a sociological theory of technology it should be the *explanandum*, not the *explanans*.

Our account would not be complete, however, without mentioning some recent developments, especially in the American history of technology. These show the emergence of a growing number of theoretical themes on which research is focused (Staudenmaier 1985; Hughes 1979). For example, the systems approach to technology,[8] consideration of the effect of labor relations on technological development,[9] and detailed studies of some not-so-successful inventions[10] seem to herald departures from the "old" history of technology. Such work promises to be valuable for a sociological analysis of technology, and we return to some of it later.

The final body of work we wish to discuss is what might be described as "sociology of technology."[11] There have been some limited attempts in recent years to launch such a sociology, using ideas developed in the history and sociology of science—studies by, for example, Johnston (1972) and Dosi (1982), who advocate the description of technological knowledge in terms of Kuhnian paradigms.[12] Such approaches certainly appear to be more promising than standard descriptive historiography, but it is not clear whether or not these authors share our understanding of technological artifacts as social constructs. For example, neither Johnston nor Dosi considers explicitly the need for a symmetric sociological explanation that treats successful and failed artifacts in an equivalent way. Indeed, by locating their discussion at the level of technological paradigms, we are not sure how the artifacts themselves are to be approached. As neither author has yet produced an empirical study using Kuhnian ideas, it is difficult to evaluate how the Kuhnian terms may be utilized.[13] Certainly this has been a pressing problem in the sociology of science, where it has not always been possible to give Kuhn's terms a clear empirical reference.

The possibilities of a more radical social constructivist view of technology have been touched on by Mulkay (1979). He argues that the success and efficacy of technology could pose a special problem for the social constructivist view of *scientific knowledge*. The argument Mulkay wishes to counter is that the practical effectiveness of technology somehow demonstrates the privileged epistemology of science and thereby exempts it from sociological explanation. Mulkay opposes this view, rightly in our opinion, by pointing out the problem of the "science discovers, technology applies" notion implicit in such claims. In a second argument against this position, Mulkay notes (following Mario Bunge (1966)) that it is possible for a false or partly false theory to be used as the basis for successful practical application: The success of the technology would not then have anything to say about the "truth" of the scientific knowledge on which it was based. We find this second point not entirely satisfactory. We would rather stress that the truth or falsity of scientific knowledge is irrelevant to sociological analysis of belief: To retreat to the argument that science may be wrong but good technology can still be based on it is missing this point. Furthermore, the success of technology is still left unexplained within such an argument. The only effective way to deal with these difficulties is to adopt a perspective that attempts to show that technology, as well as science, can be understood as a social construct.

Mulkay seems to be reluctant to take this step because, as he points out, "there are very few studies...which consider how the technical meaning of hard technology is socially constructed" (Mulkay 1979, p. 77). This situation however, is starting to change: A number of such studies have recently emerged. For example, Michel Callon, in a pioneering study, has shown the effectiveness of focusing on technological controversies. He draws on an extensive case study of the electric vehicle in France (1960–75) to demonstrate that almost everything is negotiable: what is certain and what is not; who is a scientist and who is a technologist; what is technological and what is social; and who can participate in the controversy (Callon 1980a, b, 1981, 1987). David Noble's study of the introduction of numerically controlled machine tools can also be regarded as an important contribution to a social constructivist view of technology (Noble 1984). Noble's explanatory goals come from a rather different (Marxist) tradition,[14] and his study has much to recommend it: He considers the development of both a successful and a failed technology and gives a symmetric account of both developments. Another intriguing study in this tradition is Lazonick's account (1979) of the introduction of the self-acting mule: He shows that aspects of this technical development can be understood in terms of the relations of production rather than any inner logic of technological development. The work undertaken by Bijker, Bönig, and Van Oost is another attempt to show how the socially constructed character of the content of some technological artifacts might be approached empirically: Six case studies were carried out, using historical sources.[15]

In summary, then, we can say that the predominant traditions in technology studies—innovation studies and the history of technology—do not yet provide much encouragement for our program. There are exceptions, however, and some recent studies in the sociology of technology present promising starts on which a unified approach could be built. We now give a more extensive account of how these ideas may be synthesized.

EPOR and SCOT

In this part we outline in more detail the concepts and methods that we wish to employ. We start by describing the "Empirical Programme of Relativism" as it was developed in the sociology of scientific knowledge. We then go on to discuss in more detail the approach taken by Bijker and his collaborators in the sociology of technology.

The Empirical Programme of Relativism (EPOR)

The EPOR is an approach that has produced several studies demonstrating the social construction of scientific knowledge in the "hard" sciences. This tradition of research has emerged from recent sociology of scientific knowledge. Its main characteristics, which distinguish it from other approaches in the same area, are the focus on the empirical study of contemporary scientific developments and the study, in particular, of scientific controversies.[16]

Three stages in the explanatory aims of the EPOR can be identified. In the *first stage* the interpretative flexibility of scientific findings is displayed; in other words, it is shown that scientific findings are open to more than one interpretation. This shifts the focus for the explanation of scientific developments from the natural world to the social world. Although this interpretative flexibility can be recovered in certain circumstances, it remains the case that such flexibility soon disappears in science; that is, a scientific consensus as to what the "truth" is in any particular instance usually emerges. Social mechanisms that limit interpretative flexibility and thus allow scientific controversies to be terminated are described in the *second stage*. A *third stage*, which has not yet been carried through in any study of contemporary science, is to relate such "closure mechanisms" to the wider social-cultural milieu. If all three stages were to be addressed in a single study, as Collins writes, "the impact of society on knowledge 'produced' at the laboratory bench would then have been followed through in the hardest possible case" (Collins 1981c, p. 7).

The EPOR represents a continuing effort by sociologists to understand the content of the natural sciences in terms of social construction. Various parts of the program are better researched than others. The third stage of the program has not yet even been addressed, but there are many excellent studies exploring the first stage. Most current research is aimed at elucidating the closure mechanisms whereby consensus emerges (the second stage). Many studies within the EPOR have been most fruitfully located in the area of scientific controversy. Controversies offer a methodological advantage in the comparative ease with which they reveal the interpretative flexibility of scientific results. Interviews conducted with scientists engaged in a controversy usually reveal strong and differing opinions over scientific findings. As such flexibility soon vanishes from science, it is difficult to recover from the textual sources with which historians usually work. Collins has highlighted the importance of the "controversy group" in science by his use of the term "core set" (Collins 1981b). These are the scientists most intimately involved in a controversial research topic. Because the core set is defined in relation to knowledge production in science (the core set constructs scientific knowledge), some of the empirical problems encountered in the identification of groups in science by purely sociometric means can be overcome. And studying the core set has another methodological advantage, in that the resulting consensus can be monitored. In other words, the group of scientists who experiment and theorize at the research frontiers and who become embroiled in scientific controversy will also reflect the growing consensus as to the outcome of that controversy. The same group of core set scientists can then be studied in both the first and second stages of the EPOR. For the purposes of the third stage, the notion of a core set may be too limited.

The Social Construction of Technology (SCOT)
Before outlining some of the concepts found to be fruitful by Bijker and his collaborators in their studies in the sociology of technology, we should point out an imbalance between the two approaches (EPOR and SCOT) we are considering. The EPOR is part of a flourishing tradition in the sociology of scientific knowledge: It is a well-established program supported by much empirical research. In contrast, the sociology of technology is an embryonic field with no well-established traditions of research, and the approach we draw on specifically (SCOT) is only in its early empirical stages, although clearly gaining momentum.[17]

In SCOT the developmental process of a technological artifact is described as an alternation of variation and selection.[18] This results in a "multidirectional" model, in contrast with the linear models used explicitly in many innovation studies and implicitly in much history of technology. Such a multidirectional view is essential to any social constructivist account of technology. Of course, with historical hindsight, it is possible to collapse the multidirectional model on to a simpler linear model; but this misses the thrust of our argument that the "successful" stages in the development are not the only possible ones.

"The Social Construction of Facts and Artifacts" 113

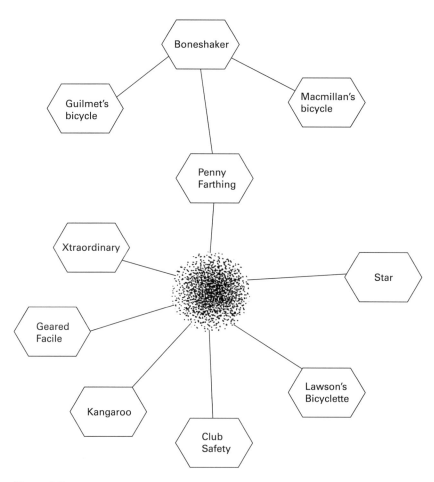

Figure 8.2
A multidirectional view of the developmental process of the Penny Farthing bicycle. The shaded area is filled in and magnified in figure 8.11. The hexagons symbolize artifacts.

Figure 8.3
A typical Penny Farthing, the Bayliss-Thomson Ordinary (1878). Photograph courtesy of the Trustees of the Science Museum, London.

Let us consider the development of the bicycle.[19] Applied to the level of artifacts in this development, this multidirectional view results in the description summarized in figure 8.2. Here we see the artifact "Ordinary" (or, as it was nicknamed after becoming less ordinary, the "Penny-farthing"; figure 8.3) and a range of possible variations. It is important to recognize that, in the view of the actors of those days, these variants were at the same time quite different from each other and equally were serious rivals. It is only by retrospective distortion that a quasi-linear development emerges, as depicted in figure 8.4. In this representation the so-called safety ordinaries (Xtraordinary (1878), Facile (1879), and Club Safety (1885)) figure only as amusing aberrations that need not be taken seriously (figure 8.5, 8.6, and 8.7). Such a retrospective description can be challenged by looking at the actual situation in the 1880s. Some of the "safety ordinaries" were produced commercially, whereas Lawson's Bicyclette, which

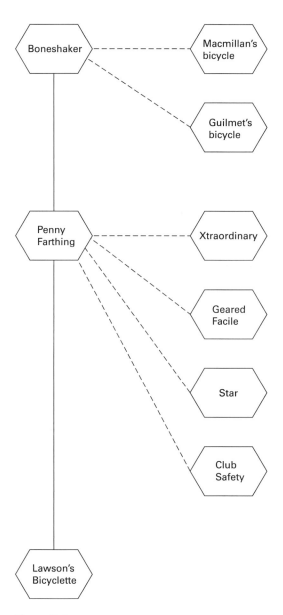

Figure 8.4
The traditional quasi-linear view of the developmental process of the Penny Farthing bicycle. Solid lines indicate successful development, and dashed lines indicate failed development.

Figure 8.5
The American Star bicycle (1885). Photograph courtesy of the Trustees of the Science Museum, London.

seems to play an important role in the linear model, proved to be a commercial failure (Woodforde 1970).

However, if a multidirectional model is adopted, it is possible to ask why some of the variants "die," whereas others "survive." To illuminate this "selection" part of the developmental processes, let us consider the problems and solutions presented by each artifact at particular moments. The rationale for this move is the same as that for focusing on scientific controversies within EPOR. In this way, one can expect to bring out more clearly the interpretative flexibility of technological artifacts.

In deciding which problems are relevant, the social groups concerned with the artifact and the meanings that those groups give to the artifact play a crucial role: A problem is defined as such only when there is a social group for which it constitutes a "problem."

The use of the concept of a relevant social group is quite straightforward. The phrase is used to denote institutions and organizations (such as the military or some specific industrial company), as well as organized or unorganized groups of individuals. The

Figure 8.6
Facile bicycle (1874). Photograph courtesy of the Trustees of the Science Museum, London.

Figure 8.7
A form of the Kangaroo bicycle (1878). Photograph courtesy of the Trustees of the Science Museum, London.

key requirement is that all members of a certain social group share the same set of meanings, attached to a specific artifact.[20] In deciding which social groups are relevant, we must first ask whether the artifact has any meaning at all for the members of the social group under investigation. Obviously, the social group of "consumers" or "users" of the artifact fulfills this requirement. But also less obvious social groups may need to be included. In the case of the bicycle, one needs to mention the "anticyclists." Their actions ranged from derisive cheers to more destructive methods. For example, Reverend L. Meadows White described such resistance to the bicycle in his book, *A Photographic Tour on Wheels*:

...but when to words are added deeds, and stones are thrown, sticks thrust into the wheels, or caps hurled into the machinery, the picture has a different aspect. All the above in certain districts are of common occurrence, and have all happened to me, especially when passing through a village just after school is closed. (Meadows, cited in Woodforde 1970, pp. 49–50)

Clearly, for the anticyclists the artifact "bicycle" had taken on meaning!

Another question we need to address is whether a provisionally defined social group is homogeneous with respect to the meanings given to the artifact—or is it more effective to describe the developmental process by dividing a rather heterogeneous group into several different social groups? Thus within the group of cycle-users we discern a separate social group of women cyclists. During the days of the high-wheeled Ordinary women were not supposed to mount a bicycle. For instance, in a magazine advice column (1885) it is proclaimed, in reply to a letter from a young lady:

The mere fact of riding a bicycle is not in itself sinful, and if it is the only means of reaching the church on a Sunday, it may be excusable. (cited in Woodforde 1970, p. 122)

Tricycles were the permitted machines for women. But engineers and producers anticipated the importance of women as potential bicyclists. In a review of the annual Stanley Exhibition of Cycles in 1890, the author observes:

From the number of safeties adapted for the use of ladies, it seems as if bicycling was becoming popular with the weaker sex, and we are not surprised at it, considering the saving of power derived from the use of a machine having only one slack. (Stanley Exhibition of Cycles, 1890, pp. 107–108)

Thus some parts of the bicycle's development can be better explained by including a separate social group of feminine cycle-users. This need not, of course, be so in other cases: For instance, we would not expect it to be useful to consider a separate social group of women users of, say, fluorescent lamps.

Once the relevant social groups have been identified, they are described in more detail. This is also where aspects such as power or economic strength enter the description, when relevant. Although the only defining property is some homogeneous meaning given to a certain artifact, the intention is not just to retreat to worn-out, general statements about "consumers" and "producers." We need to have a detailed description of the relevant social groups in order to define better the function of the artifact with respect to each group. Without this, one could not hope to be able to give any explanation of the developmental process. For example, the social group of cyclists riding the high-wheeled Ordinary consisted of "young men of means and nerve: they might be professional men, clerks, schoolmasters or dons" (Woodforde 1970, p. 47). For this social group the function of the bicycle was primarily for sport. The following comment in the *Daily Telegraph* (September 7, 1877) emphasizes sport, rather than transport:

Bicycling is a healthy and manly pursuit with much to recommend it, and, unlike other foolish crazes, it has not died out. (cited in Woodforde 1970, p. 122)

Let us now return to the exposition of the model. Having identified the relevant social groups for a certain artifact (figure 8.8), we are especially interested in the problems each group has with respect to that artifact (figure 8.9). Around each problem, several variants of solution can be identified (figure 8.10). In the case of the bicycle, some relevant problems and solutions are shown in figure 8.11, in which the shaded area of

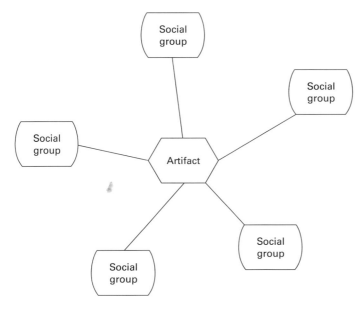

Figure 8.8
The relationship between an artifact and the relevant social groups.

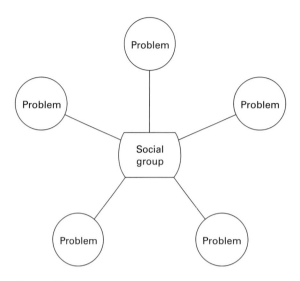

Figure 8.9
The relationship between one social group and the perceived problems.

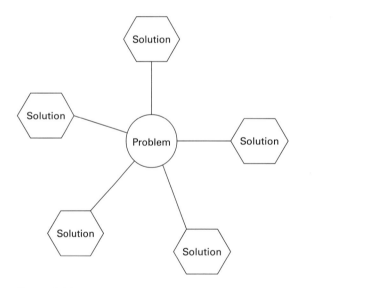

Figure 8.10
The relationship between one problem and its possible solutions.

figure 8.2 has been filled. This way of describing the developmental process brings out clearly all kinds of conflicts: conflicting technical requirements by different social groups (for example, the speed requirement and the safety requirement); conflicting solutions to the same problem (for example, the safety low-wheelers and the safety ordinaries); and moral conflicts (for example, women wearing skirts or trousers on high-wheelers; figure 8.12). Within this scheme, various solutions to these conflicts and problems are possible—not only technological ones but also judicial or even moral ones (for example, changing attitudes toward women wearing trousers).

Following the developmental process in this way, we see growing and diminishing degrees of stabilization of the different artifacts.[21] In principle, the degree of stabilization is different in different social groups. By using the concept of stabilization, we see that the "invention" of the safety bicycle was not an isolated event (1884), but a nineteen-year process (1879–98). For example, at the beginning of this period the relevant groups did not see the "safety bicycle" but a wide range of bi- and tricycles—and, among those, a rather ugly crocodilelike bicycle with a relatively low front wheel and rear chain drive (Lawson's Bicyclette; figure 8.13). By the end of the period, the phrase "safety bicycle" denoted a low-wheeled bicycle with rear chain drive, diamond frame, and air tires. As a result of the stabilization of the artifact after 1898, one did not need to specify these details: They were taken for granted as the essential "ingredients" of the safety bicycle.

We want to stress that our model is not used as a mold into which the empirical data have to be forced, *coûte que coûte*. The model has been developed from a series of case studies and not from purely philosophical or theoretical analysis. Its function is primarily heuristic—to bring out all the aspects relevant to our purposes. This is not to say that there are no explanatory and theoretical aims, analogous to the different stages of the EPOR (Bijker 1984, 1987). And indeed, as we have shown, this model already does more than merely describe technological development: It highlights its multidirectional character. Also, as will be indicated, it brings out the interpretative flexibility of technological artifacts and the role that different closure mechanisms may play in the stabilization of artifacts.

The Social Construction of Facts and Artifacts

Having described the two approaches to the study of science and technology we wish to draw on, we now discuss in more detail the parallels between them. As a way of putting some flesh on our discussion we give, where appropriate, empirical illustrations drawn from our own research.

Interpretative Flexibility

The first stage of the EPOR involves the demonstration of the interpretative flexibility of scientific findings. In other words, it must be shown that different interpretations of nature are available to scientists and hence that nature alone does not provide a determinant outcome to scientific debate.[22]

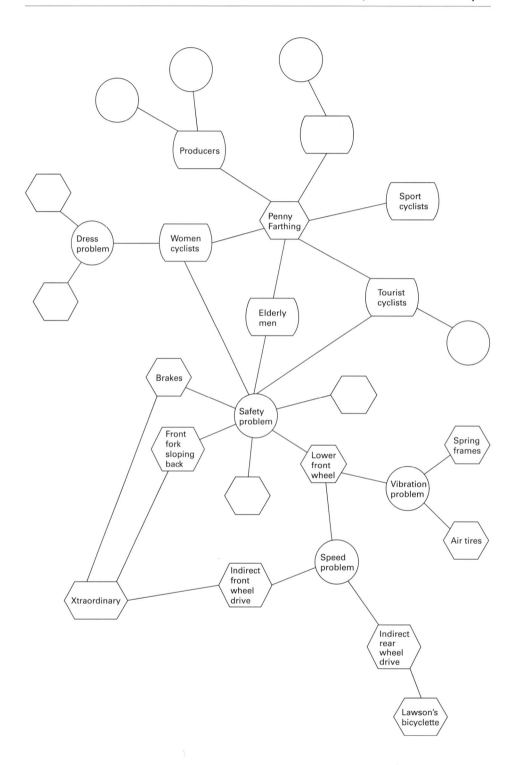

In SCOT, the equivalent of the first stage of the EPOR would seem to be the demonstration that technological artifacts are culturally constructed and interpreted; in other words, the interpretative flexibility of a technological artifact must be shown. By this we mean not only that there is flexibility in how people think of or interpret artifacts but also that there is flexibility in how artifacts are *designed*. There is not just one possible way or one best way of designing an artifact. In principle, this could be demonstrated in the same way as for the science case, that is, by interviews with technologists who are engaged in a contemporary technological controversy. For example, we can imagine that, if interviews had been carried out in 1890 with the cycle engineers, we would have been able to show the interpretative flexibility of the artifact "air tyre." For some, this artifact was a solution to the vibration problem of small-wheeled vehicles:

[The air tire was] devised with a view to afford increased facilities for the passage of wheeled vehicles—chiefly of the lighter class such for instance as velocipedes, invalid chairs, ambulances—over roadways and paths, especially when these latter are of rough or uneven character. (Dunlop 1888, p. 1).

For others, the air tire was a way of going faster (this is outlined in more detail later). For yet another group of engineers, it was an ugly looking way of making the low-wheeler even less safe (because of side-slipping) than it already was. For instance, the following comment, describing the Stanley Exhibition of Cycles, is revealing:

The most conspicuous innovation in the cycle construction is the use of pneumatic tires. These tires are hollow, about 2 in. diameter, and are inflated by the use of a small air pump. They are said to afford most luxurious riding, the roughest macadam and cobbles being reduced to the smoothest asphalte. Not having had the opportunity of testing these tires, we are unable to speak of them from practical experience; but looking at them from a theoretical point of view, we opine that considerable difficulty will be experienced in keeping the tires thoroughly inflated. Air under pressure is a troublesome thing to deal with. From the reports of those who have used these tires, it seems that they are prone to slip on muddy roads. If this is so, we fear their use on rear-driving safeties—which are all more or less addicted to side-slipping—is out of the question, as any improvement in this line should be to prevent side slip and not to increase it. Apart from these defects, the appearance of the tires destroys the symmetry and graceful appearance of a cycle, and this alone is, we think, sufficient to prevent their coming into general use. (Stanley Exhibition of Cycles, 1890, p. 107)

Figure 8.11
Some relevant social groups, problems, and solutions in the developmental process of the Penny Farthing bicycle. Because of lack of space, not all artifacts, relevant social groups, problems, and solutions are shown.

Figure 8.12
A solution to the women's dressing problem with respect to the high-wheeled Ordinary. This solution obviously has technical and athletic aspects. Probably, the athletic aspects prevented the solution from stabilizing. The set-up character of the photograph suggests a rather limited practical use. Photograph courtesy of the Trustees of the Science Museum, London.

Figure 8.13
Lawson's Bicyclette (1879). Photograph courtesy of the Trustees of the Science Museum, London.

And indeed, other artifacts were seen as providing a solution for the vibration problem, as the following comment reveals:

With the introduction of the rear-driving safety bicycle has arisen a demand for antivibration devices, as the small wheels of these machines are conducive to considerable vibration, even on the best roads. Nearly every exhibitor of this type of machine has some appliance to suppress vibration. (Stanley Exhibition of Cycles, 1889, pp. 157–158)

Most solutions used various spring constructions in the frame, the saddle, and the steering-bar (figure 8.14). In 1896, even after the safety bicycle (and the air tire with it) achieved a high degree of stabilization, "spring frames" were still being marketed.

It is important to realize that this demonstration of interpretative flexibility by interviews and historical sources is only one of a set of possible methods. At least in the study of technology, another method is applicable and has actually been used. It can be shown that different social groups have radically different interpretations of one technological artifact. We call these differences "radical" because the *content* of the artifact seems to be involved. It is something more than what Mulkay rightly claims to be rather easy—"to show that the social meaning of television varies with and depends upon the social context in which it is employed." As Mulkay notes: "It is much more difficult to show what is to count as a 'working television set' is similarly context-dependent in any significant respect" (Mulkay 1979, p. 80).

We think that our account—in which the different interpretations by social groups of the content of artifacts lead by means of different chains of problems and solutions

Figure 8.14
Whippet spring frame (1885). Photograph courtesy of the Trustees of the Science Museum, London.

to different further developments—involves the content of the artifact itself. Our earlier example of the development of the safety bicycle is of this kind. Another example is variations within the high-wheeler. The high-wheeler's meaning as a virile, high-speed bicycle led to the development of larger front wheels—for with a fixed angular velocity one way of getting a higher translational velocity over the ground was by enlarging the radius. One of the last bicycles resulting from this strand of development was the Rudge Ordinary of 1892, which had a 56-inch wheel and air tire. But groups of women and of elderly men gave quite another meaning to the high-wheeler. For them, its most important characteristic was its lack of safety:

Owing to the disparity in wheel diameters and the small weight of the backbone and trailing wheel, also to the rider's position practically over the centre of the wheel, if the large front wheel hit a brick or large stone on the road, and the rider was unprepared, the sudden check to the wheel usually threw him over the handlebar. For this reason the machine was regarded as dangerous, and however enthusiastic one may have been about the ordinary—

Figure 8.15
Singer Xtraordinary bicycle (1878). Photograph courtesy of the Trustees of the Science Museum, London.

and I was an enthusiastic rider of it once—there is no denying that it was only possible for comparatively young and athletic men. (Grew 1921, p. 8)

This meaning gave rise to lowering the front wheel, moving back the saddle, and giving the front fork a less upright position. Via another chain of problems and solutions (see figure 8.7), this resulted in artifacts such as Lawson's Bicyclette (1879) and the Xtraordinary (1878; figure 8.15). Thus there was not *one* high-wheeler; there was the *macho* machine, leading to new designs of bicycles with even higher front wheels, and there was the *unsafe* machine, leading to new designs of bicycles with lower front wheels, saddles moved backward, or reversed order of small and high wheel. Thus the interpretative flexibility of the artifact Penny-farthing is materialized in quite different design lines.

Closure and Stabilization
The second stage of the EPOR concerns the mapping of mechanisms for the closure of debate—or, in SCOT, for the stabilization of an artifact. We now illustrate what we

Figure 8.16
Geared Facile bicycle (1888). Photograph courtesy of the Trustees of the Science Museum, London.

mean by a closure mechanism by giving examples of two types that seem to have played a role in cases with which we are familiar. We refer to the particular mechanisms on which we focus as rhetorical closure and closure by redefinition of problem.

Rhetorical Closure Closure in technology involves the stabilization of an artifact and the "disappearance" of problems. To close a technological "controversy," one need not *solve* the problems in the common sense of that word. The key point is whether the relevant social groups *see* the problem as being solved. In technology, advertising can play an important role in shaping the meaning that a social group gives to an artifact.[23] Thus, for instance, an attempt was made to "close" the "safety controversy" around the high-wheeler by simply claiming that the artifact was perfectly safe. An advertisement for the "Facile" (*sic!*) Bicycle (figure 8.16) reads:

Bicyclists! Why risk your limbs and lives on high Machines when for road work a 40 inch or 42 inch "Facile" gives all the advantages of the other, together with almost absolute safety. (*Illustrated London News*, 1880; cited in Woodforde 1970, p. 60)

This claim of "almost absolute safety" was a rhetorical move, considering the height of the bicycle and the forward position of the rider, which were well known to engineers at the time to present problems of safety.

Closure by Redefinition of the Problem We have already mentioned the controversy around the air tire. For most of the engineers it was a theoretical and practical monstrosity. For the general public, in the beginning it meant an aesthetically awful accessory:

Messenger boys guffawed at the sausage tyre, factory ladies squirmed with merriment, while even sober citizens were sadly moved to mirth at a comicality obviously designed solely to lighten the gloom of their daily routine. (Woodforde 1970, p. 89)

For Dunlop and the other protagonists of the air tire, originally the air tire meant a solution to the vibration problem. However, the group of sporting cyclists riding their high-wheelers did not accept that as a problem at all. Vibration presented a problem only to the (potential) users of the low-wheeled bicycle. Three important social groups were therefore opposed to the air tire. But then the air tire was mounted on a racing bicycle. When, for the first time, the tire was used at the racing track, its entry was hailed with derisive laughter. This was, however, quickly silenced by the high speed achieved, and there was only astonishment left when it outpaced all rivals (Croon 1939). Soon handicappers had to give racing cyclists on high-wheelers a considerable start if riders on air-tire low-wheelers were entered. After a short period no racer of any pretensions troubled to compete on anything else (Grew 1921).

What had happened? With respect to two important groups, the sporting cyclists and the general public, closure had been reached, but not by convincing those two groups of the feasibility of the air tire in its meaning as an antivibration device. One can say, we think, that the meaning of the air tire was translated[24] to constitute a solution to quite another problem: the problem of how to go as fast as possible. And thus, by redefining the key problem with respect to which the artifact should have the meaning of a solution, closure was reached for two of the relevant social groups. How the third group, the engineers, came to accept the air tire is another story and need not be told here. Of course, there is nothing "natural" or logically necessary about this form of closure. It could be argued that speed is not the most important characteristic of the bicycle or that existing cycle races were not appropriate tests of a cycle's "real" speed (after all, the idealized world of the race track may not match everyday road conditions, any more than the Formula-1 racing car bears on the performance requirements of the average family sedan). Still, bicycle races have played an important role in the development of the bicycle, and because racing can be viewed as a specific form of testing, this observation is much in line with Constant's recent plea to pay more attention to testing procedures in studying technology (Constant 1983).

The Wider Context
Finally, we come to the third stage of our research program. The task here in the area of technology would seem to be the same as for science—to relate the content of a technological artifact to the wider sociopolitical milieu. This aspect has not yet been demonstrated for the science case,[25] at least not in contemporaneous sociological studies.[26]

However, the SCOT method of describing technological artifacts by focusing on the meanings given to them by relevant social groups seems to suggest a way forward. Obviously, the sociocultural and political situation of a social group shapes its norms and values, which in turn influence the meaning given to an artifact. Because we have shown how different meanings can constitute different lines of development, SCOT's descriptive model seems to offer an operationalization of the relationship between the wider milieu and the actual content of technology. To follow this line of analysis, see Bijker (1987).

Conclusion

In this chapter we have been concerned with outlining an integrated social constructivist approach to the empirical study of science and technology. We reviewed several relevant bodies of literature and strands of argument. We indicated that the social constructivist approach is a flourishing tradition within the sociology of science and that it shows every promise of wider application. We reviewed the literature on the science-technology relationship and showed that here, too, the social constructivist approach is starting to bear fruit. And we reviewed some of the main traditions in technology studies. We argued that innovation studies and much of the history of technology are unsuitable for our sociological purposes. We discussed some recent work in the sociology of technology and noted encouraging signs that a new wave of social constructivist case studies is beginning to emerge.

We then outlined in more detail the two approaches—one in the sociology of scientific knowledge (EPOR) and one in the field of sociology of technology (SCOT)—on which we base our integrated perspective. Finally, we indicated the similarity of the explanatory goals of the two approaches and illustrated these goals with some examples drawn from technology. In particular, we have seen that the concepts of interpretative flexibility and closure mechanism and the notion of social group can be given empirical reference in the social study of technology.

As we have noted throughout this chapter, the sociology of technology is still underdeveloped, in comparison with the sociology of scientific knowledge. It would be a shame if the advances made in the latter field could not be used to throw light on the study of technology. On the other hand, in our studies of technology it appeared to be fruitful to include several social groups in the analysis, and there are some indications that this method may also bear fruit in studies of science. Thus our integrated approach to the social study of science and technology indicates how the sociology of science and the sociology of technology might benefit each other.

But there is another reason, and perhaps an even more important one, to argue for such an integrated approach. And this brings us to a question that some readers might have expected to be dealt with in the first paragraph of this chapter, namely, the question of how to distinguish science from technology. We think that it is rather unfruitful to make such an a priori distinction. Instead, it seems worthwhile to start with commonsense notions of science and technology and to study them in an integrated way, as we have proposed. Whatever interesting differences may exist will gain contrast within such a program. This would constitute another concrete result of the integrated study of the social construction of facts and artifacts.

Notes

This chapter is a shortened and updated version of Pinch and Bijker (1984).

We are grateful to Henk van den Belt, Ernst Homburg, Donald MacKenzie, and Steve Woolgar for comments on an earlier draft of this chapter. We would like to thank the Stiftung Volkswagen, Federal Republic of Germany, the Twente University of Technology, The Netherlands, and the UK SSRC (under grant G/00123/0072/1) for financial support.

1. See, for example, Schumpeter (1928, 1942), Schmookler (1966, 1972), Freeman (1974, 1977), and Scholz (1977).

2. See, for example, Rosenberg (1982), Nelson and Winter (1977, 1982), and Dosi (1982, 1984). A study that preceded these is Rosenberg and Vincenti (1978).

3. Adapted from Uhlmann (1978), p. 45.

4. For another critique of these linear models, see Kline (1985).

5. Shapin writes that "a proper perspective of the uses of science might reveal that sociology of knowledge and history of technology have more in common than is usually thought" (1980, p. 132). Although we are sympathetic to Shapin's argument, we think the time is now ripe for asking more searching questions of historical studies.

6. Manuals describing resinous materials do mention Bakelite but not with the amount of attention that, retrospectively, we would think to be justified. Professor Max Bottler, for example, devotes only one page to Bakelite in his 228-page book on resins and the resin industry (Bottler 1924). Even when Bottler concentrates in another book on the *synthetic* resinous materials, Bakelite does not receive an indisputable "first place." Only half of the book is devoted to phenol/formaldehyde condensation products, and roughly half of that part is devoted to Bakelite (Bottler 1919). See also Matthis (1920).

7. For an account of other aspects of Bakelite's success, see Bijker (1987).

8. See, for example, Constant (1980), Hughes (1983), and Hanieski (1973).

9. See, for example, Noble (1979), Smith (1977), and Lazonick (1979).

10. See, for example, Vincenti (1986).

11. There is an American tradition in the sociology of technology. See, for example, Gilfillan (1935), Ogburn (1945), Ogburn and Meyers Nimkoff (1955), and Westrum (1983). A fairly comprehensive view of the present state of the art in German sociology of technology can be obtained from Jokisch (1982). Several studies in the sociology of technology that attempt to break with the traditional approach can be found in Krohn et al. (1978).

12. Dosi uses the concept of technological trajectory, developed by Nelson and Winter (1977); see also van den Belt and Rip (1987). Other approaches to technology based on Kuhn's idea of the community structure of science are mentioned by Bijker (1987). See also Constant (1987) and the collection edited by Laudan (1984).

13. One is reminded of the first blush of Kuhnian studies in the sociology of science. It was hoped that Kuhn's "paradigm" concept might be straightforwardly employed by sociologists in their studies of science. Indeed there were a number of studies in which attempts were made to identify phases in science, such as preparadigmatic, normal, and revolutionary. It soon became apparent, however, that Kuhn's terms were loosely formulated, could be subject to a variety of interpretations, and did not lend themselves to operationalization in any straightforward manner. See, for example, the inconclusive discussion over whether a Kuhnian analysis applies to psychology in Palermo (1973). A notable exception is Barnes's contribution to the discussion of Kuhn's work (Barnes 1982).

14. For a valuable review of Marxist work in this area, see MacKenzie (1984).

15. For a provisional report of this study, see Bijker et al. (1984). The five artifacts that are studied are Bakelite, fluorescent lighting, the safety bicycle, the Sulzer loom, and the transistor. See also Bijker (1987).

16. Work that might be classified as falling within the EPOR has been carried out primarily by Collins, Pinch, and Travis at the Science Studies Centre, University of Bath, and by Harvey and Pickering at the Science Studies Unit, University of Edinburgh. See, for example, Collins (1975), Wynne (1976), Pinch (1977, 1986), Pickering (1984), and the studies by Pickering, Collins, Travis, and Pinch in Collins (1981a).

17. See, for example, Bijker and Pinch (1983) and Bijker (1984, 1987). Studies by van den Belt (1985), Schot (1985, 1986), Jelsma and Smit (1986), and Elzen (1985, 1986) are also based on SCOT.

18. Constant (1980) used a similar evolutionary approach. Both Constant's model and our model seem to arise out of the work in evolutionary epistemology; see, for example, Toulmin (1972) and Campbell (1974). Elster (1983) gives a review of evolutionary models of technical change. See also van den Belt and Rip (1987).

19. It may be useful to state explicitly that we consider bicycles to be as fully fledged a technology as, for example, automobiles or aircraft. It may be helpful for readers from outside notorious cycle countries such as The Netherlands, France, and Great Britain to point out that both the automobile and the aircraft industries are, in a way, descendants from the bicycle industry. Many names occur in the histories of both the bicycle and the autocar: Triumph, Rover, Humber, and Raleigh, to mention but a few (Caunter 1955, 1957). The Wright brothers both sold and manufactured bicycles before they started to build their flying machines—mostly made out of bicycle parts (Gibbs-Smith 1960).

20. There is no cookbook recipe for how to identify a social group. Quantitative instruments using citation data may be of some help in certain cases. More research is needed to develop operationalizations of the notion of "relevant social group" for a variety of historical and sociological research sites. See also Law (1987) on the demarcation of networks and Bijker (1987).

21. Previously, two concepts have been used that can be understood as two distinctive concepts within the broader idea of stabilization (Bijker et al. 1984). *Reification* was used to denote social existence—existence in the consciousness of the members of a certain social group. *Economic stabilization* was used to indicate the economic existence of an artifact—its having a market. Both concepts are used in a continuous and relative way, thus requiring phrases such as "the *degree* of reification of the high-wheeler is *higher* in the group of young men of means and nerve than in the group of elderly men."

22. The use of the concepts of interpretative flexibility and rhetorical closure in science cases is illustrated by Pinch and Bijker (1984).

23. Advertisements seem to constitute a large and potentially fruitful data source for empirical social studies of technology. The considerations that professional advertising designers give to differences among various "consumer groups" obviously fit our use of different relevant groups. See, for example, Schwartz Cowan (1983) and Bijker (1987).

24. The concept of translation is fruitfully used in an extended way by Callon (1980b, 1981, 1986), Callon and Law (1982), and Latour (1983, 1984).

25. A model of such a "stage 3" explanation is offered by Collins (1983).

26. Historical studies that address the third stage may be a useful guide here. See, for example, MacKenzie (1978), Shapin (1979, 1984), and Shapin and Schaffer (1985).

References

Baekeland, L. H. 1909a. "On soluble, fusible, resinous condensation products of phenols and formaldehyde." *Journal of Industrial and Engineering Chemistry* 1:345–349.

Baekeland, L. H. 1909b. "The synthesis, constitution, and use of Bakelite." *Journal of Industrial and Engineering Chemistry* 1:149–161.

Barnes, B. 1982. *T. S. Kuhn and Social Science*. London: Macmillan.

Bijker, W. E. 1984. "Collectifs technologiques et styles technologiques: Eléments pour un modèle explicatif de la construction sociale des artefacts techniques," in *Travailleur collectif et relations science-production*, J. H. Jacot, ed. Paris: Editions du CNRS, 113–120.

Bijker, W. E. 1987. "The Social Construction of Bakelite: Toward a Theory of Invention," in *The Social Construction of Technological Systems*, Wiebe E. Bijker, Thomas P. Hughes, and Trevor Pinch, eds. Cambridge, Mass.: MIT Press, 159–187.

Bijker, W. E., and Pinch, T. J. 1983. "La construction sociale de faits et d'artefacts: Impératifs stratégiques et méthodologiques pour une approche unifiée de l'étude des sciences et de la technique." Paper presented to L'atelier de recherche (III) sur les problèmes stratégiques et méthodologiques en milieu scientifique et technique. Paris, March.

Bijker, W. E., Bönig, J., and van Oost, E. C. J. 1984. "The Social Construction of Technological Artefacts." Paper presented at the EASST Conference. A shorter version of this paper is published in *Zeitschrift für Wissenschaftsorschung*, Special Issue 2, 3: 39–52 (1984).

Bottler, M. 1919. *Uber Herstellung und Eigenschaften von Kunstharzen und deren Verwendung in der Lack- und Firnisindustrie und zu elektrotechnischen und industriellen Zwecken*. Munich: Lehmanns.

Bottler, M. 1924. *Harze und Harzinustrie*. Leipzig: Max Janecke.

Bunge, M. 1966. "Technology as applied science." *Technology and Culture* 7:329–347.

Callon, M. 1980a. "The state and technical innovation: A case study of the electrical vehicle in France." *Research Policy* 9:358–376.

Callon, M. 1980b. "Struggles and negotiations to define what is problematic and what is not: The sociologic of translation," in *The Social Process of Scientific Investigation*, K. Knorr, R. Krohn, and R. Whitley, eds. Dordrecht and Boston: Reidel, vol. 4, 197–219.

Callon, M. 1981. "Pour une sociologie des controverses technologiques." *Fundamenta Scientiae* 2:381–399.

Callon, M. 1986. "Some elements of a sociology of translation: Domestication of the scallops and the fishermen of St. Brieuc Bay," in *Power, Action, and Belief: A New Sociology of Knowledge?* J. Law, ed. London: Routledge and Kegan Paul, 196–233.

Callon, M. 1987. "Society in the Making: The Study of Technology as a Tool for Sociological Analysis," in *The Social Construction of Technological Systems*, Wiebe E. Bijker, Thomas P. Hughes, and Trevor Pinch, eds. Cambridge, Mass.: MIT Press, 83–110.

Callon, M., and Law, J. 1982. "On interests and their transformation: Enrollment and counter-enrollment." *Social Studies of Science* 12:615–625.

Campbell, D. T. 1974. "Evolutionary epistemology," in *The Philosophy of Karl Popper, The Library of Living Philosophers*, P. A. Schlipp, ed. La Salle, Ill.: Open Court, vol. 14-I, 413–463.

Caunter, C. F. 1955. *The History and Development of Cycles (as Illustrated by the Collection of Cycles in the Science Museum); Historical Survey*. London: HMSO.

Caunter, C. F. 1957. *The History and Development of Light Cars*. London: HMSO.

Collins, H. M. 1975. "The seven sexes: A study in the sociology of a phenomenon, or the replication of experiments in physics." *Sociology* 9:205–224.

Collins, H. M., ed. 1981a. "Knowledge and controversy." *Social Studies of Science* 11:3–158.

Collins, H. M. 1981b. "The place of the core-set in modern science: Social contingency with methodological propriety in science." *History of Science* 19:6–19.

Collins, H. M. 1981c. "Stages in the empirical programme of relativism." *Social Studies of Science* 11:3–10.

Collins, H. M. 1983. "An empirical relativist programme in the sociology of scientific knowledge," in *Science Observed: Perspectives on the Social Study of Science*, K. D. Knorr-Cetina and M. J. Mulkay, eds. Beverly Hills: Sage, 85–113.

Constant, E. W., III. 1980. *The Origins of the Turbojet Revolution*. Baltimore: Johns Hopkins University Press.

Constant, E. W., III. 1983. "Scientific theory and technological testability: Science, dynamometers, and water turbines in the 19th century." *Technology and Culture* 24:183–198.

Constant, E. W., III. 1987. "The Social Locus of Technological Practice: Community, System, or Organization?" in *The Social Construction of Technological Systems*, Wiebe E. Bijker, Thomas P. Hughes, and Trevor Pinch, eds. Cambridge, Mass.: MIT Press, 223–242.

Croon, L. 1939. *Das Fahrrad und seine Entwicklung*. Berlin: VDI-Verlag.

Dosi, G. 1982. "Technological paradigms and technological trajectories: A suggested interpretation of the determinants and directions of technical change." *Research Policy* 11:147–162.

Dosi, G. 1984. *Technical Change and Industrial Transformation*. London: Macmillan.

Dunlop, J. B. 1888. "An improvement in tyres of wheels for bicycles, tricycles, or other road cars." British Patent 10607. Filed on July 23, 1888.

Elster, J. 1983. *Explaining Technical Change*. Cambridge: Cambridge University Press.

Elzen, B. 1985. "De ultracentrifuge: op zoek naar patronen in technologische ontwikkeling door een vergelijkin van twee case-studies." *Jaarboek voor de Geschiedenis van Bedrijf en Techniek* 2:250–278.

Elzen, B. 1986. "Two ultracentrifuges: A comparative study of the social construction of artefacts." *Social Studies of Science* 16:621–662.

Ferguson, E. 1974. "Toward a discipline of the history of technology." *Technology and Culture* 15:13–30.

Freeman, C. 1974. *The Economics of Industrial Innovation*. Harmondsworth: Penguin. Reprinted by Frances Pinter, London, 1982.

Freeman, C. 1977. "Economics of research and development," in *Science, Technology and Society. A Cross-Disciplinary Perspective*, I. Spiegel-Rösing and D. de Solla Price, eds. London and Beverly Hills: Sage, 223–275.

Gibbs-Smith, C. H. 1960. *The Aeroplane: An Historical Survey of Its Origins and Development*. London: HSMO.

Gilfillan, S. G. 1935. *The Sociology of Invention*. Cambridge, Mass.: MIT Press.

Grew, W. 1921. *The Cycle Industry, Its Origin, History and Latest Developments*. London: Pitman & Sons.

Hanieski, J. F. 1973. "The airplane as an economic variable: Aspects of technological change in aeronautics, 1903–1955." *Technology and Culture* 14:535–552.

Haynes, W. 1954. *American Chemical Industry*, Vol. 2. New York: Van Nostrand.

Hughes, T. P. 1979. "Emerging themes in the history of technology." *Technology and Culture* 20:697–711.

Hughes, T. P. 1983. *Networks of Power: Electrification in Western Society, 1880–1930*. Baltimore: Johns Hopkins University Press.

Jelsma, J., and Smit, W. A. 1986. "Risks of recombinant DNA research: From uncertainty to certainty," in *Impact Assessment Today*, H. A. Becker and A. L. Porter, eds. Van Arkel: Utrecht, 715–741.

Johnston, R. 1972. "The internal structure of technology," in *The Sociology of Science*, P. Halmos, ed. Keele: University of Keele, 117–130.

Jokisch, R., ed. 1982. *Techniksoziologie*. Frankfurt am Main: Suhrkamp.

Kaufman, M. 1963. *The First Century of Plastics: Celluloid and Its Sequel*. London: The Plastics Institute.

Kline, S. 1985. "Research, invention, innovation, and production: Models and reality." *Research Management* 28 (4):36–45.

Krohn, W., Layton, E. T., and Weingart, P., eds. 1978. *The Dynamics of Science and Technology, Sociology of the Sciences Yearbook*, vol. 2. Dordrecht: Reidel.

Latour, B. 1983. "Give me a laboratory and I will raise the world," in *Science Observed: Perspectives on the Social Study of Science*, K. D. Knorr-Cetina and M. J. Mulkay, eds. London and Beverly Hills: Sage, 141–170.

Latour, B. 1984. *Les microbes, querre et paix, suivi de irréductions*. Paris: Métaillé. English translation: *The Pasteurization of French Society, Followed by Irréductions. A Politico-Scientific Essay*, Cambridge, Mass.: Harvard University Press, 1987 (forthcoming).

Laudan, R., ed. 1984. *The Nature of Technological Knowledge: Are Models of Scientific Change Relevant?* Dordrecht: Reidel.

Law, John. 1987. "Technology and Heterogeneous Engineering: The Case of Portuguese Expansion," in *The Social Construction of Technological System*, Wiebe E. Bijker, Thomas P. Hughes, and Trevor Pinch, eds. Cambridge, Mass.: MIT Press, 111–134.

Layton, E. 1977. "Conditions of technological development," in *Science, Technology and Society: A Cross-Disciplinary Perspective*, I. Spiegel-Rösing and D. de Solla Price, eds. London and Beverly Hills: Sage, 197–222.

Lazonick, W. 1979. "Industrial relations and technical change: The case of the self-acting mule." *Cambridge Journal of Economics* 3:231–262.

MacKenzie, D. 1978. "Statistical theory and social interest: A case study." *Social Studies of Science* 8:35–83.

MacKenzie, D. 1984. "Marx and the machine." *Technology and Culture* 25:473–502.

Matthis, A. R. c. 1920. *Insulating Varnishes in Electrotechnics*. London: John Heywood.

Mulkay, M. J. 1979. "Knowledge and utility: implications for the sociology of knowledge." *Social Studies of Science* 9:63–80.

Nelson, R. R., and Winter, S. G. 1977. "In search of a useful theory of innovation." *Research Policy* 6:36–76.

Nelson, R. R., and Winter, S. G. 1982. *An Evolutionary Theory of Economic Change*. Cambridge, Mass.: The Belknap Press of Harvard University Press.

Noble, D. F. 1979. "Social choice in machine design: The case of automatically controlled machine tools," in *Case Studies on the Labour Process*, A. Zimbalist, ed. New York: Monthly Review Press, 18–50.

Noble, D. F. 1984. *Forces of Production: A Social History of Industrial Automation*. New York: Knopf.

Ogburn, W. F. 1945. *The Social Effects of Aviation*. Boston: Houghton Mifflin.

Ogburn, W. F., and Meyers Nimkoff, F. 1955. *Technology and the Changing Family*. Boston: Houghton Mifflin.

Palermo, D. S. 1973. "Is a scientific revolution taking place in psychology?" *Science Studies* 3:211–244.

Pickering, A. 1984. *Constructing Quarks—A Sociological History of Particle Physics*. Chicago and Edinburgh: University of Chicago Press and Edinburgh University Press.

Pinch, T. J. 1977. "What does a proof do if it does not prove? A study of the social conditions and metaphysical divisions leading to David Bohm and John von Neumann failing to communicate in quantum physics," in *The Social Production of Scientific Knowledge*, E. Mendelsohn, P. Weingart, and R. Whitley, eds. Dordrecht: Reidel, 171–215.

Pinch, T. J. 1986. *Confronting Nature: The Sociology of Solar-Neutrino Detection*. Dordrecht: Reidel.

Pinch, T. J., and Bijker, W. E. 1984. "The social construction of facts and artefacts: Or how the sociology of science and the sociology of technology might benefit each other." *Social Studies of Science* 14:399–441. Also published in Serbo-Croatian as "Društveno Proizvodenje Činjenica I Tvorevina: O Cjelovitom Pristupu Izučavanju Znanosti I Tehnologije," *Gledišta, časopis za društvenu kritiku i teoriju*, March–April, 25:21–57.

Rosenberg, N. 1982. *Inside the Black Box: Technology and Economics*. Cambridge: Cambridge University Press.

Rosenberg, N., and Vincenti, W. G. 1978. *The Britannia Bridge: The Generation and Diffusion of Knowledge*. Cambridge, Mass.: MIT Press.

Schmookler, J. 1966. *Invention and Economic Growth*. Cambridge, Mass.: Harvard University Press.

Schmookler, J. 1972. *Patents, Invention and Economic Change, Data and Selected Essays*. Z. Griliches and L. Hurwicz, eds. Cambridge, Mass.: Harvard University Press.

Scholz, L. with a contribution by G. von L. Uhlmann, 1977. *Technik-Indikatoren, Ansätze zur Messung des Standes der Technik in der industriellen Produktion*. Berlin and München: Duncker & Humblot.

Schot, J. 1985. "De ontwikkeling van de techniek als een variatieen selectieproces. De meekrapteelt en-bereiding in het licht van een alternatieve techniekopvatting." Master's thesis, Erasmus University of Rotterdam, unpublished.

Schot, J. 1986. "De meekrapnijverheid: de ontwikkeling van de techniek als een proces van variatie en selectie," in *Jaarboek voor de Geschiedenis van Bedrijf en Techniek*, E. S. A. Bloemen, W. E. Bijker, W. van den Brocke, et al., eds. Utrecht: Stichting, vol. 3, 43–62.

Schumpeter, J. 1971 [1928]. "The instability of capitalism," in *The Economics of Technological Change*, N. Rosenberg, ed. Harmondsworth: Penguin, 13–42.

Schumpeter, J. 1942. *Capitalism, Socialism and Democracy*. New York: Harper & Row. Reprinted 1974 by Unwin University Books, London.

Schwartz Cowan, R. 1983. *More Work for Mother: The Ironies of Household Technology from the Open Hearth to the Microwave*. New York: Basic Books.

Shapin, S. 1979. "The politics of observation: Cerebral anatomy and social interests in the Edinburgh phrenology disputes," in *On the Margins of Science: The Social Construction of Rejected Knowledge*, R. Wallis, ed. Keele: University of Keele, 139–178.

Shapin, S. 1980. "Social uses of science," in *The Ferment of Knowledge*, G. S. Rousseau and R. Porter, eds. Cambridge: Cambridge University Press, 93–139.

Shapin, S. 1984. "Pump and circumstance: Robert Boyle's literary technology." *Social Studies of Science* 14:481–520.

Shapin, S., and Schaffer, S. 1985. *Leviathan and the Air-Pump: Hobbes, Boyle and the Experimental Life*. Princeton: Princeton University Press.

Smith, M. Roe. 1977. *Harpers Ferry Armory and the New Technology: The Challenge of Change*. Ithaca, N.Y.: Cornell University Press.

"The Stanley Exhibition of Cycles." 1889. *The Engineer* 67:157–158.

"The Stanley Exhibition of Cycles." 1890. *The Engineer* 69:107–108.

Staudenmaier, J. M., SJ. 1983. "What SHOT hath wrought and what SHOT hath not: Reflections on 25 years of the history of technology." Paper presented to the Twenty-fifth Annual Meeting of SHOT.

Staudenmaier, J. M., SJ. 1985. *Technology's Storytellers: Reweaving the Human Fabric*. Cambridge, Mass.: MIT Press.

Toulmin, S. 1972. *Human Understanding*, vol. 1. Oxford: Oxford University Press.

Uhlmann, L. 1978. *Der Innovationsprozess in westeuropäischen Industrieländern. Band 2: Den Ablauf industriellen Innovationsprozesses*. Berlin and München: Duncker and Humblot.

van den Belt, H., and Rip, A. 1987. "The Nelson-Winter-Dosi Model and Synthetic Dye Chemistry," in *The Social Construction of Technological Systems*, Wiebe E. Bijker, Thomas P. Hughes, and Trevor Pinch, eds. Cambridge, Mass.: MIT Press, 135–158.

van den Belt, H. 1985. "A. W. Hofman en de Franse Octrooiprocessen rond anilinerood: demarcatie als sociale constructie." *Jaarboek voor de Geschiedenis van Bedrijf en Techniek* 2:64–86.

Vincenti, W. G. 1986. "The Davis wing and the problem of airfoil design: Uncertainty and growth in engineering knowledge." *Technology and Culture* 26.

Westrum, R. 1983. "What happened to the old sociology of technology?" Paper presented to the Eighth Annual Meeting of the Society for Social Studies of Science, Blacksburg, Virginia, November.

Whitley, R. D. 1972. "Black boxism and the sociology of science: A discussion of the major developments in the field," in *The Sociology of Science*, P. Halmos, ed. Keele: University of Keele, 62–92.

Woodforde, J. 1970. *The Story of the Bicycle*. London: Routledge and Kegan Paul.

Wynne, B. 1976. "C. G. Barkla and the J phenomenon: A case study of the treatment of deviance in physics." *Social Studies of Science* 6:307–347.

9 "Technological Momentum"

Thomas P. Hughes

In this chapter, Hughes argues that both technological determinists and social constructivists have done interesting work, but neither group has provided the full picture. He argues that rather than adhering to one or the other theory, one should examine how society and technology both exert influence. Hughes acknowledges that people—in the form of individuals, governments, corporations, etc.—direct the development of new technologies. But he also claims that large sociotechnical systems can gain "momentum." By this he means that at times it may appear as though certain large technological systems have a mind of their own and cannot be stopped. But Hughes maintains that this is simply because a large number of social groups (including corporations, governments, industries, and consumers) have financial, capital, infrastructure, and ideological reasons for keeping such systems going. Once certain large systems are in place, it is much easier to keep them going and innovate "around the edges" than to radically change or abandon them altogether. In this way, Hughes offers a compromise of sorts in the social/technological determinism debate that helps to explain how both people and technological systems influence and shape each other. He argues that the investment of money, effort, and resources to develop technological systems can make subsequent efforts to change those systems very difficult.

The concepts of technological determinism and social construction provide agendas for fruitful discussion among historians, sociologists, and engineers interested in the nature of technology and technological change. Specialists can engage in a general discourse that subsumes their areas of specialization. In this essay I shall offer an additional concept—technological momentum—that will, I hope, enrich the discussion. Technological momentum offers an alternative to technological determinism and social construction. Those who in the past espoused a technological determinist approach to history offered a needed corrective to the conventional interpretation of history that virtually ignored the role of technology in effecting social change. Those who more recently advocated a social construction approach provided an invaluable corrective to an interpretation of history that encouraged a passive attitude toward an overwhelming technology. Yet both approaches suffer from a failure to encompass the complexity of technological change.

All three concepts present problems of definition. Technological determinism I define simply as the belief that technical forces determine social and cultural changes.

From Leo Marx and Merritt Roe Smith, eds., *Does Technology Drive History? The Dilemma of Technological Determinism* (Cambridge, Mass.: MIT Press, 1994), pp. 101–113. Reprinted with permission.

Social construction presumes that social and cultural forces determine technical change. A more complex concept than determinism and social construction, technological momentum infers that social development shapes and is shaped by technology. Momentum also is time dependent. Because the focus of this essay is technological momentum, I shall define it in detail by resorting to examples.

"Technology" and "technical" also need working definitions. Proponents of technological determinism and of social construction often use "technology" in a narrow sense to include only physical artifacts and software. By contrast, I use "technical" in referring to physical artifacts and software. By "technology" I usually mean technological or sociotechnical systems, which I shall also define by examples.

Discourses about technological determinism and social construction usually refer to society, a concept exceedingly abstract. Historians are wary of defining society other than by example because they have found that twentieth-century societies seem quite different from twelfth-century ones and that societies differ not only over time but over space as well. Facing these ambiguities, I define the social as the world that is not technical, or that is not hardware or technical software. This world is made up of institutions, values, interest groups, social classes, and political and economic forces. As the reader will learn, I see the social and the technical as interacting within technological systems. Technological system, as I shall explain, includes both the technical and the social. I name the world outside of technological systems that shapes them or is shaped by them the "environment." Even though it may interact with the technological system, the environment is not a part of the system because it is not under the control of the system as are the system's interacting components.

In the course of this essay the reader will discover that I am no technological determinist. I cannot associate myself with such distinguished technological determinists as Karl Marx, Lynn White, and Jacques Ellul. Marx, in moments of simplification, argued that waterwheels ushered in manorialism and that steam engines gave birth to bourgeois factories and society. Lenin added that electrification was the bearer of socialism. White elegantly portrayed the stirrup as the prime mover in a train of cause and effect culminating in the establishment of feudalism. Ellul finds the human-made environment structured by technical systems, as determining in their effects as the natural environment of Charles Darwin. Ellul sees the human-made as steadily displacing the natural—the world becoming a system of artifacts, with humankind, not God, as the artificer.[1]

Nor can I agree entirely with the social constructivists. Wiebe Bijker and Trevor Pinch have made an influential case for social construction in their essay "The Social Construction of Facts and Artifacts."[2] They argue that social, or interest, groups define and give meaning to artifacts. In defining them, the social groups determine the designs of artifacts. They do this by selecting for survival the designs that solve the problems they want solved by the artifacts and that fulfill desires they want fulfilled by the artifacts. Bijker and Pinch emphasize the interpretive flexibility discernible in the evolution of artifacts: they believe that the various meanings given by social groups to, say, the bicycle result in a number of alternative designs of that machine.

The various bicycle designs are not fixed; closure does not occur until social groups believe that the problems and desires they associate with the bicycle are solved or fulfilled.

In summary, I find the Bijker-Pinch interpretation tends toward social determinism, and I must reject it on these grounds. The concept of technological momentum avoids the extremism of both technological determinism and social construction by presenting a more complex, flexible, time-dependent, and persuasive explanation of technological change.

Technological Systems

Electric light and power systems provide an instructive example of technological systems. By 1920 they had taken on a messy complexity because of the heterogeneity of their components. In their diversity, their complexity, and their large scale, such mature technological systems resemble the megamachines that Lewis Mumford described in *The Pentagon of Power*.[3] The actor networks of Bruno Latour and Michel Callon[4] also share essential characteristics with technological systems. An electric power system consists of inanimate electrons and animate regulatory boards, both of which, as Latour and Callon suggest, can be intractable if not brought in line or into the actor network.

The Electric Bond and Share Company (EBASCO), an American electric utility holding company of the 1920s, provides an example of a mature technological system. Established in 1905 by the General Electric Company, EBASCO controlled through stock ownership a number of electric utility companies, and through them a number of technical subsystems—namely electric light and power networks, or grids.[5] EBASCO provided financial, management, and engineering construction services for the utility companies. The inventors, engineers, and managers who were the system builders of EBASCO saw to it that the services related synergistically. EBASCO management recommended construction that EBASCO engineering carried out and for which EBASCO arranged financing through sale of stocks or bonds. If the utilities lay in geographical proximity, then EBASCO often physically interconnected them through high-voltage power grids. The General Electric Company founded EBASCO and, while not owning a majority of stock in it, substantially influenced its policies. Through EBASCO General Electric learned of equipment needs in the utility industry and then provided them in accord with specifications defined by EBASCO for the various utilities with which it interacted. Because it interacted with EBASCO, General Electric was a part of the EBASCO system. Even though I have labeled this the EBASCO system, it is not clear that EBASCO solely controlled the system. Control of the complex systems seems to have resulted from a consensus among EBASCO, General Electric, and the utilities in the systems.

Other institutions can also be considered parts of the EBASCO system, but because the interconnections were loose rather than tight[6] these institutions are usually not recognized as such. I refer to the electrical engineering departments in engineering colleges, whose faculty and graduate students conducted research or consulted for EBASCO. I am also inclined to include a few of the various state regulatory authorities as parts of the EBASCO system, if their members were greatly influenced by it. If the regulatory authorities were free of this control, then they should be considered a part of the EBASCO environment, not of the system.

Because it had social institutions as components, the EBASCO system could be labeled a sociotechnical system. Since, however, the system had a technical (hardware and software) core, I prefer to name it a technological system, to distinguish it from social systems without technical cores. This privileging of the technical in a technological system is justified in part by the prominent roles played by engineers, scientists, workers, and technical-minded managers in solving the problems arising during the creation and early history of a system. As a system matures, a bureaucracy of managers and white-collar employees usually plays an increasingly prominent role in maintaining and expanding the system, so that it then becomes more social and less technical.

EBASCO as a Cause and an Effect

From the point of view of technological—better, technical—determinists, the determined is the world beyond the technical. Technical determinists considering EBASCO as a historical actor would focus on its technical core as a cause with many effects. Instead of seeing EBASCO as a technological system with interacting technical and social components, they would see the technical core as causing change in the social components of EBASCO and in society in general. Determinists would focus on the way in which EBASCO's generators, by energizing electric motors on individual production machines, made possible the reorganization of the factory floor in a manner commonly associated with Fordism. Such persons would see street, workplace, and home lighting changing working and leisure hours and affecting the nature of work and play. Determinists would also cite electrical appliances in the home as bringing less— and more—work for women,[7] and the layout of EBASCO's power lines as causing demographic changes. Electrical grids such as those presided over by EBASCO brought a new decentralized regionalism, which contrasted with the industrial, urban-centered society of the steam age.[8] One could extend the list of the effects of electrification enormously.

Yet, contrary to the view of the technological determinists, the social constructivists would find exogenous technical, economic, political, and geographical forces, as well as values, shaping with varying intensity the EBASCO system during its evolution. Social constructivists see the technical core of EBASCO as an effect rather than a cause. They could cite a number of instances of social construction. The spread of alternating (polyphase) current after 1900, for instance, greatly affected, even determined, the history of the early utilities that had used direct current, for these had to change their generators and related equipment to alternating current or fail in the face of competition. Not only did such external technical forces shape the technical core of the utilities;

economic forces did so as well. With the rapid increase in the United States' population and the concentration of industry in cities, the price of real estate increased. Needing to expand their generating capacity, EBASCO and other electric utilities chose to build new turbine-driven power plants outside city centers and to transmit electricity by high-voltage lines back into the cities and throughout the area of supply. Small urban utilities became regional ones and then faced new political or regulatory forces as state governments took over jurisdiction from the cities. Regulations also caused technical changes. As the regional utilities of the EBASCO system expanded, they conformed to geographical realities as they sought cooling water, hydroelectric sites, and mine-mouth locations. Values, too, shaped the history of EBASCO. During the Great Depression, the Roosevelt administration singled out utility holding-company magnates for criticism, blaming the huge losses experienced by stock and bond holders on the irresponsible, even illegal, machinations of some of the holding companies. Partly as a result of this attack, the attitudes of the public toward large-scale private enterprise shifted so that it was relatively easy for the administration to push through Congress the Holding Company Act of 1935, which denied holding companies the right to incorporate utilities that were not physically contiguous.[9]

Gathering Technological Momentum

Neither the proponents of technical determinism nor those of social construction can alone comprehend the complexity of an evolving technological system such as EBASCO. On some occasions EBASCO was a cause; on others it was an effect. The system both shaped and was shaped by society. Furthermore, EBASCO's shaping society is not an example of purely technical determinism, for EBASCO, as we have observed, contained social components. Similarly, social constructivists must acknowledge that social forces in the environment were not shaping simply a technical system, but a technological system, including—as systems invariably do—social components.

The interaction of technological systems and society is not symmetrical over time. Evolving technological systems are time dependent. As the EBASCO system became larger and more complex, thereby gathering momentum, the system became less shaped by and more the shaper of its environment. By the 1920s the EBASCO system rivaled a large railroad company in its level of capital investment, in its number of customers, and in its influence upon local, state, and federal governments. Hosts of electrical engineers, their professional organizations, and the engineering schools that trained them were committed by economic interests and their special knowledge and skills to the maintenance and growth of the EBASCO system. Countless industries and communities interacted with EBASCO utilities because of shared economic interests. These various human and institutional components added substantial momentum to the EBASCO system. Only a historical event of large proportions could deflect or break the momentum of an EBASCO, the Great Depression being a case in point.

Characteristics of Momentum

Other technological systems reveal further characteristics of technological momentum, such as acquired skill and knowledge, special-purpose machines and processes, enormous physical structures, and organizational bureaucracy. During the late nineteenth century, for instance, mainline railroad engineers in the United States transferred their acquired skill and knowledge to the field of intra-urban transit. Institutions with specific characteristics also contributed to this momentum. Professors in the recently founded engineering schools and engineers who had designed and built the railroads organized and rationalized the experience that had been gathered in preparing roadbeds, laying tracks, building bridges, and digging tunnels for mainline railroads earlier in the century. This engineering science found a place in engineering texts and in the curricula of the engineering schools, thus informing a new generation of engineers who would seek new applications for it.

Late in the nineteenth century, when street congestion in rapidly expanding industrial and commercial cities such as Chicago, Baltimore, New York, and Boston threatened to choke the flow of traffic, extensive subway and elevated railway building began as an antidote. The skill and the knowledge formerly expended on railroad bridges were now applied to elevated railway structures; the know-how once invested in tunnels now found application in subways. A remarkably active period of intra-urban transport construction began about the time when the building of mainline railways reached a plateau, thus facilitating the movement of know-how from one field to the other. Many of the engineers who played leading roles in intra-urban transit between 1890 and 1910 had been mainline railroad builders.[10]

The role of the physical plant in the buildup of technological momentum is revealed in the interwar history of the Badische Anilin und Soda Fabrik (BASF), one of Germany's leading chemical manufacturers and a member of the I.G. Farben group. During World War I, BASF rapidly developed large-scale production facilities to utilize the recently introduced Haber-Bosch technique of nitrogen fixation. It produced the nitrogen compounds for fertilizers and explosives so desperately needed by a blockaded Germany. The high-technology process involved the use of high-temperature, high-pressure, complex catalytic action. Engineers had to design and manufacture extremely costly and complex instrumentation and apparatus. When the blockade and the war were over, the market demand for synthetic nitrogen compounds did not match the large capacity of the high-technology plants built by BASF and other companies during the war. Numerous engineers, scientists, and skilled craftsmen who had designed, constructed, and operated these plants found their research and development knowledge and their construction skills underutilized. Carl Bosch, chairman of the managing board of BASF and one of the inventors of the Haber-Bosch process, had a personal and professional interest in further development and application of high-temperature, high-pressure, catalytic processes. He and other managers, scientists, and engineers at BASF sought additional ways of using the plant and the knowledge created during the war years. They first introduced a high-temperature, high-pressure catalytic process for manufacturing synthetic methanol in the early 1920s. The momentum of the now-generalized process next showed itself in management's decision in the mid 1920s to invest in research and development aimed at using high-temperature, high-pressure catalytic chemistry for the production of synthetic gasoline from coal. This project became the largest investment in research and development by BASF during the Weimar era. When the National Socialists took power, the government contracted for large amounts of the synthetic product. Momentum swept BASF and I.G. Farben into the Nazi system of economic autarky.[11]

When managers pursue economies of scope, they are taking into account the momentum embodied in large physical structures. Muscle Shoals Dam, an artifact of considerable size, offers another example of this aspect of technological momentum. As the loss of merchant ships to submarines accelerated during World War I, the United States also attempted to increase its indigenous supply of nitrogen compounds. Having selected a process requiring copious amounts of electricity, the government had to construct a hydroelectric dam and power station. This was located at Muscle Shoals, Alabama, on the Tennessee River. Before the nitrogen-fixation facilities being built near the dam were completed, the war ended. As in Germany, the supply of synthetic nitrogen compounds then exceeded the demand. The U.S. government was left not only with process facilities but also with a very large dam and power plant.

Muscle Shoals Dam (later named Wilson Dam), like the engineers and managers we have considered, became a solution looking for a problem. How should the power from the dam be used? A number of technological enthusiasts and planners envisioned the dam as the first of a series of hydroelectric projects along the Tennessee River and its tributaries. The poverty of the region spurred them on in an era when electrification was seen as a prime mover of economic development. The problem looking for a solution attracted the attention of an experienced problem solver, Henry Ford, who proposed that an industrial complex based on hydroelectric power be located along 75 miles of the waterway that included the Muscle Shoals site. An alliance of public power and private interests with their own plans for the region frustrated his plan. In 1933, however, Muscle Shoals became the original component in a hydroelectric, flood-control, soil-reclamation, and regional development project of enormous scope sponsored by Senator George Norris and the Roosevelt administration and presided over by the Tennessee Valley Authority. The technological momentum of the Muscle Shoals Dam had carried over from World War I to the New Deal. This durable artifact acted over time like a magnetic field, attracting plans and projects suited to its characteristics. Systems of artifacts are not neutral forces; they tend to shape the environment in particular ways.[12]

Using Momentum

System builders today are aware that technological momentum—or whatever they may call it—provides the durability and the propensity for growth that were associated more commonly in the past with the spread of bureaucracy. Immediately after World

War II, General Leslie Groves displayed his system-building instincts and his awareness of the critical importance of technological momentum as a means of ensuring the survival of the system for the production of atomic weapons embodied in the wartime Manhattan Project. Between 1945 and 1947, when others were anticipating disarmament, Groves expanded the gaseous-diffusion facilities for separating fissionable uranium at Oak Ridge, Tennessee; persuaded the General Electric Company to operate the reactors for producing plutonium at Hanford, Washington; funded the new Knolls Atomic Power Laboratory at Schenectady, New York; established the Argonne and Brookhaven National Laboratories for fundamental research in nuclear science; and provided research funds for a number of universities. Under his guiding hand, a large-scale production system with great momentum took on new life in peacetime. Some of the leading scientists of the wartime project had confidently expected production to end after the making of a few bombs and the coming of peace.[13]

More recently, proponents of the Strategic Defense Initiative (SDI), organized by the Reagan administration in 1983, have made use of momentum. The political and economic interests and the organizational bureaucracy vested in this system were substantial—as its makers intended. Many of the same industrial contractors, research universities, national laboratories, and government agencies that took part in the construction of intercontinental ballistic missile systems, National Air and Space Administration projects, and atomic weapon systems have been deeply involved in SDI. The names are familiar: Lockheed, General Motors, Boeing, TRW, McDonnell Douglas, General Electric, Rockwell, Teledyn, MIT, Stanford, the University of California's Lawrence Livermore Laboratory, Los Alamos, Hanford, Brookhaven, Argonne, Oak Ridge, NASA, the U.S. Air Force, the U.S. Navy, the CIA, the U.S. Army, and others. Political interests reinforced the institutional momentum. A number of congressmen represent districts that receive SDI contracts, and lobbyists speak for various institutions drawn into the SDI network.[14] Only the demise of the Soviet Union as a military threat allowed counter forces to build up sufficient momentum to blunt the cutting edge of SDI.

Conclusion

A technological system can be both a cause and an effect; it can shape or be shaped by society. As they grow larger and more complex, systems tend to be more shaping of society and less shaped by it. Therefore, the momentum of technological systems is a concept that can be located somewhere between the poles of technical determinism and social constructivism. The social constructivists have a key to understanding the behavior of young systems; technical determinists come into their own with the mature ones. Technological momentum, however, provides a more flexible mode of interpretation and one that is in accord with the history of large systems.

What does this interpretation of the history of technological systems offer to those who design and manage systems or to the public that might wish to shape them through a democratic process? It suggests that shaping is easiest before the system has acquired political, economic, and value components. It also follows that a system with great technological momentum can be made to change direction if a variety of its components are subjected to the forces of change.

For instance, the changeover since 1970 by U.S. automobile manufacturers from large to more compact automobiles and to more fuel-efficient and less polluting ones came about as a result of pressure brought on a number of components in the huge automobile production and use system. As a result of the oil embargo of 1973 and the rise of gasoline prices, American consumers turned to imported compact automobiles; this, in turn, brought competitive economic pressure to bear on the Detroit manufacturers. Environmentalists helped persuade the public to support, and politicians to enact, legislation that promoted both anti-pollution technology and gas-mileage standards formerly opposed by American manufacturers. Engineers and designers responded with technical inventions and developments.

On the other hand, the technological momentum of the system of automobile production and use can be observed in recent reactions against major environmental initiatives in the Los Angeles region. The host of institutions and persons dependent politically, economically, and ideologically on the system (including gasoline refiners, automobile manufacturers, trade unions, manufacturers of appliances and small equipment using internal-combustion engines, and devotees of unrestricted automobile usage) rallied to frustrate change.

Because social and technical components interact so thoroughly in technological systems and because the inertia of these systems is so large, they bring to mind the iron-cage metaphor that Max Weber used in describing the organizational bureaucracies that proliferated at the beginning of the twentieth century.[15] Technological systems, however, are bureaucracies reinforced by technical, or physical, infrastructures which give them even greater rigidity and mass than the social bureaucracies that were the subject of Weber's attention. Nevertheless, we must remind ourselves that technological momentum, like physical momentum, is not irresistible.

Notes

1. Lynn White, Jr., *Medieval Technology and Social Change* (Clarendon, 1962); Jacques Ellul, *The Technological System* (Continuum, 1980); Karl Marx, *Capital: A Critique of Political Economy*, ed. F. Engels; *Electric Power Development in the U.S.S.R.*, ed. B. I. Weitz (Moscow: INRA, 1936).

2. The essay is found in *The Social Construction of Technological Systems: New Directions in the Sociology and History of Technology*, ed. W. E. Bijker et al. (MIT Press, 1987) (and is partially reprinted as chapter 8 in this book).

3. Lewis Mumford, *The Myth of the Machine: II. The Pentagon of Power* (Harcourt Brace Jovanovich, 1970).

4. Bruno Latour, *Science in Action: How to Follow Scientists and Engineers through Society* (Harvard University Press, 1987); Michel Callon, "Society in the Making: The Study of Technology as a Tool for Sociological Analysis," in *The Social Construction of Technological Systems*.

5. Before 1905, General Electric used the United Electric Securities Company to hold its utility securities and to fund its utility customers who purchased GE equipment. See Thomas P. Hughes, *Networks of Power: Electrification in Western Society, 1880–1930* (Johns Hopkins University Press, 1983), pp. 395–396.

6. The concept of loosely and tightly coupled components in systems is found in Charles Perrow's *Normal Accidents: Living with High Risk Technology* (Basic Books, 1984).

7. Ruth Schwartz Cowan, "The 'Industrial Revolution' in the Home," *Technology and Culture* 17 (1976): 1–23.

8. Lewis Mumford, *The Culture of Cities* (Harcourt Brace Jovanovich, 1970), p. 378.

9. More on EBASCO's history can be found on pp. 392–399 of *Networks of Power*.

10. Thomas Parke Hughes, "A Technological Frontier: The Railway," in *The Railroad and the Space Program*, ed. B. Mazlish (MIT Press, 1965).

11. Thomas Parke Hughes, "Technological Momentum: Hydrogenation in Germany 1900–1933," *Past and Present* (August 1969): 106–132.

12. On Muscle Shoals and the TVA, see Preston J. Hubbard's *Origins of the TVA: The Muscle Shoals Controversy, 1920–1932* (Norton, 1961).

13. Richard G. Hewlett and Oscar E. Anderson, Jr., *The New World, 1939–1946* (Pennsylvania State University Press, 1962), pp. 624–638.

14. Charlene Mires, "The Strategic Defense Initiative" (unpublished essay, History and Sociology of Science Department, University of Pennsylvania, 1990).

15. Max Weber, *The Protestant Ethic and the Spirit of Capitalism*, tr. T. Parsons (Unwin-Hyman, 1990), p. 155.

10 "Where Are the Missing Masses? The Sociology of a Few Mundane Artifacts"

Bruno Latour

One of the most popular and powerful ways of resolving the technological determinism/social constructivism dichotomy in technology studies is the actor network approach. Those advocating the actor network approach agree with the social constructivist claim that sociotechnical systems are developed through negotiations between people, institutions, and organizations. But they make the additional interesting argument that artifacts are part of these negotiations as well. This is not to say that machines think like people do and decide how they will act, but their behavior or nature often has a comparable role. Actor network theorists argue that the material world pushes back on people because of its physical structure and design. People are free to interpret the precise meaning of an artifact, but they can't simply tell an automobile engine that it should get 100 miles per gallon. The laws of nature and the capacities of a particular design limit the ways in which artifacts can be integrated into a sociotechnical system. In this chapter, one of the foremost contributors to the actor network approach, Bruno Latour, explores how artifacts can be deliberately designed to both replace human action and constrain and shape the actions of other humans. His study demonstrates how people can "act at a distance" through the technologies they create and implement and how, from a user's perspective, a technology can appear to determine or compel certain actions. He argues that even technologies that are so commonplace that we don't even think about them can shape the decisions we make, the effects our actions have, and the way we move through the world. Technologies play such an important role in mediating human relationships, Latour argues, that we cannot understand how societies work without an understanding of how technologies shape our everyday lives. Latour's study of the relationship between producers, machines, and users demonstrates how certain values and political goals can be achieved through the construction and employment of technologies.

To Robert Fox, Again, might not the glory of the machines consist in their being without this same boasted gift of language? "Silence," it has been said by one writer, "is a virtue which render us agreeable to our fellow-creatures."
Samuel Butler (*Erewhon*, chap. 23)

Early this morning, I was in a bad mood and decided to break a law and start my car without buckling my seat belt. My car usually does not want to start before I buckle the belt. It first flashes a red light "FASTEN YOUR SEAT BELT!," then an alarm sounds; it is

---

From Wiebe E. Bijker and John Law, eds., *Shaping Technology/Building Society: Studies in Sociotechnical Change* (Cambridge, Mass.: MIT Press, 1992), pp. 225–258. Reprinted with permission.

so high pitched, so relentless, so repetitive, that I cannot stand it. After ten seconds I swear and put on the belt. This time, I stood the alarm for twenty seconds and then gave in. My mood had worsened quite a bit, but I was at peace with the law—at least with that law. I wished to break it, but I could not. Where is the morality? In me, a human driver, dominated by the mindless power of an artifact? Or in the artifact forcing me, a mindless human, to obey the law that I freely accepted when I get my driver's license? Of course, I could have put on my seat belt before the light flashed and the alarm sounded, incorporating in my own self the good behavior that everyone—the car, the law, the police—expected of me. Or else, some devious engineer could have linked the engine ignition to an electric sensor in the seat belt, so that I could not even have started the car before having put it on. Where would the morality be in those two extreme cases? In the electric currents flowing in the machine between the switch and the sensor? Or in the electric currents flowing down my spine in the automatism of my routinized behavior? In both cases the result would be the same from an outside observer—say a watchful policeman: this assembly of a driver and a car obeys the law in such a way that it is impossible for a car to be at the same time moving AND to have the driver without the belt on. *A law of the excluded middle* has been built, rendering logically inconceivable as well as morally unbearable a driver without a seat belt. Not quite. Because I feel so irritated to be forced to behave well that I instruct my garage mechanics to unlink the switch and the sensor. The excluded middle is back in! There is at least one car that is both on the move and without a seat belt on its driver—mine. This was without counting on the cleverness of engineers. They now invent a seat belt that politely makes way for me when I open the door and then straps me as politely but very tightly when I close the door. Now there is no escape. The only way not to have the seat belt on is to leave the door wide open, which is rather dangerous at high speed. Exit the excluded middle. The program of action[1] "IF a car is moving, THEN the driver has a seat belt" is enforced. It has become logically—no, it has become sociologically—impossible to drive without wearing the belt. I cannot be bad anymore. I, plus the car, plus the dozens of patented engineers, plus the police are making me be moral (figure 10.1).

According to some physicists, there is not enough mass in the universe to balance the accounts that cosmologists make of it. They are looking everywhere for the "missing mass" that could add up to the nice expected total. It is the same with sociologists. They are constantly looking, somewhat desperately, for social links sturdy enough to tie all of us together or for moral laws that would be inflexible enough to make us behave properly. When adding up social ties, all does not balance. Soft humans and weak moralities are all sociologists can get. The society they try to recompose with bodies and norms constantly crumbles. Something is missing, something that should be strongly social and highly moral. Where can they find it? Everywhere, but they too often refuse to see it in spite of much new work in the sociology of artifacts.[2]

I expect sociologists to be much more fortunate than cosmologists, because they will soon discover their missing mass. To balance our accounts of society, we simply

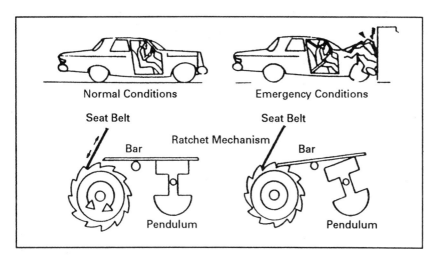

Figure 10.1
The designers of the seat belt take on themselves and then shift back to the belt contradictory programs; the belt should be lenient and firm, easy to put on and solidly fastened while ready to be unbuckled in a fraction of a second; it should be unobtrusive and strap in the whole body. The object does not reflect the social. It does more. It transcribes and displaces the contradictory interests of people and things.

have to turn our exclusive attention away from humans and look also at nonhumans. Here they are, the hidden and despised social masses who make up our morality. They knock at the door of sociology, requesting a place in the accounts of society as stubbornly as the human masses did in the nineteenth century. What our ancestors, the founders of sociology, did a century ago to house the human masses in the fabric of social theory, we should do now to find a place in a new social theory for the nonhuman masses that beg us for understanding.

Description of a Door

I will start my inquiry by following a little script written by anonymous hands.[3] On a freezing day in February, posted on the door of La Halle aux Cuirs at La Villette, in Paris, where Robert Fox's group was trying to convince the French to take up social history of science, could be seen a small handwritten notice: "The Groom Is On Strike, For God's Sake, Keep The Door Closed" ("groom" is Frenglish for an automated door-closer or butler). This fusion of labor relations, religion, advertisement, and technique in one insignificant fact is exactly the sort of thing I want to describe[4] in order to discover the missing masses of our society. As a technologist teaching in the School of Mines, an engineering institution, I want to challenge some of the assumptions sociologists often hold about the social context of machines.

Walls are a nice invention, but if there were no holes in them there would be no way to get in or out—they would be mausoleums or tombs. The problem is that if you make holes in the walls, anything and anyone can get in and out (cows, visitors, dust, rats, noise—La Halle aux Cuirs is ten meters from the Paris ring road—and, worst of all, cold—La Halle aux Cuirs is far to the north of Paris). So architects invented this hybrid: a wall hole, often called a *door*, which although common enough has always struck me as a miracle of technology. The cleverness of the invention hinges upon the hingepin: instead of driving a hole through walls with a sledgehammer or a pick, you simply gently push the door (I am supposing here that the lock has not been invented—this would overcomplicate the already highly complex story of La Villette's door); furthermore—and here is the real trick—once you have passed through the door, you do not have to find trowel and cement to rebuild the wall you have just destroyed: you simply push the door gently back (I ignore for now the added complication of the "pull" and "push" signs).

So, to size up the work done by hinges, you simply have to imagine that every time you want to get in or out of the building you have to do the same work as a prisoner trying to escape or as a gangster trying to rob a bank, plus the work of those who rebuild either the prison's or the bank's walls. If you do not want to imagine people destroying walls and rebuilding them every time they wish to leave or enter a building, then imagine the work that would have to be done to keep inside or outside all the things and people that, left to themselves, would go the wrong way.[5] As Maxwell never said, imagine his demon working *without* a door. Anything could escape from or penetrate into La Halle aux Cuirs, and soon there would be complete equilibrium between the depressing and noisy surrounding area and the inside of the building. Some technologists, including the present writer in *Material Resistance, A Textbook* (1984), have written that techniques are always involved when asymmetry or irreversibility are the goal; it might appear that doors are a striking counterexample because they maintain the wall hole in a reversible state; the allusion to Maxwell's demon clearly shows, however, that such is not the case; the reversible door is the only way to trap irreversibly inside La Halle aux Cuirs a differential accumulation of warm historians, knowledge, and also, alas, a lot of paperwork; the hinged door allows a selection of what gets in and what gets out so as to locally increase order, or information. If you let the drafts get inside (these renowned "courants d'air" so dangerous to French health), the paper drafts may never get outside to the publishers.

Now, draw two columns (if I am not allowed to give orders to the reader, then I offer it as a piece of strongly worded advice): in the right-hand column, list the work people would have to do if they had no door; in the left-hand column write down the gentle pushing (or pulling) they have to do to fulfill the same tasks. Compare the two columns: the enormous effort on the right is balanced by the small one on the left, and this is all thanks to hinges. I will define this transformation of a major effort into a minor one by the words *displacement* or *translation* or *delegation* or *shifting*;[6] I will say that we have delegated (or translated or displaced or shifted down) to the hinge the work of reversibly solving the wall-hole dilemma. Calling on Robert Fox, I do not have to do this work nor even think about it; it was delegated by the carpenter to a character, the hinge, which I will call a *nonhuman*. I simply enter La Halle aux Cuirs. As a more general descriptive rule, every time you want to know what a nonhuman does, simply imagine what other humans or other nonhumans would have to do were this character not present. This imaginary substitution exactly sizes up the role, or function, of this little character.

Before going on, let me point out one of the side benefits of this table: in effect, we have drawn a scale where tiny efforts balance out mighty weights; the scale we drew reproduces the very leverage allowed by hinges. That the small be made stronger than the large is a very moral story indeed (think of David and Goliath); by the same token, it is also, since at least Archimedes' days, a very good definition of a lever and of power: what is the minimum you need to hold and deploy astutely to produce the maximum effect? Am I alluding to machines or to Syracuse's King? I don't know, and it does not matter, because the King and Archimedes fused the two "minimaxes" into a single story told by Plutarch: the defense of Syracuse through levers and war machines.[7] I contend that this reversal of forces is what sociologists should look at in order to understand the social construction of techniques, and not a hypothetical "social context" that they are not equipped to grasp. This little point having been made, let me go on with the story (we will understand later why I do not really need your permission to go on and why, nevertheless, you are free not to go on, although only *relatively* so).

Delegation to Humans

There is a problem with doors. Visitors push them to get in or pull on them to get out (or vice versa), but then the door remains open. That is, instead of the door you have a gaping hole in the wall through which, for instance, cold rushes in and heat rushes out. Of course, you could imagine that people living in the building or visiting the Centre d'Histoire des Sciences et des Techniques would be a well-disciplined lot (after all, historians are meticulous people). They will learn to close the door behind them and retransform the momentary hole into a well-sealed wall. The problem is that discipline is not the main characteristic of La Villette's people; also you might have mere sociologists visiting the building, or even pedagogues from the nearby Centre de Formation. Are they all going to be so well trained? Closing doors would appear to be a simple enough piece of know-how once hinges have been invented, but, considering the amount of work, innovations, sign-posts, and recriminations that go on endlessly everywhere to keep them closed (at least in northern regions), it seems to be rather poorly disseminated.

This is where the age-old Mumfordian choice is offered to you: either to discipline the people or to substitute for the unreliable people another delegated human character whose only function is to open and close the door. This is called a groom or a porter (from the French word for door), or a gatekeeper, or a janitor, or a concierge, or a turnkey, or a jailer. The advantage is that you now have to discipline only one human and may safely leave the others to their erratic behavior. No matter who it is and where it comes from, the groom will always take care of the door. A nonhuman (the hinges) plus a human (the groom) have solved the wall-hole dilemma.

Solved? Not quite. First of all, if La Halle aux Cuirs pays for a porter, they will have no money left to buy coffee or books, or to invite eminent foreigners to give lectures. If they give the poor little boy other duties besides that of porter, then he will not be present most of the time and the door will stay open. Even if they had money to keep him there, we are now faced with a problem that two hundred years of capitalism has not completely solved: how to discipline a youngster to reliably fulfill a boring and underpaid duty? Although there is now only one human to be disciplined instead of hundreds, the weak point of the tactic can be seen: if this *one* lad is unreliable, then the whole chain breaks down; if he falls asleep on the job or goes walkabout, there will be no appeal: the door will stay open (remember that locking it is no solution because this would turn it into a wall, and then providing everyone with the right key is a difficult task that would not ensure that key holders will lock it back). Of course, the porter may be punished. But disciplining a groom—Foucault notwithstanding—is an enormous and costly task that only large hotels can tackle, and then for other reasons that have nothing to do with keeping the door properly closed.

If we compare the work of disciplining the groom with the work he substitutes for, according to the list defined above, we see that this delegated character has the opposite effect to that of the hinge: a simple task—forcing people to close the door—is now performed at an incredible cost; the minimum effect is obtained with maximum spending and discipline. We also notice, when drawing the two lists, an interesting difference: in the first relationship (hinges vis-à-vis the work of many people), you not only had a reversal of forces (the lever allows gentle manipulations to displace heavy weights) but also a modification of *time schedule*: once the hinges are in place, nothing more has to be done apart from maintenance (oiling them from time to time). In the second set of relations (groom's work versus many people's work), not only do you fail to reverse the forces but you also fail to modify the time schedule: nothing can be done to prevent the groom who has been reliable for two months from failing on the sixty-second day; at this point it is not maintenance work that has to be done but the *same* work as on the first day—apart from the few habits that you might have been able to *incorporate* into his body. Although they appear to be two similar delegations, the first one is concentrated at the time of installation, whereas the other is continuous; more exactly, the first one creates clear-cut distinctions between production, installation, and maintenance, whereas in the other the distinction between training and keeping in operation is either fuzzy or nil. The first one evokes the past perfect ("once hinges had been installed..."), the second the present tense ("when the groom is at his post..."). There is a built-in inertia in the first that is largely lacking in the second. The first one is Newtonian, the second Aristotelian (which is simply a way of repeating that the first is nonhuman and the other human). A profound temporal shift takes place when nonhumans are appealed to; time is *folded*.

Delegation to Nonhumans

It is at this point that you have a relatively new choice: either to discipline the people or to *substitute* for the unreliable humans a *delegated nonhuman character* whose only function is to open and close the door. This is called a door-closer or a groom ("groom" is a French trademark that is now part of the common language). The advantage is that you now have to discipline only one nonhuman and may safely leave the others (bellboys included) to their erratic behavior. No matter who they are and where they come from—polite or rude, quick or slow, friends or foes—the nonhuman groom will always take care of the door in any weather and at any time of the day. A nonhuman (hinges) plus another nonhuman (groom) have solved the wall-hole dilemma.

Solved? Well, not quite. Here comes the deskilling question so dear to social historians of technology: thousands of human grooms have been put on the dole by their nonhuman brethren. Have they been replaced? This depends on the kind of action that has been translated or delegated to them. In other words, when humans are displaced and deskilled, nonhumans have to be upgraded and reskilled. This is not an easy task, as we shall now see.

We have all experienced having a door with a powerful spring mechanism slam in our faces. For sure, springs do the job of replacing grooms, but they play the role of a very rude, uneducated, and dumb porter who obviously prefers the wall version of the door to its hole version. They simply slam the door shut. The interesting thing with such impolite doors is this: if they slam shut so violently, it means that you, the visitor, have to be very quick in passing through and that you should not be at someone else's heels, otherwise your nose will get shorter and bloody. An unskilled nonhuman groom thus presupposes a skilled human user. It is always a trade-off. I will call, after Madeleine Akrich's paper (Akrich 1992), the behavior imposed back onto the human by nonhuman delegates *prescription*.[8] Prescription is the moral and ethical dimension of mechanisms. In spite of the constant weeping of moralists, no human is as relentlessly moral as a machine, especially if it is (she is, he is, they are) as "user friendly" as my Macintosh computer. We have been able to delegate to nonhumans not only force as we have known it for centuries but also values, duties, and ethics. It is because of this morality that we, humans, behave so ethically, no matter how weak and wicked we feel we are. The sum of morality does not only remain stable but increases enormously with the population of nonhumans. It is at this time, funnily enough, that moralists who focus on isolated socialized humans despair of us—us meaning of course humans and their retinue of nonhumans.

How can the prescriptions encoded in the mechanism be brought out in words? By replacing them by strings of sentences (often in the imperative) that are uttered (silently and continuously) by the mechanisms for the benefit of those who are mechanized: do this, do that, behave this way, don't go that way, you may do so, be allowed to go there. Such sentences look very much like a programming language. This substitution of words for silence can be made in the analyst's thought experiments, but also by instruction booklets, or explicitly, in any training session, through the voice of a demonstrator or instructor or teacher. The military are especially good at shouting them out through the mouthpiece of human instructors who delegate back to themselves the task of explaining, in the rifle's name, the characteristics of the rifle's ideal user. Another way of hearing what the machines silently did and said are the accidents. When the space shuttle exploded, thousands of pages of transcripts suddenly covered every detail of the silent machine, and hundreds of inspectors, members of congress, and engineers retrieved from NASA dozens of thousands of pages of drafts and orders. This description of a machine—whatever the means—retraces the steps made by the engineers to transform texts, drafts, and projects into things. The impression given to those who are obsessed by human behavior that there is a missing mass of morality is due to the fact that they do not follow this path that leads from text to things and from things to texts. They draw a strong distinction between these two worlds, whereas the job of engineers, instructors, project managers, and analysts is to continually cross this divide. Parts of a program of action may be delegated to a human, or to a nonhuman.

The results of such *distribution of competences*[9] between humans and nonhumans is that competent members of La Halle aux Cuirs will safely pass through the slamming door at a good distance from one another while visitors, unaware of the local cultural condition, will crowd through the door and get bloody noses. The nonhumans take over the selective attitudes of those who engineered them. To avoid this discrimination, inventors get back to their drawing board and try to imagine a nonhuman character that will not *prescribe* the same rare local cultural skills to its human users. A weak spring might appear to be a good solution. Such is not the case, because it would substitute for another type of very unskilled and undecided groom who is never sure about the door's (or his own) status: is it a hole or a wall? Am I a closer or an opener? If it is both at once, you can forget about the heat. In computer parlance, a door is an exclusive OR, not an AND gate.

I am a great fan of hinges, but I must confess that I admire hydraulic door closers much more, especially the old heavy copper-plated one that slowly closed the main door of our house in Aloxe-Corton. I am enchanted by the addition to the spring of a hydraulic piston, which easily draws up the energy of those who open the door, retains it, and then gives it back slowly with a subtle type of implacable firmness that one could expect from a well-trained butler. Especially clever is its way of extracting energy from each unwilling, unwitting passerby. My sociologist friends at the School of Mines call such a clever extraction an "obligatory passage point," which is a very fitting name for a door. No matter what you feel, think, or do, you have to leave a bit of your energy, literally, at the door. This is as clever as a toll booth.[10]

This does not quite solve all of the problems, though. To be sure, the hydraulic door closer does not bang the noses of those unaware of local conditions, so its prescriptions may be said to be less restrictive, but it still leaves aside segments of human populations: neither my little nephews nor my grandmother could get in unaided because our groom needed the force of an able-bodied person to accumulate enough energy to close the door later. To use Langdon Winner's classic motto (1980): Because of their prescriptions, these doors *discriminate* against very little and very old persons. Also, if there is no way to keep them open for good, they discriminate against furniture removers and in general everyone with packages, which usually means, in our late capitalist society, working- or lower-middle-class employees. (Who, even among those from higher strata, has not been cornered by an automated butler when they had their hands full of packages?)

There are solutions, though: the groom's delegation may be written off (usually by blocking its arm) or, more prosaically, its delegated action may be opposed by a foot (salesmen are said to be expert at this). The foot may in turn be delegated to a carpet or anything that keeps the butler in check (although I am always amazed by the number of objects that *fail* this trial of force and I have very often seen the door I just wedged open politely closing when I turned my back to it).

Anthropomorphism

As a technologist, I could claim that provided you put aside the work of installing the groom and maintaining it, and agree to ignore the few sectors of the population that are discriminated against, the hydraulic groom does its job well, closing the door behind you, firmly and slowly. It shows in its humble way how three rows of delegated nonhuman actants[11] (hinges, springs, and hydraulic pistons) replace, 90 percent of the time, either an undisciplined bellboy who is never there when needed or, for the general public, the program instructions that have to do with remembering-to-close-the-door-when-it-is-cold.

The hinge plus the groom is the technologist's dream of efficient action, at least until the sad day when I saw the note posted on La Villette's door with which I started this meditation: "The groom is on strike." So not only have we been able to delegate the act of closing the door from the human to the nonhuman, we have also been able to delegate the human lack of discipline (and maybe the union that goes with it). On strike...[12] Fancy that! Nonhumans stopping work and claiming what? Pension payments? Time off? Landscaped offices? Yet it is no use being indignant, because it is very true that nonhumans are not so reliable that the irreversibility we would like to grant them is always complete. We did not want ever to have to think about this door again—apart from regularly scheduled routine maintenance (which is another way of saying that we did not have to bother about it)—and here we are, worrying again about how to keep the door closed and drafts outside.

What is interesting in this note is the humor of attributing a human characteristic to a failure that is usually considered "purely technical." This humor, however, is more profound than in the notice they could have posted: "The groom is not working." I constantly talk with my computer, who answers back; I am sure you swear at your old car; we are constantly granting mysterious faculties to gremlins inside every conceivable home appliance, not to mention cracks in the concrete belt of our nuclear plants. Yet, this behavior is considered by sociologists as a scandalous breach of natural barriers. When you write that a groom is "on strike," this is only seen as a "projection," as they say, of a human behavior onto a nonhuman, cold, technical object, one by nature impervious to any feeling. This is *anthropomorphism*, which for them is a sin akin to zoophily but much worse.

It is this sort of moralizing that is so irritating for technologists, because the automatic groom is already anthropomorphic through and through. It is well known that the French like etymology; well, here is another one: *anthropos* and *morphos* together mean either that which *has* human shape or that which *gives shape* to humans. The groom is indeed anthropomorphic, in three senses: first, it has been made by humans; second, it substitutes for the actions of people and is a delegate that permanently occupies the position of a human; and third, it shapes human action by prescribing back what sort of people should pass through the door. And yet some would forbid us to ascribe feelings to this thoroughly anthropomorphic creature, to delegate labor relations, to "project"—that is, to translate—*other* human properties to the groom. What of those many other innovations that have endowed much more sophisticated doors with the ability to see you arrive in advance (electronic eyes), to ask for your identity (electronic passes), or to slam shut in case of danger? But anyway, who are sociologists to decide the real and final shape (*morphos*) of humans (*anthropos*)? To trace with confidence the boundary between what is a "real" delegation and what is a "mere" projection? To sort out forever and without due inquiry the three different kinds of anthropomorphism I listed above? Are we not shaped by nonhuman grooms, although I admit only a very little bit? Are they not our brethren? Do they not deserve consideration? With your self-serving and self-righteous social studies of technology, you always plead against machines and for deskilled workers—are you aware of *your* discriminatory biases? You discriminate between the human and the inhuman. I do not hold this bias (this one at least) and see only actors—some human, some nonhuman, some skilled, some unskilled—that exchange their properties. So the note posted on the door is accurate; it gives with humor an exact rendering of the groom's behavior: it is not working, it is on strike (notice, that the word "strike" is a rationalization carried from the nonhuman repertoire to the human one, which proves again that the divide is untenable).

Built-in Users and Authors

The debates around anthropomorphism arise because we believe that there exist "humans" and "nonhumans," without realizing that this attribution of roles and action is also a *choice*.[13] The best way to understand this choice is to compare machines with texts, since the inscription of builders and users in a mechanism is very much the same as that of authors and readers in a story. In order to exemplify this point I have now to confess that I am *not* a technologist. I built in my article a made-up author, and I also invented possible readers whose reactions and beliefs I anticipated. Since the beginning I have many times used the "you" and even "you sociologists." I even asked you to draw up a table, and I also asked your permission to go on with the story. In doing so, I built up an inscribed reader to whom I prescribed qualities and behavior, as surely as a traffic light or a painting prepare a position for those looking at them. Did you *underwrite* or *subscribe* this definition of yourself? Or worse, is there any one at all to read this text and occupy the position prepared for the reader? This question is a source of constant difficulties for those who are unaware of the basics of semiotics or of technology. *Nothing in a given scene* can prevent the inscribed user or reader from behaving differently from what was expected (nothing, that is, until the next paragraph). The reader in the flesh may totally ignore my definition of him or her. The user of the traffic light may well cross on the red. Even visitors to La Halle aux Cuirs may never show up because it is too complicated to find the place, *in spite* of the fact that their behavior and trajectory have been perfectly anticipated by the groom. As for the computer user input, the cursor might flash forever without the user being there or knowing what to do. There might be an enormous gap between the prescribed user and the user-in-the-flesh, a difference as big as the one between the "I" of a novel and the novelist.[14] It is exactly this difference that upset the authors of the anonymous appeal on which I comment. On other occasions, however, the gap between the two may be nil: the prescribed user is so well anticipated, so carefully nested inside the scenes, so exactly dovetailed, that it does what is expected.[15]

The problem with scenes is that they are usually well prepared for anticipating users or readers who are at close quarters. For instance, the groom is quite good in its anticipation that people will push the door open and give it the energy to reclose it. It is very bad at doing anything to help people arrive there. After fifty centimeters, it is helpless and cannot act, for example, on the maps spread around La Villette to explain where La Halle aux Cuirs is. Still, no scene is prepared without a preconceived idea of what sort of actors will come to occupy the prescribed positions.

This is why I said that although *you* were free not to go on with this paper, *you* were only "relatively" so. Why? Because I know that, because you bought this book, you are hard-working, serious, English-speaking technologists or readers committed to understanding new development in the social studies of machines. So my injunction to "read the paper, you sociologist" is not very risky (but I would have taken no chance with a French audience, especially with a paper written in English). This way of counting on earlier distribution of skills to help narrow the gap between built-in users or readers and users- or readers-in-the-flesh is like a *pre*-inscription.[16]

The fascinating thing in text as well as in artifact is that they have to thoroughly organize the relation between what is inscribed in them and what can/could/should be pre-inscribed in the users. Each setup is surrounded by various arenas interrupted by different types of walls. A text, for instance, is clearly *circumscribed*[17]—the dust cover, the title page, the hard back—but so is a computer—the plugs, the screen, the disk drive, the user's input. What is nicely called "interface" allows any setup to be connected to another through so many carefully designed entry points. Sophisticated mechanisms build up a whole gradient of concentric circles around themselves. For instance, in most modern photocopy machines there are troubles that even rather incompetent users may solve themselves like "ADD PAPER;" but then there are trickier ones that require a bit of explanation: "ADD TONER. SEE MANUAL, PAGE 30." This instruction might be backed up by homemade labels: "DON'T ADD THE TONER YOURSELF, CALL THE SECRETARY," which limit still further the number of people able to troubleshoot. But then other more serious crises are addressed by labels like "CALL THE TECHNICAL STAFF AT THIS NUMBER," while there are parts of the machine that are sealed off entirely with red labels such as "DO NOT OPEN—DANGER, HIGH VOLTAGE, HEAT" or "CALL THE POLICE." Each of these messages addresses a different audience, from the widest (everyone with the rather largely disseminated competence of using photocopying machines) to the narrowest (the rare bird able to troubleshoot and who, of course, is never there).[18] Circumscription only defines how a setup itself has built-in plugs and interfaces; as the name indicates, this tracing of circles, walls, and entry points inside the text or the machine does not prove that readers and users will obey. There is nothing sadder that an obsolete computer with all its nice interfaces, but no one on earth to plug them in.

Drawing a side conclusion in passing, we can call *sociologism* the claim that, given the competence, pre-inscription, and circumscription of human users and authors, you can read out the scripts nonhuman actors have to play; and *technologism* the symmetric claim that, given the competence and pre-inscription of nonhuman actors, you can easily read out and deduce the behavior prescribed to authors and users. From now on, these two absurdities will, I hope, disappear from the scene, because the actors at any point may be human or nonhuman, and the displacement (or translation, or transcription) makes impossible the easy reading out of one repertoire and into the next. The bizarre idea that society might be made up of human relations is a mirror image of the other no less bizarre idea that techniques might be made up of nonhuman relations. We deal with characters, delegates, representatives, lieutenants (from the French "lieu" plus "tenant," i.e., holding the place of, for, someone else)—some figurative, others nonfigurative; some human, others nonhuman; some competent, others incompetent. Do you want to cut through this rich diversity of delegates and artificially create two heaps of refuse, "society" on one side and "technology" on the other? That is your privilege, but I have a less bungled task in mind.

A scene, a text, an automatism can do a lot of things to their prescribed users at the range—close or far—that is defined by the circumscription, but most of the effect finally ascribed[19] to them depends on lines of other setups being aligned. For instance, the groom closes the door only if there are people reaching the Centre d'Histoire des Sciences; these people arrive in front of the door only if they have found maps (another delegate, with the built-in prescription I like most: "*you* are here" circled in red on the map) and only if there are roads leading under the Paris ring road to the Halle (which is a condition not always fullfilled); and of course people will start bothering about reading the maps, getting their feet muddy and pushing the door open only if they are convinced that the group is worth visiting (this is about the only condition in La Villette that is fulfilled). This gradient of aligned setups that endow actors with the pre-inscribed competences to find its users is very much like Waddington's "chreod":[20] people effortlessly flow through the door of La Halle aux Cuirs and the groom, hundreds of times a day, recloses the door—when it is not stuck. The result of such an alignment of setups[21] is to decrease the number of occasions in which words are used; most of the actions are silent, familiar, incorporated (in human or in nonhuman bodies)—making the analyst's job so much harder. Even the classic debates about freedom, determination, predetermination, brute force, or efficient will—debates that are the twelfth-century version of seventeenth-century discussions on grace—will be slowly eroded. (Because *you* have reached this point, it means I was right in saying that you were not at all free to stop reading the paper: positioning myself cleverly along a chreod, and adding a few other tricks of my own, I led you *here*...or did I? Maybe you skipped most of it, maybe you did not understand a word of it, o you, undisciplined readers.)

Figurative and Nonfigurative Characters

Most sociologists are violently upset by this crossing of the sacred barrier that separate human from nonhumans, because they confuse this divide with another one between *figurative* and *nonfigurative* actors. If I say that Hamlet is the figuration of "depression among the aristocratic class," I move from a personal figure to a less personal one—that is, class. If I say that Hamlet stands for doom and gloom, I use less figurative entities, and if I claim that he represents western civilization, I use nonfigurative abstractions. Still, they all are equally actors, that is, entities that *do* things, either in Shakespeare's artful plays or in the commentators' more tedious tomes. The choice of granting actors figurativity or not is left entirely to the authors. It is exactly the same for techniques. Engineers are the authors of these subtle plots and scenarios of dozens of delegated and interlocking characters so few people know how to appreciate. The label "inhuman" applied to techniques simply overlooks translation mechanisms and the many choices that exist for figuring or defiguring, personifying or abstracting, embodying or disembodying actors. When we say that they are "mere automatisms," we project as much as when we say that they are "loving creatures;" the only difference is that the latter is an anthropomorphism and the former a technomorphism or phusimorphism.

For instance, a meat roaster in the Hôtel-Dieu de Beaune, the little groom called "le Petit Bertrand," is the delegated author of the movement (figure 10.2). This little man is as famous in Beaune as is the Mannekenpis in Brussels. Of course, he is not the one who does the turning—a hidden heavy stone collects the force applied when the human demonstrator or the cook turn a heavy handle that winds up a cord around a drum equipped with a ratchet. Obviously "le Petit Bertrand" believes he is the one doing the job because he not only smiles but also moves his head from side to side with obvious pride while turning his little handle. When we were kids, even though we had seen our father wind up the machine and put away the big handle, we liked to believe that the little guy was moving the spit. The irony of the "Petit Bertrand" is that, although the delegation to mechanisms aims at rendering any human turnspit useless, the mechanism is ornamented with a constantly exploited character "working" all day long.

Although this turnspit story offers the opposite case from that of the door closer in terms of figuration (the groom on the door does not look like a groom but really does the same job, whereas "le Petit Bertrand" does look like a groom but is entirely

Figure 10.2
Le Petit Bertrand is a mechanical meat roaster from the sixteenth century that ornaments the kitchen of the Hôtel-Dieu de Beaune, the hospital where the author was born. The big handle (bottom right) is the one that allows the humans to wind up the mechanism; the small handle (top right) is made to allow a little nonhuman anthropomorphic character to move the whole spit. Although the movement is prescribed back by the mechanism, since the Petit Bertrand smiles and turns his head from left to right, it is believed that it is at the origin of the force. This secondary mechanism—to whom is ascribed the origin of the force—is unrelated to the primary mechanism, which gathers a large-scale human, a handle, a stone, a crank, and a brake to regulate the movement.

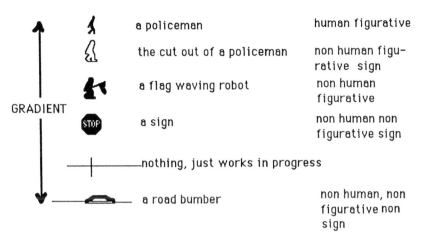

Figure 10.3
Students of technology are wary of anthropomorphism that they see as a projection of human characters to mere mechanisms, but mechanisms to another "morphism," a non-figurative one that can also be applied to humans. The difference between "action" and "behavior" is not a primary, natural one.

passive), they are similar in terms of delegation (you no longer need to close the door, and the cook no longer has to turn the skewer). The "enunciator" (a general word for the author of a text or for the mechanics who devised the spit) is free to place or not a representation of him or herself in the script (texts or machines). "Le Petit Bertrand" is a delegated version of whoever is responsible for the mechanism. This is exactly the same operation as the one in which I pretended that the author of this article was a hardcore technologist (when I really am a mere sociologist—which is a second localization of the text, as wrong as the first because really I am a mere philosopher...). If I say "we the technologists," I propose a picture of the author of the text as surely as if we place "le Petit Bertrand" as the originator of the scene. But it would have been perfectly possible for me and for the mechanics to position *no figurated character* at all as the author *in* the scripts *of* our scripts (in semiotic parlance there would be no *narrator*). I would just have had to say things like "recent developments in sociology of technology have shown that..." instead of "I," and the mechanics would simply have had to take out "le Petit Bertrand," leaving the beautiful cranks, teeth, ratchets, and wheels to work alone. The point is that removing the "Petit Bertrand" does not turn the mechanism into a "mere mechanism" where no actors are acting. It is just a different choice of style.

The distinctions between humans and nonhumans, embodied or disembodied skills, impersonation or "machination," are less interesting than the complete chain along which competences and actions are distributed. For instance, on the freeway the other day I slowed down because a guy in a yellow suit and red helmet was waving

Table 10.1
The distinction between words and things is impossible to make for technology because it is the gradient allowing engineers to shift down—from words to things—or to shift up—from things to signs—that enables them to enforce their programs of actions

	Figurative	Non-figurative
Human	"I"	"Science shows that" …
Non-human	"le Petit Bertrand"	a door-closer

a red flag. Well, the guy's moves were so regular and he was located so dangerously and had such a pale though smiling face that, when I passed by, I recognized it to be a machine (it failed the Turing test, a cognitivist would say). Not only was the red flag delegated; not only was the arm waving the flag also delegated; but the body appearance was also added to the machine. We road engineers (see? I can do it again and carve out another author) could move much further in the direction of figuration, although at a cost: we could have given him electronic eyes to wave only when a car approaches, or have regulated the movement so that it is faster when cars do not obey. We could also have added (why not?) a furious stare or a recognizable face like a mask of Mrs. Thatcher or President Mitterand—which would have certainly slowed drivers very efficiently.²² But we could also have moved the other way, to a *less* figurative delegation: the flag by itself could have done the job. And why a flag? Why not simply a sign "work in progress?" And why a sign at all? Drivers, if they are circumspect, disciplined, and watchful will see for themselves that there is work in progress and will slow down. But there is another radical, nonfigurative solution: the road bumper, or a speed trap that we call in French "un gendarme couché," a laid policeman. It is impossible for us not to slow down, or else we break our suspension. Depending on where we stand along this chain of delegation, we get classic moral human beings endowed with self-respect and able to speak and obey laws, or we get stubborn and efficient machines and mechanisms; halfway through we get the usual power of signs and symbols. It is the complete chain that makes up the missing masses, not either of its extremities. The paradox of technology is that it is thought to be at one of the extremes, whereas it is the ability of the engineer to travel easily along the whole gradient and substitute one type of delegation for another that is inherent to the job.²³

From Nonhumans to Superhumans

The most interesting (and saddest) lesson of the note posted on the door at La Villette is that people are not circumspect, disciplined, and watchful, especially not French drivers doing 180 kilometers an hour on a freeway on a rainy Sunday morning when the speed limit is 130 (I inscribe the legal limit in this article because this is about the only place where you could see it printed in black and white; no one else seems to bother, except the mourning families). Well, that is exactly the point of the note:

"The groom is on strike, *for God's sake*, keep the door closed." In our societies there are two systems of appeal: nonhuman and superhuman—that is, machines and gods. This note indicates how desperate its anonymous frozen authors were (I have never been able to trace and honor them as they deserved). They first relied on the inner morality and common sense of humans; this failed, the door was always left open. Then they appealed to what we technologists consider the supreme court of appeal, that is, to a nonhuman who regularly and conveniently does the job in place of unfaithful humans; to our shame, we must confess that it also failed after a while, the door was again left open. How poignant their line of thought! They moved up and backward to the oldest and firmest court of appeal there is, there was, and ever will be. If humans and nonhumans have failed, certainly God will not deceive them. I am ashamed to say that when I crossed the hallway this February day, the door *was* open. Do not accuse God, though, because the note did not make a direct appeal; God is not accessible without mediators—the anonymous authors knew their catechisms well—so instead of asking for a direct miracle (God holding the door firmly closed or doing so through the mediation of an angel, as has happened on several occasions, for instance when Saint Peter was delivered from his prison) they appealed to the respect for God in human hearts. This was their mistake. In our secular times, this is no longer enough.

Nothing seems to do the job nowadays of disciplining men and women to close doors in cold weather. It is a similar despair that pushed the road engineer to add a golem to the red flag to force drivers to beware—although the only way to slow French drivers is still a good traffic jam. You seem to need more and more of these figurated delegates, aligned in rows. It is the same with delegates as with drugs; you start with soft ones and end up shooting up. There is an inflation for delegated characters, too. After a while they weaken. In the old days it might have been enough just to have a door for people to know how to close it. But then, the embodied skills somehow disappeared; people had to be reminded of their training. Still, the simple inscription "keep the door closed" might have been sufficient in the good old days. But you know people, they no longer pay attention to the notice and need to be reminded by stronger devices. It is then that you install automatic grooms, since electric shocks are not as acceptable for people as for cows. In the old times, when quality was still good, it might have been enough just to oil it from time to time, but nowadays even automatisms go on strike.

It is not, however, that the movement is always from softer to harder devices, that is, from an autonomous body of knowledge to force through the intermediary situation of worded injunctions, as the La Villette door would suggest. It goes also the other way. It is true that in Paris no driver will respect a sign (for instance, a white or yellow line forbidding parking), nor even a sidewalk (that is a yellow line plus a fifteen centimeter curb); so instead of embodying in the Parisian consciousness an *intrasomatic* skill, authorities prefer to align yet a third delegate (heavy blocks shaped like truncated pyramids and spaced in such a way that cars cannot sneak through); given the results, only a complete two-meter high continuous Great Wall could do the job, and even this might not make the sidewalk safe, given the very poor sealing efficiency of

China's Great Wall. So the deskilling thesis appears to be the general case: always go from intrasomatic to *extrasomatic* skills; never rely on undisciplined people, but always on safe, delegated nonhumans. This is far from being the case, even for Parisian drivers. For instance, red lights are usually respected, at least when they are sophisticated enough to integrate traffic flows through sensors; the delegated policeman standing there day and night is respected even though it has no whistles, gloved hands, and body to *enforce* this respect. Imagined collisions with other cars or with the absent police are enough to keep the drivers in check. The thought experiment "what would happen if the delegated character was not there" is the same as the one I recommended above to size up its function. The same *incorporation* from written injunction to body skills is at work with car manuals. No one, I guess, casts more than a cursory glance at the manual before starting the engine of an unfamiliar car. There is a large *body* of skills that we have so well embodied or incorporated that the mediations of the written instructions are useless.[24] From extrasomatic, they have become intrasomatic. Incorporation in human or "excorporation" in nonhuman bodies is also one of the choices left to the designers.

The only way to follow engineers at work is not to look for extra- or intrasomatic delegation, but only at their work of *re-inscription*.[25] The beauty of artifacts is that they take on themselves the contradictory wishes or needs of humans and nonhumans. My seat belt is supposed to strap me in firmly in case of accident and thus impose on me the respect of the advice DON'T CRASH THROUGH THE WINDSHIELD, which is itself the translation of the unreachable goal DON'T DRIVE TOO FAST into another less difficult (because it is a more selfish) goal: IF YOU DO DRIVE TOO FAST, AT LEAST DON'T KILL YOURSELF. But accidents are rare, and most of the time the seat belt should not tie me firmly. I need to be able to switch gears or tune my radio. The car seat belt is not like the airplane seat belt buckled only for landing and takeoff and carefully checked by the flight attendants. But if auto engineers invent a seat belt that is completely elastic, it will not be of any use in case of accident. This first contradiction (be firm and be lax) is made more difficult by a second contradiction (you should be able to buckle the belt very fast—if not, no one will wear it—but also unbuckle it very fast, to get out of your crashed car). Who is going to take on all of these contradictory specifications? The seat belt mechanism—if there is no other way to go, for instance, by directly limiting the speed of the engine, or having roads so bad that no one can drive fast on them. The safety engineers have to re-inscribe in the seat belt all of these contradictory usages. They pay a price, of course: the mechanism is *folded* again, rendering it more complicated. The airplane seat belt is childish by comparison with an automobile seat belt. If you study a complicated mechanism without seeing that it reinscribes contradictory specifications, you offer a dull description, but every piece of an artifact becomes fascinating when you see that every wheel and crank is the possible answer to an objection. The program of action is in practice the answer to an *antiprogram* against which the mechanism braces itself. Looking at the mechanism alone is like watching half the court during a tennis game; it appears as so many meaningless moves. What analysts of artifacts have to do is similar to what we all did when studying scientific texts: we added the other half of the court.²⁶ The scientific literature looked dull, but when the agonistic field to which it reacts was brought back in, it became as interesting as an opera. The same with seat belts, road bumpers, and grooms.

Texts and Machines

Even if it is now obvious that the missing masses of our society are to be found among the nonhuman mechanisms, it is not clear how they get there and why they are missing from most accounts. This is where the comparison between texts and artifacts that I used so far becomes misleading. There is a crucial distinction between stories and machines, between narrative programs and programs of action, a distinction that explains why machines are so hard to retrieve in our common language. In storytelling, one calls *shifting out* any displacement of a character to another space time, or character. If I tell you "Pasteur entered the Sorbonne amphitheater," I translate the present setting—you and me—and shift it to another space (middle of Paris), another time (mid-nineteenth century), and to other characters (Pasteur and his audience). "I" the enunciator may decide to appear, disappear, or be represented by a narrator who tells the story ("that day, I was sitting on the upper row of the room"); "I" may also decide to position you and any reader inside the story ("had you been there, you would have been convinced by Pasteur's experiments"). There is no limit to the number of shiftings out with which a story may be built. For instance, "I" may well stage a dialogue inside the amphitheater between two characters who are telling a story about what happened at the Académie des Sciences between, say, Pouchet and Milnes-Edwards. In that case, the room becomes the place *from which* narrators shift out to tell a story about the Academy, and they may or not shift *back in* the amphitheater to resume the first story about Pasteur. "I" may also *shift in* the entire series of nested stories to close mine and come back to the situation I started from—you and me. All these displacements are well known in literature departments (Latour 1988) and make up the craft of talented writers.

No matter how clever and crafty are our novelists, they are no match for engineers. Engineers constantly shift out characters in other spaces and other times, devise positions for human and nonhuman users, break down competences that they then redistribute to many different actors, and build complicated narrative programs and subprograms that are evaluated and judged by their ability to stave off antiprograms. Unfortunately, there are many more literary critics than technologists, and the subtle beauties of technosocial imbroglios escape the attention of the literate public. One of the reasons for this lack of concern may be the peculiar nature of the shifting-out that generates machines and devices. Instead of sending the listener of a story into another world, the technical shifting-out inscribes the words into *another matter*. Instead of allowing the reader of the story to be *at the same time* away (in the story's frame of reference) and here (in an armchair), the technical shifting-out forces the reader to choose *between* frames of reference. Instead of allowing enunciators and enunciatees a sort of simultaneous presence and communion to other actors, techniques allow both to *ignore* the delegated actors and walk away without even feeling their presence. This is the profound meaning of Butler's sentence I placed at the beginning of this chapter: machines are not talking actors, not because they are unable to do so, but because they might have chosen to remain silent to become agreeable to their fellow machines and fellow humans.

To understand this difference in the two directions of shifting out, let us venture once more onto a French freeway; for the umpteenth time I have screamed at my son Robinson, "Don't sit in the middle of the rear seat; if I brake too hard, you're dead." In an auto shop further along the freeway I come across a device *made for* tired-and-angry-parents-driving-cars-with-kids-between-two-and-five (too old for a baby seat and not old enough for a seat belt) and-from-small-families (without other persons to hold them safely) with-cars-with-two-separated-front-seats-and-head-rests. It is a small market, but nicely analyzed by the German manufacturers and, given the price, it surely pays off handsomely. This description of myself and the small category into which I am happy to belong is transcribed in the device—a steel bar with strong attachments connecting the head rests—and in the advertisement on the outside of the box; it is also pre-inscribed in about the only place where I could have realized that I needed it, the freeway. (To be honest and give credit where credit is due, I must say that Antoine Hennion has a similar device in his car, which I had seen the day before, so I really looked for it in the store instead of "coming across" it as I wrongly said; which means that a) there is some truth in studies of dissemination by imitation; b) if I describe this episode in as much detail as the door I will never have been able to talk about the work done by the historians of technology at La Villette.) Making a short story already too long, I no longer scream at Robinson, and I no longer try to foolishly stop him with my extended right arm: he firmly holds the bar that protects him against my braking. I have delegated the continuous injunction of my voice and extension of my right arm (with diminishing results, as we know from Feschner's law) to a reinforced, padded, steel bar. Of course, I had to make two detours: one to my wallet, the second to my tool box; 200 francs and five minutes later I had fixed the device (after making sense of the instructions encoded with Japanese ideograms).

We may be able to follow these detours that are characteristic of the technical form of delegation by adapting a linguistic tool. Linguists differentiate the *syntagmatic* dimension of a sentence from the *paradigmatic* aspect. The syntagmatic dimension is the possibility of *associating* more and more words in a grammatically correct sentence: for instance, going from "the barber" to "the barber goes fishing" to the "barber goes fishing with his friend the plumber" is what linguists call moving through the syntagmatic dimension. The number of elements tied together increases, and nevertheless the sentence is still meaningful. The paradigmatic dimension is the possibility, in a sentence of a given length, of *substituting* a word for another while still maintaining a grammatically correct sentence. Thus, going from "the barber goes fishing" to the "plumber goes fishing" to "the butcher goes fishing" is tantamount to moving through the paradigmatic dimension.[27]

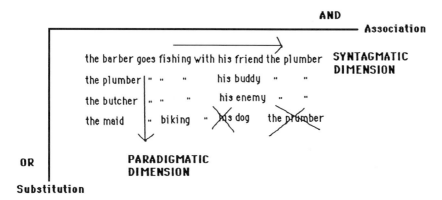

Figure 10.4
Linguists define meaning as the intersection of a horizontal line of association—the syntagm—and a vertical line of substitution—the paradigm. The touchstone in linguistics is the decision made by the competent speaker that a substitution (OR) or an association (AND) is grammatically correct in the language under consideration. For instance, the last sentence is incorrect.

Linguists claim that these two dimensions allow them to describe the system of any language. Of course, for the analysis of artifacts we do not have a structure, and the definition of a grammatically correct expression is meaningless. But if, by substitution, we mean the technical shifting to another *matter*, then the two dimensions become a powerful means of describing the dynamic of an artifact. The syntagmatic dimension becomes the AND dimension (how many elements are tied together), and the paradigmatic dimension becomes the OR dimension (how many translations are necessary in order to move through the AND dimension). I could not tie Robinson to the order, but through a detour and a translation I now hold together my will and my son.

The detour, plus the translation of words and extended arm into steel, is a shifting out to be sure, but not of the same type as that of a story. The steel bar has now taken over my competence as far as keeping my son at arm's length is concerned. From speech and words and flesh it has become steel and silence and extrasomatic. Whereas a narrative program, no matter how complicated, always remains a text, the program of action substitutes part of its character to other nontextual elements. This divide between text and technology is at the heart of the myth of Frankenstein (Latour 1992). When Victor's monster escapes the laboratory in Shelley's novel, is it a metaphor of fictional characters that seem to take up a life of their own? Or is it the metaphor of technical characters that do take up a life of their own because they cease to be texts and become flesh, legs, arms, and movements? The first version is not very interesting because in spite of the novelist's cliché, a semiotic character in a text always needs the reader to offer it an "independent" life. The second version is not very interesting

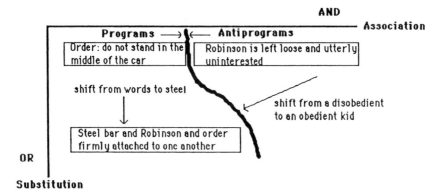

Figure 10.5
The translation diagram allows one to map out the story of a script by following the two dimensions: AND, the association (the latitude, so to speak), and OR, the substitution (the longitude). The plot is defined by the line that separates the programs of action chosen for the analysis and the antiprograms. The point of the story is that it is impossible to move in the AND direction without paying the price of the OR dimension, that is renegotiating the sociotechnical assemblage.

either, because the "autonomous" thrust of a technical artifact is a worn-out commonplace made up by bleeding-heart moralists who have never noticed the throngs of humans necessary to keep a machine alive. No, the beauty of Shelley's myth is that we cannot choose between the two versions: parts of the narrative program are still texts, others are bits of flesh and steel—and this mixture is indeed a rather curious monster.

To bring this chapter to a close and differentiate once again between texts and artifacts, I will take as my final example not a flamboyant Romantic monster but a queer little surrealist one: the Berliner key:[28]

Yes, this is a key and not a surrealist joke (although this is *not* a key, because it is a picture and a text about a key). The program of action in Berlin is almost as desperate a plea as in La Villette, but instead of begging CLOSE THE DOOR BEHIND YOU PLEASE it is slightly more ambitious and *orders*: RELOCK THE DOOR BEHIND YOU. Of course the preinscription is much narrower: only people endowed with the competence of living in the house can use the door; visitors should ring the doorbell. But even with such a limited group the antiprogram in Berlin is the same as everywhere: undisciplined tenants forget to lock the door behind them. How can you force them to lock it? A normal key[29] endows you with the *competence* of opening the door—it proves you are *persona grata*—but nothing in it entails the *performance* of actually using the key again once you have opened the door and closed it behind you. Should you put up a sign? We know that signs are never forceful enough to catch people's attention for long. Assign a police officer to every doorstep? You could do this in East Berlin, but not in reunited

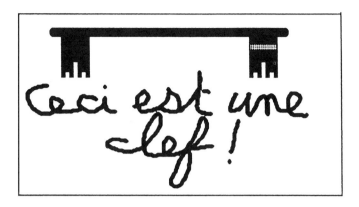
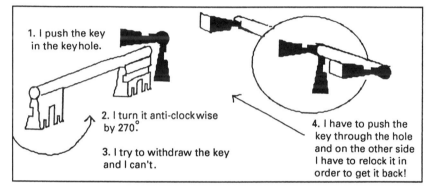
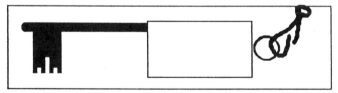

Figure 10.6
The key, its usage, and its holder.

Berlin. Instead, Berliner blacksmiths decided to re-inscribe the program of action in the very shape of the key and its lock—hence this surrealist form. They in effect sunk the contradiction and the lack of discipline of the Berliners in a more "realist" key. The program, once translated, appears innocuous enough: UNLOCK THE DOOR. But here lies the first novelty: it is impossible to remove the key in the normal way; such a move is "proscribed" by the lock. Otherwise you have to break the door, which is hard as well as impolite; the only way to retrieve the key is to push the whole key through the door to the other side—hence its symmetry—but then it is still impossible to retrieve the key. You might give up and leave the key in the lock, but then you lose the competence of the tenant and will never again be able to get in or out. So what do you do? You rotate the key one more turn and, yes, you have in effect relocked the door and then, only then, are you able to retrieve the precious "sesame." This is a clever translation of a possible program relying on morality into a program relying on dire necessity: you might not want to relock the key, but you cannot do otherwise. The distance between morality and force is not as wide as moralists expect; or more exactly, clever engineers have made it smaller. There is a price to pay of course for such a shift away from morality and signs; you have to replace most of the locks in Berlin. The pre-inscription does not stop here, however, because you now have the problem of keys that no decent key holder can stack into place because they have no hole. On the contrary, the new sharp key is going to poke holes in your pockets. So the blacksmiths go back to the drawing board and invent specific key holders adapted to the Berliner key!

The key in itself is not enough to fulfill the program of action. Its effects are very severely circumscribed, because it is only when you have a Berliner endowed with the double competence of being a tenant and knowing how to use the surrealist key that the relocking of the door may be enforced. Even such an outcome is not full proof, because a really bad guy may relock the door without closing it! In that case the worst possible antiprogram is in place because the lock stops the door from closing. Every passerby may see the open door and has simply to push it to enter the house. The setup that prescribed a very narrow segment of the human population of Berlin is now so lax that it does not even discriminate against nonhumans. Even a dog knowing nothing about keys, locks, and blacksmiths is now allowed to enter! No artifact is idiot-proof because any artifact is only a portion of a program of action and of the fight necessary to win against many antiprograms.

Students of technology are never faced with people on the one hand and things on the other, they are faced with programs of action, sections of which are endowed to *parts* of humans, while other sections are entrusted to parts of nonhumans. In practice they are faced with the front line of figure 10.7. This is the only thing they can *observe*: how a negotiation to associate dissident elements requires more and more elements to be tied together and more and more shifts to other matters. We are now witnessing in technology studies the same displacement that has happened in science studies during the last ten years. It is not that society and social relations invade the certainty of science or the efficiency of machines. It is that society itself is to be rethought from

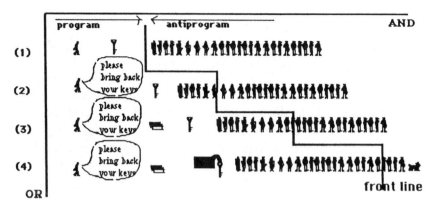

Figure 10.7
The hotel manager successively adds keys, oral notices, written notices, and finally weights; each time he thus modifies the attitude of some part of the "hotel customers" group while he extends the syntagmatic assemblage of elements. From Madeleine Akrich and Bruno Latour, "A Summary of a Convenient Vocabulary for the Semiotics of Human and Nonhuman Assemblies," in Wiebe E. Bijker and John Law, eds., *Shaping Technology/Building Society: Studies in Sociotechnical Change* (Cambridge, Mass.: MIT Press, 1992), p. 263.

top to bottom once we add to it the facts and the artifacts that make up large sections of our social ties. What appears in the place of the two ghosts—society and technology—is not simply a hybrid object, a little bit of efficiency and a little bit of sociologizing, but a *sui generis* object: the collective thing, the trajectory of the front line between programs and anti-programs. It is too full of humans to look like the technology of old, but it is too full of nonhumans to look like the social theory of the past. The missing masses are in our traditional social theories, not in the supposedly cold, efficient, and inhuman technologies.

Notes

This chapter owes to many discussions held at the Centre de Sociologie de l'Innovation, especially with John Law, the honorary member from Keele, and Madeleine Akrich. It is particularly indebted to Françoise Bastide, who was still working on these questions of semiotics of technology a few months before her death.

I had no room to incorporate a lengthy dispute with Harry Collins about this chapter (but see Collins and Yearley 1992, and Callon and Latour, 1992).

Trevor Pinch and John Law kindly corrected the English.

1. The program of action is the set of written instructions that can be substituted by the analyst to any artifact. Now that computers exist, we are able to conceive of a text (a programming language) that is at once words and actions. How to do things with words and then turn words into things is now clear to any programmer. A program of action is thus close

to what Pinch et al. (1992) call "a social technology," except that all techniques may be made to be a program of action....

2. In spite of the crucial work of Diderot and Marx, careful description of techniques is absent from most classic sociologists—apart from the "impact of technology on society" type of study—and is simply black-boxed in too many economists' accounts. Modern writers like Leroi-Gourhan (1964) are not often used. Contemporary work is only beginning to offer us a more balanced account. For a reader, see MacKenzie and Wacjman 1985; for a good overview of recent developments, see Bijker et al. (1987). A remarkable essay on how to describe artifacts—an iron bridge compared to a Picasso portrait—is offered by Baxandall (1985). For a recent essay by a pioneer of the field, see Noble 1984. For a remarkable and hilarious description of a list of artifacts, see Baker 1988.

3. Following Madeleine Akrich's lead (Akrich 1992), we will speak only in terms of *scripts* or scenes or scenarios, or setups as John Law says, played by human or nonhuman actants, which may be either figurative or nonfigurative.

4. After Akrich, I will call the retrieval of the script from the situation *de-scription*. They define actants, endow them with competences, make them do things, and evaluate the sanction of these actions like the *narrative program* of semioticians.

5. Although most of the scripts are in practice silent, either because they are intra- or extrasomatic, the written descriptions are not an artifact of the analyst (technologist, sociologist, or semiotician), because there exist many states of affairs in which they are *explicitly* uttered. The gradient going from intrasomatic to extrasomatic skills through discourse is never fully stabilized and allows many entries revealing the process of translation: user manuals, instruction, demonstration or drilling situations, practical thought experiments ("what would happen if, instead of the red light, a police officer were there"). To this should be added the innovator's workshop, where most of the objects to be devised are still at the stage of *projects* committed to paper ("if we had a device doing this and that, we could then do this and that"); market analysis in which consumers are confronted with the new device; and, naturally, the exotic situation studied by anthropologists in which people faced with a foreign device talk to themselves while trying out various combinations ("what will happen if I attach this lead here to the mains?"). The analyst has to empirically capture these situations to write down the scripts. When none is available, the analyst may still make a thought experiment by comparing prescence/absence tables and collating the list of all the actions taken by actors ("if I take this one away, this and that other action will be modified"). There are dangers in such a counterfactual method, as Collins has pointed out (Collins and Yearley 1992), but it is used here only to outline the semiotics of artifacts. In practice, as Akrich (1992) shows, the scripts are explicit and accountable.

6. We call the translation of any script from one repertoire to a *more durable* one transcription, inscription, or encoding. This definition does *not* imply that the direction always goes from soft bodies to hard machines, but simply that it goes from a provisional, less reliable one to a longer-lasting, more faithful one. For instance, the embodiment in cultural tradition of the user manual of a car is a transcription, but so is the replacement of a police officer by a traffic light; one goes from machines to bodies, whereas the other goes the opposite way. Specialists of robotics have abandoned the pipe dream of total automation; they

learned the hard way that many skills are better delegated to humans than to nonhumans, whereas others may be taken away from incompetent humans.

7. See Authier 1989 on Plutarch's Archimedes.

8. We call prescription whatever a scene presupposes from its *transcribed* actors and authors (this is very much like "role expectation" in sociology, except that it may be inscribed or encoded in the machine). For instance, a Renaissance Italian painting is designed to be viewed from a specific angle of view prescribed by the vanishing lines, exactly like a traffic light expects that its users will watch it from the street and not sideways (French engineers often hide the lights directed toward the side street so as to hide the state of the signals, thus preventing the strong temptation to rush through the crossing at the first hint that the lights are about to be green; this prescription of who is allowed to watch the signal is very frustrating). "User input" in programming language, is another very telling example of this inscription in the automatism of a living character whose behavior is both free and predetermined.

9. In this type of analysis there is no effort to attribute forever certain competences to humans and others to nonhumans. The attention is focused on following how *any* set of competences is *distributed* through various entities.

10. Interestingly enough, the oldest Greek engineering myth, that of Daedalus, is about cleverness, deviousness. "Dedalion" means something that goes away from the main road, like the French word "bricole." In the mythology, science is represented by a straight line and technology by a detour, science by *epistémè* and technology by the *métis*. See the excellent essay of Frontisi-Ducroux (1975) on the semantic field of the name Daedalus.

11. We use *actant* to mean anything that acts and *actor* to mean what is made the source of an action. This is a semiotician's definition that is not limited to humans and has no relation whatsoever to the sociological definition of an actor by opposition to mere behavior. For a semiotician, the act of attributing "inert force" to a hinge or the act of attributing it "personality" are comparable in principle and should be studied symmetrically.

12. I have been able to document a case of a five-day student strike at a French school of management (ESSEC) to urge that a door closer be installed in the student cafeteria to keep the freezing cold outside.

13. It is of course another choice to decide who makes such a choice: a man? a spirit? no one? an automated machine? The *scripter* or designer of all these scripts is itself (himself, herself, themselves) negotiated.

14. This is what Norman (1988) calls the Gulf of Execution. His book is an excellent introduction to the study of the tense relations between inscribed and real users. However, Norman speaks only about dysfunction in the interfaces with the final user and never considers the shaping of the artifact by the engineer themselves.

15. To stay within the same etymological root, we call the way actants (human or nonhuman) tend to extirpate themselves from the prescribed behavior *de-inscription* and the way they accept or happily acquiesce to their lot *subscription*.

16. We call *pre-inscription* all the work that has to be done upstream of the scene and all the things assimilated by an actor (human or nonhuman) before coming to the scene as a user or an author. For instance, how to drive a car is basically preinscribed in any (Western) youth years before it comes to passing the driving test; hydraulic pistons were also pre-inscribed for slowly giving back the energy gathered, years before innovators brought them to bear on automated grooms. Engineers can bet on this predetermination when they draw up their prescriptions. This is what is called "articulation work" (Fujimura 1987).

17. We call *circumscription* the organization in the setting of its own limits and of its own demarcation (doors, plugs, hall, introductions).

18. See Suchman for a description of such a setting (1987).

19. We call *ascription* the attribution of an effect to one aspect of the setup. This new decision about attributing efficiency—for instance, to a person's genius, to workers' efforts, to users, to the economy, to technology—is as important as the others, but it is derivative. It is like the opposition between the primary mechanism—who is allied to whom—and the secondary mechanism—whose leadership is recognized—in history of science (Latour 1987).

20. Waddington's term for "necessary paths"—from the Greek *creos* and *odos*.

21. We call *conscription* this mobilization of well-drilled and well-aligned resources to render the behavior of a human or a nonhuman predictable.

22. Trevor Pinch sent me an article from the *Guardian* (2 September 1988) titled "Cardboard coppers cut speeding by third."

A Danish police spokesman said an advantage of the effigies, apart from cutting manpower costs, was that they could stand for long periods undistracted by other calls of duty. Additional assets are understood to be that they cannot claim overtime, be accused of brutality, or get suspended by their chief constable without explanation. "For God's sake, don't tell the Home Office," Mr. Tony Judge, editor of the Police Review Magazine in Britain, said after hearing news of the [Danish] study last night. "We have enough trouble getting sufficient men already." The cut-outs have been placed beside notorious speeding blackspots near the Danish capital. Police said they had yielded "excellent" results. Now they are to be erected at crossings where drivers often jump lights. From time to time, a spokesman added, they would be replaced by real officers.

23. Why did the (automatic) groom go on strike? The answers to this are the same as for the question posed earlier of why no one showed up at La Halle aux Cuirs: it is not because a piece of behavior is prescribed by an inscription that the predetermined characters will show up on time and do the job expected of them. This is true of humans, but it is truer of nonhumans. In this case the hydraulic piston did its job, but not the spring that collaborated with it. Any of the words employed above may be used to describe a setup at any level and not only at the simple one I chose for the sake of clarity. It does not have to be limited to the case where a human deals with a series of nonhuman delegates; it can also be true of relations among nonhumans (yes, you sociologists, there are also relations among things, and *social* relations at that).

24. For the study of user's manual, see Norman 1988 and Boullier, Akrich, and Le Goaziou 1990.

25. Re-inscription is the same thing as inscription or translation or delegation, but seen in its movement. The aim of sociotechnical study is thus to follow the *dynamic* of re-inscription transforming a silent artifact into a *polemical* process. A lovely example of efforts at re-inscription of what was badly pre-inscribed outside of the setting is provided by Orson Welles in *Citizen Kane*, where the hero not only bought a theater for his singing wife to be applauded in, but also bought the journals that were to do the reviews, bought off the art critics themselves, and paid the audience to show up—all to no avail, because the wife eventually quit. Humans and nonhumans are very undisciplined no matter what you do and how many predeterminations you are able to control inside the setting.

For a complete study of this dynamic on a large technical system, see Law (1992) and Latour (1992).

26. The study of scientific text is now a whole industry: see Callon, Law, and Rip (1986) for a technical presentation and Latour (1987) for an introduction.

27. The linguistic meaning of a paradigm is unrelated to the Kuhnian usage of the word. For a complete description of these diagrams, see Latour, Mauguin, and Teil (1992).

28. I am grateful to Berward Joerges for letting me interview his key and his key holder. It alone was worth the trip to Berlin.

29. Keys, locks, and codes are of course a source of marvelous fieldwork for analysts. You may for instance replace the key (excorporation) by a memorized code (incorporation). You may lose both, however, since memory is not necessarily more durable than steel.

References

Akrich, Madeleine. 1992. "The De-Scription of Technical Objects," in Wiebe E. Bijker and John Law, eds., *Shaping Technology/Building Society*. Cambridge, MA: MIT Press, 205–224.

Authier, M. 1989. "Archimède, le canon du savant," in *Eléments d'Histoire des Sciences*, Michel Serres, ed. Paris: Bordas, 101–127.

Baker, N. 1988. *The Mezzanine*. New York: Weidenfeld and Nicholson.

Baxandall, Michael. 1985. *Patterns of Intention. On the Historical Explanation of Pictures*. New Haven, CT: Yale University Press.

Bijker, Wiebe E., Hughes, T. P., and Pinch, Trevor J. 1987. *The Social Construction of Technological Systems: New Directions in the Sociology and History of Technology*. Cambridge, MA: MIT Press.

Boullier, D., Akrich, M., and Le Goaziou, V. 1990. *Représentation de l'utilisateur final et genèse des modes d'emploi*. Miméo, Ecole des Mines.

Butler, Samuel 1872 (paperback edition 1970). *Erewhon*. Harmondsworth: Penguin.

Callon, Michel, Law, John, and Rip, Arie, eds. 1986. *Mapping the Dynamics of Science and Technology*. Basingstoke: Macmillan.

Callon, Michel, and Latour, Bruno. 1992. "Don't Throw Out the Baby with the Bath School: Reply to Collins and Yearley," in *Science as Practice and Culture*, A. Pickering, ed. Chicago: Chicago University Press.

Collins, H. M., and Yearley, Steven. 1992. "Epistemological Chicken," in *Science in Practice and Culture*, A Pickering, ed. Chicago: Chicago University Press.

Frontisi-Ducroux, F. 1975. *Dédale, Mythologie de l'artisan en Grèce Ancienne*. Paris: Maspéro-La Découvertè.

Fujimura, Joan. 1987. "Constructing 'Do-able' Problems in Cancer Research: Articulating Alignment." *Social Studies of Science* 17: 257–293.

Latour, Bruno. 1987. *Science in Action. How to follow scientists and engineers through society*. Milton Keynes: Open University Press; and Cambridge, MA: Harvard University Press.

Latour, Bruno. 1988. "A Relativist Account of Einstein's Relativity." *Social Studies of Science* 18: 3–45.

Latour, Bruno. 1992. *Aramis ou l'amour des techniques*. Paris: La Découvertè.

Latour, Bruno, Mauguin, P., and Teil, Genevieve. 1992. "A Note on Socio-Technical Graphs." *Social Studies of Science* 22: 33–57.

Law, John. 1992. "The Olympus 320 Engine: A Case Study in Design, Autonomy and Organizational Control," in *Technology and Culture*.

MacKenzie, Donald, and Wajcman, Judy, eds. 1985. *The Social Shaping of Technology: How the Refrigerator Got Its Hum*. Bristol, PA: Open University Press.

Noble, David. 1984. *Forces of Production: A Social History of Industrial Automation*. New York: Knopf.

Norman, David. 1988. *The Psychology of Everyday Things*. New York: Basic Books.

Pinch, Trevor, Ashmore, Malcolm, and Mulkay, Michael. 1992. "Technology, Testing, Text: Clinical Budgeting in the U.K. National Health Service," in Wiebe E. Bijker and John Law, eds., *Shaping Technology/Building Society*. Cambridge, MA: MIT Press, 265–289.

Suchman, Lucy. 1987. *Plans and Situated Actions. The Problem of Human Machine Communication*. Cambridge: Cambridge University Press.

Winner, Langdon. 1980. "Do Artefacts Have Politics?" *Daedalus* 109: 121–136.

11 "Code Is Law"

Lawrence Lessig

Like the other authors in this section, Lawrence Lessig believes that technology and society are intertwined. Instead of only examining the past, however, Lessig is especially concerned about the present and future. In tackling a relatively new and still evolving technology—the Internet—Lessig goes as far as to say that computer code will not simply affect our lives, but that it will be as powerful as law in enabling and limiting our actions. He argues that an individual's behavior is influenced by four forms of regulation—law, norms, markets, and what he calls architecture. Architecture is the term he uses for computer code though what he claims applies equally well and more broadly to the built environment. He urges us to act while the technology is still developing so that we can direct it to create a world we want rather than one dictated by those involved in writing code. Lessig's analysis illustrates just one more way in which the social and technical are intertwined and reveals an important implication: In order to get the kind of future we want, we can't focus only on laws, norms and markets. We also must pay attention to technology, technological design, and technical details. Some of Lessig's arguments seem deterministic, but Lessig recognizes that humans have power to control the technology that will ultimately shape us.

Code Is Law

A decade ago, in the spring of 1989, communism in Europe died—collapsed, as a tent would fall if its main post were removed. No war or revolution brought communism to its end. Exhaustion did. Born in its place across Central and Eastern Europe was a new political regime, the beginnings of a new political society.

For constitutionalists (as I am), this was a heady time. I had just graduated from law school in 1989, and in 1991 I began teaching at the University of Chicago. Chicago had a center devoted to the study of the emerging democracies in Central and Eastern Europe. I was a part of that center. Over the next five years I spent more hours on airplanes, and more mornings drinking bad coffee, than I care to remember.

Eastern and Central Europe were filled with Americans telling former Communists how they should govern. The advice was endless and silly. Some of these visitors literally sold constitutions to the emerging constitutional republics; the balance had innumerable half-baked ideas about how the new nations should be governed. These Americans came from a nation where constitutionalism had worked, yet apparently had no clue why.

From *Code: And Other Laws of Cyberspace* (New York: Basic Books, 1999), pp. 3–8; 85–90, 241–242, 254–255. Reprinted by permission of Basic Books, a member of Perseus Books Group.

The center's mission, however, was not to advise. We knew too little to guide. Our aim was to watch and gather data about the transitions and how they progressed. We wanted to understand the change, not direct it.

What we saw was striking, if understandable. Those first moments after communism's collapse were filled with antigovernmental passion—with a surge of anger directed against the state and against state regulation. Leave us alone, the people seemed to say. Let the market and nongovernmental organizations—a new society— take government's place. After generations of communism, this reaction was completely understandable. What compromise could there be with the instrument of your repression?

A certain American rhetoric supported much in this reaction. A rhetoric of libertarianism. Just let the market reign and keep the government out of the way, and freedom and prosperity would inevitably grow. Things would take care of themselves. There was no need, and could be no place, for extensive regulation by the state.

But things didn't take care of themselves. Markets didn't flourish. Governments were crippled, and crippled governments are no elixir of freedom. Power didn't disappear—it simply shifted from the state to mafiosi, themselves often created by the state. The need for traditional state functions—police, courts, schools, health care— didn't magically go away. Private interests didn't emerge to fill the need. Instead, needs were unmet. Security evaporated. A modern if plodding anarchy replaced the bland communism of the previous three generations: neon lights flashed advertisements for Nike; pensioners were swindled out of their life savings by fraudulent stock deals; bankers were murdered in broad daylight on Moscow streets. One system of control had been replaced by another, but neither system was what Western libertarians would call freedom.

At just about the time when this post-communist euphoria was waning—in the mid-1990s—there emerged in the West another "new society," to many just as exciting as the new societies promised in post-communist Europe. This was cyberspace. First in universities and centers of research, and then within society generally, cyberspace became the new target of libertarian utopianism. Here freedom from the state would reign. If not in Moscow or Tblisi, then here in cyberspace would we find the ideal libertarian society.

The catalyst for this change was likewise unplanned. Born in a research project in the Defense Department, cyberspace too arose from the displacement of a certain architecture of control. The tolled, single-purpose network of telephones was displaced by the untolled and multipurpose network of packet-switched data. And thus the old one-to-many architectures of publishing (television, radio, newspapers, books) were supplemented by a world where everyone could be a publisher. People could communicate and associate in ways that they had never done before. The space promised a kind of society that real space could never allow—freedom without anarchy, control without government, consensus without power. In the words of a manifesto that will define our generation: "We reject: kings, presidents and voting. We believe in: rough consensus and running code."[1]

As in post-Communist Europe, first thoughts about cyberspace tied freedom to the disappearance of the state. But here the bond was even stronger than in post-Communist Europe. The claim now was that government *could not* regulate cyberspace, that cyberspace was essentially, and unavoidably, free. Governments could threaten, but behavior could not be controlled; laws could be passed, but they would be meaningless. There was no choice about which government to install—none could reign. Cyberspace would be a society of a very different sort. There would be definition and direction, but built from the bottom up, and never through the direction of a state. The society of this space would be a fully self-ordering entity, cleansed of governors and free from political hacks.

I taught in Central Europe during the summers of the early 1990s; I witnessed the transformation in attitudes about communism that I described at the start of this chapter. And so I felt a bit of déjà vu when in the spring of 1995, I began to teach the law of cyberspace, and saw in my students these very same post-communist thoughts about freedom and government. Even at Yale—not known for libertarian passions—the students seemed drunk with what James Boyle would later call the "libertarian gotcha":[2] no government could survive without the Internet's riches, yet no government could control what went on there. Real-space governments would become as pathetic as the last Communist regimes. It was the withering of the state that Marx had promised, jolted out of existence by trillions of gigabytes flashing across the ether of cyberspace. Cyberspace, the story went, could *only* be free. Freedom was its nature.

But why was never made clear. That *cyberspace* was a place that governments could not control was an idea that I never quite got. The word itself speaks not of freedom but of control. Its etymology reaches beyond a novel by William Gibson (*Neuromancer*, published in 1984) to the world of "cybernetics," the study of control at a distance.[3] Cybernetics had a vision of perfect regulation. Its very motivation was finding a better way to direct. Thus, it was doubly odd to see this celebration of non-control over architectures born from the very ideal of control.

As I said, I am a constitutionalist. I teach and write about constitutional law. I believe that these first thoughts about government and cyberspace are just as misguided as the first thoughts about government after communism. Liberty in cyberspace will not come from the absence of the state. Liberty there, as anywhere, will come from a state of a certain kind.[4] We build a world where freedom can flourish not by removing from society any self-conscious control; we build a world where freedom can flourish by setting it in a place where a particular kind of self-conscious control survives. We build liberty, that is, as our founders did, by setting society upon a certain *constitution*.

But by "constitution" I don't mean a legal text. Unlike my countrymen in Eastern Europe, I am not trying to sell a document that our framers wrote in 1787. Rather, as the British understand when they speak of their constitution, I mean an *architecture*—not just a legal text but a way of life—that structures and constrains social and legal power, to the end of protecting fundamental *values*—principles and ideals that reach beyond the compromises of ordinary politics.

Constitutions in this sense are built, they are not found. Foundations get laid, they don't magically appear. Just as the founders of our nation learned from the anarchy that followed the revolution (remember: our first constitution, the Articles of Confederation, was a miserable failure of do-nothingness), so too are we beginning to see in cyberspace that this building, or laying, is not the work of an invisible hand. There is no reason to believe that the grounding for liberty in cyberspace will simply emerge. In fact, as I will argue, quite the opposite is the case. As our framers learned, and as the Russians saw, we have every reason to believe that cyberspace, left to itself, will not fulfill the promise of freedom. Left to itself, cyberspace will become a perfect tool of control.[5]

Control. Not necessarily control by government, and not necessarily control to some evil, fascist end. But the argument of this book is that the invisible hand of cyberspace is building an architecture that is quite the opposite of what it was at cyberspace's birth. The invisible hand, through commerce, is constructing an architecture that perfects control—an architecture that makes possible highly efficient regulation. As Vernor Vinge warned in 1996, a distributed architecture of regulatory control; as Tom Maddox added, an axis between commerce and the state.[6]

This book is about that change, and about how we might prevent it. When we see the path that cyberspace is on...we see that much of the "liberty" present at cyberspace's founding will vanish in its future. Values that we now consider fundamental will not necessarily remain. Freedoms that were foundational will slowly disappear.

If the original cyberspace is to survive, and if values that we knew in that world are to remain, we must understand how this change happens, and what we can do in response.... Cyberspace presents something new for those who think about regulation and freedom. It demands a new understanding of how regulation works and of what regulates life there. It compels us to look beyond the traditional lawyer's scope—beyond laws, regulations, and norms. It requires an account of a newly salient regulator.

That regulator is the obscurity in the book's title—*Code*. In real space we recognize how laws regulate—through constitutions, statutes, and other legal codes. In cyberspace we must understand how code regulates—how the software and hardware that make cyberspace what it is *regulate* cyberspace as it is. As William Mitchell puts it, this code is cyberspace's "law."[7] *Code is law.*

This code presents the greatest threat to liberal or libertarian ideals, as well as their greatest promise. We can build, or architect, or code cyberspace to protect values that we believe are fundamental, or we can build, or architect, or code cyberspace to allow those values to disappear. There is no middle ground. There is no choice that does not include some kind of *building*. Code is never found; it is only ever made, and only ever made by us. As Mark Stefik puts it, "Different versions of [cyberspace] support different kinds of dreams. We choose, wisely or not."[8]

My argument is not for some top-down form of control; my claim is not that regulators must occupy Microsoft. A constitution envisions an environment; as Justice

Holmes said, it "call[s] into life a being the development of which [can not be] foreseen."[9] Thus, to speak of a constitution is not to describe a one-hundred-day plan. It is instead to identify the values that a space should guarantee. It is not to describe a "government"; it is not even to select (as if a single choice must be made) between bottom-up or top-down control. In speaking of a constitution in cyberspace we are simply asking: What values are protected there? What values will we build into the space to encourage certain forms of life?

The "values" here are of two sorts—substantive and structural. In the American tradition, we worried about the second first. The framers of the Constitution of 1787 (enacted without a Bill of Rights) were focused on structures of government. Their aim was to ensure that a particular government (the federal government) did not become too powerful. And so they built into its design checks on the power of the federal government and limits on its reach over the states.

Opponents of that Constitution insisted that more checks were needed, that the Constitution needed to impose substantive limits on government's power as well as structural limits. And thus the Bill of Rights was born. Ratified in 1791, the Bill of Rights promised that the federal government will not remove certain protections—of speech, privacy, and due process. And it guaranteed that the commitment to these substantive values will remain despite the passing fancy of normal government. These values were to be entrenched, or embedded, in our constitutional design; they can be changed, but only by changing the Constitution's design.

These two kinds of protection go together in our constitutional tradition. One would have been meaningless without the other. An unchecked structure could easily have overturned the substantive protections expressed in the Bill of Rights, and without substantive protections, even a balanced and reflective government could have violated values that our framers thought fundamental.

We face the same questions in constituting cyberspace, but we have approached them from an opposite direction. Already we are struggling with substance: Will cyberspace promise privacy or access? Will it preserve a space for free speech? Will it facilitate free and open trade? These are choices of substantive value.

But structure matters as well. What checks on arbitrary regulatory power can we build into the design of the space? What "checks and balances" are possible? How do we separate powers? How do we ensure that one regulator, or one government, doesn't become too powerful? . . .

Theorists of cyberspace have been talking about these questions since its birth.[10] But as a culture, we are just beginning to get it. We are just beginning to see why the architecture of the space matters—in particular, why the *ownership* of that architecture matters. If the code of cyberspace is owned, . . . it can be controlled; if it is not owned, control is much more difficult. The lack of ownership, the absence of property, the inability to direct how ideas will be used—in a word, the presence of a commons—is key to limiting, or checking, certain forms of governmental control.

One part of this question of ownership is at the core of the current debate between open and closed source software. In a way that the American founders would have instinctively understood, "free software" or "open source software"—or "open code," to (cowardly) avoid taking sides in a debate I describe later—is itself a check on arbitrary power. A structural guarantee of constitutionalized liberty, it functions as a type of separation of powers in the American constitutional tradition. It stands alongside substantive protections, like freedom of speech or of the press, but its stand is more fundamental.... The first intuition of our founders was right: structure builds substance. Guarantee the structural (a space in cyberspace for open code), and (much of) the substance will take care of itself.

. . .

Given our present tradition in constitutional law and our present faith in representative government, are we able to respond collectively to the changes I will have described?

My strong sense is that we are not. We are at a stage in our history when we urgently need to make fundamental choices about values, but we trust no institution of government to make such choices. Courts cannot do it, because as a legal culture we don't want courts choosing among contested matters of values, and Congress should not do it because, as a political culture, we so deeply question the products of ordinary government.

Change is possible. I don't doubt that revolutions remain in our future; the open code movement is just such a revolution. But I fear that it is too easy for the government to dislodge these revolutions, and that too much will be at stake for it to allow the revolutionaries to succeed. Our government has already criminalized the core ethic of this movement, transforming the meaning of *hacker* into something quite alien to its original sense. This, I argue, is only the start.

Things could be different. They are different elsewhere. But I don't see how they could be different for us just now. This no doubt is a simple confession of the limits of my own imagination. I would be grateful to be proven wrong. I would be grateful to watch as we relearn—as the citizens of the former Communist republics are learning—how to escape our disabling ideas about the possibilities for governance.

. . .

What Things Regulate

John Stuart Mill was an Englishman, though one of the most influential political philosophers in America in the nineteenth century. His writings ranged from important work on logic to a still striking text, *The Subjection of Women*. But his continuing influence comes from a relatively short book titled *On Liberty*. Published in 1859, this powerful argument for individual liberty and diversity of thought represents an important view of liberal and libertarian thinking in the second half of the nineteenth century.

"Libertarian," however, has a specific meaning for us. It associates with arguments against government.[11] Government, in the modern libertarian's view, is the threat to liberty; private action is not. Thus, the good libertarian is focused on reducing government's power. Curb the excesses of government, the libertarian says, and you will have ensured freedom for your society.

Mill's view was not so narrow. He was a defender of liberty and an opponent of forces that suppressed it. But those forces were not confined to government. Liberty, in Mill's view, was threatened as much by norms as by government, as much by stigma and intolerance as by the threat of state punishment. His objective was to argue against these private forces of coercion. His work was a defense against liberty-suppressing norms, because in England at the time these were the real threat to liberty.

Mill's method is important, and it should be our own. It asks, What is the threat to liberty, and how can we resist it? It is not limited to asking, What is the threat to liberty *from government?* It understands that more than government can threaten liberty, and that sometimes this something more can be private rather than state action. Mill was not so concerned with the source. His concern was with liberty.

Threats to liberty change. In England norms may have been the problem in the late nineteenth century; in the United States in the first two decades of the twentieth century it was state suppression of speech.[12] The labor movement was founded on the idea that the market is sometimes a threat to liberty—not just because of low wages, but also because the market form of organization itself disables a certain kind of freedom.[13] In other societies, at other times, the market is the key, not the enemy, to liberty.

Thus, rather than think of an enemy in the abstract, we should understand the particular threat to liberty that exists in a particular time and place. And this is especially true when we think about liberty in cyberspace. For my argument is that cyberspace teaches a new threat to liberty. Not new in the sense that no theorist has conceived of it before. Others have.[14] But new in the sense of newly urgent. We are coming to understand a newly powerful regulator in cyberspace, and we don't yet understand how best to control it.

This regulator is code—or more generally, the "built environment" of social life, its architecture.[15] And if in the middle of the nineteenth century it was norms that threatened liberty, and at the start of the twentieth state power that threatened liberty, and during much of the middle twentieth century the market that threatened liberty, my argument is that we understand how in the late twentieth century, and into the twenty-first, it is a different regulator—code—that should be our concern.

But it is not my aim to say that this should be our new single focus. My argument is not that there is a new single enemy different from the old. Instead, I believe we need a more general understanding of how regulation works. One that focuses on more than the single influence of any one force such as government, norms, or the market, and instead integrates these factors into a single account.

This chapter is a step toward that more general understanding.[16] It is an invitation to think beyond the narrow threat of government. The threats to liberty have never come solely from government, and the threats to liberty in cyberspace certainly will not.

A Dot's Life
There are many ways to think about constitutional law and the limits it may impose on government regulation. I want to think about it from the perspective of someone who is regulated or constrained. That someone regulated is represented by this (pathetic) dot—a creature (you or me) subject to the different constraints that might regulate it. By describing the various constraints that might bear on this individual, I hope to show you something about how these constraints function together.

Here then is the dot.

How is this dot "regulated"?

Let's start with something easy: smoking. If you want to smoke, what constraints do you face? What factors *regulate* your decision to smoke or not?

One constraint is legal. In some places at least, laws regulate smoking—if you are under eighteen, the law says that cigarettes cannot be sold to you. If you are under twenty-six, cigarettes cannot be sold to you unless the seller checks your ID. Laws also regulate where smoking is permitted—not in O'Hare Airport, on an airplane, or in an elevator, for instance. In these two ways at least, laws aim to direct smoking behavior. They operate as a kind of constraint on an individual who wants to smoke.[17]

But laws are not the most significant constraints on smoking. Smokers in the United States certainly feel their freedom regulated, even if only rarely by the law. There are no smoking police, and smoking courts are still quite rare. Rather, smokers in America are regulated by norms. Norms say that one doesn't light a cigarette in a private car without first asking permission of the other passengers. They also say, however, that one needn't ask permission to smoke at a picnic. Norms say that others can ask you to stop smoking at a restaurant, or that you never smoke during a meal.

European norms are savagely different. There the presumption is in the smoker's favor; vis-à-vis the smoker, the norms are laissez-faire. But in the States the norms effect a certain constraint, and this constraint, we can say, *regulates* smoking behavior.

Law and norms are still not the only forces regulating smoking behavior. The market too is a constraint. The price of cigarettes is a constraint on your ability to smoke. Change the price, and you change this constraint. Likewise with quality. If the market supplies a variety of cigarettes of widely varying quality and price, your ability to select the kind of cigarette you want increases; increasing choice here reduces constraint.

Finally, there are the constraints created, we might say, by the technology of cigarettes, or by the technologies affecting their supply.[18] Unfiltered cigarettes present a greater constraint on smoking than filtered cigarettes if you are worried about your health. Nicotine-treated cigarettes are addictive and therefore create a greater constraint on smoking than untreated cigarettes. Smokeless cigarettes present less of a constraint because they can be smoked in more places. Cigarettes with a strong odor present more of a constraint because they can be smoked in fewer places. In all of these ways, how the cigarette *is* affects the constraints faced by a smoker. How it is, how it is designed, how it is built—in a word, its *architecture*.

Thus, four constraints regulate this pathetic dot—the law, social norms, the market, and architecture—and the "regulation" of this dot is the sum of these four constraints. Changes in any one will affect the regulation of the whole. Some constraints will support others; some may undermine others. A complete view, however, should consider them together.

So think of the four together like this:

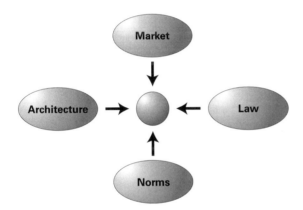

In this drawing, each oval represents one kind of constraint operating on our pathetic dot in the center. Each constraint imposes a different kind of cost on the dot for engaging in the relevant behavior—in this case, smoking. The cost from norms is different from the market cost, which is different from the cost from law and the cost from the (cancerous) architecture of cigarettes.

The constraints are distinct, yet they are plainly interdependent. Each can support or oppose the others. Technologies can undermine norms and laws; they can also support them. Some constraints make others possible; others make some impossible. Constraints work together, though they function differently and the effect of each is distinct. Norms constrain through the stigma that a community imposes; markets constrain through the price that they exact; architectures constrain through the physical burdens they impose; and law constrains through the punishment it threatens.

We can call each constraint a "regulator," and we can think of each as a distinct modality of regulation. Each modality has a complex nature, and the interaction among these four is hard to describe. I've worked through this complexity more completely in the appendix. But for now, it is enough to see that they are linked and that, in a sense, they combine to produce the regulation to which our pathetic dot is subject in any given area.

The same model describes the regulation of behavior in cyberspace.

Law regulates behavior in cyberspace. Copyright law, defamation law, and obscenity laws all continue to threaten *ex post* sanction for the violation of legal rights. How well law regulates, or how efficiently, is a different question: in some cases it does so more efficiently, in some cases less. But whether better or not, law continues to threaten a certain consequence if it is defied. Legislatures enact;[19] prosecutors threaten;[20] courts convict.[21]

Norms also regulate behavior in cyberspace. Talk about democratic politics in the alt.knitting newsgroup, and you open yourself to flaming; "spoof" someone's identity in a MUD, and you may find yourself "toaded";[22] talk too much in a discussion list, and you are likely to be placed on a common bozo filter. In each case, a set of understandings constrain behavior, again through the threat of *ex post* sanctions imposed by a community.

Markets regulate behavior in cyberspace. Pricing structures constrain access, and if they do not, busy signals do. (AOL learned this quite dramatically when it shifted from an hourly to a flat rate pricing plan.)[23] Areas of the Web are beginning to charge for access, as online services have for some time. Advertisers reward popular sites; online services drop low-population forums. These behaviors are all a function of market constraints and market opportunity. They are all, in this sense, regulations of the market.

And finally, an analog for architecture regulates behavior in cyberspace—*code*. The software and hardware that make cyberspace what it is constitute a set of constraints on how you can behave. The substance of these constraints may vary, but they are experienced as conditions on your access to cyberspace. In some places (online services such as AOL, for instance) you must enter a password before you gain access; in other places you can enter whether identified or not.[24] In some places the transactions you engage in produce traces that link the transactions (the "mouse droppings") back to you; in other places this link is achieved only if you want it to be.[25] In some places you can choose to speak a language that only the recipient can hear (through encryption);[26] in other places encryption is not an option.[27] The code or software or architecture or protocols set these features; they are features selected by code writers; they constrain some behavior by making other behavior possible, or impossible. The code embeds certain values or makes certain values impossible. In this sense, it too is regulation, just as the architectures of real-space codes are regulations.

As in real space, then, these four modalities regulate cyberspace. The same balance exists. As William Mitchell puts it (though he omits the constraint of the market):

Architecture, laws, and customs maintain and represent whatever balance has been struck [in real space]. As we construct and inhabit cyberspace communities, we will have to make and maintain similar bargains—though they will be embodied in software structures and electronic access controls rather than in architectural arrangements.[28]

Laws, norms, the market, and architectures interact to build the environment that "Netizens" know. The code writer, as Ethan Katsh puts it, is the "architect."[29]

But how can we "make and maintain" this balance between modalities? What tools do we have to achieve a different construction? How might the mix of real-space values be carried over to the world of cyberspace? How might the mix be changed if change is desired?

Notes

1. Paulina Borsook, "How Anarchy Works," *Wired* 110 (October 1995): 3.10, available at http://www.wired.com/wired/archive/3.10/ietf_pr.html (visited May 30, 1999), quoting Netlander, David Clark.

2. James Boyle, talk at Telecommunications Policy Research Conference (TPRC), Washington, D.C., September 28, 1997. David Shenk discusses the libertarianism that cyberspace inspires (as well as other, more fundamental problems with the age) in a brilliant cultural how-to book that responsibly covers both the technology and the libertarianism; see *Data Smog: Surviving the Information Glut* (San Francisco, Harper Edge, 1997), esp. 174–77. The book also describes technorealism, a responsive movement that advances a more balanced picture of the relationship between technology and freedom.

3. See Kevin Kelley, *Out of Control: The New Biology of Machines, Social Systems, and the Economic World* (Reading, Mass.: Addison-Wesley, 1994), 119.

4. As Stephen Holmes has put it, "Rights depend upon the competent exercise of ... legitimate public power.... The largest and most reliable human rights organization is the liberal state.... Unless society is politically well organized, there will be no individual liberties and no civil society"; "What Russia Teaches Us Now: How Weak States Threaten Freedom," *American Prospect* 33 (1997): 30, 33.

5. This is a dark picture, I confess, and it contrasts with the picture of control drawn by Andrew Shapiro in *The Control Revolution* (New York: Public Affairs, 1999). As I discuss later, however, the difference between Shapiro's view and my own turns on the extent to which architectures enable top-down regulation. In my view, one highly probable architecture would enable greater regulation than Shapiro believes is likely.

6. See "We Know Where You Will Live," Computers, Freedom, and Privacy Conference, March 30, 1996, audio link available at http://www-swiss.ai.mit.edu/projects/mac/cfp96/plenary-sf.html.

7. See William J. Mitchell, *City of Bits: Space, Place, and the Infobahn* (Cambridge, Mass.: MIT Press, 1995), 111. In much of this book, I work out Mitchell's idea, though I drew the metaphor from others as well. Ethan Katsh discusses this notion of software worlds in "Software

Worlds and the First Amendment: Virtual Doorkeepers in Cyberspace," *University of Chicago Legal Forum* (1996): 335, 338. Joel Reidenberg discusses the related notion of "lex informatica" in "Lex Informatica: The Formulation of Information Policy Rules Through Technology," *Texas Law Review* 76 (1998): 553. I have been especially influenced by James Boyle's work in the area. I discuss his book in chapter 9, but see also "Foucault in Cyberspace: Surveillance, Sovereignty, and Hardwired Censors," *University of Cincinnati Law Review* 66 (1997): 177. For a recent and powerful use of the idea, see Shapiro, *The Control Revolution*. Mitch Kapor is the father of the meme "architecture is politics" within cyberspace talk. I am indebted to him for this.

8. Mark Stefik, "Epilogue: Choices and Dreams," in *Internet Dreams: Archetypes, Myths, and Metaphors*, edited by Mark Stefik (Cambridge, Mass.: MIT Press, 1996), 390.

9. *Missouri v Holland*, 252 US 416, 433 (1920).

10. Richard Stallman, for example, organized resistance to the emergence of passwords at MIT. Passwords are an architecture that facilitates control by excluding users not "officially sanctioned." Steven Levy, *Hackers* (Garden City, N.Y.: Anchor Press/Doubleday, 1984), 416–17.

11. Or more precisely, against a certain form of government regulation. The more powerful libertarian arguments against regulation in cyberspace are advanced, for example, by Peter Huber in *Law and Disorder in Cyberspace: Abolish the FCC and Let Common Law Rule the Telecosm*. Huber argues against agency regulation and in favor of regulation by the common law. See also Thomas Hazlett in "The Rationality of U.S. Regulation of the Broadcast Spectrum," *Journal of Law and Economics* 33 (1990): 133, 133–39. For a lawyer, it is hard to understand precisely what is meant by "the common law." The rules of the common law are many, and the substantive content has changed. There is a common law process, which lawyers like to mythologize, in which judges make policy decisions in small spaces against the background of binding precedent. It might be this that Huber has in mind, and if so, there are, of course, benefits to this system. But as he plainly understands, it is a form of *regulation* even if it is constituted differently.

12. The primary examples are the convictions under the 1917 Espionage Act; see, for example, *Schenck v United States*, 249 US 47 (1919) (upholding conviction for distributing a leaflet attacking World War I conscription); *Frohwerk v United States*, 249 US 204 (1919) (upholding conviction based on newspaper alleged to cause disloyalty); *Debs v United States*, 249 US 211 (1919) (conviction upheld for political speech said to cause insubordination and disloyalty).

13. See, for example, the work of John R. Commons, *Legal Foundations of Capitalism* (1924), 296–98, discussed in Herbert Hovenkamp, *Enterprise and American Law, 1836–1937* (Cambridge, Mass.: Harvard University Press, 1991), 235; see also John R. Commons, *Institutional Economics: Its Place in Political Economy* (1934).

14. The general idea is that the tiny corrections of space enforce a discipline, and that this discipline is an important regulation. Such theorizing is a tiny part of the work of Michel Foucault; see *Discipline and Punish: The Birth of the Prison* (1979), 170–77, though his work generally inspires this perspective. It is what Oscar Gandy speaks about in *The Panoptic Sort: A Political Economy of Personal Information* (Boulder, Colo.: Westview Press, 1993), 23. David

Brin makes the more general point that I am arguing—that the threat to liberty is broader than a threat by the state; see David Brin, *The Transparent Society: Will Technology Force Us to Choose between Privacy and Freedom?* (New York: Basic Books, 1990), 110.

15. See, for example, *The Built Environment: A Creative Inquiry into Design and Planning*, edited by Tom J. Bartuska and Gerald L. Young (Menlo Park, Calif.: Crisp Publications, 1994); *Preserving the Built Heritage: Tools for Implementation*, edited by J. Mark Schuster et al. (Hanover, N.H.: University Press of New England, 1997). In design theory, the notion I am describing accords with the tradition of Andres Duany and Elizabeth Plater-Zyberk; see, for example, William Lennertz, "Town-Making Fundamentals," in *Towns and Town-Making Principles*, edited by Andres Duany and Elizabeth Plater-Zyberk (New York: Rizzoli, 1991): "The work of ... Duany and ... Plater-Zyberk begins with the recognition that design affects behavior. [They] see the structure and function of a community as interdependent. Because of this, they believe a designer's decisions will permeate the lives of residents not just visually but in the way residents live. They believe design structures functional relationships, quantitatively and qualitatively, and that it is a sophisticated tool whose power exceeds its cosmetic attributes" (21).

16. Elsewhere I've called this the "New Chicago School"; see Lawrence Lessig, "The New Chicago School," *Journal of Legal Studies* 27 (1998): 661. It is within the "tools approach" to government action (see John de Monchaux and J. Mark Schuster, "Five Things to Do," in Schuster, *Preserving the Built Heritage*, 3), but it describes four tools whereas Schuster describes five. I develop the understanding of the approach in the appendix to this book.

17. See generally *Smoking Policy: Law, Politics, and Culture*, edited by Robert L. Rabin and Stephen D. Sugarman (New York: Oxford University Press, 1993); Lawrence Lessig, "The Regulation of Social Meaning," *University of Chicago Law Review* 62 (1995): 943, 1025–34; Cass R. Sunstein, "Social Norms and Social Roles," *Columbia Law Review* 96 (1996): 903.

18. These technologies are themselves affected, no doubt, by the market. Obviously, these constraints could not exist independently of each other but affect each other in significant ways.

19. The ACLU lists twelve states that passed Internet regulations between 1995 and 1997; see http://www.aclu.org/issues/cyber/censor/stbills.html#bills (visited May 31, 1999).

20. See, for example, the policy of the Minnesota attorney general on the jurisdiction of Minnesota over people transmitting gambling information into the state; available at http://www.ag.state.mn.us/home/consumer/consumernews/OnlineScams/memo.html (visited May 31, 1999).

21. See, for example, *Playboy Enterprises v Chuckleberry Publishing, Inc.*, 939 FSupp 1032 (SDNY 1996); *United States v Thomas*, 74 F3d 1153 (6th Cir 1996); *United States v Miller*, 166 F3d 1153 (11th Cir 1999); *United States v Lorge*, 166 F3d 516 (2d Cir 1999); *United States v Whiting*, 165 F3d 631 (8th Cir 1999); *United States v Hibbler*, 159 F3d 233 (6th Cir 1998); *United States v Fellows*, 157 F3d 1197 (9th Cir 1998); *United States v Simpson*, 152 F3d 1241 (10th Cir 1998); *United States v Hall*, 142 F3d 988 (7th Cir 1998); *United States v Hockings*, 129 F3d 1069 (9th Cir 1997); *United States v Lacy*, 119 F3d 742 (9th Cir 1997); *United States v Smith*, 47 MJ 588 (CrimApp 1997); *United States v Ownby*, 926 FSupp 558 (WDVa 1996).

22. See Julian Dibbell, "A Rape in Cyberspace," *Village Voice*, December 23, 1993, 36.

23. See, for example, "AOL Still Suffering but Stock Price Rises," *Network Briefing*, January 31, 1997; David S. Hilzenrath, "'Free' Enterprise, Online Style; AOL, CompuServe, and Prodigy Settle FTC Complaints," *Washington Post*, May 2, 1997, G1; "America Online Plans Better Information About Price Changes," *Wall Street Journal*, May 29, 1998, B2; see also Swisher, *Aol.com*, 206–8.

24. USENET postings can be anonymous; see Henry Spencer and David Lawrence, *Managing USENET* (Sebastopol, Calif.: O'Reilly & Associates, 1998), 366–67.

25. Web browsers make this information available, both in real time and archived in a cookie file; see http://www.cookiecentral.com/faq.htm (visited May 31, 1999). They also permit users to turn this tracking feature off.

26. PGP is a program to encrypt messages that is offered both commercially and free.

27. Encryption, for example, is illegal in some international contexts; see Stewart A. Baker and Paul R. Hurst, *The Limits of Trust: Cryptography, Governments, and Electronic Commerce* (Boston: Kluwer Law International, 1998) 130–6.

28. William J. Mitchell, *City of Bits: Space, Place, and the Infobahn* (Cambridge, MA: MIT Press, 1995), 159.

29. See M. Ethan Katsh, "Software Worlds and the First Amendment: Virtual Doorkeepers in Cyberspace," *University of Chicago Legal Forum* (1996): 335, 340. "If a comparison to the physical world is necessary, one might say that the software designer is the architect, the builder, and the contractor, as well as the interior decorator."

12 "The Intersection of Culture, Gender, and Technology"

Patrick D. Hopkins

In recent years the idea that gender and technology shape one another has come to the fore. The gender relations prevailing in a society may influence the design and direction of technology and vice versa, the introduction of a new technology may affect gender relations. In this chapter, Hopkins outlines several different types of relationships between gender and technology. He argues that specific technologies are often associated with women or men; that technology can reinforce a gender system; that technology can allow individuals or groups to subvert an existing gender arrangement; and that technology can alter the very nature of sex and gender. Using gender as a social category, Hopkins demonstrates a variety of complex ways in which the social and the technological are interwoven and shows how misleading it can be to force such distinctions. Hopkins's analysis is similar to Lessig's in the sense that it also suggests that in order to get a future world that we want, we will have to pay attention to technology. Ideas about gender are so intertwined with technology that if we want a more equitable gender system, we will have to change technology, and vice versa. Although Hopkins limits his analysis to gender, his approach could be extended to other categories that affect our society such as race, class, politics, or religion.

- In a small Tennessee city, a divorcing couple argues about abortion. One claims that their embryos are unborn children with a right to life; the other argues they are just lumps of disposable tissue. Standard legal reasoning about abortion, privacy, and the right to control one's body doesn't help much in the argument, however—because the embryos are not inside anyone's body. They never have been. They sit frozen in a small cylinder on the other side of town. The abortion debate is raging and no one is even pregnant.
- In Sri Lanka, well-meaning innovators import water pumps to ease the drudgery of women's long, hot walks to wells—but they only teach men how to repair the devices. When the pumps break, the actual users, women, don't know how to fix them. So the pumps sit there unused while women lug water back and forth.
- Medical technologists figure out a way to choose the sex of a baby—and make it available to a culture which prefers their firstborn children to be male. Could an entire generation of firstborn male children make a difference?
- Women try to take advantage of a new invention, the automobile—but they find out that only electric cars are considered appropriate for women. Gasoline cars are for men.

From Patrick D. Hopkins (ed.), *Sex/Machine: Readings in Culture, Gender, and Technology* (Bloomington: Indiana University Press, 1998), pp. 1–9. Reprinted with permission.

- The entire abortion controversy might be put to rest, some feminists argue—if only we could find a way to get machines pregnant, rather than women.
- Healthy babies are born all the time that are neither male nor female, or that are perhaps both—but they don't get out of the hospital that way. Someone chooses what sex they will be.
- A lesbian feminist is called an "intruder" and an "oppressor" by other lesbian feminists—because she used to be a man.
- A researcher suggests using amniocentesis to test fetuses for homosexuality—and then "curing" them with androgen injections.
- On the Internet, you can fall in love with a clever, articulate, beautiful young woman—and then find out that her personality was generated by a dull, laconic, unattractive, middle-aged man. Have you been lied to? Or does your version of reality serve you poorly?
- Scientists come up with a way to get men pregnant—and are swamped with requests.
- A theorist looks toward science fiction novels for inspiration—and argues that the best way to be a feminist is to become a cyborg.

These situations only hint at the degree to which issues of gender and technology are complex, far-reaching, and fascinating. As powerful interacting social and physical forces, gender and technology shape our experiences, cultures, and identities—sometimes in such comfortable and subtle ways that it takes effort to appreciate them; sometimes in such conspicuous and explosive ways that everyone recognizes their importance. Delving into these issues is an opportunity to discover how technology promises or threatens to rewrite our ideas about sex, sexuality, and gender identity. It is an opportunity to debate ethical and legal issues at the very core of human experiences—procreation, labor, sex, our bodies. It is the chance to find out how sex role restrictions prevent each of us from using certain technologies, or require us to use others.

Examining these topics can be both illuminating and unsettling, particularly because we discover how our own lives are and will be affected by shifts in ideas of gender and by changes in technology. In my classes on these issues, students often remark that they never realized how much their daily lives, their career choices, their thoughts on ethical and social issues, and even their self-concepts have been affected by assumptions about technology, sex, and gender. What seemed like little things before (so little they were ignored)—why a student's husband automatically gets into the driver's side of their car, or why she tends to think of hunting as a technological activity, but not cooking—take on larger significance. Topics that previously attracted little attention or seemed like science fiction—sex selection, ectogenesis, cloning, or concepts of personal identity on the Internet—now have the potential to produce culture shock.

The issues in this book, then, are both global and personal. Like race, age, religion, science, culture, and politics, gender and technology form and transform society and individuals. Questions about these forces and their interactions get at the multiple hearts of major philosophical and social problems—questions of ethics, social justice, epistemic constraints, personal and social identity, economics and labor, realism and irrealism, and ideas of human nature. Since these sorts of questions have generated such exciting, interdisciplinary work, it is time to create a single text large enough to give readers a taste of the issues and methods that exist at the intersection of gender studies and technology studies. This book attempts to meet that goal, showcasing the variety of perspectives that inform this diverse field of study. Although approaches and topics are varied, there is enough information here for me to generate a useful classification for considering the ways in which technology and gender interact.

I begin with the givenness of both technology and gender. Humans (like many other animals) are a technology-using and technology-producing species. Technology is always present in variegated forms, both subdued and obvious, and is always fundamental to the basic structure and activity of society. Similarly, humans are always already embedded in some sex/gender system, some ideological framework that varies in significant detail from culture to culture and time to time, but which layers vast cultural meaning on the evolved sexual dimorphism of the human organism, setting different roles, expectations, assessments, and values for members of different sexes.

Neither technology nor gender is static, of course. They are both dynamic, though material technology has a way of building upon itself so that its kind of dynamism is often seen as "progressive," not necessarily in the sense of getting continually better (even cancer can be diagnosed as "progressive"), but in the sense of developing finer, greater, and different kinds of manipulability without losing earlier effectiveness. Typically then (though not always), technology increases, and does not merely change into other forms. As such, there is a strong tendency (at least in historical spurts) for technology to "arrive," for technology to be "new" (whether or not "improved"). This important, ever-present association of newness with technology has itself grown stronger. While technology has always been around, it has increased in power and capability exponentially in the nineteenth and twentieth centuries. In the late twentieth century, at least in "high"-tech areas of the world, people have come to expect newer and newer technologies, faster and faster. Much of the debate over technology in general, and technology's effects on gender roles and identity in particular, is generated by the fear that new technologies are moving too fast, or too far, or in the wrong direction from traditional, or at least temporarily established, gender norms.

Gender, on the other hand, though certainly dynamic and in some ways capable of being developed into new and improved forms by inventive experts, does not have the material quality that most technologies do. Changes in gender, even if parallel to changes in technology in many substantial ways, are typically not subject to the kinds of economic distribution or production that technologies are. Changes in gender (as an ideological system) are less likely to be available through a catalog, or at a local factory, or at a trade show. They are less likely to be instantly upgradable or purchasable. The upshot of this is that new technologies are at least somewhat more likely to arrive in existing gender systems than new gender ideologies are to arrive in existing technology systems (though the latter can and does happen). For my purposes here, this means that a significant part of the study of technology and gender is the study of how new technologies are evaluated through the lens of an existing gender system and how new technologies alter existing concepts and practices of that system for better or worse. There are at least four ways in which these sorts of evaluations and alterations can occur—either separately or in various combinations—all of which are examined in this volume.

Technology's Association with Gender

The very concept of technology, as well as its practices, may be more or less strongly embedded in a gender framework itself. Since most gender arrangements are dichotomous, with fairly fixed categories of masculine and feminine, it is not surprising that various other concepts and phenomena get associated with one or the other of these poles. In Western culture men have historically been associated with technology, while women are more typically associated with "nature," perceived (incorrectly, I would argue) as the opposite of technology. Layering these dichotomies on top of one another—man/woman, nature/technology, nature/culture—tends to influence assessments of technology and gender in particular and often contrary ways.

For instance, if men are associated with technology and masculine psychology is considered technology-oriented, then technological development may be interpreted as predominantly a masculine act, the outcome of a masculine drive. This is rather weighty when cashed out in historical, anthropological, and moral terms. As a matter of historical appraisal, when advances in technology such as spears, knives, hammers, and other hunting and building devices are understood as innovations of male hunters, then major leaps forward in human cultural evolution are attributed to men. Male psychology itself is seen as pushing culture ahead. Women, on the other hand, are often assumed to be absent from technological history and cultural evolution, stuck in their primeval homes giving birth, raising children, and gathering food while males roam the countryside, building, warring, changing, and disrupting human society.

As a matter of moral discourse, these associations can lead to conflicting and influential judgments. While some may valorize men's purported technological drive as a positive force, indispensable to intellectual and cultural progress, others may vilify it, claiming that male technophilia is dangerous, and is responsible for environmental destruction, war, the nuclear threat, and alienation from "nature." Some feminists have argued, for example, that men's obsession with technology is a form of "womb envy," a degraded and inferior attempt to imitate women's more purely "natural" creativity. Women, closer to "nature" and more concerned with healthy, authentic lives, may be idealized as moral exemplars in this view, which counters the androcentric dichotomies of *male/technology/cultural/progress* and *female/procreation/cultural/stasis* with the gynocentric dichotomies of *male/technology/bad* and *female/nature/good*.

Whatever the consequences of these long-standing associations, they need to be deeply and critically examined. In fact, the very definition of technology needs to be scrutinized because cultural ideas about gender and technology may detrimentally constrain and bias historical and moral assessments on all sides. The technological character of traditionally "feminine" activities may be ignored—as with technologies of gathering, cooking, and sewing. Female inventors may be disregarded or their inventions attributed to men, gestures that further the idea that women are not technologists. Potentially beneficial technologies created by men (such as some reproductive technologies) may be reflexively rejected because of the assumption that the technology is tainted by male worldviews or the drive to conquer "nature." Technologies that blur the distinctions between the sexes may be automatically interpreted as "threatening" because of their blurring effects.

Technology Reinforcing Gender Systems

New technologies that arrive in existing gender systems (which are almost always hierarchical and typically male-dominated) may be used to shore up those in power, entrench current standards, or extend the ideals of the system. This can happen in several ways, sometimes consciously, sometimes reflexively. Most explicitly and crudely, technology may be used simply to enforce gender roles and restrictions. A chastity belt is a simple device for controlling women's sexual activity and extending the property status of wives. Cosmetic surgeries may be aggressively marketed to magnify sexual differences and ideals—breast implants for increasing women's sexual attractiveness to heterosexual men, or pectoral and bicep implants for increasing men's apparent physical strength. Sex-specific toys encourage children to model particular sex roles—family and dating for girls, war for boys. In more extreme forms, technologies may be used to ensure other cultural gender ideals, such as identifying and eradicating homosexuality through genetic testing or guaranteeing firstborn male children through sex selection techniques.

Another kind of reinforcement occurs when one's gendered social position limits access to technology (new or old). This is likely to happen when there is a strong sexual division of labor or when cultural roles are sharply divided along gender lines. For example, if women are not permitted to work outside the home, particularly in management positions, and a technology like the telephone or computer is marketed primarily as a business machine, then their access to these technologies will be limited. If cooking and sewing are seen as feminine tasks, then microwave ovens and computerized sewing machines may be seen as frivolous expenses by male heads-of-household, or if such devices are purchased, men may be so reluctant to learn how to use them that they are unable to perform the simple tasks of stitching up a seam or baking a potato. If women's social spheres are limited to church and home, they may be refused access to the "family" automobile because it is thought they simply have no reason to drive.

Unequal access is also sometimes the result of beliefs about the "natural" abilities of the sexes. Some technologies are seen as psychologically or physically inappropriate for members of a particular sex—something they could not operate or could not understand. For example, if women are perceived as passive, physically weak, and technically inept, it may be seen as inappropriate for them to use guns, and thus their experience with guns will be limited, despite the fact that lightweight materials and automatic innovations have made guns easy to use or that women might have the need for guns in the first place (e.g., for defending themselves against dangerous men). If men are seen as clumsy, hotheaded, and useful mainly for brute-force manual labor, particularly in cheap export labor markets, they may be excluded from high-tech electronics manufacturing jobs which are thought to need the "delicate" fingers, patience, and task precision of women.

Restricted access also occurs when particular technologies retain their gender connections in the form of cultural prohibitions, even after formal obstacles have been dropped or previous rationales for restrictions (such as physical strength) have been surmounted. The "masculine" or "feminine" aura may still linger, making it more difficult for someone to approach a technology. For instance, cars are still often seen as men's machines and responsibility, even though the computerization of cars has left many formerly "mechanically-minded" men as ignorant of how to fix them as supposedly "non-mechanically-minded" women. A woman who knows how to fix her sewing machine may not be lauded as "technically-minded" even though contemporary sewing machines are complex, computerized devices. Men, on the other hand, can be praised for their technical know-how for replacing the blade on a power mower, a task that requires minimal technical knowledge. Female fighter pilots are still a rarity, even though the old military concerns about upper-body strength and hand-to-hand combat are hardly relevant.

Technology Subverting Gender Systems

While technology can be used to reinforce particular gender roles, it can also be used to subvert them. Technologies can open up options for challenging sex-based restrictions, allowing people to "break out" of proscribed roles and limited spheres of action. This can occur when technology permits people to enter labor markets and professions from which they had previously been excluded because of an actual or perceived sex-based lack of ability. For example, in the military and police, on assembly lines and farms, in construction and landscaping, and in other professions, brute strength was often a (sometimes specious) requirement excluding women from participation. When machines begin to perform most of the hard labor, or in the case of the military, when they vastly decrease hand-to-hand combat in favor of machine-to-machine or machine-to-soldier combat, then the job of the humans involved is to assist, manage, program, or take care of the machines rather than to labor directly themselves. This allows women either to enter professions that no longer require (if they ever really did) assumed male-specific strength, or to extend their previous roles—today's female technician "nurses" fighting machines as the female nurse of yesteryear "repaired" fighting men.

Another kind of alteration occurs when technology changes or eliminates a profession outright, including sex-segregated ones. Sometimes mechanization or other technological shifts eradicate specialized positions, such as typists, gas station attendants, and blacksmiths. If not eradicated, sometimes jobs are changed in ways that eliminate their masculine and feminine associations. For example, the difficult and time-consuming procedure of carrying water from a well often gets cast as a woman's job, because much of the water's use is for domestic chores. However, when indoor plumbing arrives on the scene, the task of water collection is simply abolished and most of its gender connotations along with it. Femininity does not make it through the transition. Turning on a faucet is not considered woman's work, though the water-using domestic chores are still likely to be hers.

Technology can also subvert gender roles by permitting activities which cross restrictive cultural, social, ethical, and interpersonal boundaries, expanding one's movements, social scope, and access to information. For example, women might use automobiles to get out of their house and see a bit of the world. Women might use guns to protect themselves while traveling (the "great equalizer"), lessening their sense of vulnerability. And while television can reinforce gender roles by bringing Donna Reeds and Carol Bradys into the household, it can also open up new possibilities by bringing Mary Tyler Moores, Cagneys and Laceys, and Captain Janeways into the household.

More extensive technologies can shake our assumptions about gender in other ways by opening up the very biological correlates of sex to alteration. Women and men can be turned into each other, at least on a certain anatomical and hormonal level, which generates the very important concept that sex and gender can be divided into kinds or levels, such as anatomical, genetic, hormonal, social, psychological, and sartorial. Women can modify their experiences of childbirth, and the various cultural values that go along with the act, by using anesthesia (considered a sinful technology early on) when giving birth, using reproductive technology to overcome infertility, scheduling C-sections for particular days to work around their calendars, or using treatments which can allow a sixty-three-year-old, post-menopausal woman to have a healthy baby.

As a matter of politics and morality, these gender-subverting uses of technologies are particularly interesting because they are both resisted and demanded. Depending on which ideals of gender are dominant, these technologies can take on an aura of perversion for allowing men and women to step out of their "natural," traditional, and socially legitimated roles, or they can take on a salvific role, offering release from toil, drudgery, and the limitations of social and biological sex.

Technology Altering the Very Nature of Gender and Sex

These last examples begin to get at issues which go beyond merely challenging gender roles and restrictions. They point toward the possibility of a more radical challenge to gender by the technological transformation of sex and of the human body itself.

For some time now, gender studies and feminist theory have been involved in a debate over the meaning of gender and sex, over the very character of gender and sex.

Divided roughly into camps of "essentialists" versus "social constructionists," the debate parallels older realist and idealist battles. The essentialist position may be oversimplified this way: some core, objective property (typically understood as biological or biopsychological) defines what it means to be a woman or a man, and the categories of male and female are thus culture-independent and mind-independent "natural" kinds. The social constructionist position may be oversimplified this way: the categories of male, female, man, and woman are not "natural" kinds but are rather culturally constructed ideals, irreducible to biological or psychological properties, which change demonstrably in meaning and practice over time and across cultures.

The political outcome of these positions is that essentialists tend to view gender differences as innate and immutable, closed at some fundamental level to modification by education, parenting, or ideological movements, with some basic differences in gender roles pragmatically and objectively justified. Social constructionists tend to view gender differences as created, learned, and alterable, with gender role divisions always historically relative, contingent, and ultimately unwarranted by appeal to an objective reality outside human culture. While both sides of this debate can marshal compelling evidence for their general claims, neither is unassailable. The dominant criticism of essentialism is that it does not account for actual observed variability in these "natural" categories and ignores a tremendous amount of conceptual fuzziness and empirical counterexample in its biologistic definitions. The dominant criticism of social constructionism is that it simply seems to rule out any influences of the physical body on behavior, social categories, and self-concepts, treating human beings as if they were only pure minds, exempt from the biological and evolutionary forces that constrain all other organisms.

Irrespective of the theoretical merits of these two positions, technology threatens or promises to circumvent the political heart of the debate by altering the connection between the premises and conclusions of both sides. Essentialists move from the belief that sex and gender differences are hardwired, largely immutable, and socially valuable to the conclusion that attempts to ignore or eradicate them are futile, harmful, and sexually confusing. Social constructionists move from the belief that sex and gender differences are culturally produced and often socially detrimental to the conclusion that they can be radically altered for the better through education, legal reform, and improved theoretical understanding.

While both sides depend for these moves on the assumption that "biological" equals "immutable," technology increasingly erodes that assumption. Taking seriously the essentialist idea that gender identity, behavior, or cognitive and personality traits may be sex-linked physical characteristics of the body does not mean that they are fixed. "Genetic," "biological," and "bodily" do not imply "unchangeable." Even if we doubt the simplified social constructionist claims that sex and gender are categories unconstrained by objective, empirical bodily facts, we have to grant that technology can nonetheless allow us to alter the body in such ways that gender's "naturalness" or "reality" no longer has any permanent sway. Categories of gender and sex, regardless of their possible "essentialist" foundations, are as open to change and difference as the categories of social constructionism.

At proximal technological levels, the "natural" or "biological" constraints of sex are already being modified as reproductive technology permits procreation without sexual intercourse, removes menopause as a barrier to pregnancy, and allows gender-disorienting or gender-ignoring personal interaction through Internet technologies and virtual reality. At slightly more distal levels, technologies such as cloning and *in vitro* gestation allow reproduction without either sexual dimorphism or pregnancy. At more speculative levels, radical bodily changes produced by genetic engineering, cybernetic implants, nanotechnological reconstruction, and artificial intelligence uploading open the possibility of a completely postgendered cyborgism and perhaps even a post-human subjectivity altogether.

As with the use of technology to more mildly subvert existing gender systems, these potential effects on gender identity and sexual being are both resisted and invited. However, these radically disruptive effects on sexual biology and gender identity seem to be more anxiety-producing and politically explosive than mere gender-role shifting technologies because altering the very physicality of sex appears to get at the heart of some cherished and previously unalterable correlates of human social and personal identity. This can be received as a great liberating step forward, or rejected as a great and dangerous loss. It is in response to these sorts of radical technological changes that familiar social and political alliances realign in odd ways. Religious conservatives and radical feminists can find themselves on the same side responding to reproductive technologies, or gender-bending virtual technologies, while gruff old male science fiction writers can find themselves being theorized as postmodern feminists.

This classification system may not exhaust the possibilities for studying technology and gender, but it does get at the core of many debates, evaluations, hopes, and anxieties.

III TECHNOLOGY AND VALUES

At first, the emphasis on sociotechnical systems in this book may seem academic. Why is it so important to understand how users shape machines, how devices cannot exist without social practices, and how institutions weave artifacts and people together into systems? This section will begin to unravel and reveal the importance of these matters. They are significant at least in part because they are connected to values. Many of us want our lives and the lives of future generations to be better than the past and present. We want the future to embody values we hold dear; to be safe, comfortable, peaceful, and secure; to consist of socially just arrangements, and sustainable practices. Understanding how values are entwined in sociotechnical systems is crucial to steering technology to a future that we want.

The idea that values can be wrapped up in sociotechnical systems may seem a bit odd at first. The traditional view is to think of technology as material objects. When technologies are abstracted out of their social and technical context and presented as chunks of matter, it is difficult to see how they can embody or propagate values. But when they are viewed as an essential part of a sociotechnical system—as part of a network of people, institutions, relationships, and interconnections, the value aspects become much clearer. When looked at in their broader social context, as Latour, Lessig, and others in this book have already shown, we can discern how artifacts and techniques shape the way we see the world, impact the way people interact with each other, and help certain values to be realized or not realized. No matter how autonomously a technology may operate, it is never separated from human beings, human endeavors, and human values. The readings in this section offer additional examples of how technologies can enable certain social relationships and constrict others, thereby reinforcing a system of values.

Asserting that technologies are value free is also a common trope in our society. When viewed through the lens of sociotechnical systems, however, this argument dissolves. Consider the claim that engineers have no ethical responsibilities because they simply manipulate matter using scientific laws. An engineer may insist that his or her job is merely to figure out what is possible and leave it to others to decide what should be done. Or the engineer may simply assume that there is one ideal and beautiful technologically sweet solution to any problem and that his or her job is to seek out this solution. The only way these claims can be made plausible, however, is by completely ignoring the social context in which technology is produced and the social consequences of adopting and using technologies. The daily life of an engineer involves much more than just methodically applying scientific laws. In order to be successful, engineers must continuously take into account the social context in which they are working and the social context in which their products, designs, and knowledge will be used. An engineer attempting to design anything must consider who will use it, how it will be used, and what effect it will have. In taking these factors into account, engineers make value judgments. In a similar way, everyone else who makes decisions about technology—from government officials to marketers to users—also take a variety of technical and social factors into account and thus make value judgments.

Yet, the idea that technology and the processes by which it is developed are neutral seems to persist. Politicians who fund projects they believe are important, corporate executives who determine which technologies should be built and how, regulators who specify the standards to which technologies must conform, manufacturers who render designs into concrete form, and users who integrate technologies into their daily lives, all may be tempted to think that a particular machine or device has no influence on values. Yet they are making, using or promoting specific technologies because they hope to shape society in a particular way; they are making decisions in an effort to advance certain values. People design and deploy technologies with certain uses (and users) in mind and once integrated into society, technologies may privilege those uses and users. For instance, a technology may make it slightly easier for one group of people to participate in an activity and slightly more difficult for another group. Once a group gains some control over a technology, they can redefine the activity as their own and continue to increase their impact through subsequent revisions.

While those involved in building a particular technology intentionally seek certain ends and design the technology to reinforce or reconfigure a particular value, in some cases values are privileged without any one individual actively pushing for it. Because of the complexities of existing sociotechnical systems, it is never completely clear what the effects of a new device will be. New technologies can have unexpected consequences and they can produce effects that, while imperceptible at first, can accumulate over time and become quite powerful. Regardless of whether a specific value was actively promoted by a group or individual, technologies can have significant effects on the values and social arrangements in a society.

Most of the authors in this section hold out hope for the future. They have written on this topic because they believe that individuals who understand the ways in which sociotechnical systems work are better able to build sociotechnical systems that make an even better world than we currently have. Recognizing that sociotechnical systems embody social values, enhance the interests of some, and constrain the interests of others, can empower those who contribute to the design of technology to use technologies to shape society for the better.

Questions to consider while reading the selections in this section:

1. How do established, functioning technologies affect social values? Consider examples not mentioned in the readings.

2. How do values influence the development of technology? Explore examples not mentioned in the readings.

3. Distinguish the notion of technological progress from that of social change. What are the values involved in each?

4. How can race and gender bias be reinforced (or countered) in sociotechnical systems?

5. What does the example of the Amish teach us about technology in any society?

13 "Do Artifacts Have Politics?"

Langdon Winner

In this chapter, Langdon Winner raises the seemingly simple question, "Do artifacts have politics?" In asking the question, Winner isn't asking whether televisions belong to one or another political party but whether artifacts alter the power relationships between people. While the traditional answer is to say that artifacts are value neutral—as in the saying "guns don't kill people, people kill people"—Winner argues to the contrary, that artifacts are political. He does this by showing how artifacts are integrated into social systems, how they are supported by social arrangements, and in turn how they embody certain values. He contends that artifacts can be political in two important ways. First, the deployment of a technology may become "a way of settling an issue in the affairs of a particular community." Second, artifacts may be inherently political in the sense that they require or are strongly compatible with particular kinds of political arrangements. Winner argues that technology is not simply associated with political order, but in some real sense *is* political order. Technology, at least in some cases, is the method by which society is constructed. Although scholars have called into question the specific facts underlying Winner's most compelling example (the Long Island bridges), the story he tells serves as a parable of the political nature of technologies. Winner's analysis points to the ways in which certain values—like racial segregation, freedom, and universal access—can be effected, promoted, or denied, either intentionally or accidentally, through the process of design, construction, and use of a technology.

No idea is more provocative in controversies about technology and society than the notion that technical things have political qualities. At issue is the claim that the machines, structures, and systems of modern material culture can be accurately judged not only for their contributions to efficiency and productivity and their positive and negative environmental side effects, but also for the ways in which they can embody specific forms of power and authority. Since ideas of this kind are a persistent and troubling presence in discussions about the meaning of technology, they deserve explicit attention.

Writing in the early 1960s, Lewis Mumford gave classic statement to one version of the theme, arguing that "from late neolithic times in the Near East, right down to our own day, two technologies have recurrently existed side by side: one authoritarian, the other democratic, the first system-centered, immensely powerful, but inherently

---

From Langdon Winner, "Do Artifacts Have Politics?" *The Whale and the Reactor: a Search for Limits in an Age of High Technology* (Chicago: University of Chicago Press, 1986), pp. 19–39. Reprinted with permission. © 1986 The University of Chicago Press.

unstable, the other man-centered, relatively weak, but resourceful and durable."[1] This thesis stands at the heart of Mumford's studies of the city, architecture, and history of technics, and mirrors concerns voiced earlier in the works of Peter Kropotkin, William Morris, and other nineteenth-century critics of industrialism. During the 1970s, antinuclear and pro-solar energy movements in Europe and the United States adopted a similar notion as the centerpiece of their arguments. According to environmentalist Denis Hayes, "The increased deployment of nuclear power facilities must lead society toward authoritarianism. Indeed, safe reliance upon nuclear power as the principal source of energy may be possible only in a totalitarian state." Echoing the views of many proponents of appropriate technology and the soft energy path, Hayes contends that "dispersed solar sources are more compatible than centralized technologies with social equity, freedom and cultural pluralism."[2]

An eagerness to interpret technical artifacts in political language is by no means the exclusive property of critics of large-scale, high-technology systems. A long lineage of boosters has insisted that the biggest and best that science and industry made available were the best guarantees of democracy, freedom, and social justice. The factory system, automobile, telephone, radio, television, space program, and of course nuclear power have all at one time or another been described as democratizing, liberating forces. David Lillienthal's *T.V.A.: Democracy on the March*, for example, found this promise in the phosphate fertilizers and electricity that technical progress was bringing to rural Americans during the 1940s.[3] Three decades later Daniel Boorstin's *The Republic of Technology* extolled television for "its power to disband armies, to cashier presidents, to create a whole new democratic world—democratic in ways never before imagined, even in America."[4] Scarcely a new invention comes along that someone doesn't proclaim it as the salvation of a free society.

It is no surprise to learn that technical systems of various kinds are deeply interwoven in the conditions of modern politics. The physical arrangements of industrial production, warfare, communications, and the like have fundamentally changed the exercise of power and the experience of citizenship. But to go beyond this obvious fact and to argue that certain technologies *in themselves* have political properties seems, at first glance, completely mistaken. We all know that people have politics; things do not. To discover either virtues or evils in aggregates of steel, plastic, transistors, integrated circuits, chemicals, and the like seems just plain wrong, a way of mystifying human artifice and of avoiding the true sources, the human sources of freedom and oppression, justice and injustice. Blaming the hardware appears even more foolish than blaming the victims when it comes to judging conditions of public life.

Hence, the stern advice commonly given those who flirt with the notion that technical artifacts have political qualities: What matters is not technology itself, but the social or economic system in which it is embedded. This maxim, which in a number of variations is the central premise of a theory that can be called the social determination of technology, has an obvious wisdom. It serves as a needed corrective to those who focus uncritically upon such things as "the computer and its social impacts" but who fail to look behind technical devices to see the social circumstances of their development, deployments, and use. This view provides an antidote to naive technological determinism—the idea that technology develops as the sole result of an internal dynamic and then, unmediated by any other influence, molds society to fit its patterns. Those who have not recognized the ways in which technologies are shaped to social and economic forces have not gotten very far.

But the corrective has its own shortcomings; taken literally, it suggests that technical *things* do not matter at all. Once one has done the detective work necessary to reveal the social origins—power holders behind a particular instance of technological change—one will have explained everything of importance. This conclusion offers comfort to social scientists. It validates what they had always suspected, namely, that there is nothing distinctive about the study of technology in the first place. Hence, they can return to their standard models of social power—those of interest-group politics, bureaucratic politics, Marxist models of class struggle, and the like—and have everything they need. The social determination of technology is, in this view, essentially no different from the social determination of, say, welfare policy or taxation.

There are, however, good reasons to believe that technology is politically significant in its own right, good reasons why the standard models of social science only go so far in accounting for what is most interesting and troublesome about the subject. Much of modern social and political thought contains recurring statements of what can be called a theory of technological politics, an odd mongrel of notions often crossbred with orthodox liberal, conservative, and socialist philosophies.[5] The theory of technological politics draws attention to the momentum of large-scale sociotechnical systems, to the response of modern societies to certain technological imperatives, and to the ways human ends are powerfully transformed as they are adapted to technical means. This perspective offers a novel framework of interpretation and explanation for some of the more puzzling patterns that have taken shape in and around the growth of modern material culture. Its starting point is a decision to take technical artifacts seriously. Rather than insist that we immediately reduce everything to the interplay of social forces, the theory of technological politics suggests that we pay attention to the characteristics of technical objects and the meaning of those characteristics. A necessary complement to, rather than a replacement for, theories of the social determination of technology, this approach identifies certain technologies as political phenomena in their own right. It points us back, to borrow Edmund Husserl's philosophical injunction, *to the things themselves*.

In what follows I will outline and illustrate two ways in which artifacts can contain political properties. First are instances in which the invention, design, or arrangement of a specific technical device or system becomes a way of settling an issue in the affairs of a particular community. Seen in the proper light, examples of this kind are fairly straightforward and easily understood. Second are cases of what can be called "inherently political technologies," man-made systems that appear to require or to be strongly compatible with particular kinds of political relationships. Arguments about cases of this kind are much more troublesome and closer to the heart of the matter. By the term "politics" I mean arrangements of power and authority in human associations as well as the activities that take place within those arrangements. For my purposes here, the term "technology" is understood to mean all of modern practical artifice, but to avoid confusion I prefer to speak of "technologies" plural, smaller or larger pieces or systems of hardware of a specific kind.[6] My intention is not to settle any of the issues here once and for all, but to indicate their general dimensions and significance.

Technical Arrangements and Social Order

Anyone who has traveled the highways of America and has gotten used to the normal height of overpasses may well find something a little odd about some of the bridges over the parkways on Long Island, New York. Many of the overpasses are extraordinarily low, having as little as nine feet of clearance at the curb. Even those who happened to notice this structural peculiarity would not be inclined to attach any special meaning to it. In our accustomed way of looking at things such as roads and bridges, we see the details of form as innocuous and seldom give them a second thought.

It turns out, however, that some two hundred or so low-hanging overpasses on Long Island are there for a reason. They were deliberately designed and built that way by someone who wanted to achieve a particular social effect. Robert Moses, the master builder of roads, parks, bridges, and other public works of the 1920s to the 1970s in New York, built his overpasses according to specifications that would discourage the presence of buses on his parkways. According to evidence provided by Moses' biographer, Robert A. Caro, the reasons reflect Moses' social class bias and racial prejudice. Automobile-owning whites of "upper" and "comfortable middle" classes, as he called them, would be free to use the parkways for recreation and commuting. Poor people and blacks, who normally used public transit, were kept off the roads because the twelve-foot tall buses could not handle the overpasses. One consequence was to limit access of racial minorities and low-income groups to Jones Beach, Moses' widely acclaimed public park. Moses made doubly sure of this result by vetoing a proposed extension of the Long Island Railroad to Jones Beach.

Robert Moses' life is a fascinating story in recent U.S. political history. His dealings with mayors, governors, and presidents; his careful manipulation of legislatures, banks, labor unions, the press, and public opinion could be studied by political scientists for years. But the most important and enduring results of his work are his technologies, the vast engineering projects that give New York much of its present form. For generations after Moses' death and the alliances he forged have fallen apart, his public works, especially the highways and bridges he built to favor the use of the automobile over the development of mass transit, will continue to shape that city. Many of his monumental structures of concrete and steel embody a systematic social inequality, a way of engineering relationships among people that, after a time, became just another part of the landscape. As New York planner Lee Koppleman told Caro about the low bridges on Wantagh Parkway, "The old son of a gun had made sure that buses would *never* be able to use his goddamned parkways."[7]

Histories of architecture, city planning, and public works contain many examples of physical arrangements with explicit or implicit political purposes. One can point to Baron Haussmann's broad Parisian thoroughfares, engineered at Louis Napoleon's direction to prevent any recurrence of street fighting of the kind that took place during the revolution of 1848. Or one can visit any number of grotesque concrete buildings and huge plazas constructed on university campuses in the United States during the late 1960s and early 1970s to defuse student demonstrations. Studies of industrial machines and instruments also turn up interesting political stories, including some that violate our normal expectations about why technological innovations are made in the first place. If we suppose that new technologies are introduced to achieve increased efficiency, the history of technology shows that we will sometimes be disappointed. Technological change expresses a panoply of human motives, not the least of which is the desire of some to have dominion over others even though it may require an occasional sacrifice of cost savings and some violation of the normal standard of trying to get more from less.

One poignant illustration can be found in the history of nineteenth-century industrial mechanization. At Cyrus McCormick's reaper manufacturing plant in Chicago in the middle 1880s, pneumatic molding machines, a new and largely untested innovation, were added to the foundry at an estimated cost of $500,000. The standard economic interpretation would lead us to expect that this step was taken to modernize the plant and achieve the kind of efficiencies that mechanization brings. But historian Robert Ozanne has put the development in a broader context. At the time, Cyrus McCormick II was engaged in a battle with the National Union of Iron Molders. He saw the addition of the new machines as a way to "weed out the bad element among the men," namely, the skilled workers who had organized the union local in Chicago.[8] The new machines, manned by unskilled laborers, actually produced inferior castings at a higher cost than the earlier process. After three years of use the machines were, in fact, abandoned, but by that time they had served their purpose—the destruction of the union. Thus, the story of these technical developments at the McCormick factory cannot be adequately understood outside the record of workers' attempts to organize, police repression of the labor movement in Chicago during that period, and the events surrounding the bombing at Haymarket Square. Technological history and U.S. political history were at that moment deeply intertwined.

In the examples of Moses' low bridges and McCormick's molding machines, one sees the importance of technical arrangements that precede the *use* of the things in question. It is obvious that technologies can be used in ways that enhance the power, authority, and privilege of some over others, for example, the use of television to sell a candidate. In our accustomed way of thinking technologies are seen as neutral tools that can be used well or poorly, for good, evil, or something in between. But we usually do not stop to inquire whether a given device might have been designed and built in such a way that it produces a set of consequences logically and temporally *prior to any of its professed uses*. Robert Moses' bridges, after all, were used to carry automobiles from one point to another; McCormick's machines were used to make metal castings; both technologies, however, encompassed purposes far beyond their immediate use. If our moral and political language for evaluating technology includes only categories having to do with tools and uses, if it does not include attention to the meaning of the designs and arrangements of our artifacts, then we will be blinded to much that is intellectually and practically crucial.

Because the point is most easily understood in the light of particular intentions embodied in physical form, I have so far offered illustrations that seem almost conspiratorial. But to recognize the political dimensions in the shapes of technology does not require that we look for conscious conspiracies or malicious intentions. The organized movement of handicapped people in the United States during the 1970s pointed out the countless ways in which machines, instruments, and structures of common use—buses, buildings, sidewalks, plumbing fixtures, and so forth—made it impossible for many handicapped persons to move freely about, a condition that systematically excluded them from public life. It is safe to say that designs unsuited for the handicapped arose more from long-standing neglect than from anyone's active intention. But once the issue was brought to public attention, it became evident that justice required a remedy. A whole range of artifacts have been redesigned and rebuilt to accommodate this minority.

Indeed, many of the most important examples of technologies that have political consequences are those that transcend the simple categories "intended" and "unintended" altogether. These are instances in which the very process of technical development is so thoroughly biased in a particular direction that it regularly produces results heralded as wonderful breakthroughs by some social interests and crushing setbacks by others. In such cases it is neither correct nor insightful to say, "Someone intended to do somebody else harm." Rather one must say that the technological deck has been stacked in advance to favor certain social interests and that some people were bound to receive a better hand than others.

The mechanical tomato harvester, a remarkable device perfected by researchers at the University of California from the late 1940s to the present offers an illustrative tale. The machine is able to harvest tomatoes in a single pass through a row, cutting the plants from the ground, shaking the fruit loose, and (in the newest models) sorting the tomatoes electronically into large plastic gondolas that hold up to twenty-five tons of produce headed for canning factories. To accommodate the rough motion of these harvesters in the field, agricultural researchers have bred new varieties of tomatoes that are hardier, sturdier, and less tasty than those previously grown. The harvesters replace the system of handpicking in which crews of farm workers would pass through the fields three or four times, putting ripe tomatoes in lug boxes and saving immature fruit for later harvest.[9] Studies in California indicate that the use of the machine reduces costs by approximately five to seven dollars per ton as compared to hand harvesting.[10] But the benefits are by no means equally divided in the agricultural economy. In fact, the machine in the garden has in this instance been the occasion for a thorough reshaping of social relationships involved in tomato production in rural California.

By virtue of their very size and cost of more than $50,000 each, the machines are compatible only with a highly concentrated form of tomato growing. With the introduction of this new method of harvesting, the number of tomato growers declined from approximately 4,000 in the early 1960s to about 600 in 1973, and yet there was a substantial increase in tons of tomatoes produced. By the late 1970s an estimated 32,000 jobs in the tomato industry had been eliminated as a direct consequence of mechanization.[11] Thus, a jump in productivity to the benefit of very large growers has occurred at the sacrifice of other rural agricultural communities.

The University of California's research on and development of agricultural machines such as the tomato harvester eventually became the subject of a lawsuit filed by attorneys for California Rural Legal Assistance, an organization representing a group of farm workers and other interested parties. The suit charged that university officials are spending tax monies on projects that benefit a handful of private interests to the detriment of farm workers, small farmers, consumers, and rural California generally and asks for a court injunction to stop the practice. The university denied these charges, arguing that to accept them "would require elimination of all research with any potential practical application."[12]

As far as I know, no one argued that the development of the tomato harvester was the result of a plot. Two students of the controversy, William Friedland and Amy Barton, specifically exonerate the original developers of the machine and the hard tomato from any desire to facilitate economic concentration in that industry.[13] What we see here instead is an ongoing social process in which scientific knowledge, technological invention, and corporate profit reinforce each other in deeply entrenched patterns, patterns that bear the unmistakable stamp of political and economic power. Over many decades agricultural research and development in U.S. land-grant colleges and universities has tended to favor the interests of large agribusiness concerns.[14] It is in the face of such subtly ingrained patterns that opponents of innovations such as the tomato harvester are made to seem "antitechnology" or "antiprogress." For the harvester is not merely the symbol of a social order that rewards some while punishing others; it is in a true sense an embodiment of that order.

Within a given category of technological change there are, roughly speaking, two kinds of choices that can affect the relative distribution of power, authority, and privilege in a community. Often the crucial decision is a simple "yes or no" choice—are we going to develop and adopt the thing or not? In recent years many local, national, and international disputes about technology have centered on "yes or no" judgments about such things as food additives, pesticides, the building of highways, nuclear reactors, dam projects, and proposed high-tech weapons. The fundamental choice about an antiballistic missile or supersonic transport is whether or not the thing is going to join society as a piece of its operating equipment. Reasons given for and against are frequently as important as those concerning the adoption of an important new law.

A second range of choices, equally critical in many instances, has to do with specific features in the design or arrangement of a technical system after the decision to go ahead with it has already been made. Even after a utility company wins permission to build a large electric power line, important controversies can remain with respect to the placement of its route and the design of its towers; even after an organization has decided to institute a system of computers, controversies can still arise with regard to the kinds of components, programs, modes of access, and other specific features the system will include. Once the mechanical tomato harvester had been developed in its basic form, a design alteration of critical social significance—the addition of electronic sorters, for example—changed the character of the machine's effects upon the balance of wealth and power in California agriculture. Some of the most interesting research on technology and politics at present focuses upon the attempt to demonstrate in a detailed, concrete fashion how seemingly innocuous design features in mass transit systems, water projects, industrial machinery, and other technologies actually mask social choices of profound significance. Historian David Noble has studied two kinds of automated machine tool systems that have different implications for the relative power of management and labor in the industries that might employ them. He has shown that although the basic electronic and mechanical components of the record/playback and numerical control systems are similar, the choice of one design over another has crucial consequences for social struggles on the shop floor. To see the matter solely in terms of cost cutting, efficiency, or the modernization of equipment is to miss a decisive element in the story.[15]

From such examples I would offer some general conclusions. These correspond to the interpretation of technologies as "forms of life" presented in the previous chapter, filling in the explicitly political dimensions of that point of view.

The things we call "technologies" are ways of building order in our world. Many technical devices and systems important in everyday life contain possibilities for many different ways of ordering human activity. Consciously or unconsciously, deliberately or inadvertently, societies choose structures for technologies that influence how people are going to work, communicate, travel, consume, and so forth over a very long time. In the processes by which structuring decisions are made, different people are situated differently and possess unequal degrees of power as well as unequal levels of awareness. By far the greatest latitude of choice exists the very first time a particular instrument, system, or technique is introduced. Because choices tend to become strongly fixed in material equipment, economic investment, and social habit, the original flexibility vanishes for all practical purposes once the initial commitments are made. In that sense technological innovations are similar to legislative acts or political foundings that establish a framework for public order that will endure over many generations. For that reason the same careful attention one would give to the rules, roles, and relationships of politics must also be given to such things as the building of highways, the creation of television networks, and the tailoring of seemingly insignificant features on new machines. The issues that divide or unite people in society are settled not only in the institutions and practices of politics proper, but also, and less obviously, in tangible arrangements of steel and concrete, wires and semiconductors, nuts and bolts.

Inherently Political Technologies

None of the arguments and examples considered thus far addresses a stronger, more troubling claim often made in writings about technology and society—the belief that some technologies are by their very nature political in a specific way. According to this view, the adoption of a given technical system unavoidably brings with it conditions for human relationships that have a distinctive political cast—for example, centralized or decentralized, egalitarian or inegalitarian, repressive or liberating. This is ultimately what is at stake in assertions such as those of Lewis Mumford that two traditions of technology, one authoritarian, the other democratic, exist side by side in Western history. In all the cases cited above the technologies are relatively flexible in design and arrangement and variable in their effects. Although one can recognize a particular result produced in a particular setting, one can also easily imagine how a roughly similar device or system might have been built or situated with very much different political consequences. The idea we must now examine and evaluate is that certain kinds of technology do not allow such flexibility, and that to choose them is to choose unalterably a particular form of political life.

A remarkably forceful statement of one version of this argument appears in Friedrich Engels's little essay "On Authority" written in 1872. Answering anarchists who believed that authority is an evil that ought to be abolished altogether, Engels launches into a panegyric for authoritarianism, maintaining, among other things, that strong authority is a necessary condition in modern industry. To advance his case in the strongest possible way, he asks his readers to imagine that the revolution has already occurred. "Supposing a social revolution dethroned the capitalists, who now exercise their authority over the production and circulation of wealth. Supposing, to adopt entirely the point of view of the anti-authoritarians, that the land and the instruments of labour had become the collective property of the workers who use them. Will authority have disappeared or will it have only changed its form?"[16]

His answer draws upon lessons from three sociotechnical systems of his day, cotton-spinning mills, railways, and ships at sea. He observes that on its way to becoming finished thread, cotton moves through a number of different operations at different locations in the factory. The workers perform a wide variety of tasks, from running the steam engine to carrying the products from one room to another. Because these tasks must be coordinated and because the timing of the work is "fixed by the authority of the steam," laborers must learn to accept a rigid discipline. They must, according to Engels, work at regular hours and agree to subordinate their individual wills to the persons in charge of factory operations. If they fail to do so, they risk the horrifying possibility that production will come to a grinding halt. Engels pulls no punches. "The automatic machinery of a big factory," he writes, "is much more despotic than the small capitalists who employ workers ever have been."[17]

Similar lessons are adduced in Engels's analysis of the necessary operating conditions for railways and ships at sea. Both require the subordination of workers to an "imperious authority" that sees to it that things run according to plan. Engels finds that far from being an idiosyncrasy of capitalist social organization, relationships of authority and subordination arise "independently of all social organization, [and] are imposed upon us together with the material conditions under which we produce and make products circulate." Again, he intends this to be stern advice to the anarchists who, according to Engels, thought it possible simply to eradicate subordination and superordination at a single stroke. All such schemes are nonsense. The roots of unavoidable authoritarianism are, he argues, deeply implanted in the human involvement with science and technology. "If man, by dint of his knowledge and inventive genius, has subdued the forces of nature, the latter avenge themselves upon him by subjecting him, insofar as he employs them, to a veritable despotism independent of all social organization."[18]

Attempts to justify strong authority on the basis of supposedly necessary conditions of technical practice have an ancient history. A pivotal theme in the *Republic* is Plato's quest to borrow the authority of *technē* and employ it by analogy to buttress his argument in favor of authority in the state. Among the illustrations he chooses, like Engels, is that of a ship on the high seas. Because large sailing vessels by their very nature need to be steered with a firm hand, sailors must yield to their captain's commands; no reasonable person believes that ships can be run democratically. Plato goes on to suggest that governing a state is rather like being captain of a ship or like practicing medicine as a physician. Much the same conditions that require central rule and decisive action in organized technical activity also create this need in government.

In Engels's argument, and arguments like it, the justification for authority is no longer made by Plato's classic analogy, but rather directly with reference to technology itself. If the basic case is as compelling as Engels believed it to be, one would expect that as a society adopted increasingly complicated technical systems as its material basis, the prospects for authoritarian ways of life would be greatly enhanced. Central control by knowledgeable people acting at the top of a rigid social hierarchy would seem increasingly prudent. In this respect his stand in "On Authority" appears to be at variance with Karl Marx's position in Volume I of *Capital*. Marx tries to show that increasing mechanization will render obsolete the hierarchical division of labor and the relationships of subordination that, in his view, were necessary during the early stages of modern manufacturing. "Modern Industry," he writes, "sweeps away by technical means the manufacturing division of labor, under which each man is bound hand and foot for life to a single detail operation. At the same time, the capitalistic form of that industry reproduces this same division of labour in a still more monstrous shape; in the factory proper, by converting the workman into a living appendage of the machine."[19] In Marx's view the conditions that will eventually dissolve the capitalist division of labor and facilitate proletarian revolution are conditions latent in industrial technology itself. The differences between Marx's position in *Capital* and Engels's in his essay raise an important question for socialism: What, after all, does modern technology make possible or necessary in political life? The theoretical tension we see here mirrors many troubles in the practice of freedom and authority that had muddied the tracks of socialist revolution.

Arguments to the effect that technologies are in some sense inherently political have been advanced in a wide variety of contexts, far too many to summarize here. My reading of such notions, however, reveals there are two basic ways of stating the case. One version claims that the adoption of a given technical system actually requires the creation and maintenance of a particular set of social conditions as the operating environment of that system. Engels's position is of this kind. A similar view is offered by a contemporary writer who holds that "if you accept nuclear power plants, you also accept a techno-scientific-industrial-military elite. Without these people in charge, you could not have nuclear power."[20] In this conception some kinds of technology require their social environments to be structured in a particular way in much the same sense that an automobile requires wheels in order to move. The thing could not exist as an effective operating entity unless certain social as well as material conditions were met. The meaning of "required" here is that of practical (rather than logical) necessity. Thus, Plato thought it a practical necessity that a ship at sea have one captain and an unquestionably obedient crew.

A second, somewhat weaker, version of the argument holds that a given kind of technology is strongly compatible with, but does not strictly require, social and political relationships of a particular stripe. Many advocates of solar energy have argued that technologies of that variety are more compatible with a democratic, egalitarian society than energy systems based on coal, oil, and nuclear power; at the same time they do not maintain that anything about solar energy requires democracy. Their case is, briefly, that solar energy is decentralizing in both a technical and political sense: technically speaking, it is vastly more reasonable to build solar systems in a disaggregated, widely distributed manner than in large-scale centralized plants; politically speaking, solar energy accommodates the attempts of individuals and local communities to manage their affairs effectively because they are dealing with systems that are more accessible, comprehensible, and controllable than huge centralized sources. In this view solar energy is desirable not only for its economic and environmental benefits, but also for the salutary institutions it is likely to permit in other areas of public life.[21]

Within both versions of the argument there is a further distinction to be made between conditions that are internal to the workings of a given technical system and those that are external to it. Engels's thesis concerns internal social relations said to be required within cotton factories and railways, for example; what such relationships mean for the condition of society at large is, for him, a separate question. In contrast, the solar advocate's belief that solar technologies are compatible with democracy pertains to the way they complement aspects of society removed from the organization of those technologies as such.

There are, then, several different directions that arguments of this kind can follow. Are the social conditions predicated said to be required by, or strongly compatible with, the workings of a given technical system? Are those conditions internal to that system or external to it (or both)? Although writings that address such questions are often unclear about what is being asserted, arguments in this general category are an important part of modern political discourse. They enter into many attempts to explain how changes in social life take place in the wake of technological innovation. More important, they are often used to buttress attempts to justify or criticize proposed courses of action involving new technology. By offering distinctly political reasons for or against the adoption of a particular technology, arguments of this kind stand apart from more commonly employed, more easily quantifiable claims about economic costs and benefits, environmental impacts, and possible risks to public health and safety that technical systems may involve. The issue here does not concern how many jobs will be created, how much income generated, how many pollutants added, or how many cancers produced. Rather, the issue has to do with ways in which choices about technology have important consequences for the form and quality of human associations.

If we examine social patterns that characterize the environments of technical systems, we find certain devices and systems almost invariably linked to specific ways of organizing power and authority. The important question is: Does this state of affairs derive from an unavoidable social response to intractable properties in the things themselves, or is it instead a pattern imposed independently by a governing body, ruling class, or some other social or cultural institution to further its own purposes?

Taking the most obvious example, the atom bomb is an inherently political artifact. As long as it exists at all, its lethal properties demand that it be controlled by a centralized, rigidly hierarchical chain of command closed to all influences that might make its workings unpredictable. The internal social system of the bomb must be authoritarian; there is no other way. The state of affairs stands as a practical necessity independent of any larger political system in which the bomb is embedded, independent of the type of regime or character of its rulers. Indeed, democratic states must try to find ways to ensure that the social structures and mentality that characterize the management of nuclear weapons do not "spin off" or "spill over" into the polity as a whole.

The bomb is, of course, a special case. The reasons very rigid relationships of authority are necessary in its immediate presence should be clear to anyone. If, however, we look for other instances in which particular varieties of technology are widely perceived to need the maintenance of a special pattern of power and authority, modern technical history contains a wealth of examples.

Alfred D. Chandler in *The Visible Hand*, a monumental study of modern business enterprise, presents impressive documentation to defend the hypothesis that the construction and day-to-day operation of many systems of production, transportation, and communication in the nineteenth and twentieth centuries require the development of particular social form—a large-scale centralized, hierarchical organization administered by highly skilled managers. Typical of Chandler's reasoning is his analysis of the growth of the railroads.[22]

Technology made possible fast, all-weather transportation; but safe, regular, reliable movement of goods and passengers, as well as the continuing maintenance and repair of locomotives, rolling stock, and track, roadbed, stations, roundhouses, and other equipment, required the creation of a sizable administrative organization. It meant the employment of a set of managers to supervise these functional activities over an extensive geographical area; and the appointment of an administrative command of middle and top executives to monitor, evaluate, and coordinate the work of managers responsible for the day-to-day operations.

Throughout his book Chandler points to ways in which technologies used in the production and distribution of electricity, chemicals, and a wide range of industrial goods "demanded" or "required" this form of human association. "Hence, the operational requirements of railroads demanded the creation of the first administrative hierarchies in American business."[23]

Were there other conceivable ways of organizing these aggregates of people and apparatus? Chandler shows that a previously dominant social form, the small traditional family firm, simply could not handle the task in most cases. Although he does not speculate further, it is clear that he believes there is, to be realistic, very little latitude in the forms of power and authority appropriate within modern sociotechnical systems. The properties of many modern technologies—oil pipelines and refineries, for example—are such that overwhelmingly impressive economies of scale and speed are possible. If such systems are to work effectively, efficiently, quickly, and safely, certain requirements of internal social organization have to be fulfilled; the material possibilities that modern technologies make available could not be exploited otherwise. Chandler acknowledges that as one compares sociotechnical institutions of different nations, one sees "ways in which cultural attitudes, values, ideologies, political systems, and social structure affect these imperatives."[24] But the weight of argument and empirical evidence in *The Visible Hand* suggests that any significant departure from the basic pattern would be, at best, highly unlikely.

It may be that other conceivable arrangements of power and authority, for example, those of decentralized, democratic worker self-management, could prove capable of administering factories, refineries, communications systems, and railroads as well as or better than the organizations Chandler describes. Evidence from automobile assembly teams in Sweden and worker-managed plants in Yugoslavia and other countries is often presented to salvage these possibilities. Unable to settle controversies over this matter here, I merely point to what I consider to be their bone of contention. The available evidence tends to show that many large, sophisticated technological systems are in fact highly compatible with centralized, hierarchical managerial control. The interesting question, however, has to do with whether or not this pattern is in any sense a requirement of such systems, a question that is not solely empirical. The matter ultimately rests on our judgments about what steps, if any, are practically necessary in the workings of particular kinds of technology and what, if anything, such measures require of the structure of human associations. Was Plato right in saying that a ship at sea needs steering by a decisive hand and that this could only be accomplished by a single captain and an obedient crew? Is Chandler correct in saying that the properties of large-scale systems require centralized, hierarchical managerial control?

To answer such questions, we would have to examine in some detail the moral claims of practical necessity (including those advocated in the doctrines of economics) and weigh them against moral claims of other sorts, for example, the notion that it is good for sailors to participate in the command of a ship or that workers have a right to be involved in making and administering decisions in a factory. It is characteristic of societies based on large, complex technological systems, however, that moral reasons other than those of practical necessity appear increasingly obsolete, "idealistic," and irrelevant. Whatever claims one may wish to make on behalf of liberty, justice, or equality can be immediately neutralized when confronted with arguments to the effect, "Fine, but that's no way to run a railroad" (or steel mill, or airline, or communication system, and so on). Here we encounter an important quality in modern political discourse and in the way people commonly think about what measures are justified in response to the possibilities technologies make available. In many instances, to say that some technologies are inherently political is to say that certain widely accepted reasons of practical necessity—especially the need to maintain crucial technological systems as smoothly working entities—have tended to eclipse other sorts of moral and political reasoning.

One attempt to salvage the autonomy of politics from the bind of practical necessity involves the notion that conditions of human association found in the internal workings of technological systems can easily be kept separate from the polity as a whole. Americans have long rested content in the belief that arrangements of power and authority inside industrial corporations, public utilities, and the like have little bearing on public institutions, practices, and ideas at large. That "democracy stops at the factory gates" was taken as a fact of life that had nothing to do with the practice of political freedom. But can the internal politics of technology and the politics of the whole community be so easily separated? A recent study of business leaders in the United States, contemporary exemplars of Chandler's "visible hand of management," found them remarkably impatient with such democratic scruples as "one man, one vote." If democracy doesn't work for the firm, the most critical institution in all of society, American executives ask, how well can it be expected to work for the government of a nation—particularly when that government attempts to interfere with the achievements of the firm? The authors of the report observe that patterns of authority that work effectively in the corporation become for businessmen "the desirable model against which to compare political and economic relationships in the rest of society."[25] While such findings are far from conclusive, they do reflect a sentiment increasingly common in the land: what dilemmas such as the energy crisis require is not a redistribution of wealth or broader public participation but, rather, stronger, centralized public and private management.

An especially vivid case in which the operational requirements of a technical system might influence the quality of public life is the debates about the risks of nuclear power. As the supply of uranium for nuclear reactors runs out, a proposed alternative fuel is the plutonium generated as a by-product in reactor cores. Well-known objections to plutonium recycling focus on its unacceptable economic costs, its risks of environmental contamination, and its dangers in regard to the international proliferation of nuclear weapons. Beyond these concerns, however, stands another less widely appreciated set of hazards—those that involve the sacrifice of civil liberties. The widespread use of plutonium as a fuel increases the chance that this toxic substance might be stolen by terrorists, organized crime, or other persons. This raises the prospect, and not a trivial one, that extraordinary measures would have to be taken to safeguard plutonium from theft and to recover it should the substance be stolen. Workers in the nuclear industry as well as ordinary citizens outside could well become subject to background security checks, covert surveillance, wiretapping, informers, and even emergency measures under martial law—all justified by the need to safeguard plutonium.

Russell W. Ayres's study of the legal ramifications of plutonium recycling concludes: "With the passage of time and the increase in the quantity of plutonium in existence will come pressure to eliminate the traditional checks the courts and legislatures place on the activities of the executive and to develop a powerful central authority better able to enforce strict safeguards." He avers that "once a quantity of plutonium had been stolen, the case for literally turning the country upside down to get it back would be overwhelming." Ayres anticipates and worries about the kinds of thinking that, I have argued, characterize inherently political technologies. It is still true that in a world in which human beings make and maintain artificial systems nothing is "required" in an absolute sense. Nevertheless, once a course of action is under way, once artifacts such as nuclear power plants have been built and put in operation, the kinds of reasoning that justify the adaptation of social life to technical requirements pop up as spontaneously as flowers in the spring. In Ayres's words, "Once recycling begins and the risks of plutonium theft become real rather than hypothetical, the case for governmental infringement of protected rights will seem compelling."[26] After a certain point, those who cannot accept the hard requirements and imperatives will be dismissed as dreamers and fools.

The two varieties of interpretation I have outlined indicate how artifacts can have political qualities. In the first instance we noticed ways in which specific features in the design or arrangement of a device or system could provide a convenient means of establishing patterns of power and authority in a given setting. Technologies of this kind have a range of flexibility in the dimensions of their material form. It is precisely because they are flexible that their consequences for society must be understood with reference to the social actors able to influence which designs and arrangements are chosen. In the second instance we examined ways in which the intractable properties of certain kinds of technology are strongly, perhaps unavoidably, linked to particular institutionalized patterns of power and authority. Here the initial choice about whether or not to adopt something is decisive in regard to its consequences. There are no alternative physical designs or arrangements that would make a significant difference; there are, furthermore, no genuine possibilities for creative intervention by different social systems—capitalist or socialist—that could change the intractability of the entity or significantly alter the quality of its political effects.

To know which variety of interpretation is applicable in a given case is often what is at stake in disputes, some of them passionate ones, about the meaning of technology for how we live. I have argued a "both/and" position here, for it seems to me that both kinds of understanding are applicable in different circumstances. Indeed, it can happen that within a particular complex of technology—a system of communication or transportation, for example—some aspects may be flexible in their possibilities for society, while other aspects may be (for better or worse) completely intractable. The two varieties of interpretation I have examined here can overlap and intersect at many points.

There are, of course, issues on which people can disagree. Thus, some proponents of energy from renewable resources now believe they have at last discovered a set of intrinsically democratic, egalitarian, communitarian technologies. In my best estimation, however, the social consequences of building renewable energy systems will surely depend on the specific configurations of both hardware and the social institutions created to bring that energy to us. It may be that we will find ways to turn this silk purse into a sow's ear. By comparison, advocates of the further development of nuclear power seem to believe that they are working on a rather flexible technology whose adverse social effects can be fixed by changing the design parameters of reactors and nuclear waste disposal systems. For reasons indicated above, I believe them to be dead wrong in that faith. Yes, we may be able to manage some of the "risks" to public health and safety that nuclear power brings. But as society adapts to the more dangerous and apparently indelible features of nuclear power, what will be the long-range toll in human freedom?

My belief that we ought to attend more closely to technical objects themselves is not to say that we can ignore the contexts in which those objects are situated. A ship at sea may well require, as Plato and Engels insisted, a single captain and obedient crew. But a ship out of service, parked at the dock, needs only a caretaker. To understand which technologies and which contexts are important to us, and why, is an enterprise that must involve both the study of specific technical systems and their history as well as a thorough grasp of the concepts and controversies of political theory. In our times people are often willing to make drastic changes in the way they live to accommodate technological innovation while at the same time resisting similar kinds of changes justified on political grounds. If for no other reason than that, it is important for us to achieve a clearer view of these matters than has been our habit so far.

Notes

1. Lewis Mumford, "Authoritarian and Democratic Technics," *Technology and Culture* 5:1–8, 1964.

2. Denis Hayes, *Rays of Hope: The Transition to a Post-Petroleum World* (New York: W. W. Norton, 1977), 71, 159.

3. David Lillienthal, *T.V.A.: Democracy on the March* (New York: Harper and Brothers, 1944), 72–83.

4. Daniel J. Boorstin, *The Republic of Technology* (New York: Harper and Row, 1978), 7.

5. Langdon Winner, *Autonomous Technology: Technics-Out-of-Control as a Theme in Political Thought* (Cambridge: MIT Press, 1977).

6. The meaning of "technology" I employ in this essay does not encompass some of the broader definitions of that concept found in contemporary literature, for example, the notion of "technique" in the writings of Jacques Ellul. My purposes here are more limited. For a discussion of the difficulties that arise in attempts to define "technology," see *Autonomous Technology*, 8–12.

7. Robert A. Caro, *The Power Broker: Robert Moses and the Fall of New York* (New York: Random House, 1974), 318, 481, 514, 546, 951–958, 952.

8. Robert Ozanne, *A Century of Labor-Management Relations at McCormick and International Harvester* (Madison: University of Wisconsin Press, 1967), 20.

9. The early history of the tomato harvester is told in Wayne D. Rasmussen, "Advances in American Agriculture: The Mechanical Tomato Harvester as a Case Study," *Technology and Culture* 9:531–543, 1968.

10. Andrew Schmitz and David Seckler, "Mechanized Agriculture and Social Welfare: The Case of the Tomato Harvester," *American Journal of Agricultural Economics* 52:569–577, 1970.

11. William H. Friedland and Amy Barton, "Tomato Technology," *Society* 13:6, September/October 1976. See also William H. Friedland, *Social Sleep-walkers: Scientific and Technological Research in California Agriculture*, University of California, Davis, Department of Applied Behavioral Sciences, Research Monograph No. 13, 1974.

12. *University of California Clip Sheet* 54:36, May 1, 1979.

13. "Tomato Technology."

14. A history and critical analysis of agricultural research in the land-grant colleges is given in James Hightower, *Hard Tomatoes, Hard Times* (Cambridge: Schenkman, 1978).

15. David F. Noble, *Forces of Production: A Social History of Machine Tool Automation* (New York: Alfred A. Knopf, 1984).

16. Friedrich Engels, "On Authority," in *The Marx-Engels Reader*, ed. 2, Robert Tucker (ed.) (New York: W. W. Norton, 1978), 731.

17. Ibid.

18. Ibid., 732, 731.

19. Karl Marx, *Capital*, vol. 1, ed. 3, translated by Samuel Moore and Edward Aveling (New York: Modern Library, 1906), 530.

20. Jerry Mander, *Four Arguments for the Elimination of Television* (New York: William Morrow, 1978), 44.

21. See, for example, Robert Argue, Barbara Emanuel, and Stephen Graham, *The Sun Builders: A People's Guide to Solar, Wind and Wood Energy in Canada* (Toronto: Renewable Energy in Canada, 1978). "We think decentralization is an implicit component of renewable energy; this implies the decentralization of energy systems, communities and of power. Renewable energy doesn't require mammoth generation sources of disruptive transmission corridors. Our cities and towns, which have been dependent on centralized energy supplies, may be able to achieve some degree of autonomy, thereby controlling and administering their own energy needs." (16)

22. Alfred D. Chandler, Jr., *The Visible Hand: The Managerial Revolution in American Business* (Cambridge: Belknap, 1977), 244.

23. Ibid.

24. Ibid., 500.

25. Leonard Silk and David Vogel, *Ethics and Profits: The Crisis of Confidence in American Business* (New York: Simon and Schuster, 1976), 191.

26. Russell W. Ayres, "Policing Plutonium: The Civil Liberties Fallout," *Harvard Civil Rights—Civil Liberties Law Review* 10 (1975): 443, 413–414, 374.

14 "Control: Human and Nonhuman Robots"

George Ritzer

This chapter by George Ritzer makes Forster's vision of an automated world seem not too far off in the future. It is taken from his book, *The McDonaldization of Society*, in which he examines how corporate management of the fast food industry uses technologies to promote its own values. Two of their most important, and interrelated, goals are efficiency and predictability. Because they do not want to rely on their employees to accept and promote these goals, fast food management has sought to design technologies that either control or bypass the need to rely on employees. The resulting machines, technologies, and techniques not only direct the minutia of how people work, but radically alter who can perform the work and, perhaps more importantly, who would *want* to do the work—thereby giving management even more control. Ritzer goes on to argue that these same techniques are used not only in the fast food industry, but in many other fields as well including education, health care, and even farming. And it is not just employees who are subject to such control. The way that customers, students, patients, and even livestock experience certain situations is carefully crafted by those who design the technologies and techniques. Ritzer's chapter leaves us with at least two important questions. First, what values are being realized when nearly every machine or technique is designed to further the interest of specific groups of people (often at the expense of other groups)? And second, what will happen as we develop technologies with new and more powerful abilities to control people?

This chapter presents the fourth dimension of McDonaldization: increased control through the replacement of human with nonhuman technology. *Technology* includes not only machines and tools but also materials, skills, knowledge, rules, regulations, procedures, and techniques. Thus, technologies include not only the obvious, such as robots and computers, but also the less obvious, such as the assembly line, bureaucratic rules, and manuals prescribing accepted procedures and techniques. A *human technology* (a screwdriver, for example) is controlled by people; *a nonhuman technology* (the order window at the drive-through, for instance) controls people.

The great source of uncertainty, unpredictability, and inefficiency in any rationalizing system is people—either those who work within it or those served by it. Hence, efforts to increase control are usually aimed at both employees and customers, although processes and products may also be the targets.

From *The McDonaldization of Society*. Revised New Century Edition (Thousand Oaks, Calif.: Pine Forge Press, 2004), pp. 106–133, 269–274. © 2004 Sage Publications Inc. Reprinted by permission of Pine Forge Press Inc.

Historically, organizations gained control over people gradually through increasingly effective technologies.[1] Eventually, they began reducing people's behavior to a series of machinelike actions. And once people were behaving like machines, they could be replaced with actual machines. The replacement of humans by machines is the ultimate stage in control over people; people can cause no more uncertainty and unpredictability because they are no longer involved, at least directly, in the process.

Control is not the only goal associated with nonhuman technologies. These technologies are created and implemented for many reasons, such as increased productivity, greater quality control, and lower cost. However, this chapter is mainly concerned with the ways nonhuman technologies have increased control over employees and consumers in a McDonaldizing society.

Controlling Employees

Before the age of sophisticated nonhuman technologies, people were largely controlled by other people. In the workplace, owners and supervisors controlled subordinates directly, face-to-face. But such direct, personal control is difficult, costly, and likely to engender personal hostility. Subordinates will likely strike out at an immediate supervisor or an owner who exercises excessively tight control over their activities. Control through a technology is easier, less costly in the long run, and less likely to engender hostility toward supervisors and owners. Thus, over time, control by people has shifted toward control by technologies.[2]

The Fast-Food Industry: From Human to Mechanical Robots

Fast-food restaurants have coped with problems of uncertainty by creating and instituting many nonhuman technologies. Among other things, they have done away with a cook, at least in the conventional sense. Grilling a hamburger is so simple that anyone can do it with a bit of training. Furthermore, even when more skill is required (as in the case of cooking an Arby's roast beef), the fast-food restaurant develops a routine involving a few simple procedures that almost anyone can follow. Cooking fast food is like a game of connect-the-dots or painting-by-numbers. Following prescribed steps eliminates most of the uncertainties of cooking.

Much of the food prepared at McDonaldized restaurants arrives preformed, precut, presliced, and "preprepared." All employees need to do, when necessary, is cook or often merely heat the food and pass it on to the customer. At Taco Bell, workers used to spend hours cooking meat and shredding vegetables. Now, the workers simply drop bags of frozen ready-cooked beef into boiling water. They have used preshredded lettuce for some time, and more recently preshredded cheese and prediced tomatoes have appeared.[3] The more that is done by nonhuman technologies before the food arrives at the restaurant, the less workers need to do and the less room they have to exercise their own judgment and skill.

McDonald's has developed a variety of machines to control its employees. The soft drink dispenser has a sensor that automatically shuts off the flow when the cup is full.

Ray Kroc's dissatisfaction with the vagaries of human judgment led to the elimination of french fry machines controlled by humans and to the development of machines that ring or buzz when the fries are done or that automatically lift the french fry baskets out of the hot oil. When an employee controls the french fry machine, misjudgment may lead to undercooked, overcooked, or even burned fries. Kroc fretted over this problem: "It was amazing that we got them as uniform as we did, because each kid working the fry vats would have his own interpretation of the proper color and so forth."[4]

At the cash register, workers once had to look at a price list and then punch the prices in by hand—so that the wrong (even lower) amount could be rung up. Computer screens and computerized cash registers forestall that possibility.[5] All the employees need do is press the image on the register that matches the item purchased; the machine then produces the correct price.

If the objective in a fast-food restaurant is to reduce employees to human robots, we should not be surprised by the spread of robots that prepare food. For example, a robot cooks hamburgers at one campus restaurant:

The robot looks like a flat oven with conveyor belts running through and an arm attached at the end. A red light indicates when a worker should slide in a patty and bun, which bob along in the heat for 1 minute 52 seconds. When they reach the other side of the machine, photo-optic sensors indicate when they can be assembled.

The computer functioning as the robot's brain determines when the buns and patty are where they should be. If the bun is delayed, it slows the patty belt. If the patty is delayed, it slows bun production. It also keeps track of the number of buns and patties in the oven and determines how fast they need to be fed in to keep up speed.[6]

Robots offer a number of advantages—lower cost, increased efficiency, fewer workers, no absenteeism, and a solution to the decreasing supply of teenagers needed to work at fast-food restaurants. The professor who came up with the idea for the robot that cooks hamburgers said, "Kitchens have not been looked at as factories, which they are.... Fast-food restaurants were the first to do that."[7]

Taco Bell developed "a computer-driven machine the size of a coffee table that...can make and seal in a plastic bag a perfect hot taco."[8] Another company worked on an automated drink dispenser that produced a soft drink in fifteen seconds: "Orders are punched in at the cash register by a clerk. A computer sends the order to the dispenser to drop a cup, fill it with ice and appropriate soda, and place a lid on top. The cup is then moved by conveyor to the customer."[9] When such technologies are refined and prove to be less expensive and more reliable than humans, fast-food restaurants will employ them widely.

McDonald's experimented with a limited program called ARCH, or Automated Robotic Crew Helper. A french fry robot fills the fry basket, cooks the fries, empties the basket when sensors tell it the fries are done, and even shakes the fries while they are being cooked. In the case of drinks, an employee pushes a button on the cash register to place an order. The robot then puts the proper amount of ice in the cup, moves the cup under the correct spigot, and allows the cup to fill. It then places the cup on a conveyor, which moves it to the employee, who passes it on to the customer.[10]

Like the military, fast-food restaurants have generally recruited teenagers because they surrender their autonomy to machines, rules, and procedures more easily than adults.[11] Fast-food restaurants also seek to maximize control over the work behavior of adults. Even managers are not immune from such efforts. Another aspect of McDonald's experimental ARCH program is a computerized system that, among other things, tells managers how many hamburgers or orders of french fries they will require at a given time (the lunch hour, for example). The computerized system takes away the need to make such judgments and decisions from managers.[12] Thus, "Burger production has become an exact science in which everything is regimented, every distance calculated and every dollop of ketchup monitored and tracked."[13]

Education: McChild Care Centers
Universities have developed a variety of nonhuman technologies to exert control over professors. For instance, class periods are set by the university. Students leave at the assigned time no matter where the professor happens to be in the lecture. Because the university requires grading, the professor must test students. In some universities, final grades must be submitted within forty-eight hours of the final exam, which may force professors to employ computer-graded, multiple-choice exams. Required evaluations by students may force professors to teach in a way that will lead to high ratings. The publishing demands of the tenure and promotion system may force professors to devote far less time to their teaching than they, and their students, would like.

An even more extreme version of this emphasis appears in the child care equivalent of the fast-food restaurant, KinderCare, which was founded in 1969, and now operates over 1,250 learning centers in the United States. Over 120,000 children between the ages of 6 weeks and 12 years attend the centers.[14] KinderCare tend to hire short-term employees with little or no training in child care. What these employees do in the "classroom" is largely determined by an instruction book with a ready-made curriculum. Staff members open the manual to find activities spelled out in detail for each day. Clearly, a skilled, experienced, creative teacher is not the kind of person that such "McChild" care centers seek to hire. Rather, relatively untrained employees are more easily controlled by the nonhuman technology of the omnipresent "instruction book."

Another example of organizational control over teachers is the franchised Sylvan Learning Center, often thought of as the "McDonald's of Education."[15] (There are over nine hundred Sylvan Learning Centers in the United States, Canada, and Asia.[16]) Sylvan Learning Centers are after-school centers for remedial education. The corporation "trains staff and tailors a McDonald's type uniformity, down to the U-shaped tables at which instructors work with their charges."[17] Through their training methods, rules, and technologies, for-profit systems such as the Sylvan Learning Center exert great control over their "teachers."

Health Care: Who's Deciding Our Fate?

As is the case with all rationalized systems, medicine has moved away from human toward nonhuman technologies. The two most important examples are the growing importance of bureaucratic rules and controls and the growth of modern medical machinery. For example, the prospective payment and DRG (diagnostic related groups) systems—not physicians and their medical judgment—tend to determine how long a patient must be hospitalized. Similarly, the doctor operating alone out of a black bag with a few simple tools has virtually become a thing of the past. Instead, doctors serve as dispatchers, sending patients on to the appropriate machines and specialists. Computer programs can diagnose illnesses.[18] Although it is unlikely that they will ever replace the physician, computers may one day be the initial, if not the prime, diagnostic agents. It is now even possible for people to get diagnoses, treatment, and prescriptions over the Internet with no face-to-face contact with a physician.

These and other developments in modern medicine demonstrate increasing external control over the medical profession by third-party payers, employing organizations, for-profit hospitals, health maintenance organizations (HMOs), the federal government, and "McDoctors"-like organizations. Even in its heyday the medical profession was not free of external control, but now the nature and extent of the control is changing and its degree and extent is increasing greatly. Instead of decisions being made by the mostly autonomous doctor in private practice, doctors are more likely to conform to bureaucratic rules and regulations. In bureaucracies, employees are controlled by their superiors. Physicians' superiors are increasingly likely to be professional managers and not other doctors. Also, the existence of hugely expensive medical technologies often mandates that they be used. As the machines themselves grow more sophisticated, physicians come to understand them less and are therefore less able to control them. Instead, control shifts to the technologies as well as to the experts who create and handle them.

An excellent recent example of increasing external control over physicians (and other medical personnel) is called "pathways."[19] a pathway is a standardized series of steps prescribed for dealing with an array of medical problems. Involved are a series of "if-then" decision points—if a certain circumstance exists, the action to follow is prescribed. What physicians do in a variety of situations is determined by the pathway and *not* the individual physician. To put it in terms of this chapter, the pathway—a nonhuman technology—exerts external control over physicians.

Various terms have been used to describe pathways—standardization, "cookbook" medicine, a series of recipes, a neat package tied together with a bow, and so on—and all describe the rationalization of medical practice. The point is that there are prescribed courses of action under a wide range of circumstances. While doctors need not, indeed should not, follow a pathway at all times, they do so most of the time. A physician who spearheads the protocol movement says he grows concerned when physicians follow a pathway more than 92% of the time. While this leaves some leeway for physicians, it is clear that what they are supposed to do is predetermined in the vast majority of instances.

Let us take, for example, an asthma patient. In this case, the pathway says that if the patient's temperature rises above 101 degrees, then a complete blood count is to be ordered. A chest X ray is to be ordered under certain circumstances—if it's the patient's initial wheezing episode or if there is chest pain, respiratory distress, or a fever of over 101 degrees. And so it goes—a series of if-then steps prescribed for and controlling what physicians and other medical personnel do. While there are undoubted advantages associated with such pathways (e.g., lower likelihood of using procedures or medicines that have been shown not to work), they do tend to take decision making away from physicians. Continued reliance on such pathways is likely to adversely affect the ability of physicians to make independent decisions.

The Workplace: Do as I Say, Not as I Do
Most workplaces are bureaucracies that can be seen as large-scale nonhuman technologies. Their innumerable rules, regulations, guidelines, positions, lines of command, and hierarchies dictate what people do within the system and how they do it. The consummate bureaucrat thinks little about what is to be done: He or she simply follows the rules, deals with incoming work, and passes it on to its next stop in the system. Employees need do little more than fill out the required forms, these days most likely right on the computer screen.

At the lowest levels in the bureaucratic hierarchy ("blue-collar work"), scientific management clearly strove to limit or replace human technology. For instance, the "one best way" required workers to follow a series of predefined steps in a mindless fashion. More generally, Frederick Taylor believed that the most important part of the work world was not the employees but, rather, the organization that would plan, oversee, and control their work.

Although Taylor wanted all employees to be controlled by the organization, he accorded managers much more leeway than manual workers. It was the task of management to study the knowledge and skills of workers and to record and tabulate them and ultimately to reduce them to laws, rules, and even mathematical formulas. In other words, managers were to take a body of human skills, abilities, and knowledge and transform them into a set of nonhuman rules, regulations, and formulas. Once human skills were codified, the organization no longer needed skilled workers. Management would hire, train, and employ unskilled workers in accord with a set of strict guidelines.

In effect, then, Taylor separated "head" work from "hand" work. Prior to Taylor's day, the skilled worker had performed both. Taylor and his followers studied what was in the heads of those skilled workers, then translated that knowledge into simple, mindless routines that virtually anyone could learn and follow. Workers were thus left with little more than repetitive "hand" work. This principle remains at the base of the movement throughout our McDonaldizing society to replace human with nonhuman technology.

Behind Taylor's scientific management, and all other efforts at replacing human with nonhuman technology, lies the goal of being able to employ human beings with minimal intelligence and ability. In fact, Taylor sought to hire people who resembled animals:

Now one of the very first requirements for a man who is fit to handle pig iron as a regular occupation is that he shall be so stupid and so phlegmatic that he more nearly resembles in his mental make-up the ox than any other type. The man who is mentally alert and intelligent is for this very reason entirely unsuited to what would, for him, be the grinding monotony of work of this character. Therefore the workman who is best suited to handling pig iron is unable to understand the real science of doing this class of work. He is so stupid that the word "percentage" has no meaning to him, and he must consequently be trained by a man more intelligent than himself into the habit of working in accordance with the laws of this science before he can be successful.[20]

Not coincidentally, Henry Ford had a similar view of the kinds of people who were to work on his assembly lines:

Repetitive labour—the doing of one thing over and over again and always in the same way—is a terrifying prospect to a certain kind of mind. It is terrifying to me. I could not possibly do the same thing day in and day out, but to other minds, perhaps I might say to the majority of minds, repetitive operations hold no terrors. In fact, to some types of mind thought is absolutely appalling. To them the ideal job is one where creative instinct need not be expressed. The jobs where it is necessary to put in mind as well as muscle have very few takers—we always need men who like a job because it is difficult. The average worker, I am sorry to say, wants a job in which he does not have to think. Those who have what might be called the creative type of mind and who thoroughly abhor monotony are apt to imagine that all other minds are similarly restless and therefore to extend quite unwanted sympathy to the labouring man who day in and day out performs almost exactly the same operation.[21]

The kind of person sought out by Taylor was the same kind of person Ford thought would work well on the assembly line. In their view, such people would more likely submit to external technological control over their work and perhaps even crave such control.

Not surprisingly, a perspective similar to that held by Taylor and Ford can be attributed to other entrepreneurs: "The obvious irony is that the organizations built by W. Clement Stone [the founder of Combined Insurance] and Ray Kroc, both highly creative and innovative entrepreneurs, depend on the willingness of employees to follow detailed routines precisely."[22]

Many workplaces have come under the control of nonhuman technologies. In the supermarket, for example, the checker once had to read the prices marked on food products and enter them into the cash register. As with all human activities, however, the process was slow, with a chance of human error. To counter these problems, many supermarkets installed optical scanners, which "read" a code preprinted on each item. Each code number calls up a price already entered into the computer that controls the cash register. This nonhuman technology has thus reduced the number and sophistication of the tasks performed by the checker. Only the less-skilled tasks remain, such as physically scanning the food and bagging it. And even those tasks are being eliminated with the development of self-scanning and having consumers bag their groceries, especially in discount supermarkets. In other words, the work performed by the supermarket checker, when it hasn't been totally eliminated, has been "de-skilled"; that is, a decline has occurred in the amount of skill required for the job.

The nonhuman technologies in telemarketing "factories" can be even more restrictive. Telemarketers usually have scripts they must follow unerringly. The scripts are designed to handle most foreseeable contingencies. Supervisors often listen in on solicitations to make sure employees follow the correct procedures. Employees who fail to meet the quotas for the number of calls made and sales completed in a given time may be fired summarily.

Similar control is exerted over the "phoneheads," or customer service representatives, who work for many companies. Those who handle reservations for the airlines (for example, United Airlines) must log every minute spent on the job and justify each moment away from the phone. Employees have to punch a "potty button" on the phone to let management know of their intentions. Supervisors sit in an elevated "tower" in the middle of the reservations floor, "observing like [prison] guards the movements of every operator in the room." They also monitor phone calls to make sure that employees say and do what they are supposed to. This control is part of a larger process of "omnipresent supervision increasingly taking hold in so many workplaces—not just airline reservations centers but customer service departments and data-processing businesses where computers make possible an exacting level of employee scrutiny."[23] No wonder customers often deal with representatives who behave like automatons. Said one employee of United Airlines, "My body became an extension of the computer terminal that I typed the reservations into. I came to feel emptied of self."[24]

Sometimes telephone service representatives are literally prisoners. Prison inmates are now used in at least 17 states in this way, and the idea is currently on the legislative table in several more states. The attractions of prisoners are obvious—they work for very little pay and they can be controlled to a far higher degree than even the "phoneheads" discussed above. Furthermore, they can be relied on to show up for work. As one manager put it, "I need people who are there every day."[25]

Many telemarketing firms are outsourcing much of their labor overseas, especially to India, where people who are desperate for well-paying jobs are willing to accept levels of control that would be found unacceptable by many in the United States. Indian call centers afford a number of advantages, including lower wage costs than in the United States; the availability of an English-speaking, computer-literate, and college-educated workforce with a strong work ethic; and significant experience and familiarity with business processes.[26]

Following the logical progression, some companies now use computer calls instead of having people solicit us over the phone.[27] Computer voices are far more predictable and controllable than even the most rigidly controlled human operator, including prisoners and those who work in Indian call centers. Indeed, in our increasingly McDonaldized society, I have had some of my most "interesting" conversations with such computer voices.

Of course, lower-level employees are not the only ones whose problem-solving skills are lost in the transition to more nonhuman technology. I have already mentioned the controls on professors and doctors. In addition, pilots flying the modern, computerized airplane (such as the Boeing 757, 767, and 777) are being controlled and, in the process, de-skilled. Instead of flying "by the seat of their pants" or using old-fashioned autopilots for simple maneuvers, modern pilots can "push a few buttons and lean back while the plane flies to its destination and lands on a predetermined runway." Said one FAA official, "We're taking more and more of these functions out of human control and giving them to machines." These airplanes are in many ways safer and more reliable than older, less technologically advanced models. However, pilots, dependent on these technologies, may lose the ability to handle emergency situations creatively. The problem, said one airline manager, is that "I don't have computers that will do that [be creative]; I just don't."[28]

Controlling Customers

Employees are relatively easy to control, because they rely on employers for their livelihood. Customers have much more freedom to bend the rules and go elsewhere if they don't like the situations in which they find themselves. Still, McDonaldized systems have developed and honed a number of methods for controlling customers.

The Fast-Food Industry: Get the Hell Out of There

Whether they go into a fast-food restaurant or use the drive-through window, customers enter a kind of conveyor system that moves them through the restaurant in the manner desired by the management. It is clearest in the case of the drive-through window (the energy for this conveyor comes from one's own automobile), but it is also true for those who enter the restaurant. Consumers know that they are supposed to line up, move to the counter, order their food, pay, carry the food to an available table, eat, gather up their debris, deposit it in the trash receptacle, and return to their cars.

Three mechanisms help to control customers:[29]

1. Customers receive cues (for example, the presence of lots of trash receptacles, especially at the exits) that indicate what is expected of them.
2. A variety of structural constraints lead customers to behave in certain ways. For example, the drive-through window, as well as the written instructions on the menu marquee at the counter (and elsewhere), gives customers few, if any, alternatives.
3. Customers have internalized taken-for-granted norms and follow them when they enter a fast-food restaurant.

When my children were young, they admonished me after we finished our meal at McDonald's (I ate in fast-food restaurants in those days before I "saw the light") for not cleaning up the debris and carting it to the trash can. My children were, in effect, serving as agents for McDonald's, teaching me the norms of behavior in such settings. I (and most others) have long-since internalized these norms, and I still dutifully follow them these days on the rare occasions that a lack of any other alternative (or the need for a clean restroom) forces me into a fast-food restaurant.

One goal of control in fast-food restaurants is to influence customers to spend their money and leave quickly. The restaurants need tables to be vacated rapidly so other diners will have a place to eat their food. A famous old chain of cafeterias, the Automat,[30] was partly undermined by people who occupied tables for hours on end. The Automat became a kind of social center, leaving less and less room for people to eat the meals they had purchased. The deathblow was struck when street people began to monopolize the Automat's tables.

Some fast-food restaurants employ security personnel to keep street people on the move or, in the suburbs, to prevent potentially rowdy teenagers from monopolizing tables or parking lots. 7-Eleven has sought to deal with loitering teenagers outside some of its stores by playing saccharine tunes such as "Some Enchanted Evening." Said a 7-Eleven spokesperson, "They won't hang around and tap their feet to Mantovani."[31]

In some cases, fast-food restaurants have put up signs limiting a customer's stay in the restaurant (and even its parking lot), say, to twenty minutes.[32] More generally, fast-food restaurants have structured themselves so that people do not need or want to linger over meals. Easily consumed finger foods make the meal itself a quick one. Some fast-food restaurants use chairs that make customers uncomfortable after about twenty minutes.[33] Much the same effect is produced by the colors used in the decor: "Relaxation isn't the point. Getting the Hell out of there is the point. The interior colours have been chosen carefully with this end in mind. From the scarlet and yellow of the logo to the maroon of the uniform; everything clashes. It's designed to stop people from feeling so comfortable they might want to stay."[34]

Other Settings: It's Like Boot Camp
In the university, students (the "consumers" of university services) are obviously even more controlled than professors. For example, universities often give students little leeway in the courses they may take. The courses themselves, often highly structured, force the students to perform in specific ways.

Control over students actually begins long before they enter the university. Grade schools in particular have developed many ways to control students. Kindergarten has been described as an educational "boot camp."[35] Students are taught not only to obey authority but also to embrace the rationalized procedures of rote learning and objective testing. More important, spontaneity and creativity tend not to be rewarded and may even be discouraged, leading to what one expert calls "education for docility."[36] Those who conform to the rules are thought of as good students; those who don't are labeled bad students. As a general rule, the students who end up in college are the ones who have successfully submitted to the control mechanisms. Creative, independent students are often, from the educational system's point of view, "messy, expensive, and time-consuming."[37]

The clock and the lesson plan also exert control over students, especially in grade school and high school. Because of the "tyranny of the clock," a class must end at the sound of the bell, even if students are just about to comprehend something important. Because of the "tyranny of the lesson plan," a class must focus on what the plan requires for the day, no matter what the class (and perhaps the teacher) may find interesting. Imagine "a cluster of excited children examining a turtle with enormous fascination and intensity. Now children, put away the turtle, the teacher insists. We're going to have our science lesson. The lesson is on crabs."[38]

In the health care industry, the patient (along with the physician) is increasingly under the control of large, impersonal systems. For example, in many medical insurance programs, patients can no longer decide on their own to see a specialist. Rather, the patient must first see a primary-care physician who must decide whether a specialist is necessary. Because of the system's great pressure on the primary physician to keep costs down, fewer patients visit specialists and primary-care physicians perform more functions formerly handled by specialists.

The supermarket scanners that control checkers also control customers. When prices were marked on all the products, customers could calculate roughly how much they were spending as they shopped. They could also check the price on each item to be sure that they were not being overcharged at the cash register. But with scanners, it is almost impossible for consumers to keep tabs on prices and on the checkers.

Supermarkets also control shoppers with food placement. For example, supermarkets take pains to put the foods that children find attractive in places where youngsters can readily grab them (for example, low on the shelves). Also, what a market chooses to feature through sale prices and strategic placement in the store profoundly affects what is purchased. Manufacturers and wholesalers battle one another for coveted display positions, such as at the front of the market or at the "endcaps" of aisles. Foods placed in these positions will likely sell far more than they would if they were relegated to their usual positions.

Malls also exert control over customers, especially children and young adults, who are programmed by the mass media to be avid consumers. Going to the mall can become a deeply ingrained habit. Some people are reduced to what Kowinski calls "zombies," shopping the malls hour after hour, weekend after weekend.[39] More specifically, the placement of food courts, escalators, and stairs force customers to traverse corridors and pass attractive shop windows. Benches are situated so that consumers might be attracted to certain sites even though they are seeking a brief respite from the labors of consumption. The strategic placement of shops, as well as goods within shops, leads people to be attracted to products in which they might not otherwise have been interested.

Computers that respond to the human voice via voice recognition systems exert great control over people. A person receiving a collect call might be asked by the computer voice whether she will accept the charges. The computer voice requests, "Please say yes or no." Although efficient and cost-saving, such a system has its drawbacks:

> The person senses that he cannot use free-flowing speech. He's being constrained. The computer is controlling him. It can be simply frustrating.... People adapt to it, but only by filing it away subconsciously as another annoyance of living in our technological world.[40]

Even religion and politics are being marketed today, and like all McDonaldizing systems, they are adopting technologies that help them control the behavior of their "customers." For example, the Roman Catholic Church has its Vatican Television (which conducts about 130 televised broadcasts each year of events inside the Vatican).[41] More generally, instead of worshiping with a human preacher, millions of worshipers now "interact" with a televised image.[42] Television permits preachers to reach far more people than they could in a conventional church, so they can exert (or so they hope) greater control over what people believe and do and, in the process, extract higher contributions. TV preachers use the full panoply of techniques developed by media experts to control their viewers. Some use a format much like that of the talk shows hosted by Jay Leno or David Letterman, complete with jokes, orchestras, singers, and guests. Here is how one observer describes Vatican television: "The big advantage to the Vatican of having its own television operation... is that they can put their own spin on anything they produce. If you give them the cameras and give them access, they are in control."[43]

A similar point can be made about politics. The most obvious example is the use of television to market politicians and manipulate voters. Indeed, most people never see a politician except on TV, most likely in a firmly controlled format designed to communicate the exact message and image desired by the politicians and their media advisers. President Ronald Reagan raised such political marketing to an art form in the 1980s. On many occasions, visits were set up and TV images arranged (the president in front of a flag or with a military cemetery behind him) so that the viewers and potential voters received precisely the visual message intended by Reagan's media advisers. Tightly controlled TV images are similarly important to President George W. Bush, as reflected, for example, in his landing as "co-pilot" of a jet plane on an aircraft carrier in order to announce (erroneously) the end of hostilities with Iraq in 2003. Conversely, as we saw earlier, President Reagan and especially President George W. Bush tended to avoid freewheeling press conferences where they were not in control.

Controlling the Process and the Product

In a society undergoing McDonaldization, people are the greatest threat to predictability. Control over people can be enhanced by controlling processes and products, but control over processes and products also becomes valued in itself.

Food Production, Cooking, and Vending: It Cooks Itself

Technologies designed to reduce uncertainties are found throughout the manufacture of food. For example, the mass manufacturing of bread is not controlled by skilled bakers who lavish love and attention on a few loaves of bread at a time. Such skilled bakers cannot produce enough bread to supply the needs of our society. Furthermore, the bread they do produce can suffer from the uncertainties involved in having humans do the work. The bread may, for example, turn out to be too brown or too doughy. To increase productivity and eliminate these unpredictabilities, mass producers of bread have developed an automated system in which, as in all automated systems, humans play a minimal role rigidly controlled by the technology:

The most advanced bakeries now resemble oil refineries. Flour, water, a score of additives, and huge amounts of yeast, sugar, and water are mixed into a broth that ferments for an hour. More flour is then added, and the dough is extruded into pans, allowed to rise for an hour, then moved through a tunnel oven. The loaves emerge after eighteen minutes, to be cooled, sliced, and wrapped.[44]

In one food industry after another, production processes in which humans play little more than planning and maintenance roles have replaced those dominated by skilled craftspeople. The warehousing and shipping of food has been similarly automated.

Further along in the food production process, other nonhuman technologies have affected how food is cooked. Technologies such as ovens with temperature probes "decide" for the cook when food is done. Many ovens, coffeemakers, and other appliances can turn themselves on and off. The instructions on all kinds of packaged foods dictate precisely how to prepare and cook the food. Premixed products, such as Mrs. Dash, eliminate the need for the cook to come up with creative combinations of seasonings. Nissin Foods' Super Boil soup—"the soup that cooks itself!"—has a special compartment in the bottom of the can. A turn of a key starts a chemical reaction that eventually boils the soup.[45] Even the cookbook was designed to take creativity away from the cook and control the process of cooking.

Some rather startling technological developments have occurred in the ways in which animals are raised for food. For instance, "aquaculture," a $57-billion-a-year business in 2000,[46] is growing dramatically because of the spiraling desire for seafood in an increasingly cholesterol-conscious population.[47] Instead of the old inefficient, unpredictable methods of harvesting fish—a lone angler casting a line or even boats catching tons of fish at a time in huge nets—we now have the much more predictable and efficient "farming" of seafood. More than 50% of the fresh salmon found in restaurants is now raised in huge sea cages off the coast of Norway.

Sea farms offer several advantages. Most generally, aquaculture allows humans to exert far greater control over the vagaries that beset fish in their natural habitat, thus producing a more predictable supply. Various drugs and chemicals increase predictability in the amount and quality of seafood. Aquaculture also permits a more predictable and efficient harvest because the creatures are confined to a limited space. In addition, geneticists can manipulate them to produce seafood more efficiently. For example, it takes a standard halibut about ten years to reach market size, but a new dwarf variety can reach the required size in only three years. Sea farms also allow for greater calculability—the greatest number of fish for the least expenditure of time, money, and energy.

Relatively small, family-run farms for raising other animals are being rapidly replaced by "factory farms."[48] The first animal to find its way into the factory farm was the chicken. Here is the way one observer describes a chicken "factory":

A broiler producer today gets a load of 10,000, 50,000, or even more day-old chicks from the hatcheries, and puts them straight into a long, windowless shed.... Inside the shed, every aspect of the birds' environment is controlled to make them grow faster on less feed. Food and water are fed automatically from hoppers suspended from the roof. The lighting is adjusted.... For instance, there may be bright light twenty-four hours a day for the first week or two, to encourage the chicks to gain [weight] quickly....

Toward the end of the eight- or nine-week life of the chicken, there may be as little as half a square foot of space per chicken—or less than the area of a sheet of quarto paper for a three-and-one-half-pound bird.[49]

Among its other advantages, such chicken farms allow one person to raise over fifty thousand chickens.

Raising chickens this way ensures control over all aspects of the business. For instance, the chickens' size and weight are more predictable than that of free-ranging chickens. "Harvesting" chickens confined in this way is also more efficient than is catching chickens that roam over large areas.

However, confining chickens in such crowded quarters creates unpredictabilities, such as violence and even cannibalism. Farmers deal with these irrational "vices" in a variety of ways, such as dimming the lights as chickens approach full size and "debeaking" chickens so they cannot harm each other.

Some chickens are allowed to mature so they can be used for egg production. However, they receive much the same treatment as chickens raised for food. Hens are viewed as little more than "converting machines" that transform raw material (feed) into a finished product (eggs). Peter Singer describes the technology employed to control egg production:

The cages are stacked in tiers, with food and water troughs running along the rows, filled automatically from a central supply. They have sloping wire floors. The slope... makes it more difficult for the birds to stand comfortably, but it causes the eggs to roll to the front of the cage where they can easily be collected... [and] in the more modern plants, carried by conveyor belt to a packing plant.... The excrement drops through [the wire floor] and can be allowed to pile up for many months until it is all removed in a single operation.[50]

This system obviously imposes great control over the production of eggs, leading to greater efficiency, to a more predictable supply, and more uniform quality than the old chicken coop.

Other animals—pigs, lambs, steers, and calves especially—are raised similarly. To prevent calves' muscles from developing, which toughens the veal, they are immediately confined to tiny stalls where they cannot exercise. As they grow, they may not even be able to turn around. Being kept in stalls also prevents the calves from eating grass, which would cause their meat to lose its pale color; the stalls are also kept free of straw, which, if eaten by the calves, would also darken the meat. "They are fed a totally liquid diet, based on nonfat milk powder with added vitamins, minerals, and growth-promoting drugs," says Peter Singer in his book, *Animal Liberation*.[51] To make sure the calves take in the maximum amount of food, they are given no water, which forces them to keep drinking their liquid food. By rigidly controlling the size of the stall and the diet, veal producers can maximize two quantifiable objectives: the production of the largest amount of meat in the shortest possible time and the creation of the tenderest, whitest, and therefore most desirable veal.

Employment of a variety of technologies obviously leads to greater control over the process by which animals produce meat, thereby increasing the efficiency, calculability, and predictability of meat production. In addition, they exert control over farm workers. Left to their own devices, ranchers might feed young steers too little or the wrong food or permit them too much exercise. In fact, in the rigidly controlled factory ranch, human ranch hands (and their unpredictabilities) are virtually eliminated.

The Ultimate Examples of Control? Birth and Death

Not just fish, chickens, and calves are being McDonaldized, but also people, especially the processes of birth and death.

Controlling Conception: Even Granny Can Conceive

Conception is rapidly becoming McDonaldized, and increasing control is being exercised over the process. For example, the problem of male impotence[52] has been attacked by the burgeoning impotence clinics, some of which have already expanded into chains,[53] and an increasingly wide array of technologies, including medicine (especially Viagra) and mechanical devices. Many males are now better able to engage in intercourse and to play a role in pregnancies that otherwise might not have occurred.

Similarly, female infertility has been ameliorated by advances in the technologies associated with artificial (more precisely, "donor"[54]) insemination, in vitro fertilization,[55] intracytoplasmic sperm injection,[56] various surgical and nonsurgical procedures associated with the Wurn technique,[57] and so on. Some fertility clinics have grown so confident that they offer a money-back guarantee if there is no live baby after three attempts.[58] For those women who still cannot become pregnant or carry to term, surrogate mothers can do the job.[59] Even postmenopausal women now have the chance of becoming pregnant ("granny pregnancies");[60] the oldest, thus far, is a sixty-five-year-old Indian woman who gave birth to a boy in April 2003.[61] These developments and many others, such as ovulation predictor home tests,[62] have made having a child far more predictable. Efficient, easy-to-use home pregnancy tests are also available to take the ambiguity out of determining whether or not a woman has become pregnant.

One of the great unpredictabilities tormenting some prospective parents is whether the baby will turn out to be a girl or a boy. Sex selection[63] clinics have opened in England, India, and Hong Kong as the first of what may eventually become a chain of "gender choice centers." The technology, developed in the early 1970s, is actually rather simple: Semen is filtered through albumen to separate sperm with male chromosomes from sperm with female chromosomes. The woman is then artificially inseminated with the desired sperm. The chances of creating a boy are 75%; a girl, 70%.[64] A new technique uses staining of sperm cells to determine which cells carry X (male) and Y (female) chromosomes. Artificial insemination or in vitro fertilization then mates the selected sperm with an egg. The U.S. lab that developed this technique is able to offer a couple an 85% chance of creating a girl; the probabilities of creating a boy are still unclear but are expected to be lower.[65] The goal is to achieve 100% accuracy in using "male" or "female" sperm to tailor the sex of the offspring to the needs and demands of the parents.

The increasing control over the process of conception delights some but horrifies others: "Being able to specify your child's sex in advance leads to nightmare visions of ordering babies with detailed specifications, like cars with automatic transmission or leather upholstery."[66] Said a medical ethicist, "Choosing a child like we choose a car is part of a consumerist mentality, the child becomes a 'product' rather than a full human being."[67] By turning a baby into just another "product" to be McDonaldized—engineered, manufactured, and commodified—people are in danger of dehumanizing the birth process.

Of course, we are just on the frontier of the McDonaldization of conception (and just about everything else). For example, the first cloned sheep, Dolly (now deceased), was created in Scotland in 1996, and other animals have since been cloned. This opened the door to the possibility of the cloning of humans. In fact, Clonaid, a Raelian sect that believes aliens populated the earth through cloning and that the destiny of humankind is to clone, recently claimed (thus far unsubstantiated) to have cloned its fifth human being.[68] Cloning involves the creation of identical copies of molecules, cells, or even entire organisms.[69] This conjures up the image of the engineering and mass production of a "cookie-cutter" race of people, all good-looking, athletic, intelligent, free of genetic defects, and so on. If everyone were to be conceived through cloning, we would be close to the ultimate in the control of this process. And a world in which everyone was the same would be a world in which they would be ready to accept a similar sameness in everything around them. Of course, this is a science fiction scenario, but the technology needed to take us down this road is already here!

Controlling Pregnancy: Choosing the Ideal Baby

Some parents wait until pregnancy is confirmed before worrying about the sex of their child. But then, amniocentesis can be used to determine whether a fetus is male or female. First used in 1968 for prenatal diagnosis, amniocentesis is a process whereby fluid is drawn from the amniotic sac, usually between the fourteenth and eighteenth weeks of pregnancy.[70] With amniocentesis, parents might choose to exert greater control over the process by aborting a pregnancy if the fetus is of the "wrong" sex. This is clearly a far less efficient technique than prepregnancy sex selection, because it occurs after conception. In fact, very few Americans (only about 5% in one study) say that they might use abortion as a method of sex selection.[71] However, amniocentesis does allow parents to know well in advance what the sex of their child will be.

Concern about a baby's sex pales in comparison to concern about the possibility of genetic defects. In addition to amniocentesis, a variety of recently developed tests can be used to determine whether a fetus carries genetic defects such as cystic fibrosis, Down syndrome, Huntington's disease, hemophilia, Tay-Sachs disease, and sickle-cell disease.[72] These newer tests include the following:

- Chorionic villus sampling (CVS)   Generally done earlier than amniocentesis, between the tenth and twelfth weeks of pregnancy, CVS involves taking a sample from the fingerlike structures projecting from the sac that later becomes the placenta. These structures have the same genetic makeup as the fetus.[73]
- Maternal serum alpha-fetoprotein (MSAFP) testing   A simple blood test done in the sixteenth to eighteenth weeks of pregnancy. A high level of alpha-fetoprotein might indicate spina bifida; a low level might indicate Down syndrome.
- Ultrasound   A technology derived from sonar that provides an image of the fetus by bouncing high-frequency energy off it. Ultrasound can reveal various genetic defects, as well as many other things (sex, gestational age, and so on).

The use of all these nonhuman technologies has increased dramatically in recent years, with some (ultrasound, MSAFP) already routine practices.[74] Many other technologies for testing fetuses are also available, and others will undoubtedly be created.

If one or more of these tests indicate the existence of a genetic defect, then abortion becomes an option. Parents who choose abortion are unwilling to inflict the pain and suffering of genetic abnormality or illness on the child and on the family. Eugenicists feel that it is not rational for a society to allow genetically disabled children to be born and to create whatever irrationalities will accompany their birth. From a cost-benefit point of view (calculability), abortion is less costly than supporting a child with serious physical or mental abnormalities or problems, sometimes for a number of years. Given such logic, it makes sense for society to use the nonhuman technologies now available to discover which fetuses are to be permitted to survive and which are not. The ultimate step would be a societal ban on certain marriages and births, something that China has considered, with the goal of such a law being the reduction of the number of sick or retarded children that burden the state.[75]

Efforts to predict and repair genetic anomalies are proceeding at a rapid rate. The Human Genome Project constructed a map of 99% of the human genome's gene-containing regions.[76] When the project began, only about 100 human disease genes were known; today we know of over 140 such genes.[77] Such knowledge will allow scientists to develop new diagnostic tests and therapeutic methods. Knowledge of where each gene is and what each does will also extend the ability to test fetuses, children, and prospective mates for genetic diseases. Prospective parents who carry problematic genes may choose not to marry or not to procreate. Another possibility (and fear) is that as the technology gets cheaper and becomes more widely available, people may be able to do the testing themselves (we already have home pregnancy tests) and then make a decision to try a risky home abortion.[78] Overall, human mating and procreation will come to be increasingly affected and controlled by these new nonhuman technologies.

Controlling Childbirth: Birth as Pathology

McDonaldization and increasing control is also manifest in the process of giving birth. One measure is the decline of midwifery, a very human and personal practice. In 1900, midwives attended about half of American births, but by 1986, they attended only 4%.[79] Today, however, midwifery has enjoyed a slight renaissance because of the dehumanization and rationalization of modern childbirth practices,[80] and 6.5% of babies in the United States are now delivered by midwives.[81] When asked why they sought out midwives, women complain about things such as the "callous and neglectful treatment by the hospital staff," "labor unnecessarily induced for the convenience of the doctor," and "unnecessary cesareans for the same reason."[82]

The flip side of the decline of midwives is the increase in the control of the birth process by professional medicine,[83] especially obstetricians. It is they who are most likely to rationalize and dehumanize the birth process. Dr. Michelle Harrison, who served as a resident in obstetrics and gynecology, is but one physician willing to admit that hospital birth can be a "dehumanized process."[84]

The increasing control over childbirth is also manifest in the degree to which it has been bureaucratized. "Social childbirth," the traditional approach, once took place largely in the home, with female relatives and friends in attendance. Now, childbirth takes place almost totally in hospitals, "alone among strangers."[85] In 1900, less than 5% of U.S. births took place in hospitals; by 1940, it was 55%; and by 1960, the process was all but complete, with nearly 100% of births occurring in hospitals.[86] In more recent years, hospital chains and birthing centers have emerged, modeled after my paradigm for the rationalization process—the fast-food restaurant.

Over the years, hospitals and the medical profession have developed many standard, routinized (McDonaldized) procedures for handling and controlling childbirth. One of the best known, created by Dr. Joseph De Lee, was widely followed through the first half of the twentieth century. De Lee viewed childbirth as a disease (a "pathologic process"), and his procedures were to be followed even in the case of low-risk births:[87]

1. The patient was placed in the lithotomy position, "lying supine with legs in air, bent and wide apart, supported by stirrups."[88]
2. The mother-to-be was sedated from the first stage of labor on.
3. An episiotomy[89] was performed to enlarge the area through which the baby must pass.
4. Forceps were used to make the delivery more efficient.
5. Describing this type of procedure, one woman wrote, "Women are herded like sheep through an obstetrical assembly line, are drugged and strapped on tables where their babies are forceps delivered."[90]

De Lee's standard practice includes not only control through nonhuman technology (the procedure itself, forceps, drugs, an assembly line approach) but most of the other elements of McDonaldization—efficiency, predictability, and the irrationality of turning the human delivery room into an inhuman baby factory. The calculability that it lacked was added later in the form of Emanuel Friedman's "Friedman Curve." This curve prescribed three rigid stages of labor. For example, the first stage was allocated exactly 8.6 hours, during which cervical dilation was to proceed from two to four centimeters.[91]

The moment that babies come into the world, they, too, are greeted by a calculable scoring system, the Apgar test. The babies receive scores of zero to two on each of five factors (for example, heart rate, color), with ten being the healthiest total score. Most babies have scores between seven and nine a minute after birth and scores of eight to ten after five minutes. Babies with scores of zero to three are considered to be in very serious trouble. Dr. Harrison wonders why medical personnel don't ask about more subjective things, such as the infant's curiosity and mood:

A baby doesn't have to be crying for us to know it is healthy. Hold a new baby. It makes eye contact. It breathes. It sighs. The baby has color. Lift it in your arms and feel whether it has good tone or poor, strong limbs or limp ones. The baby does not have to be on a cold table to have its condition measured.[92]

The use of various nonhuman technologies in the delivery of babies has tended to ebb and flow. The use of forceps, invented in 1588, reached a peak in the United States in the 1950s, when as many as 50% of all births involved their use. However, forceps fell out of vogue, and in the 1980s, only about 15% of all births employed forceps. Many methods of drugging mothers-to-be have also been widely used. The electronic fetal monitor became popular in the 1970s. Today, ultrasound is a popular technology.

Another worrisome technology associated with childbirth is the scalpel. Many doctors routinely perform episiotomies during delivery so that the opening of the vagina does not tear or stretch unduly during pregnancy. Often done to enhance the pleasure of future sex partners and to ease the passage of the infant, episiotomies are quite debilitating and painful for the woman. Dr. Harrison expresses considerable doubt about episiotomies. "I want those obstetricians to stop cutting open women's vaginas. Childbirth is not a surgical procedure."[93]

The scalpel is also a key tool in cesarean sections. Birth, a perfectly human process, has come to be controlled by this technology (and those who wield it) in many cases.[94] The first modern "C-section" took place in 1882, but as late as 1970, only 5% of all births involved cesarean. Its use skyrocketed in the 1970s and 1980s, reaching 25% of all births in 1987 in what has been described as a "national epidemic."[95] By the mid-1990s, the practice had declined slightly, to 21%.[96] However, as of August 2002, 25% of births were once again by cesarean, first-time C-sections were at an all-time high of almost 17%, and the rate of vaginal births after a previous cesarean was down to 16.5%.[97] This latter occurred even though the American College of Obstetricians formally abandoned the time-honored idea, "once a cesarean, always a cesarean." That is, it no longer supports the view that once a mother has a cesarean section, all succeeding births must be cesarean.

In addition, many people believe that cesareans are often performed unnecessarily. The first clue is historical data: Why do we see a sudden need for so many more cesareans? Weren't cesareans just as necessary a few decades ago? The second clue is data indicating that private patients who can pay are more likely to get cesareans than those on Medicaid (which reimburses far less) and are twice as likely as indigent patients to get cesareans.[98] Are those in higher social classes and with more income really more likely to need cesareans than those with less income and from the lower social classes?[99]

One explanation for the dramatic increase in cesareans is that they fit well with the idea of the substitution of nonhuman for human technology, but they also mesh with the other elements of the increasing McDonaldization of society:

- They are more *predictable* than the normal birth process that can occur a few weeks (or even months) early or late. It is frequently noted that cesareans generally seem to be performed before 5:30 P.M. so that physicians can be home for dinner. Similarly, well-heeled women may choose a cesarean so that the unpredictabilities of natural childbirth do not interfere with careers or social demands.
- As a comparatively simple operation, the cesarean is more *efficient* than natural childbirth, which may involve many more unforeseen circumstances.
- Cesareans births are more *calculable*, normally involving no less than twenty minutes and no more than forty-five minutes. The time required for a normal birth, especially a first birth, may be far more variable.
- Irrationalities exist, including the risks associated with surgery—anesthesia, hemorrhage, blood replacement. Compared with those who undergo a normal childbirth, women who have cesareans seem to experience more physical problems and a longer period of recuperation, and the mortality rate can be as much as twice as high. Then there are the higher costs associated with cesareans. One study indicated that physicians' costs were 68% higher and hospital costs 92% higher for cesareans compared with natural childbirth.[100]

- Cesareans are dehumanizing because a natural human process is transformed, often unnecessarily, into a nonhuman or even inhuman process in which women endure a surgical procedure. At the minimum, many of those who have cesareans are denied unnecessarily the very human experience of vaginal birth. The wonders of childbirth are reduced to the routines of a minor surgical procedure.

Controlling the Process of Dying: Designer Deaths

The months or years of decline preceding most deaths involve a series of challenges irresistible to the forces of McDonaldization. In the natural order of things, the final phase of the body's breakdown can be hugely inefficient, incalculable, and unpredictable. Why can't all systems quit at once instead of, say, the kidneys going and then the intellect and then the heart? Many a dying person has confounded physicians and loved ones by rallying and persisting longer than expected or, conversely, giving out sooner than anticipated. Our seeming lack of control in face of the dying process is pointed up in the existence of powerful death figures in myth, literature, and film.

But now we have found ways to rationalize the dying process, giving us at least the illusion of control. Consider the increasing array of nonhuman technologies designed to keep people alive long after they would have expired had they lived at an earlier time in history. In fact, some beneficiaries of these technologies would not want to stay alive under those conditions that allow them to survive (a clear irrationality). Unless the physicians are following an advance directive (a living will) that explicitly states "do not resuscitate," or "no heroic measures," people lose control over their own dying process. Family members, too, in the absence of such directives, must bow to the medical mandate to keep people alive as long as possible.

At issue is who should be in control of the process of dying. It seems increasingly likely that the decision about who dies and when will be left to the medical bureaucracy. Of course, we can expect bureaucrats to focus on rational concerns. For instance, the medical establishment is making considerable progress in maximizing the number of days, weeks, or years a patient remains alive. However, it has been slower to work on improving the quality of life during the extra time. This focus on calculability is akin to the fast-food restaurant telling people how large its sandwiches are but saying nothing about their quality.

We can also expect an increasing reliance on nonhuman technologies. For example, computer systems may be used to assess a patient's chances of survival at any given point in the dying process—90%, 50%, 10%, and so on. The actions of medical personnel are likely to be influenced by such assessments. Thus, whether a person lives or dies may come to depend increasingly on a computer program.

As you can see, death has followed much the same path as birth. That is, the dying process has been moved out of the home and beyond the control of the dying and their families and into the hands of medical personnel and hospitals.[101] Physicians have gained a large measure of control over death just as they won control over birth, and death, like birth, is increasingly likely to take place in the hospital. In 1900, only 20% of deaths took place in hospitals; by 1977, it had reached 70%. By 1993, the number of hospital deaths was down slightly to 65%, but to that percentage must be added the increasing number of people who die in nursing homes (11%) and residences such as hospices (22%).[102] The growth of hospital chains and chains of hospices, using principles derived from the fast-food restaurant, signals death's bureaucratization, rationalization, even McDonaldization.

The McDonaldization of the dying process, as well as of birth, has spawned a series of counterreactions, efforts to cope with the excesses of rationalization. For example, as a way to humanize birth, interest in midwives has grown. However, the greatest counterreaction has been the search for ways to regain control over our own deaths. Advance directives and living wills tell hospitals and medical personnel what they may or may not do during the dying process. Suicide societies and books such as Derek Humphry's *Final Exit*[103] give people instructions on how to kill themselves. Finally, there is the growing interest in and acceptance of euthanasia,[104] most notably the work of "Dr. Death," Jack Kevorkian, whose goal is to give back to people control over their own deaths.

However, these counterreactions themselves have elements of McDonaldization. For example, Dr. Kevorkian (now serving a ten- to twenty-five-year prison term for second-degree murder) uses a nonhuman technology, a "machine," to help people kill themselves. More generally, and strikingly, he is an advocate of a "rational policy" for the planning of death.[105] Thus, the rationalization of death is found even in the efforts to counter it. Dr. Kervorkian's opponents have noted the limitations of his rational policy:

It is not so difficult... to envision a society of brave new benignity and rationality, in which a sort of humane disposal system would tidy up and whisk away to dreamland the worst-case geezers and crones. They are, after all, incredibly expensive and non-productive.... And they are a terrible inconvenience to that strain of the American character that has sought to impose rational control on all aspects of life....

Would it be so farfetched to envision a society that in the name of efficiency and convenience... practiced Kevorkianism as a matter of routine in every community?[106]

Conclusion

The fourth dimension of McDonaldization is control, primarily through the replacement of human with nonhuman technology. Among the many objectives guiding the development of nonhuman technologies, the most important here is increased control over the uncertainties created by people—especially employees and consumers. The ultimate in control is reached when employees are replaced by nonhuman technologies such as robots. Nonhuman technologies are also employed to control the uncertainties created by customers. The objective is to make them more pliant participants in McDonaldized processes. In controlling employees and consumers, nonhuman technologies also lead to greater control over work-related processes and finished products. However, the ultimate examples of the efforts at obtaining greater control through the use of nonhuman technologies are found in the realms of birth and death.

Clearly, the future will bring with it an increasing number of nonhuman technologies with greater ability to control people and processes. Even today, listening to audiotapes rather than reading books, for example, shifts control to those who do the reading on the tape: "The mood, pace and intonation of the words are decided for you. You can't linger or rush headlong into them anymore."[107] Military hardware such as "smart bombs" (for example, the "jdams"—joint direct attack munitions— used so frequently and with such great effect in the 2003 war with Iraq) adjust their trajectories without human intervention, but in the future, smart bombs may be developed that scan an array of targets and "decide" which one to hit. Perhaps the next great step will be the refinement of artificial intelligence, which gives machines the apparent ability to think and to make decisions as humans do.[108] Artificial intelligence promises many benefits in a wide range of areas (medicine, for example). However, it also constitutes an enormous step in de-skilling. In effect, more and more people will lose the opportunity, and perhaps the ability, to think for themselves.

Notes

1. Richard Edwards. *Contested Terrain: The Transformation of the Workplace in the Twentieth Century*. New York: Basic Books, 1979.

2. Richard Edwards. *Contested Terrain: The Transformation of the Workplace in the Twentieth Century*. New York: Basic Books, 1979.

3. Michael Lev. "Raising Fast Food's Speed Limit." *Washington Post*, August 7, 1991, pp. D1, D4.

4. Ray Kroc. *Grinding It Out*. New York: Berkeley Medallion, 1977, pp. 131–132.

5. Eric A. Taub. "The Burger Industry Takes a Big Helping of Technology." *New York Times*, October 8, 1998, pp. 13Gff.

6. William R. Greer. "Robot Chef's New Dish: Hamburgers." *New York Times*, May 27, 1987, p. C3.

7. William R. Greer. "Robot Chef's New Dish: Hamburgers." *New York Times*, May 27, 1987, p. C3.

8. Michael Lev. "Taco Bell Finds Price of Success (59 cents)." *New York Times*, December 17, 1990, p. D9.

9. Calvin Sims. "Robots to Make Fast Food Chains Still Faster." *New York Times*, August 24, 1988, p. 5.

10. Chuck Murray. "Robots Roll from Plant to Kitchen." *Chicago Tribune–Business*, October 17, 1993, pp. 3ff; "New Robots Help McDonald's Make Fast Food Faster." Business Wire, August 18, 1992.

11. In recent years, the shortage of a sufficient number of teenagers to keep turnover-prone fast-food restaurants adequately stocked with employees has led to a widening of the traditional labor pool of fast-food restaurants.

12. Chuck Murray. "Robots Roll from Plant to Kitchen." *Chicago Tribune–Business*, October 17, 1993, pp. 3ff.

13. Eric A. Taub. "The Burger Industry Takes a Big Helping of Technology." *New York Times*, October 8, 1998, pp. 13Gff.

14. KinderCare Web site: www.kindercare.com/about_6.php3.

15. "The McDonald's of Teaching." *Newsweek*, January 7, 1985, p. 61.

16. Sylvan Learning Center Web site: www.educate.com/about.html.

17. "The McDonald's of Teaching." *Newsweek*, January 7, 1985, p. 61.

18. William Stockton. "Computers That Think." *New York Times Magazine*, December 14, 1980, p. 48.

19. Bernard Wysocki, Jr. "Follow the Recipe: Children's Hospital in San Diego Has Taken the Standardization of Medical Care to an Extreme." *Wall Street Journal* April 22, 2003, p. R4ff.

20. Frederick W. Taylor. *The Principles of Scientific Management*. New York: Harper & Row, 1947, p. 59.

21. Henry Ford. *My Life and Work*. Garden City, NY: Doubleday, 1922, p. 103.

22. Robin Leidner. *Fast Food, Fast Talk: Service Work and the Routinization of Everyday Life*. Berkeley: University of California Press, 1993, p. 105.

23. Virginia A. Welch. "Big Brother Flies United." *Washington Post–Outlook*, March 5, 1995, p. C5.

24. Virginia A. Welch. "Big Brother Flies United." *Washington Post–Outlook*, March 5, 1995, p. C5.

25. StopJunkCalls Web site: www.stopjunkcalls.com/convict.htm.

26. Staff. "Call Centres Become Bigger." Global News Wire, *India Business Insight*, September 30, 2002.

27. Gary Langer. "Computers Reach Out, Respond to Human Voice." *Washington Post*, February 11, 1990, p. H3.

28. Carl H. Lavin. "Automated Planes Raising Concerns." *New York Times*, August 12, 1989, pp. 1, 6.

29. Robin Leidner. *Fast Food, Fast Talk: Service Work and the Routinization of Everyday Life*. Berkeley: University of California Press, 1993.

30. L. B. Diehl and M. Hardart. *The Automat: The History, Recipes, and Allure of Horn and Hardart's Masterpiece*. New York: Clarkson Potter, 2002.

31. "Disenchanted Evenings." *Time*, September 3, 1990, p. 53.

32. Ester Reiter. *Making Fast Food*. Montreal and Kingston: McGill-Queens University Press, p. 86.

33. Stan Luxenberg. *Roadside Empires: How the Chains Franchised America*. New York: Viking, 1985.

34. Martin Plimmer. "This Demi-Paradise: Martin Plimmer Finds Food in the Fast Lane Is Not to His Taste." *Independent* (London), January 3, 1998, p. 46.

35. Harold Gracey. "Learning the Student Role: Kindergarten as Academic Boot Camp." In Dennis Wrong and Harold Gracey, eds., *Readings in Introductory Sociology*. New York: Macmillan, 1967, pp. 243–254.

36. Charles E. Silberman. *Crisis in the Classroom: The Remaking of American Education*. New York: Random House, 1970, p. 122.

37. Charles E. Silberman. *Crisis in the Classroom: The Remaking of American Education*. New York: Random House, 1970, p. 137.

38. Charles E. Silberman. *Crisis in the Classroom: The Remaking of American Education*. New York: Random House, 1970, p. 125.

39. William Severini Kowinski. *The Malling of America: An Inside Look at the Great Consumer Paradise*. New York: William Morrow, 1985, p. 359.

40. Gary Langer. "Computers Reach Out, Respond to Human Voice." *Washington Post*, February 11, 1990, p. H3.

41. Vatican Web site: www.vatican.va/news_services/television.

42. Jeffrey Hadden and Charles E. Swann. *Prime Time Preachers: The Rising Power of Televangelism*. Reading, MA: Addison-Wesley, 1981.

43. E. J. Dionne, Jr. "The Vatican Is Putting Video to Work." *New York Times*, August 11, 1985, sec. 2, p. 27.

44. William Serrin. "Let Them Eat Junk." *Saturday Review*, February 2, 1980, p. 23.

45. "Super Soup Cooks Itself." *Scholastic News*, January 4, 1991, p. 3.

46. AquaSol, Inc. Web site: www.fishfarming.com.

47. Martha Duffy. "The Fish Tank on the Farm." *Time*, December 3, 1990, pp. 107–111.

48. Peter Singer. *Animal Liberation: A New Ethic for Our Treatment of Animals*. New York: Avon, 1975.

49. Peter Singer. *Animal Liberation: A New Ethic for Our Treatment of Animals*. New York: Avon, 1975, pp. 96–97.

50. Peter Singer. *Animal Liberation: A New Ethic for Our Treatment of Animals*. New York: Avon, 1975, pp. 105–106.

51. Peter Singer. *Animal Liberation: A New Ethic for Our Treatment of Animals*. New York: Avon, 1975, p. 123.

52. Lenore Tiefer. "The Medicalization of Impotence: Normalizing Phallocentrism." *Gender and Society* 8(1994):363–377.

53. Cheryl Jackson. "Impotence Clinic Grows into Chain." *Tampa Tribune–Business and Finance*, February 18, 1995, p. 1.

54. Annette Baran and Reuben Pannor. *Lethal Secrets: The Shocking Consequences and Unresolved Problems of Artificial Insemination*. New York: Warner, 1989.

55. Paula Mergenbagen DeWitt. "In Pursuit of Pregnancy." *American Demographics*, May 1993, pp. 48ff.

56. Eric Adler. "The Brave New World: It's Here Now, Where In Vitro Fertilization Is Routine and Infertility Technology Pushes Back All the Old Limitations." *Kansas City Star*, October 25, 1998, pp. G1ff.

57. Clear Passage Web site: www.clearpassage.com/about_infertility_therapy.htm.

58. "No Price for Little Ones." *Financial Times*, September 28, 1998, pp. 17ff.

59. Diederika Pretorius. *Surrogate Motherhood: A Worldwide View of the Issues*. Springfield, IL: Charles C Thomas, 1994.

60. Korky Vann. "With In-Vitro Fertilization, Late-Life Motherhood Becoming More Common." *Hartford Courant*, July 7, 1997, pp. E5ff.

61. Ian MacKinnon. "Mother of Newborn Child Says She Is 65." *The Times* (London), April 10, 2003, Overseas News sec., p. 28.

62. Angela Cain. "Home Test Kits Fill an Expanding Health Niche." *Times Union-Life and Leisure* (Albany, NY), February 12, 1995, p. 11.

63. Neil Bennett, ed. *Sex Selection of Children*. New York: Academic Press, 1983.

64. "Selecting Sex of Child." *South China Morning Post*, March 20, 1994, p. 15.

65. Rick Weiss. "Va. Clinic Develops System for Choosing Sex of Babies." *Washington Post*, September 10, 1998, pp. A1ff; Randeep Ramesh. "Can You Trust That Little Glow When You Choose Sex?" *Guardian* (London), October 6, 1998, pp. 14ff; Abigail Trafford. "Is Sex Selection Wise?" *Washington Post*, September 22, 1998, pp. Z6ff.

66. Janet Daley. "Is Birth Ever Natural?" *The Times* (London), March 16, 1994, p. 18.

67. Matt Ridley. "A Boy or a Girl: Is It Possible to Load the Dice?" *Smithsonian* 24(June 1993):123.

68. Gina Kolata and Kenneth Chang. "For Clonaid, a Trail of Unproven Claims." *New York Times*, January 1, 2003, p. A13.

69. Roger Gosden. *Designing Babies: The Brave New World of Reproductive Technology*. New York: W. H. Freeman, 1999, p. 243.

70. Rayna Rapp. "The Power of 'Positive' Diagnosis: Medical and Maternal Discourses on Amniocentesis." In Donna Bassin, Margaret Honey, and Meryle Mahrer Kaplan, eds., *Representations of Motherhood*. New Haven, CT: Yale University Press, 1994, pp. 204–219.

71. Aliza Kolker and B. Meredith Burke. *Prenatal Testing: A Sociological Perspective*. Westport, CT: Bergin & Garvey, 1994, p. 158.

72. Jeffrey A. Kuller and Steven A. Laifer. "Contemporary Approaches to Prenatal Diagnosis." *American Family Physician 52*(December 1996):2277ff.

73. Aliza Kolker and B. Meredith Burke. *Prenatal Testing: A Sociological Perspective*. Westport, CT: Bergin & Garvey, 1994; Ellen Domke and Al Podgorski. "Testing the Unborn: Genetic Test Pinpoints Defects, But Are There Risks?" *Chicago Sun-Times*, April 17, 1994, p. C5.

74. However, some parents do resist the rationalization introduced by fetal testing. See Shirley A. Hill. "Motherhood and the Obfuscation of Medical Knowledge." *Gender and Society* 8(1994):29–47.

75. Mike Chinoy. *CNN News*. February 8, 1994.

76. Joan H. Marks. "The Human Genome Project: A Challenge in Biological Technology." In Gretchen Bender and Timothy Druckery, eds., *Culture on the Brink: Ideologies of Technology*. Seattle, WA: Bay Press, 1994, pp. 99–106; R. C. Lewontin. "The Dream of the Human Genome." In Gretchen Bender and Timothy Druckery, eds., *Culture on the Brink: Ideologies of Technology*. Seattle, WA: Bay Press, 1994, pp. 107–127; Staff. "Genome Research: International Consortium Completes Human Genome Project." *Genomics & Genetics Weekly*, May 9, 2003, p. 32.

77. Staff. "Genome Research: International Consortium Completes Human Genome Project." *Genomics & Genetics Weekly*, May 9, 2003, p. 32.

78. Matt Ridley. "A Boy or a Girl: Is It Possible to Load the Dice?" *Smithsonian 24*(June 1993):123.

79. Jessica Mitford. *The American Way of Birth*. New York: Plume, 1993.

80. For a critique of midwifery from the perspective of rationalization, see Charles Krauthammer. "Pursuit of a Hallmark Moment Costs a Baby's Life." *Tampa Tribune*, May 27, 1996, p. 15.

81. Judy Foreman. "The Midwives' Time Has Come—Again." *Boston Globe*, November 2, 1998, pp. C1ff.

82. Jessica Mitford. *The American Way of Birth*. New York: Plume, 1993, p. 13.

83. Catherine Kohler Riessman. "Women and Medicalization: A New Perspective." In P. Brown, ed., *Perspectives in Medical Sociology*. Prospect Heights, IL: Waveland, 1989, pp. 190–220.

84. Michelle Harrison. *A Woman in Residence*. New York: Random House, 1982, p. 91.

85. Judith Walzer Leavitt. *Brought to Bed: Childbearing in America, 1750–1950*. New York: Oxford University Press, 1986, p. 190.

86. Judith Walzer Leavitt. *Brought to Bed: Childbearing in America, 1750–1950*. New York: Oxford University Press, 1986, p. 190.

87. Paula A. Treichler. "Feminism, Medicine, and the Meaning of Childbirth." In Mary Jacobus, Evelyn Fox Keller, and Sally Shuttleworth, eds., *Body Politics: Women and the Discourses of Science*. New York: Routledge, 1990, pp. 113–138.

88. Jessica Mitford. *The American Way of Birth*. New York: Plume, 1993, p. 59.

89. An episiotomy is an incision from the vagina toward the anus to enlarge the opening needed for a baby to pass.

90. Jessica Mitford. *The American Way of Birth*. New York: Plume, 1993, p. 61.

91. Jessica Mitford. *The American Way of Birth*. New York: Plume, 1993, p. 143.

92. Michelle Harrison. *A Woman in Residence*. New York: Random House, 1982, p. 86.

93. Michelle Harrison. *A Woman in Residence*. New York: Random House, 1982, p. 113.

94. Jeanne Guillemin. "Babies by Cesarean: Who Chooses, Who Controls?" In P. Brown, ed., *Perspectives in Medical Sociology*. Prospect Heights, IL: Waveland, 1989, pp. 549–558.

95. L. Silver and S. M. Wolfe. *Unnecessary Cesarean Sections: How to Cure a National Epidemic*. Washington, DC: Public Citizen Health Research Group, 1989.

96. Joane Kabak. "C Sections." *Newsday*, November 11, 1996, pp. B25ff.

97. Susan Brink. "Too Posh to Push?" *U.S. News & World Report*, August 5, 2002, Health and Medicine sec., p. 42.

98. Randall S. Stafford. "Alternative Strategies for Controlling Rising Cesarean Section Rates." *JAMA*, February 2, 1990, pp. 683–687.

99. Jeffrey B. Gould, Becky Davey, and Randall S. Stafford. "Socioeconomic Differences in Rates of Cesarean Sections." *New England Journal of Medicine*, 321(4)(July 27, 1989):233–239; F. C. Barros et al. "Epidemic of Caesarean Sections in Brazil." *The Lancet*, July 20, 1991, pp. 167–169.

100. Randall S. Stafford. "Alternative Strategies for Controlling Rising Cesarean Section Rates." *JAMA*, February 2, 1990, pp. 683–687.

101. Although, more recently, insurance and hospital practices have led to more deaths in nursing homes or even at home.

102. Sherwin B. Nuland. *How We Die: Reflections on Life's Final Chapter*. New York: Knopf, 1994, p. 255; National Center for Health Statistics. *Vital Statistics of the United States, 1992–1993, Volume II—Mortality, Part A*. Hyattsville, MD: Public Health Service, 1995.

103. Derek Humphry. *Final Exit: The Practicalities of Self-Deliverance and Assisted Suicide for the Dying*, 3rd ed. New York: Delta, 2002.

104. Richard A. Knox. "Doctors Accepting of Euthanasia, Poll Finds: Many Would Aid in Suicide Were It Legal." *Boston Globe*, April 23, 1998, pp. A5ff.

105. Ellen Goodman. "Kevorkian Isn't Helping 'Gentle Death.'" *Newsday*, August 4, 1992, p. 32.

106. Lance Morrow. "Time for the Ice Floe, Pop: In the Name of Rationality, Kevorkian Makes Dying—and Killing—Too Easy." *Time*, December 7, 1998, pp. 48ff.

107. Amir Muhammad. "Heard Any Good Books Lately?" *New Straits Times*, October 21, 1995, pp. 9ff.

108. Raymond Kurzweil. *The Age of Intelligent Machines*. Cambridge: MIT Press, 1990.

15   White

Richard Dyer

Richard Dyer's study of photography and film technology from the nineteenth and early twentieth centuries provides a powerful example of technological decision making that was both shaped by and, in turn, shaped ideas about race and racial social arrangements. Without Dyer digging into the details of technological innovation, we might simply have assumed that the refinement of photographic techniques was based solely on chemistry, mechanics, theories of light, and the development of lenses; that is, we might have presumed they were technologically determined. Yet while these were all part of the history of photographic technology, Dyer shows that this knowledge was deployed in specific ways to enhance the features of white people and even to make them appear whiter than they might appear in real life. While we might think that cameras "don't lie" and, therefore, can't discriminate against or privilege certain races, Dyer suggests otherwise. Because of the values that were inherent in technical decisions, photographic equipment was developed in a particular way that enhanced white but not black faces. The technologies were motivated by and subsequently reinforced certain ideas about race.

All technologies are at once technical in the most limited sense (to do with their material properties and functioning) and also always social (economic, cultural, ideological). Cultural historians sometimes ride roughshod over the former, unwilling to accept the stubborn resistance of matter, the sheer time and effort expended in the trial and error processes of technological discovery, the internal dynamics of technical knowledge. Yet the technically minded can also underestimate, or even entirely discount, the role of the social in technology. Why a technology is even explored, why that exploration is funded, what is actually done with the result (out of all the possible things that could be done with it), these are not determined by purely technical considerations. Given tools and media do set limitations to what can be done with them, but these are very broad; in the immediacy and instantaneity of using technologies we don't stop to consider them culturally, we just use them as we know how—but the history, the social inscription, is there all the same.

Several writers have traced the interplay of factors involved in the development of the photographic media (e.g. Altman 1984, Coleman 1985, Neale 1985, Williams 1974) and this chapter is part of that endeavour. I am trying to add two things, in addition to the specificity of a focus on light. The first is a sense of the racial character of technologies, supplementing the emphasis on class and gender in previous work. Thus just as perspective as an artistic technique has been argued to be implicated in an

From *White* (New York: Routledge, 1997), pp. 83–94, 229, 242–246. © 1997 Richard Dyer. Reprinted by permission of Taylor & Francis Books UK.

individualistic world view that privileges both men and the bourgeoisie, so I want to argue that photography and cinema, as media of light, at the very least lend themselves to privileging white people. Second, I also want to insist on the aesthetic, on the technological construction of beauty and pleasure, as well as on the representation of the world. Much historical work on media technology is concerned with how media construct images of the world. This is generally too sophisticatedly conceptualised to be concerned with anything so vulgar as whether a medium represents the world accurately (though in practice, and properly, this lingers as an issue) but is concerned with how an ideology—a way of seeing the world that serves particular social interests is implicated in the mode of representation. I have no quarrel with this as such, but I do want to recognise that cultural media are only sometimes concerned with reality and are at least as much concerned with ideals and indulgence, that are themselves socially constructed. It is important to understand this too and, indeed, to understand how representation is actually implicated in inspirations and pleasures.

. . .

Lighting for Whiteness

The photographic media and, *a fortiori*, movie lighting assume, privilege and construct whiteness. The apparatus was developed with white people in mind and habitual use and instruction continue in the same vein, so much so that photographing non-white people is typically construed as a problem.

All technologies work within material parameters that cannot be wished away. Human skin does have different colours which reflect light differently. Methods of calculating this differ, but the degree of difference registered is roughly the same: Millerson (1972: 31), discussing colour television, gives light skin 43 per cent light reflectance and dark skin 29 per cent; Malkiewicz (1986: 53) states that "a Caucasian face has about 35 percent reflectance but a black face reflects less than 16 percent." This creates problems if shooting very light and very dark people in the same frame. Writing in *Scientific American* in 1921, Frederick Mills, "electrical illuminating engineer at the Lasky Studios," noted that

> when there are two persons in [a] scene, possibly a star and a leading player, if one has a dark make-up and the other a light, much care must be exercised in so regulating the light that it neither 'burns up' the light make-up nor is of insufficient strength to light up the dark make-up. (1921: 148)

The problem is memorably attested in a racial context in school photos where either the black pupils' faces look like blobs or the white pupils have theirs bleached out.

The technology at one's disposal also sets limits. The chemistry of different stocks registers shades and colours differently. Cameras offer varying degrees of flexibility with regard to exposure (effecting their ability to take a wide lightness/darkness range).

Different kinds of lighting have different colours and degrees of warmth, with concomitant effects on different skins. However, what is at one's disposal is not all that could exist. Stocks, cameras and lighting were developed taking the white face as the touchstone. The resultant apparatus came to be seen as fixed and inevitable, existing independently of the fact that it was humanly constructed. It may be—certainly was—true that photo and film apparatuses have seemed to work better with light-skinned peoples, but that is because they were made that way, not because they could be no other way.

All this is complicated still further by the habitual practices and uses of the apparatus. Certain exposures and lighting set-ups, as well as make-ups and developing processes, have become established as normal. They are constituted as the way to use the medium. Anything else becomes a departure from the norm, or even a problem. In practice, such normality is white.

The question of the relationship between the variously coloured human subject and the apparatus of photography is not simply one of accuracy. This is certainly how it is most commonly discussed, in accounts of innovation or advice to photographers and film-makers. There are indeed parameters to be recognised. If someone took a photo of me and made it look as if I had olive skin and black hair, I should be grateful but have to acknowledge that it was inaccurate. However, we also find acceptable considerable departures from how we 'really' look in what we regard as accurate photos, and this must be all the more so with photography of people whom we don't know, such as celebrities, stars and models. In the history of photography and film, getting the right image meant getting the one which conformed to prevalent ideas of humanity. This included ideas of whiteness, of what colour—what range of hue—white people wanted white people to be.

The rest of this section is concerned with the way the aesthetic technology of photography and film is involved in the production of images of whiteness. I look first at the assumption of whiteness as norm at different moments of technical innovation in film history, before looking at examples of that assumption in standard technical guides to the photographic media. The section ends with a discussion of how lighting privileges white people in the image and begins to open up the analysis of the construction of whiteness through light.

Innovation in the photographic media has generally taken the human face as its touchstone, and the white face as the norm of that. The very early experimenters did not take the face as subject at all, but once they and their followers turned to portraits, and especially once photographic portraiture replaced painted portraits in popularity (from the 1840s on), the issue of the "right" technology (apparatus, consumables, practice) focused on the face and, given the clientele, the white face. Experiment with, for instance, the chemistry of photographic stock, aperture size, length of development and artificial light all proceeded on the assumption that what had to be got right was the look of the white face. This is where the big money lay, in the everyday practices of professional portraiture and amateur snapshots. By the time of film (some sixty years after the first photographs), technologies and practices were already well established. Film borrowed these, gradually and selectively, carrying forward the assumptions that had gone into them. In turn, film history involves many refinements, variations and innovations, always keeping the white face central as a touchstone and occasionally revealing this quite explicitly, when it is not implicit within such terms as "beauty," "glamour" and "truthfulness." Let me provide some instances of this.

The interactions of film stock, lighting and make-up illustrate the assumption of the white face at various points in film history. Film stock repeatedly failed to get the whiteness of the white face. The earliest stock, orthochromatic, was insensitive to red and yellow, rendering both colours dark. Charles Handley, looking back in 1954, noted that with orthochromatic stock, "even a reasonably light-red object would photograph black" (1967: 121). White skin is reasonably light-red. Fashion in make-up also had to be guarded against, as noted in one of the standard manuals of the era, Carl Louis Gregory's *Condensed Course in Motion Picture Photography* (1920):

Be very sparing in the use of lip rouge. Remember that red photographs black and that a heavy application of rouge shows an unnaturally black mouth on the screen. (316)

Yellow also posed problems. One derived from theatrical practices of make-up, against which Gregory inveighs in a passage of remarkable racial resonance:

Another myth that numerous actors entertain is the yellow grease-paint theory. Nobody can explain why a performer should make-up in chinese yellow.... The objections to yellow are that it is non-actinic and if the actor happens to step out of the rays of the arcs for a moment or if he is shaded from the distinct force of the light by another actor, his face photographs BLACK instantly. (ibid.: 317; emphasis in original)

The solution to these problems was a "dreadful white make-up" (actress Geraldine Farrar, interviewed in Brownlow 1968: 418) worn under carbon arc lights so hot that they made the make-up run, involving endless retouching. This was unpleasant for performers and exacerbated by fine dust and ultraviolet light from the arcs, making the eyes swollen and pink (so-called "Klieg eyes" after the Kliegl company which was the main supplier of arc lights at the time (Salt 1983: 136)). These eyes filmed big and dark, in other words, not very "white," and involved the performers in endless "trooping down to the infirmary" (Brownlow 1968: 418), constantly interrupting shooting for their well-being and to avoid the (racially) wrong look.

It would have been possible to use incandescent tungsten light instead of carbon arcs; this would have been easier to handle, cheaper (requiring fewer people to operate and using less power) and pleasanter to work with (much less hot). It would also have suited one of the qualities of orthochromatic stock, its preference for subtly modulated lighting rather than high contrast of the kind created by arcs. But incandescent tungsten light has a lot of red and yellow in it and thus tends to bring out those colours in all subjects, including white faces, with consequent blacking effect on orthochromatic stock. This was a reason for sticking with arcs, for all the expense and discomfort.

The insensitivity of orthochromatic stock to yellow also made fair hair look dark "unless you specially lit it" (cinematographer Charles Rosher, interviewed in Brownlow 1968: 262). Gregory similarly advised:

Yellow blonde hair photographs dark... the more loosely [it] is arranged the lighter it photographs, and different methods of studio lighting also affect the photographic values of hair. (1920: 317)

One of the principal benefits of the introduction of backlighting, in addition to keeping the performer clearly separate from the background, was that it ensured that blonde hair looked blonde:

The use of backlighting on blonde hair was not only spectacular but *necessary*—it was the only way filmmakers could get blonde hair to look light-coloured on the yellow-insensitive orthochromatic stock. (Bordwell *et al.* 1985: 226; my emphasis)

Backlighting became part of the basic vocabulary of movie lighting. As the cinematographer Joseph Walker put it in his memoirs:

We found [backlighting] necessary to keep the actors from blending into the background. [It] also adds a halo of highlights to the hair and brilliance to the scene. (Walker and Walker 1984: 218)

From 1926, the introduction of panchromatic stock, more sensitive to yellow, helped with some of the problems of ensuring white people looked properly white, as well as permitting the use of incandescent tungsten, but posed its own problems of make-up. It was still not so sensitive to red, but much more to blue. Max Factor recognised this problem, developing a make-up that would "add to the face sufficient blue coloration in proportion to red... in order to prevent excessive absorption of light by the face" (Factor 1937: 54); faces that absorb light "excessively" are of course dark ones.

Colour brought with it a new set of problems, explored in Brian Winston's article on the invention of "colour film that more readily photographs Caucasians than other human types" (1985: 106). Winston argues that at each stage the search for a colour film stock (including the development process, crucial to the subtractive systems that have proved most workable) was guided by how it rendered white flesh tones. Not long after the introduction of colour in the mid-1930s, the cinematographer Joseph Valentine commented that "perhaps the most important single factor in dramatic cinematography is the relation between the colour sensitivity of an emulsion and the reproduction of pleasing flesh tones" (1939: 54). Winston looks at one such example of the search for "pleasing flesh tones" in researches undertaken by Kodak in the early 1950s. A series of prints of "a young lady" were prepared and submitted to a panel, and a report observed:

Optimum reproduction of skin colour is not "exact" reproduction... "exact reproduction" is rejected almost unanimously as "beefy." On the other hand, when the print of highest acceptance is masked and compared with the original subject, it seems quite pale. (David L. MacAdam 1951, quoted in Winston 1985: 120)

As noted above, white skin is taken as a norm but what that means in terms of colour is determined not by how it is but by how, as Winston puts it, it is "preferred—a whiter shade of white" (ibid.: 121). Characteristically too, it is a woman's skin which provides the litmus test.

Colour film was a possibility from 1896 (when R. W. Paul showed his hand-tinted prints), with Technicolor, the "first entirely successful colour process used in the cinema," available from 1917 (Coe 1981: 112–39). Yet it did not become anything like a norm until the 1950s, for a complex of economic, technological and aesthetic reasons (cf. Kindem 1979), among which was a sense that colour film was not realistic. As Gorham Kindem suggests, this may have been partly due to a real limitation of the processes adopted from the late 1920s, in that they "could not reproduce all the colours of the visible spectrum" (1979: 35) but it also had to do with an early association with musicals and spectacle. The way Kindem elaborates this point is racially suggestive:

While flesh tones, the most important index of accuracy and consistency, might be carefully controlled through heavy make-up, practically dictating the overall colour appearance, it is quite likely that other colour in the set or location had to be sacrificed and appeared unnatural or "gaudy." (ibid.)

As noted elsewhere, accurate flesh tones are again the key issue in innovation. The tones involved here are evidently white, for it was lighting the compensatory heavy make-up with sufficient force to ensure a properly white look that was liable to make everything else excessively bright and "gaudy." Kindem relates a resistance to such an excess of colour with growing pessimism and cynicism through the 1930s as the weight of the Depression took a hold, to which black and white seemed more appropriate. Yet this seems to emphasise the gangster and social problem films of the 1930s over and above the comedies, musicals, fantasies and adventure films (think screwball, Fred and Ginger, Tarzan) that were, all the same, made in black and white. May it not be that what was not acceptable was escapism that was visually too loud and busy, because excess colour, and the very word "gaudy," was associated with, indeed, coloured people?

A last example of the operation of the white face as a control on media technology comes from professional television production in the USA.[1] In the late 1970s the WGBH Educational Foundation and the 3M Corporation developed a special television signal, to be recorded on videotape, for the purpose of evaluating tapes. This signal, know as "skin," was of a pale orange colour and was intended to duplicate the appearance on a television set of white skin. The process of scanning was known as "skinning." Operatives would watch the blank pale orange screen produced by tapes prerecorded with the "skin" signal, making notes whenever a visible defect appeared. The fewer defects, the greater the value of the tape (reckoned in several hundreds of dollars) and thus when and by whom it was used. The whole process centred on blank images representing nothing, and yet founded in the most explicit way on a particular human flesh colour.

The assumption that the normal face is a white face runs through most published advice given on photo- and cinematography.[2] This is carried above all by illustrations which invariably use a white face, except on those rare occasions when they are discussing the "problem" of dark-skinned people. The aesthetic technology of photography, as it has been invented, refined and elaborated, and the dominant uses of that technology, as they have become fixed and naturalised, assume and privilege the white subject. They also construct that subject, that is, draw on and contribute to a perception of what it means to be white. They do this as part of a much more general culture of light. This has produced both an astonishing set of technologies of light and certain fundamental philosophical, scientific and aesthetic perceptions of the nature of light. White people are central to it, to the extent that they come to seem to have a special relationship to light.

Notes

1. I am grateful to William Spurlin of the Visual and Environmental Studies Department at Harvard University for telling me of this, explaining it and commenting on drafts of this paragraph.

2. This observation is based on analysis of a random cross-section of such books published this century.

Bibliography

Altman, Rick (1984) "Towards a Theory of the History of Representational Technologies," *Iris* 2(2): 111–124.

Bordwell, David, Staiger, Janet and Thompson, Kristin (1985) *The Classical Hollywood Cinema: Film Style and Mode of Production to 1960*, New York: Columbia University Press.

Brownlow, Kevin (1968) *The Parade's Gone By*, London: Secker & Warburg.

Coe, Brian (1981) *The History of Movie Photography*, London: Ash & Grant.

Coleman, A. D. (1985) "Lentil Soup," *Et cetera* Spring: 19–31.

Factor, M. (1937) "Standardization of Motion Picture Make-up," *Journal of the Society of Motion Picture Engineers* 28(1): 52–62.

Gregory, Carl Louis, editor (1920) *A Condensed Course in Motion Picture Photography*, New York: New York Institute of Photography.

Handley, C. W. (1967) "History of Motion-Picture Studio Lighting" in Fielding 1967: 120–124. (First published in the *Journal of the Society of Motion Picture and Television Engineers*, October 1954.)

Kindem, Gorham A. (1979) "Hollywood's Conversion to Color: The Technological, Economic and Aesthetic Factors," *Journal of the University Film Association* 31(2): 29–36.

Malkiewicz, Kris (1986) *Film Lighting*, New York: Prentice-Hall.

Millerson, Gerald (1972) *The Technique of Lighting for Television and Motion Pictures*, London: Focal Press.

Mills, Frederick S. (1921) "Film Lighting as a Fine Art: Explaining Why the Fireplace Glows and Why Films Stars Wear Halos," *Scientific American* 124: 148, 157–158.

Neale, Steve (1985) *Cinema and Technology: Image, Sound, Colour*, London: Macmillan/British Film Institute.

Salt, Barry (1983) *Film Style and Technology: History and Analysis*, London: Starword.

Valentine, Joseph (1939) "Make-up and Set Painting Aid New Film," *American Cinematographer* February: 54–56, 85.

Walker, Joseph B. and Walker, Juanita (1984) *The Light on Her Face*, Hollywood: ASC Press.

Williams, Raymond (1974) *Television: Technology and Cultural Form*, London: Fontana.

Winston, Brian (1985) "A Whole Technology of Dyeing: A Note on Ideology and the Apparatus of the Chromatic Moving Image," *Daedalus* 114(4): 105–123.

16 "Manufacturing Gender in Commercial and Military Cockpit Design"

Rachel N. Weber

The preceding readings indicate that values can be designed into technology; however, this is sometimes done deliberately and other times not. In this chapter, Weber provides a case in which the design of a technology denied access to a majority of women. In large part this was not intentional. Designers often work with a specific set of users in mind by using data and specifications on that group or by simply assuming that most users will be similar to themselves. In so doing, however, the designed object may discriminate against other users that don't fit this vision. (Winner makes a reference to this when he mentions that buses, sidewalks, and buildings have often been designed in ways that discriminate against people in wheelchairs.) Once in place, a designed artifact can be very difficult to change and will continue to be a barrier to certain groups. But in Weber's example, the "momentum" of the sociotechnical system was overcome. She describes how a concerned group of people successfully redesigned airplane cockpits in order to accommodate a larger number of users including a significantly greater percentage of women. They were able to counter arguments that such a change would cost too much money, would take too much time, and would be resisted by some of those previously discriminated against because they did not want to appear to be getting special treatment. In this respect, Weber's account illustrates both unintentional incorporation of values into a design and subsequent deliberate embedding of values in a revised version of the technology.

Technological Bias in Existing Aircraft

Civilian and defense aircraft have traditionally been built to male specifications (Binkin 1993). Since women tend to be shorter, have smaller limbs and less upper-body strength, some may not be accommodated by such systems and may experience difficulty in reaching controls and operating certain types of equipment (McDaniel 1994). To understand how women's bodies become excluded by design and how difference becomes technologically embodied, it is necessary to examine how current military systems are designed with regard to the physical differences of their human operators.

To integrate the user into current design practices, engineers rely on the concepts of ergonomics and anthropometrics (McCormick and Sanders 1982). Ergonomics, also called "human factors," addresses the human characteristics, expectations, and behaviors in the design of items which people use. During World War II, ergonomics became a distinct discipline, practiced predominantly by the U.S. military. Ergonomic theories were first implemented when it became obvious that new and more complicated types

From *Science, Technology, & Human Values* 22, no. 2 (Spring 1997): 235–253. © 1997 Sage Publications Inc. Reprinted by permission of Sage Publications Inc.

of military equipment could not be operated safely or effectively or maintained adequately even by well-trained personnel. The term "human engineering" was coined and efforts were made to design equipment that would be more suitable for human use.

Anthropometrics refers to the measurement of dimensions and physical characteristics of the body as it occupies space, moves, and applies energy to physical objects as a function of age, sex, occupation, and ethnic origin and other demographic variables. Engineers at the Pentagon and at commercial airframe manufacturers rely on the U.S. Army Natick Research Development and Engineering Center's "1988 Anthropometric Survey of Army Personnel," in which multiple body dimensions are measured and categorized to standardize the design of systems. The Natick survey contains data on more than 180 body and head dimension measurements of a population of more than 9,000 soldiers. Age and race distributions match those of the June 1988 active duty Army, but minority groups were intentionally oversampled to accommodate anticipated demographic shifts in Army population (Richman-Loo and Weber 1996).

Technological Bias within Defense Aircraft
Department of Defense acquisition policy mandates that human considerations be integrated into design efforts to improve total system performance by focusing attention on the capabilities and limitations of the human operator. In other words, the Defense Department recognizes that the best defense technology is useless if it is incompatible with the capabilities and limitations of its users. In the application of anthropometric data, systems designers commonly rely on Military Standard 1472, "Human Engineering Design Criteria for Military Systems, Equipment and Facilities." Like the use of military specifications in the procurement process, these guidelines are critical in developing standards; they embody decisions made which reflect the military's needs and goals and are ultimately embodied in the technology (Roe Smith 1985).

These guidelines suggest the use of 95th and 5th percentile male dimensions in designing weapons systems. Use of this standard implies that only 10 percent of men in the population will not be accommodated by a given design feature. If the feature in question is sitting height, the 5 percent of men who are very short and the 5 percent who are very tall will not be accommodated.

Accommodation becomes more difficult when more than one physical dimension is involved, and several dimensions need to be considered in combination. The various dimensions often have low correlations with each other (e.g., sitting height and arm length). For example, approximately 52 percent of Naval aviators would not be accommodated by a particular cockpit specification if both the 5th and 95th percentiles were used for each of the thirteen dimensions.

Because women are often smaller in all physical dimensions than men, the gap between a 5th percentile woman and a 95th percentile man can be very large (Richman-Loo and Weber 1996). Women who do not meet requirements are deemed ineligible to use a variety of military systems.

The case of the Joint Primary Aircraft Training System (JPATS) has been the most publicized case of military design bias against women.[1] Engineers and human factors specialists considered minimum anthropometric requirements needed by an individual to operate the JPATS effectively and wrote specifications to reflect such requirements. For example, "the ability to reach and operate leg and hand controls, see cockpit gauges and displays, and acquire external vision required for safe operation" was considered critical to the safe and efficient operation of the system. Navy and Air Force engineers determined the five critical anthropometry design "drivers" to be sitting height, functional arm reach, leg length, buttock-knee length, and weight (Department of Defense 1993, 2).

Original JPATS specifications included a 34-inch minimum sitting height requirement in order to safely operate cockpit controls and eject. This specification is based on sitting height minimums in the current aircraft fleet and reflects a 5th percentile male standard. However, at 34 inches, anywhere from 50 to 65 percent of the American female population is excluded because female sitting heights are generally smaller than male. Therefore, JPATS, as originally intended, accommodated the 5th through 95th percentile male, but only approximately the 65th through 95th percentile female.

After successful completion of mandatory JPATS training, student pilots advance to intermediate trainers and then to aircraft-specific training. Therefore, if women cannot "fit" into the JPATS cockpit or if the cockpit does not "fit" women pilots, they will be unable to pursue aviation careers in the Navy or Air Force. In other words, design bias has far-reaching implications for gender equity in the military.

Technological Bias within Commercial Aircraft

Engineers design commercial cockpits based on military specifications, aiming to accommodate a population ranging from 25th percentile military women to 99th percentile military men. The methods used by human factors practitioners in the commercial world to determine accommodation are quite similar to those used by the military, many having been developed by internal defense divisions or borrowed directly from the public sector research laboratories (Weber 1995). Using computerized human modeling packages, engineers are able to analyze visibility and reach in a proposed cockpit design. Such programs create three-dimensional graphic representations of pilots which can be adjusted to different body sizes and proportions based on accumulated anthropometric data from the Army surveys, such as those published by the U.S. Army Natick Research Development and Engineering Center.

Although military and commercial engineers use similar methods and data, their pilot populations may differ. Commercial aviation relies on anthropometric data representative only of military populations, even though a different pool of pilots may be flying commercial planes. Many of the human factors engineers interviewed maintained that one of the obstacles to overcoming design bias against commercial women pilots is the lack of comprehensive anthropometric data for civilian female populations. The only available civilian anthropometric data are very old; for female measurements, some manufacturers still use a 1940 Department of Agriculture survey conducted for clothing dimensions. Human factors engineers agree that these data are not extensive enough for use in designing large, complex interfaces such as cockpits.

Commercial manufacturers do not possess conclusive data regarding the total population of women commercial pilots, let alone their body dimensions.[2] Approximately 3 percent of all pilots in the U.S. are women, and the percentage is significantly lower worldwide (Gilmartin 1992). In 1990 the Air Line Pilots Association (ALPA) estimated that there were approximately 900 women pilots (out of a total of 43,000) at forty-four of the airlines where it had members at that time. However, the number of women earning their air transport rating in the United States has increased by 325 percent since 1980.

Human systems specialists suspect that the civilian pilot population is more varied than the military because civilian airlines have less restrictive eligibility requirements. Commercial airlines do not maintain the same limits on body weight and height as the military. Moreover, in the military most pilots are between twenty-one and thirty-five years old, whereas commercial airlines employ an older population, often composed of retired servicemen.[3] This results in a less standardized commercial pilot population, one that might not be represented in the anthropometric data culled by the military.

Principal airframe manufacturers, such as Boeing and McDonnell Douglas, contract with both the government and private airlines. Much of the technology base, supplier base, skills, and processes used by defense and civil aircraft are common even though the divisions responsible for military and civilian work are organizationally and physically separate (Markusen and Yudken 1992). Whereas the defense division responds to a single client—the Pentagon—whose main concern is the performance characteristics, the concerns of the commercial division focus primarily on production costs or marketing (Melman 1983; Markusen 1985).

Despite a similar technological base, the cockpit technology encountered in civilian aviation differs from that found in the military. The role of the human being and the control processes available to him or her also will differ. For example, the extreme rates of acceleration experienced in military cockpits require elaborate restraining devices. Such restraints must be designed to fit the anthropometric characteristics of the intended users. Ejection is also an issue limited to military cockpit design. Much of the JPATS controversy centers on ejection seats and the need to provide safe ejection to lighter individuals.

In contrast, commercial aircraft do not reach the same high speeds as military planes, nor do they contain ejection seats. The seats in a commercial cockpit are adjustable to meet the varied comfort and safety requirements of the users. Thus certain anthropometrics such as height, weight, and strength do not have the same valence in commercial aviation as they do in the military. Many argue that commercial aircraft can accommodate a more variable population because the operating requirements are not as stringent as in the military.

However, the location of various controls on the commercial flight deck has been found to disadvantage women and smaller-statured men (Sexton 1988). Although the seats are more adjustable, individuals with smaller functional arm reach and less upper-body strength may still experience difficulties manipulating controls and reaching pedals. When smaller women are sitting in the co-pilot seat, some complain that they are not able to reach controls on the right side of the control panel. Reach concerns become increasingly important during manual reversion (when the system reverts to manual operation) even though electrical and hydraulic systems require smaller forces to actuate.

Cockpit design specifications have protected what has traditionally been a male occupation. Because both commercial and defense aircraft have been built for use by male pilots, the physical differences between men and women serve as very tangible rationales for gender-based exclusion. Although technology certainly is not the only "cause" of exclusion and segregation, biased aircraft act as symbolic markers, used to delineate the boundaries between men's and women's social space. Reppy (1993, 6) notes that

it is not that women are not physically capable of flying these particular aircraft or that they are not equally exposed to danger in other aircraft; rather denying women access to combat aircraft is a way of protecting a distinctly male arena. The technical artifact...has functioned to delineate the "other."

Regulating Accommodation in Defense Aircraft

The decision to standardize any technology is often contested, occurring within a space where social, economic, and political factors vie for position. In this case, standardization involved altering technologies in order to adjust to a changed sociopolitical environment. In the military, cockpit technology had to be adjusted to the entry of women into the armed forces and their new roles within the services. The process of design accommodation in the military became a process of negotiation between various social groups who held different stakes in and interpretations of the technology in question (Pinch and Bijker 1984).

One could argue that negotiations over accommodation arose as a result of changes made in policies regarding women in combat. Former Secretary of Defense Les Aspin publicly recognized that women should play a greater role in the military when he issued a directive in April 1993 on the assignment of women in the armed forces. The directive states that

the services shall permit women to compete for assignments in aircraft, including aircraft engaged in combat missions.

The Army and Marine Corps shall study opportunities for women to serve in additional assignments, including, but not limited to, field artillery and air defense artillery. (Aspin 1993, 1)

Although the new policy gave women a greater combat aviation role and was intended to allow for their entry into many new assignments, the aircraft associated with these assignments precluded the directive from being implemented. The realization that existing systems could contain a technological bias against women's bodies despite the Congressional mandate for accessibility alarmed policy specialists at the Pentagon. This contradiction would potentially embarrass a new administration which was reeling from its handling of the gays in the military debacle and desperately trying to define a working relationship with an antagonistic Pentagon.

In May 1993 the Under Secretary of Defense (Acquisition) directed the Assistant Secretary of Defense (Personnel and Readiness) to develop a new JPATS sitting height threshold which would accommodate at least 80 percent of eligible women. He delayed release of the JPATS draft Request for Proposal until a new threshold could be documented. This move led to the establishment of the JPATS Cockpit Accommodation Working Group which included representatives from the Air Force and Navy JPATS Program Offices as well as from service acquisition, personnel, human factors, and flight surgeon organizations. After months of deliberation, the Working Group determined that a reduction of the sitting height requirement by 3 inches would accommodate approximately 82 percent of the eligible female population (Department of Defense 1993).

Reducing the operational requirements would entail modifying existing cockpit specifications. Significant modifications were needed because the requirement for an ejection seat restricts the possibility of making the seat adjustable. In addition, the aircraft nose, rudder, and other flight controls would also need to be substantially modified to accommodate a smaller person. Further, since ejections at smaller statures and corresponding body weights had yet to be certified for safety, test articles and demonstrators had to be developed to ensure safe ejection (Dorn 1993).

After the May 1993 directive, many procurement specialists at the Pentagon were perplexed: a design which would accommodate the 5th percentile female through the 95th percentile male would have to incorporate a very wide variability of human dimensions. Some senior defense officials opposed such a change because they believed that such alterations would delay the development of the JPATS, would raise the price of training, and would be prohibitively expensive.

In opposition to these officials, pragmatists within the Pentagon—including most members of the Working Group—argued that it was both efficient and economical to integrate human factors into acquisition. Pragmatists felt that the technologies built for the military, as opposed to civilian markets, tended to privilege capability over maintenance and operability and hardware over personnel. They argued that with decreasing budgets, this could no longer be the case. Design changes, they claimed, would not only benefit women assigned to weapons systems originally designed for male operators, but would benefit smaller men as well. Studies have shown that smaller men also have difficulty operating hatches, damage control equipment, and scuttles on ships (Key, Fleischer, and Gauthier 1993). Shrinking personnel resources and a changing demographic pool from which the military recruits also mandated that defense technologies be more closely matched to human capabilities. The pragmatists were quick to emphasize that the inclusion and accommodation of smaller men would be necessary given changes in the ethnic and racial makeup of the nation (Stiehm 1985).

Pragmatists also pointed to the prospect of foreign military sales to countries with smaller-sized populations, which would make design accommodation an important economic consideration as well. Edwin Dorn (1993), the Assistant Secretary of Defense, in a memorandum to the Under Secretary of Defense (Acquisition), stressed that

a reduced JPATS sitting height threshold will also expand the accommodation of shorter males who may have previously been excluded from pilot training. For potential foreign military sales, this enhances its marketability in countries where pilot populations are of smaller average stature.

The pragmatists emphasized that cockpit accommodation would benefit all soldiers because it required the acquisition process to consider differences concerning capabilities and limitations. In pursuing this line of argument, they essentially neutered the discourse, erasing the specificities of women's bodies. By refusing to engage in a gendered discourse and instead emphasizing economic benefits, they hoped to appeal to a broader segment of the population and to a Pentagon traditionally hostile to women's issues.

In contrast to the Pentagon pragmatists, women's groups both within the military and outside supported the decision to alter the JPATS sitting height requirement on more ideological grounds. The fact that women were being excluded by the operational requirements and by the technology was central to their decision to support the changes. In general, feminism in the contemporary military environment is organized around ideals of parity and equal opportunity regarding career opportunities (Katzenstein 1993). Insisting that career advancement be based on qualifications, not biology, many argued that physical restrictions which disqualified women would unfairly limit women's mobility in the services.

Through informal networks and more formal associations such as the Defense Advisory Committee on Women in the Service (DACOWITS), new groups of activists set about to influence policy decisions about career opportunities for women.[4] Women aviators organized around the issue of female accommodation and found a receptive audience in some of the new Clinton appointees, such as Edwin Dorn, Assistant Secretary of Defense. Unlike other changes imposed from the top, the decision to alter JPATS was part of a low-level process that began with limited intervention from high-ranking administrators (Brundage 1993).

Although the media spectacle of the Tailhook scandal provided the necessary momentum for feminist groups in the military and brought gender issues to the forefront of national debates, the decision to accommodate more women in the JPATS cockpit was not without dissension.[5] Some women officers—many of whom also considered themselves feminists—believed that, as one of the people I have interviewed told me, "shrill cries for accommodation could be used against women politically." They insisted that demanding special treatment would single women out in an institution which, on the surface, seeks to eradicate differences between the sexes. In a sense, they were asking women to ignore their difference and prove themselves on gender-neutral terms.

A few women pilots questioned the construction of the operational requirements and thresholds but insisted that the existing cockpits were not biased. Is it really necessary, some asked, to possess a sitting height of 34 inches to fly defense aircraft? Women with smaller sitting heights had flown during wartime, and many believed that pilots at shorter sitting heights were no less capable of flying safely. One woman claimed that "the whole issue of height in aircraft is overstated, and just ignorance on the part of the Navy."

As debates raged in the press and within the Working Group during 1993, the possibilities for technological variety began to close down. The Pentagon pragmatists attempted to stabilize the debate, but the public spectacle of the issue facilitated closure by broadening the deliberative arena. With the JPATS case, "administrative" closure was achieved when the 1994 Defense Authorization Bill was passed. The bill included a provision which prevented the Air Force, the lead agency in the purchase of the JPATS, from spending $40 million of its $41.6 million trainer budget unless the Pentagon altered the cockpit design. John Deutsch (1992), then the Under Secretary of Defense, wrote a memo legitimizing the problem of accommodation of women in defense aircraft, stating:

I believe the Office of the Secretary of Defense (OSD) should continue to take the lead in addressing this problem. Other platforms in addition to aircraft should be considered as well. We must determine what changes are practical and cost effective in support of Secretary of Defense policy to expand combat roles for females. I request that you take the lead in determining specification needs. Further, you should determine the impact of defense platforms already in production and inventory. (Deutsch 1992, 1)

After Working Group deliberations, the Air Force issued a revised JPATS Draft Request for Proposal that included a 32.8-inch sitting height threshold. The RFP identified crew accommodation as a key source selection criterion so that during the selection process, prospective contractors would be required to submit cockpit mock-ups which would be evaluated for their adherence to the revised JPATS anthropometric requirements. Candidates who adhered to and even exceeded these requirements stood the best chance of winning the contract.

As the preceding case reveals, the relevant social groups who had a stake in changing the technology were able to voice their interests in quasi-public fora: in legislative committees, in the JPATS Working Group meetings, and in the popular media. The debates surrounding accommodation exposed the interpretive flexibility of cockpit design but also demonstrated how the more powerful and pragmatic groups were able to push forth their agenda. Able to increase momentum because of intersecting debates on "women in combat," the Working Group cast the issue of altering military technologies in terms of accommodating all types of operators and emphasized the political accountability of a public consumer to these operators....

Notes

1. The JPATS is the aircraft used by both the Navy and the Air Force to train its pilot candidates.

2. The FAA Statistics and Forecast Branch maintains information on the number of women pilots who have a current medical certificate and a pilot license. In 1993, 39,460 women held both the certificate and license out of a total of 665,069 pilots (Office of Aviation Policy, Plans and Management 1993). However these figures do not reflect the number of women actually employed as commercial pilots.

3. In the past, commercial pilots received their training in the military, whereas now the trend is to filter through private flight training schools.

4. Mary Katzenstein provided me with these insights. See also Enloe (1993, 208–14).

5. The Tailhook scandal refers to the annual Tailhooker's (Navy carrier pilots) convention of 1991 where several women were sexually harassed by servicemen and later went public with their charges. As a result, three admirals were disciplined, although none of the servicemen were officially charged.

References

Aspin, L. 1993. *Policy on the assignment of women in the armed forces*. 28 April. Washington, D.C.: Department of Defense.

Binkin, M. 1993. *Who will fight the next war? The changing face of the American military*. Washington, D.C.: Brookings Institute.

Brundage, W. 1993. The changing self-definitions of the military and women's occupational specializations. Paper presented at the workshop on "Institutional change and the U.S. military: The changing role of women," October, Cornell University.

Department of Defense. 1993. JPATS cockpit accommodation working group report. Unpublished report, 3 May.

Deutsch, J. 1992. Memorandum on JPATS cockpit accommodation working group report. Unpublished report, 2 December.

Dorn, E. 1993. Memorandum on JPATS cockpit accommodation working group report. Unpublished report, 19 October.

Enloe, C. 1993. *The morning after: Sexual politics and the end of the cold war*. Berkeley: University of Califorma Press.

Gilmartin, P. 1992. Women pilots' performance in Desert Storm helps lift barriers in military, civilian market. *Aviation Week and Space Technology*, 13 January.

Katzenstein, M. 1993. The formation of feminism in the military environment. Paper presented at the workshop on "Institutional change and the U.S. military: The changing role of women," October, Cornell University.

Key, E., E. Fleischer, and E. Gauthier. 1993. Women at sea: Design considerations. Paper presented at the Association of Scientists and Engineers, 30th Annual Technological Symposium, March, Houston, TX.

Markusen, A. 1985. *The economic and regional consequences of military innovation*. Berkeley, CA: Institute of Urban and Regional Development.

Markusen, A., and J. Yudken. 1992. *Dismantling the Cold War economy*. New York: Basic Books.

McCormick, E., and M. Sanders. 1982. *Human factors in engineering and design*. New York: McGraw-Hill.

McDaniel, J. 1994. Strength capability for operating aircraft controls. In *Advances in industrial ergonomics and safety*, vol. 6, edited by E. Aghazadeh, 58–73. Bristol, PA: Taylor and Francis.

Melman, S. 1983. *Profits without production*. New York: Knopf.

Office of Aviation Policy, Plans and Management. 1993. *U.S. civil airmen statistics*. FAA APO-94-6. Washington, D.C.: Federal Aviation Administration.

Pinch, T., and W. Bijker. 1984. The social construction of facts and artifacts. *Social Studies of Science* 14:399–441.

Reppy, J. 1993. New technologies in the gendered workplace. Paper presented at the workshop on "Institutional change and the U.S. military: The changing role of women," October, Cornell University.

Richman-Loo, N., and R. Weber. 1996. Gender and weapons design: Are military technologies biased against women? In *It's our military too! Women and the U.S. military*, edited by J. Stiehm, 136–155. Philadelphia: Temple University Press.

Roe Smith, M., ed. 1985. *Military enterprise and technological change*. Cambridge: MIT Press.

Sexton, G. 1988. Cockpit-crew systems design and integration. In *Human factors in aviation*, edited by E. Weiner and T. Nagle, 495–526. San Diego: Academic Press.

Stiehm, J. 1985. Women's biology and the U.S. military. In *Women, biology, and public policy*, edited by J. Stiehm, 205–34. Beverly Hills, CA: Sage.

Weber, R. 1995. Accommodating difference: Gender and cockpit design in military and civilian aviation. *Transportation Research Record* 1480:51–6.

17 "Pas de Trois: Science, Technology, and the Marketplace"

Daniel Sarewitz

One of the most powerful versions of technological determinism is the, often implicit, claim that technology brings not just social change but progress. In fact, some have argued that the easiest way to chart the progress of a civilization is simply to catalogue the inventions and technologies that it has produced. In this article, Daniel Sarewitz examines and ultimately rejects the idea that technological progress will necessarily lead to social progress. He acknowledges that technology has both catalyzed economic growth and directly solved certain human problems. But he maintains that these have not necessarily increased what he believes to be the true measure of progress: quality of life. He gives evidence for how this is true in countries like the United States and argues that the problems are several fold worse in developing nations. Sarewitz's examination of progress, economic wellbeing, and quality of life compels us to ask about, rather than just assume we understand the connections between technology and values.

Nothing obscures our social vision as effectively as the economic prejudice.
Karl Polanyi, *The Great Transformation*

The products of the scientific imagination are responsible for shaping, to some considerable degree, the character of modern existence. The standard litany of discoveries and innovations made over the past century—plastics, vaccines, transistors, lasers, fiber optics, recombinant DNA—serves as one measure of societal progress, progress apparently derived from the astonishing and ever-expanding variety of new products and processes that have permitted us to conquer diseases, reduce the burdens of physical labor, increase the ease and speed of transport and communication, expand the options for leisure-time activities, widen access to information and data of all types, and create the seemingly infinite number of conveniences that consumers throughout the industrialized world take for granted. One may envision even greater progress in the future, measured by advances in such areas as artificial intelligence, biotechnology, high temperature superconductivity, molecular-scale manufacturing, advanced materials, and hydrogen power.

The very idea of progress in science and technology implies a sort of directionality that seems, when viewed retrospectively, to be logical, linear, and incremental. From semaphore to telegraph to telephone to computer network; from stethoscope to X ray

---

From *Frontiers of Illusion: Science, Technology, and the Politics of Progress* (Philadelphia: Temple University Press, 1996), pp. 117–140. Used by permission of Temple University Press. © 1996 Temple University. All Rights Reserved.

to electroencephalograph to magnetic resonance imagery; from wood to coal to oil to natural gas to photovoltaic cell—these are all apparently unidirectional trends: things become faster, stronger, more precise, more efficient, more concentrated, more comprehensive. Such sequences can be interpreted as measures of progress in society; they imply a path of change defined by the resolution of a particular succession of related problems. Thus it seems that humankind is moving forward as well, borne upon the shoulders of the scientists and technologists who make such progress possible.

Such a perspective suggests that the relationship between progress in the laboratory and progress in society is largely unmediated—that the beneficial impacts of science and technology on society derive from the intrinsic attributes of new products and processes. This is the familiar rationale behind the R&D system: that scientific and technological progress is necessary to solve a wide range of societal problems, that in solving such problems the cause of human progress will be advanced, and that human progress can therefore be facilitated by generous government support for research and development.

Of course new products and processes are not automatically introduced into or spontaneously assimilated by society. In most industrialized nations, a diversity of paths from laboratory to society are created by the mechanism of the marketplace. This mechanism suggests a different perspective on the relationship between science, technology, and societal progress. Consider a hypothetical biotechnology company that pioneers a new technique for identifying people who are genetically predisposed to develop a certain type of cancer. Production and application of this new technology generate revenues, attract investment, create jobs that permit more consumption of other goods and services—all of which helps to stimulate more competition, productivity, innovation, and job creation. The result is economic growth. While this scenario presupposes a demand for the product that ultimately emerges from the innovation process—in this case, a new health-care technology—it does not presuppose any particular characteristics of the product. Video games and assault weapons would be equally suitable examples. Furthermore, continued technological innovation must allow all such products to be manufactured at lower cost over time, or it must permit continued improvement of the original products or their replacement by new products, in order to boost demand, consumption, and profits.

In 1957, the economist Robert M. Solow published a paper showing that technological innovation acted as the driving force behind rising productivity and economic growth.[1] In 1987, Solow won the Nobel Prize for this work, by which time his conclusion had become dogma, transcending even partisan politics.[2] A budget document released by the Bush Administration in 1992 stated: "It is now widely recognized that a key to enhancing long-term economic growth in America is improving productivity.... The Bush Administration has proposed, over the past three fiscal years, a pattern of investment in areas of research and development that will help to boost productivity and improve economic performance. This budget continues the pattern of aggressive investment in both basic and applied R&D."[3] Less than a year later, one of the first public documents released by the new Clinton Administration sent a similar message: "Technology is the engine of economic growth. In the United States, technological advance has been responsible for as much as two-thirds of the productivity growth since the Depression.... [Manufacturers] depend on the continuous generation of new technological innovations and the rapid transformation of these innovations into commercial products the world wants to buy."[4]

These economic perspectives put a different slant on the idea of societal progress and what such progress owes to science and technology. Rather than viewing progress as deriving from the direct application of science and technology to societal problems, the economic frame of reference interprets progress as a function of the enhanced ability of individuals to pursue options, opportunities, and desires. This ability is encapsulated by the term "standard of living," an elusive but widely used concept that commonly refers to the purchasing power available to an individual, quantified by such measures as per capita gross national product.[5] In market economies, rising standards of living are therefore interpreted by economists, policy makers, and voters alike as a concrete indication of societal progress: more people have more financial resources to pursue a wider range of choices in their lives, to meet a wider range of needs, purchase a wider range of goods, indulge in a wider range of leisure-time activities.

In the modern free-market state, science and technology are thus conventionally viewed as offering two pathways to societal progress—one by direct application to societal problems, and the other by catalysis of economic growth and rising standards of living: "Through scientific discovery and technological innovation, we enlist the forces of the natural world to solve many of the uniquely human problems we face—feeding and providing energy to a growing population, improving human health, taking responsibility for protecting the environment.... Technology—the engine of economic growth—creates jobs, builds new industries, and improves our standard of living. Science fuels technology's engine."[6]

Such notions of progress are incomplete, of course. Civic ideals such as justice, freedom (in its many forms), and community, as well as ideals of individual welfare such as emotional, spiritual, and intellectual satisfaction and fulfillment, may be conceptually slippery, but they are nevertheless crucial elements of the quality of life. These elements are not well captured by anecdotal descriptions of revolutionary technological innovations or quantitative measures of standard of living. Moreover, a view of the relationship between quality of life and progress in science and technology is particularly difficult to bring into focus because the two distinctive roles that such progress plays in society—catalyzing economic growth and directly solving human problems—are not necessarily complementary.

For example, in the 1930s, a middle-class family might have owned a radio, a Victrola, and a telephone; today it might have home entertainment systems, cellular phones, and personal computers. Where does the progress lie? The fidelity of sound reproduction has greatly increased, but we cannot know whether the analog Louis Armstrong in mono at 78 rpm gave less pleasure to listeners in 1938 than the digital Wynton Marsalis on the CD player does today. We do know, however, that more people buy a wider variety of sound reproduction products that are manufactured more efficiently than ever before, that manufacturers add progressively more value to the cost of raw materials, that they employ progressively more people and pay them progressively better wages. In other words, more people in the 1990s can stock their homes with state-of-the-art consumer products than could do so in the 1930s because the size of the economy has grown and standards of living have risen. Is the progress of society between the 1930s and the 1990s more appropriately measured by the increased technological sophistication and diversity of consumer products or by the increased number of people who are able to purchase them?

In the case of medical care, access to sophisticated diagnostic and treatment technologies is certainly less important for physical and mental well-being than having the financial wherewithal to live in sanitary, uncrowded conditions, maintain a healthful diet, escape urban violence, and pursue an occupation that is not physically or emotionally deleterious. In fact, new medical technologies need not even create direct societal benefit in order to create economic benefit. Consider the accelerating rush to develop advanced technologies that can identify the genetic origins of a variety of noninfectious diseases such as cancers, heart conditions, and mental disorders. These technologies promise—and are already beginning to deliver—profits for the health care industry and generous federal funding for biomedical research.[7] But their capacity to improve public health is far from proven. Most noninfectious diseases are not caused simply by a defect at a single genetic location but in fact reflect complex and poorly understood interactions between multiple genetic elements and the outside environment. This complexity has not subdued the passion of researchers, technologists, investors, and entrepreneurs striving to develop new genetic tests, nor has it undermined public enthusiasm and demand for new diagnostic and predictive technologies. The result may well be the flowering of an economically vibrant sector of the biomedical industry, one that is capable of augmenting the average standard of living yet may make little ascertainable direct contribution to societal welfare.[8] At the same time, genetic testing technologies could plausibly contribute to a range of consequences that adversely affect quality of life, including rising medical costs, declining availability of medical and other types of insurance, and even the creation of a potentially new "scientific" justification for eugenic social policies.*

A high standard of living—more purchasing power—affords one the flexibility to pursue one's personal aspirations more freely. To the extent that personal aspirations include the use and consumption of new technologies, then the existence of the aspirations presupposes the existence of the technologies: we do not desire products that do not exist. Indeed, at any given time the connection between the current state of technological know-how and one's perception of an acceptable standard of living is largely arbitrary: people in the 1950s, no matter how wealthy they might have been, could not bake their potatoes in a microwave oven. While the evolving products of

---

*The latter possibility is hardly unrealistic, judging by the attention lavished by the public, press, and politicians on such recent eugenic approaches to social policy as Richard J. Herrnstein and Charles Murray's *The Bell Curve* (New York: Free Press, 1994).

science and technology may define an ever-advancing horizon of material aspirations, at any given moment in time these aspirations are fixed by whatever happens to be both available and desirable. The capacity to fulfill *current* aspirations gives standard of living its meaning, but the material products that are associated with, and in some ways define, a certain standard of living are obviously very different today than they were in the past. In 1965, a middle-class American family could not imagine life with a personal computer; the absence of the technology was not a source of deprivation. Today, a socioeconomically equivalent family may be unable to imagine life *without* a personal computer; its absence is felt as a deprivation. Because the horizon of material aspirations is constantly moving—fueled by the products of scientific and technological advance—we may feel as though we are better off today than yesterday not merely because the economy has expanded and our standard of living has risen but also because we possess things today that yesterday we could not even imagine.

Indeed, a belief in the intrinsic positive value of technological innovation is essential to the health of the market economy. As one analysis of innovation and industrialization explained: "The long growth in scientific and technological knowledge could not have been transformed into continuing economic growth had Western society not enjoyed a social consensus that favored the everyday use of the products of innovation."[9] This consensus means that daily life must be continually modified by the assimilation of new processes and products, as indeed it has been throughout this century. Because it is the job of the marketplace to foster economic growth, the principal filters on scientific and technological progress are consumer demand and potential profitability. Other filters—curing a certain disease, generating power more efficiently, producing tasty vegetables in winter—are of the second order; technologies may pass through these second-order filters but still fail to pass through the filters of profitability and desirability. Conversely, the creation of adverse effects, such as elimination of jobs, generation of pollution, erosion of personal privacy, or facilitation of violence, need not prevent a given technology from passing into society so long as there is a demand and a potential for profit—and government regulations do not prohibit its use.

Science and technology may therefore make a crucial contribution to economic growth and standard of living without necessarily making a net positive noneconomic contribution to the quality of human life or the welfare of society. I do not argue that there is no such positive contribution; but I do suggest that the marketplace does not provide a mechanism to ensure that this contribution will occur. Rather, the generalized societal benefit created by economic growth may obscure the negative impacts of science and technology on quality of life. While it is likely that a greater percentage of people living in the industrialized world today is free from abject poverty than was ever the case during the past several thousand years, we cannot know whether this group enjoys a higher quality of life than similarly fortunate people enjoyed in the past, or what the direct contribution of scientific and technological progress has been to the emotional, intellectual, and spiritual fulfillment of the average person. Speculating upon historical levels of personal satisfaction or happiness is probably fruitless,[10] but the record of literature suggests that the heights of joy and the nadir of despair have been fairly constant over time and that most people float somewhere in between the extremes. If life is in any sense "better" in the present than it was in the past—if levels of societal welfare are higher—this may not be a reflection of the direct contribution of telephones and computers to personal happiness or satisfaction, but of their contribution to an economic system that helps to shield us from elemental want while affording us a wider range of options in pursuing our aspirations. The fact that a modern urbanite would be miserable living in a cave, hunting big game, and wearing animal skins by no means implies that a well-fed paleolithic hunter was miserable too (at least any more so than today's average city dweller).[11]

The Sound of Invisible Hands Clapping

The underlying relationship between economic growth, scientific and technological progress, and societal well-being is perhaps most clearly illustrated by the role of government in supporting basic research—research that is not obviously connected to particular applications. The ideologies of basic research and market economics in fact hold much in common. During the 1980s, the supply-side, anti-interventionist economic-policy makers of the Reagan Administration looked with favor upon government sponsorship of basic research even as they sought to reduce spending for a broad range of domestic programs (including civilian applied research). The policy rationale behind basic research, founded on a belief in the gradual, serendipitous, and unpredictable diffusion of scientific knowledge into the marketplace, was philosophically compatible with "trickle-down" economics. Government support of the creation of knowledge for its own sake was ideologically acceptable because such knowledge was supposed to translate into more generic potential for innovation, more opportunities for new products and higher productivity, and more generation of wealth without requiring lawmakers or bureaucrats to make choices about specific directions of innovation—choices that could instead be made in the marketplace. The Republican Congress that came to power in 1994 is pursuing a similar approach—slashing civilian applied research and development budgets but maintaining a commitment to basic research.

The raison d'être of both the basic research system and capitalism is the pursuit of growth: growth of knowledge and insight in the one case, and of productivity and wealth in the other. And the key to growth in each case is the self-interested motivations of the individual—of individual scientists pursuing their curiosity and individual consumers maximizing their utility. The cumulative effect of all this selfish action is progress for all. But the analogy goes deeper, in that the rhetoric of both basic research and the free market is rooted in an efficiency ethic that gives primacy to magnitude of growth while viewing direction of growth as intrinsically unpredictable and thus outside the domain of government control. From this perspective it is the job of the government to encourage growth of the knowledge base and of the economy but not to try to influence the character of this growth in any way. "[The] pursuit of science by independent self-coordinated initiatives assures the most efficient possible organization of scientific progress," argued Michael Polanyi in a seminal 1962 article on the philosophy of basic research. "Such self-coordination ... leads to a joint result which is unpremeditated by any of those who bring it about. Their coordination is guided as by 'an invisible hand.' ... [Any] attempt at guiding scientific research toward a purpose other than its own is an attempt to deflect it from the advancement of science."[12] These considerations are explicitly analogous to the position of free-market absolutists who view any government interference in the marketplace as ill conceived and counter to the purpose of creating new wealth.

Economists have attempted to quantify the generalized contribution of basic research to economic growth. Such efforts have not yielded precise results—is it really possible to calculate the long-term value of the discovery of the transistor effect or penicillin?*—but few would disagree that, in particular cases, economic payoffs can be very high indeed. General economic arguments are less politically compelling than specific success stories, however, and the language typically used to promote and justify federal expenditures on basic research tends to be concrete, focusing on examples of how such research leads to the resolution of particular problems or the creation of specific avenues of technological innovation: "Rosenberg's research on the potential effects of electric fields on cell division led to the discovery of an important cancer drug; Kendall's work on the hormones of the adrenal gland led to an anti-inflammatory substance; Carothers' work on giant molecules led to the invention of Nylon; Bloch and Purcell's fundamental work on the absorption of radio frequency by atomic nuclei in a magnetic field led to MRI [magnetic resonance imaging].... [Research] designed to answer problems posed by nature, or questions based on scientific hypotheses, may often lead to important technological progress."[13]

These sorts of retrospective, anecdotal portraits—restatements of the myth of unfettered research—introduce a significant distortion into the effort to understand how the research and development enterprise creates most of its value for society. By singling out the practical, positive consequences of particular scientific discoveries, basic-research advocates create the illusion that there is a connection between the serendipitous course of basic research and the specific problems that society most needs or wants to have solved—as if consumers were demanding that the electronics industry invent transistors so that their radios and TVs could warm up faster. But if progress in basic research is truly unpredictable, then there can be no way to anticipate what particular problem a given scientific discovery might ultimately address. Conversely, there will be no reason to think that any particular problem—say the cure for cancer—will have a higher likelihood of being solved than any other problem—the invention of a new type of weapon, for example. From an economic perspective, this assertion of unpredictability is not problematical because there are no priorities in the marketplace save the creation of economic growth. Faith in the serendipitous progress of basic research is rewarded because of the nature of economic markets—it doesn't

---

*Nor, for that matter, would it be feasible to calculate the long-term costs resulting from the discovery of the law of special relativity and the consequent development of nuclear weapons.

matter what gets created or discovered as long as it leads to something that someone wants to buy. In retrospect, this looks to us like progress.

That there are both positive and negative dimensions to scientific and technological progress is obvious. Attempts to balance them may be futile, but an excessive faith in the promise of benefits may dull the sensitivity of society as a whole to the potential for negative consequences. The experience of television—once envisioned as a revolutionary cultural and educational medium, now reviled as the opiate of the masses and the force that transformed democratic dialogue into thirty-second sound bites—does not seem to have dampened anyone's ardor for the coming nationwide high-speed computer network—the so-called information superhighway that will revolutionize data transmission and communication in the United States and create huge new economic opportunities for an array of businesses ranging from phone companies to publishers. As science writer James Gleick explains it: "Once the new highway is in place consumers will be able to send or receive vast amounts of information with great speed—transmitting entire, up-to-the-minute electronic encyclopedias, receiving messages worldwide, seeing current movies as well as the full range of cable and network TV, sending out color images, hearing the sounds of La Scala from Milan."[14] Alternatively, the commercial imperatives and manipulative capabilities of the superhighway may overwhelm its potential cultural and political benefits, as they did for television, and saturate the world with dreck. In any case, there will be benefits, there will be negative consequences, and above all, there will be economic growth.[15] An illuminating example of such a tradeoff was reported in a *New York Times* article about Japan's relative backwardness in establishing its own computer networks. One of the explanations offered for the reluctance of Japan to embrace network technology was cultural—that the Japanese "value face-to-face communications [above indirect ones]."[16] What are the costs of failing to overcome this barrier to innovation? "Japan is now waking up to the fact that it is far behind the United States [in computer networking], at a great risk to its economy." If this risk is to be averted, face-to-face communications may have to go.

Whereas the collective action of large numbers of people seeking to maximize their utility in the marketplace leads to economic benefit for all, there is no invisible hand ministering to the wise application of scientific and technological progress. Individual preferences combined with the potential for profit can result in the assimilation of products and processes that do not increase, and may even undermine, societal welfare. Individuals may rationally decide to seek the most advanced medical technologies because they are sick, to buy gas-guzzling cars when fuel prices are low, to log onto the information superhighway because everyone else is doing it. Millions of individuals making such choices—solving individual problems—may lead not only to economic growth but to an increasingly expensive and decreasingly equitable medical system, to increased emissions of greenhouse gases and decreased energy independence, to a decline in face-to-face communication that undermines social comity. Economists often label such impacts "externalities," but another word for them is "reality." Indeed, as commercially successful technologies become inextricably woven into the fabric of society, their negative effects become integral to this fabric as well; often, these threads cannot be removed or even avoided without creating unthinkable disruption. Industrialized society will not, for example, abandon automobiles and electricity in order to reduce the emission of greenhouse gases.

The societal implications of scientific and technological progress are further complicated by the never-ending quest of human beings to satisfy their desire for material goods, wealth, and status. If people weren't constantly eager to trade in last year's model for this year's; if they weren't demanding the most advanced medical treatments, the fastest computers, the smallest videocams; if they weren't seeking to increase their incomes and expand their businesses; if they did not, on the whole, perceive a strong positive correlation between standard of living and quality of life, then economic growth would not occur. Science and technology are thus faced with an economic task that is inherently Sisyphean: to nourish the human need to consume. From the perspective of the consumer, there is no such thing as progress; there is only the infinitely ascending ladder of material and social aspirations. Because everyone cannot be on top, the desire for upward mobility—broadly defined—cannot be staunched. There can be no end to growth. In the industrialized nations, where population has stabilized and 85 percent of the world's wealth resides, private consumption continues to rise exponentially, at a rate of about 3 percent a year.[17]

The implications of continuous, technology-fueled economic growth for the future of humanity, however, are by no means clear. In particular, the question of whether the planet can sustain another century or so of the resource consumption and waste generation that accompany such growth, without also undergoing profound and perhaps irreversible environmental change, is not resolved, but most informed observers acknowledge that such change is at least a reasonable possibility.[18] The optimistic scenario goes something like this: progress in science and technology will lead to increases in the efficiency of energy and material use, production of food, and productivity of manufacturing. This progress will allow the world economy to keep growing without seriously compromising the integrity of the earth's environment, thus permitting a more or less continual worldwide increase in standard of living.[19] The pessimistic scenario does not assume that science and technology will have the time or the capability to save humanity from itself and from the insatiable essence of market economics; it anticipates that present trajectories of development could lead to accelerating environmental and economic crises and mounting political and social chaos, the results of uncontrolled, exponentially increasing resource exploitation, pollution, and population.[20]

The widespread and deeply held belief that scientific and technological advance is directly linked to societal progress creates a strong policy bias in favor of the optimistic scenario. But if this progress is attributable less to specific and direct contributions of science and technology to quality of life than to economic growth and the consumption patterns of market economies, such optimism may not be entirely warranted. The optimistic scenario is further reinforced in modern society because scientific and technological progress are often promoted as necessary to counter the negative impacts of existing technologies—oil can save us from coal; nuclear fission can save us from oil; nuclear fusion can save us from fission. In this manner, the history of technology can be viewed as a series of advances that have allowed humanity progressively to escape the constraints of its environment—to grow more food, extract more minerals, synthesize artificial substitutes for natural resources, produce more energy. This sort of progress is driven not just by technical ingenuity, however, but by resource scarcity, consumer preference, and government action as well. In the absence of economic incentives, technology may not evolve in necessary directions. Indeed, persistently low oil prices in the United States have demonstrated that there is no reason to expect the marketplace to respond spontaneously to the rising curve of atmospheric carbon dioxide. Furthermore, technology can reasonably be viewed as the mechanism by which the constraints of the environment are created in the first place: Without the internal combustion engine there would be no need to find energy alternatives for oil. The question is not whether the world is better off with the internal combustion engine than it was without it; the question is whether this technological house of cards can be sustained indefinitely.

The objection may be raised that the economic effects of science and technology cannot be separated from the problem-solving effects—that more people are better off today not just because they are prospering financially but because scientific and technological progress has boosted agricultural and industrial productivity to meet the needs of a growing middle class for food, transport, and energy. If the Malthusian dilemma—population growth outstripping resource availability—has thus far been avoided, it is certainly because progress in research and development, in combination with the operations of the marketplace, has resulted in more efficient exploitation, distribution, and utilization of natural resources. Without internal combustion engines, plant hybridization, and electricity, among many other (and generally lesser) advances, not only would economies have been unable to grow but the hardship of enormous numbers of human lives could never have been eased. This is especially true in the workplace, where the physical strength of an individual human has been rendered irrelevant to most types of productivity.

But those advances that have contributed to meeting the elemental needs of increasingly large numbers of people do not comprise the major portion of economic activity in the industrialized world; in fact, a principal indicator of a modernizing economy is its decreasing economic dependence on agriculture, natural resource extraction, and low-technology manufacturing of clothing and other goods and an increasing dependence on high-technology manufacturing with its capacity to augment hugely the value of raw materials.[21] An average of about 30 percent of the economic output of low-income developing countries comes from agriculture; in the industrialized world, this value is about 3 percent. In Tanzania, 75 percent of all household consumption is devoted to food and clothing; in the United States, the figure is 16 percent.[22]

If we begin to strip away the camouflage provided by the mechanisms of the marketplace, technology may look less like an agent of inevitable progress and more like a loose cannon. The potential for major destructive impacts cannot be discounted. Environmental effects are only one type of impact; of perhaps even greater immediacy is the proliferation of nuclear weapons in the developing world. Other possibilities are more subtle and perhaps less subject to overt societal intervention. For example, scientific progress and technology development may significantly transform the character and viability of democratic institutions.[23] These changes may come from many directions and include the growing influence and importance of technical experts in the democratic decision-making process; the trivializing influence of mass communication technologies on the quality of political debate; the effects of consumerism on democratic values and sense of community; and the threat to privacy created by computerized compilations of personal information such as credit rating, medical history, and consumer preferences. More generally, science and technology may exert an intrinsically undemocratic influence on society because they require people to adapt to change over which they have no control: society must follow where science and technology lead; the relationship is not one of mutual consent, and there is no going back.

Finally, because the creations of the marketplace are preferentially and inevitably (and virtually by definition) a response to the demands of wealth, rather than to poverty or want, the problem-solving capacity of science and technology will preferentially serve those who already have a high standard of living. Research and development priorities in an affluent society may have little bearing on those human needs or goals that cannot be effectively expressed or served through the marketplace. In this sense, the direct positive impacts of R&D activities on quality of life may tend to become more marginal with time, as science and technology increasingly contribute to superfluity and excess rather than fundamental human welfare. The implications of this dynamic are most striking when considered in the context of the developing world.

Nobody's Partner
The affluence of the industrialized nations—on display for all the world to see through television and other mass communication technologies—has created a remarkably universal vision of material well-being. Science and technology are widely understood to be necessary tools for pursuing this affluence: "All major developmental goals... depend to a large degree on the ability of countries to absorb and use science and technology."[24]

Two very pronounced but conflicting global development trends have become apparent over the past several decades. The first trend is positive and constitutes a narrowing of the disparity between industrialized nations and the developing world in basic human development indicators such as life expectancy, literacy, nutrition, infant and child mortality, and access to safe drinking water. For example, between 1960 and 1992 the disparity in life expectancy between the developing and industrialized nations fell from twenty-three years to twelve years, due mostly to a decrease in infant and child mortality. Overall, and on average, basic human needs are increasingly being met throughout the world, although severe deprivation in the South has in no way been eliminated, with over 1.3 billion people still living in absolute poverty, a similar number lacking access to safe drinking water, nearly two billion people without adequate sanitation facilities or electricity, and infant mortality rates still five times greater than in the North.[25]

The second trend is negative. Between 1960 and 1991, those nations comprising the richest 20 percent of the world's population increased their share of the global gross national product from 70 percent to 84 percent. This growing concentration of wealth in the industrialized world has been accompanied by increasing disparity between industrialized and developing nations in such areas as per capita mean years of schooling, per capita enrollment in higher education, scientists and engineers per capita, total investment in research and development, availability of computers, and proliferation of communications technologies such as telephones and radios.[26] In 1990, the United States had 545 telephones per thousand population; Brazil had 63, the Philippines 10; Indonesia 6.[27]

Progress is being made precisely in those areas of global development that are *not* greatly dependent on the generation of new scientific knowledge and technological innovation. Advances in agriculture (the Green Revolution) and medicine (especially vaccinations) have played an important role in meeting basic needs, but continued development progress has been less dependent on new science than on the ability of economic, political, and social institutions to deliver such services as clean water, education, and rudimentary health care. Furthermore, if the human development gap between the industrialized and developing world is slowly narrowing, this is occurring in large part because most basic development indicators have natural maxima and the industrialized world has already reached or is approaching many of these maxima—literacy rates and access to clean water cannot exceed 100 percent; average life expectancy probably runs into a natural limit of considerably less than a hundred years.

In contrast, the industrialized nations are rapidly outdistancing the developing world in measures of human development and standard of living that are dependent on continued scientific advance and technological innovation—especially those based on the creation of wealth and the acquisition of material goods. A nation with many scientists and engineers, many telephones, many computers, many universities, and many high-technology companies will generate more ideas, more opportunities, more productivity, and more economic growth than a nation that lacks these assets. This kind of growth perpetuates itself by demanding more technical expertise, more research, and more information and communication technology, which in turn leads to more innovation, greater productivity, and increased economic growth. Few nations that are not already part of this loop find it possible to join in. In general, the world's wealthiest nations have benefited from several decades of conspicuous increases in concentration of global income, wealth, savings, foreign and domestic investment, trade, and bank lending. Meanwhile, income disparities within individual developing nations have typically increased, as those few who have access to the benefits of advanced education and technologies and to foreign capital are able to command greater proportions of their nations' wealth.[28]

Matters are compounded by large-scale trends in the global economic marketplace. The profitability of modern, high technology manufacturing has become progressively less dependent on those strengths that the South can bring to the marketplace —inexpensive labor and abundant raw materials—and more dependent on those attributes that the South lacks—technological sophistication and a highly skilled and educated workforce. Erosion of the South's comparative advantage in labor and natural resources is further exacerbated because the most highly trained scientists and engineers from developing nations often emigrate to the North, where job opportunities are more plentiful and salaries are higher. This "brain drain" robs the developing world of indigenous expertise that is a crucial prerequisite for economic development.[29]

These trends are accelerated still further by the global emphasis on increased privatization of technological innovation. Efforts to strengthen the world patent system reinforce the gap in economic opportunity that exists between the developing and industrialized worlds. Patents offer a strong inducement to creative scientists and engineers, not to mention entrepreneurs and investors, but they freeze out poorer nations that cannot afford access to patent-protected innovations. The negotiation of scientific cooperation agreements between the United States and developing nations has frequently bogged down over the issue of intellectual property rights. The United States has been unwilling to enter into such agreements without formal assurance that marketable scientific and technological knowledge developed under U.S.-funded programs will, in fact, belong to the United States. Similarly, opposition in the United States to international treaties governing biodiversity and commercial exploitation of sea-floor resources has come in part "because these treaties contained provisions that were perceived as attempts by developing nations to misappropriate the benefits of technological investment by developed nations."[30]

Because science-based technological innovation is widely understood to be the dynamo of economic growth, and fostering economic growth is widely viewed as the principal function of the modern industrialized nation, the products of research and development are inevitably treated as proprietary. A widely distributed 1992 report on science and U.S. foreign policy argued that "incentives for invention and innovation, such as patent laws and intellectual property rights, must be extended and protected around the world."[31] Yet on the same page this report stated that the prospects of developing nations "will depend in large part on the evolution and diffusion of technologies." These two goals are not necessarily compatible. To the extent that nations insist upon maximizing the return on their science and technology investments through patents and other mechanisms, developing nations will be at a disadvantage in acquiring new technologies, modernizing their industrial base, and fueling economic expansion. The economic "cycle of concentration," as one United Nations report termed it, can only be exacerbated.[32]

Science does not provide equal benefit for all people; scientific knowledge can be —and is—appropriated for the economic and political gain of specific nations, trading blocs, and political ideologies. Although basic research is in many ways an international activity, with much cooperation between nations and unrestricted availability of many types of data and research results, the beneficiaries of this research are those nations that already have sophisticated scientific and technological infrastructures and can therefore capitalize on new scientific knowledge, whether created at home or abroad. Thus, in the early and mid-twentieth century, it was the United States—not China or Egypt—that became the preeminent industrial power on earth, in part because it could exploit new knowledge emerging from the research laboratories of Europe. Similarly, it is often noted that the Japanese technological juggernaut of the latter part of this century benefited from the results of research conducted in the United States.[33] More to the point, however, is that very few developing nations have either the physical, intellectual, or economic resources necessary to turn new scientific knowledge and technology (whether created in the United States or elsewhere) into economic growth. It is therefore not surprising that few nations have managed to join the ranks of the industrialized world in the past century.[34] The great industrial and economic successes of recent decades—Korea, Taiwan, Hong Kong, and Singapore—comprise about 70 million people, or less than 2 percent of the developing world's total population. Although many other nations are said to be "industrializing," the economic gaps between these nations and those of the already industrialized North continue, with few exceptions, to grow wider. Of more than one hundred developing nations for which data are available, eighty-one show declining real per capita GDP relative to the North over the past several decades.[35]

Although the gradual creation of efficient domestic economies that can exploit science and technology to create economic growth may offer a route to relative well-being for a few countries of the South, this is not a viable near-term option for most nations. About 80 percent of the world's population lives in the developing world, and 95 percent of the world's population growth occurs there as well. Not only must these nations struggle to create a reasonable quality of life for their people (30 percent of whom live in abject poverty, a proportion comparable to that of eighteenth-century Europe),[36] but they must do so in the face of high rates of population growth—much higher, in fact, than those experienced by the North during the industrial revolution—and the consequent rapid acceleration in demand for basic resources and services. Furthermore, the great majority of people in the South still live in rural areas and still depend directly on exploitation of the local natural-resource base for their income and sustenance. This diminishing resource base must therefore support not only all future increases in standard of living but also a population whose size will double over the next thirty years or so. The industrialized world reached its current state of development in fits and starts and over many centuries, with no fixed vision of societal evolution, no absolute limitations on economic growth, relatively small populations, and virtually unlimited access to the huge—and cheap—resource base provided by the rest of the world. In contrast, the South (that is, the rest of the world) confronts a future in which the earth's capacity for supporting human activity may well become an impassible roadblock to development.[37] As one analyst explained it: "The *increase* in the population of the 40 low-income countries [by the year 2050 will] be about equal to the *total* population in the world in 1960. Thus, the equivalent of a life-support infrastructure superior to the one that took thousands of years to develop [in the North] would have to be put in place within 60 years in order to improve the quality of life in those regions."[38]

The question here is not whether science and technology are capable of contributing substantially to this goal but whether the science and technology *system* now operating in the industrialized world is capable of doing so. The answer seems to be "no," at least in the absence of major institutional and political change. Today, the research agenda of the industrialized world is far removed from the immediate objectives of developing nations. Although in a very general sense there is a shared need to foster economic development while conserving resources and preserving environmental quality, the goals of these two worlds—and the specific strategies open to them in pursuing these goals—are entirely different. The science and technology agenda of the United States government, for example, is dominated by the development of military technologies that are, for the most part, of no conceivable benefit to the developing world. Indeed, the worldwide proliferation of modern weaponry peddled by the United States and other industrialized nations continues to add to the woes of the South. The U.S. space program, though more benign than weapons programs, similarly has little connection to the developmental needs of the South.[39] Yet defense and space together constitute almost 70 percent of the federal R&D budget. Health research in the United States is focused on diseases of specific concern to societies with long life spans, especially cancer and heart disease. Energy research is dominated by programs in fossil fuels, nuclear fusion, and nuclear fission and focused on large-scale energy production. Federal information-technology programs emphasize the establishment of the national high-speed computer network. Environmental research is concentrating on large-scale modeling of global climate change. Basic research in the traditional natural sciences nourishes academic institutions and a high technology industrial base that exist only in the wealthiest nations. Some or all of these research efforts may have the potential to contribute to the needs of the developing world, but such contributions will usually be marginal and fortuitous.

If the developing nations were setting research priorities for the world, they might emphasize such problems as: improving the efficiency, productivity, and environmental soundness of subsistence and production farming and low-technology manufacturing; devising energy-efficient "end-use" technologies for basic needs such as cooking, lighting, and transport; creating small-scale, nonpolluting, decentralized energy-supply technologies; preventing tropical diseases, such as malaria and cholera, and developing better diagnostics and treatments for respiratory infections; reducing the consequences of natural disasters such as floods and typhoons; increasing the effectiveness of reforestation. But developing nations, with less than 15 percent of the world's scientists and engineers and less than 5 percent of the world's total research and development funding,[40] do not have the resources to maintain major programs in areas such as these, while industrialized nations lack the economic and political motivation to pursue aggressively such research goals.

Transfer of technology from North to South has proven to be a complex and unpredictable mechanism for redressing these imbalances. Not only are the products of northern R&D often discordant with the needs of the South, but poor nations generally lack the human resources and physical infrastructure necessary to assimilate high technology products and processes successfully. Although one can envision optimistic scenarios by which poor nations "leapfrog" into the world of high technology, there is not much precedent for miracles on a large scale. Forty years of North-to-South "technical assistance" has been rife with failures, especially in terms of building economic capacity.[41] From the perspective of science and technology policy, the disappointing performance of technology transfer reflects the false assumption that the utility of scientific and technological know-how is context independent—that an idea or process or machine that works in America will work in much the same fashion in Burundi or Bangladesh. In the absence of healthy market economies, the direct negative impacts of new exogenous technologies are often readily apparent and often quite surprising. Who could have anticipated, for example, that a variety of Green Revolution agricultural projects would have the indirect effect of *decreasing* the economic and social welfare of women in many areas of Africa, by eliminating incentives and flexibility that had previously allowed them to cultivate surplus crops?[42] Who could have predicted that high technology weapons exported to developing nations—ostensibly for defense against external aggression—would increase the prestige and viability of armies relative to other institutions, thus making them "attractive to ambitious but impecunious young men, so the military diverted talent from business, education, and civilian public administration?"[43] Introduced technologies ranging in scale from wheelchairs to hydroelectric dams have often created as many or more problems than they have solved,[44] while projects initiated by foreign experts and supported by foreign aid expenditures are often abandoned as soon as they are turned over to local control.[45]

As long as the global R&D agenda primarily reflects the economic and geopolitical interests of the industrialized world, the capacity of science and technology to serve the development needs of the South will not greatly improve and may in fact deteriorate. Consider, for example, the problem of vaccinations. Eradication of smallpox and global efforts to vaccinate children against such diseases as polio, measles, whooping cough, and diphtheria can justifiably be portrayed as one of the great successes of development aid and technology transfer. But the research and development necessary to develop vaccinations against these diseases was not motivated by global concern for public health in the developing world; rather, it arose from the fact that these were diseases of the North as well as the South, so that affluent markets created both a political and an economic incentive for vaccine development. An analogous situation exists today with AIDS, where the ongoing search for a vaccine and other potential treatments reflects the course of the disease in the North even though the greatest long-term public health threat lies in the South. But apart from AIDS, most of the serious diseases of the developing world today are not significant public health threats in the North. Thus, a biomedical research agenda aligned with the disease priorities of the industrialized world has moved progressively farther from the needs of the South. According to one estimate, 90 percent of the global burden of disease is borne by developing nations, while just 5 percent of biomedical R&D expenditures worldwide are devoted to the tropical diseases that cause most of this burden.[46] In the South today, the potential market for malaria, tuberculosis, and other much-needed vaccines offers the promise of meager profits because these diseases occur predominantly among poor people in poor nations; in contrast, the industrialized world offers huge opportunities for profit from new drugs and other medical technologies that are largely irrelevant to the health priorities of the developing world.[47] A stark illustration of this problem is provided by the case of invasive pneumococcal disease, a major cause of pneumonia and childhood mortality in the South. The streptococcus bacteria that causes this disease also occurs in the North—but primarily as a cause of earaches. The pharmaceuticals industry, therefore, is supporting research and development aimed at finding a vaccine that can protect Americans from earaches; in this direction lies the potential for profit. Such a vaccine would neither be designed to prevent, nor necessarily be effective against, the pneumonia-causing form of the virus.[48]

While it may well be true that science and technology are crucial to the development prospects of the South, it is equally true that the world's research enterprise primarily serves the needs of the North through its contribution to the growth of technology-intensive economies. Continued concentration of wealth, scientific know-how, and technological advance in the North dictates that the priorities of the global R&D enterprise will continue to diverge from the knowledge and innovation needs of the South. Thus, the intimate connection between the economic marketplace and progress in science and technology may be an inherent obstacle to the reduction of social and economic inequity at the international level. To the extent that such inequity is a catalyst for environmental degradation and military conflict, the current organization of knowledge production may be seen as contributing to these consequences as well.

A twofold problem emerges. First, the marketplace—which is the principal venue through which the products of science and technology pass into society—provides no first-order mechanism for evaluating or assessing the intrinsic impact of a given technology on society, save for its potential contribution to profitability and growth. Thus, there is no a priori reason to expect that the direct net consequences of science and technology on society will be positive. Second, the symbiosis of science, technology, and the marketplace may skew the R&D agenda away from society's most urgent problems and toward the relatively less compelling needs of those who have already achieved a decent standard of living. If these observations are at least partly valid, then science and technology carried out in industrialized nations in the future may have a progressively diminishing capacity to provide a net benefit to humanity as a whole.

Notes

1. Robert M. Solow, "Technical Change and the Aggregate Production Function," *Review of Economics and Statistics* (August 1957): 214–31.

2. This is not to say that everyone believes it. For a dissenting view see David F. Noble, "Automation Madness, or the Unautomatic History of Automation," in *Science, Technology, and Social Progress*, ed. Steven L. Goldman (Bethlehem, Pa.: Lehigh University Press, 1989), pp. 65–91.

3. Office of Management and Budget, *The Budget of the United States Government, Fiscal Year 1993* (Washington, D.C.: Executive Office of the President, 1992), part 1, p. 87.

4. William J. Clinton and Albert Gore, Jr., *Technology for America's Economic Growth: A New Direction to Build Strength* (Washington, D.C.: U.S. Government Printing Office, February 22, 1993), p. 7.

5. For a discussion of different ways to view standard of living, see essays by Amartya Sen and others in *The Standard of Living: The Tanner Lectures, Clare Hall, Cambridge, 1985*, ed. Geoffrey Hawthorn (New York: Cambridge University Press, 1987).

6. William J. Clinton and Albert Gore, Jr., *Science in the National Interest* (Washington, D.C.: Executive Office of the President, August 1994), introductory letter.

7. Rachel Nowak, "Genetic Testing Set for Takeoff," *Science* 265 (July 22, 1994): 464–67; Sherman Elias and George J. Annas, "Generic Consent for Genetic Screening," *New England Journal of Medicine* 330 (June 2, 1994): 1611–13.

8. Richard Strohman, "Epigenesis: The Missing Beat in Biotechnology," *Bio/Technology* 12 (February 1994): 156–64; Charles C. Mann, "The Prostate-Cancer Dilemma," *Atlantic Monthly* (November 1993): 102–18; "Is a Little Knowledge a Dangerous Thing?" *The Economist* (August 21, 1993): 67–68; Ruth Hubbard and Elijah Wald, *Exploding the Gene Myth: How Genetic Information Is Produced and Manipulated by Scientists, Physicians, Employers, Insurance Companies, Educators, and Law Enforcers* (Boston: Beacon Press, 1993); also see essays on medical promise and social implications of genetic testing in *The Code of Codes: Scientific and Social Issues in the Human Genome Project*, ed. Daniel J. Kevles and Leroy Hood (Cambridge, Mass.: Harvard University Press, 1992).

9. Nathan Rosenberg and L. E. Birdzell, Jr., *How the West Grew Rich: The Economic Transformation of the Industrial World* (New York: Basic Books, 1986), p. 264.

10. For example, see Herbert A. Simon, *The Sciences of the Artificial* (Cambridge, Mass.: MIT Press, 1981), pp. 183–84.

11. The anthropologist Marvin Harris suggests that the earliest big-game hunters enjoyed "relatively high standards of comfort and security" and "an enviable standard of living," working only a few hours a day and enjoying the privilege of life at the top of the food chain: Marvin Harris, *Cannibals and Kings: The Origins of Cultures* (New York: Vintage Books, 1991), pp. 11–12.

12. Michael Polanyi, "The Republic of Science: Its Political and Economic Theory," *Minerva* 1 (Autumn 1962): 56, 55, 62.

13. Ernest L. Eliel, *Science and Serendipity: The Importance of Basic Research* (Washington, D.C.: American Chemical Society, March 1993), p. 3; also see National Science Foundation,

Benefits of Basic Research, NSF 83–81 (Washington, D.C.: National Science Foundation, 1983).

14. James Gleick, "The Telephone Transformed—Into Almost Everything," *New York Times Magazine*, May 16, 1993, 26–29, 50–64.

15. The superhighway metaphor is perhaps more precise than its coiners intended. The interstate highway system, originally justified in terms of civil defense, has had an enormous but mixed effect on the evolution of America: solidifying the place of the automobile as the dominant transportation mode, opening up vast areas of the nation to development, compromising the economic viability of thousands of small towns, locking the nation into its dependence on oil, increasing the personal convenience of long-distance travel, destroying the commercial potential of long-distance passenger railroads.

For a variety of perspectives on the social and technological implications of the information superhighway, see "Seven Thinkers in Search of the Information Superhighway," *Technology Review* (August/September 1994): 42–52.

16. Andrew Pollack, "Now It's Japan's Turn to Play Catch-Up," *New York Times*, November 21, 1993, section 3, pp. 1, 6; also see "Second Time Around," *The Economist* (August 12, 1995): 51–52.

17. World Bank, *World Development Report 1994* (New York: Oxford University Press, 1994), p. 177.

18. Even the mainstream and risk-averse National Academy of Sciences has associated itself with this perspective; see Cheryl Simon Silver and Ruth S. De-Fries, *One Earth, One Future: Our Changing Global Environment* (Washington, D.C.: National Academy Press, 1990).

19. For example, see Julian L. Simon and Herman Kahn, eds., *The Resourceful Earth* (Cambridge, Mass.: Basil Blackwell, 1984).

20. For example, see the annual *State of the World* reports (New York: W. W. Norton) issued by the Worldwatch Institute; also see Robert D. Kaplan, "The Coming Anarchy," *Atlantic Monthly*, February 1994, 44–76.

21. These trends are also reflected in R&D priorities, of course. Before World War II, most government research was aimed specifically at meeting society's basic needs for commodities. In 1920, 40 percent of the federal civilian R&D budget was devoted to agriculture, and 25 percent went to exploration for and exploitation of natural resources; similar proportions held throughout the period between the world wars. Today, those values stand at about 4 and 6 percent, respectively (A. Hunter Dupree, *Science in the Federal Government: A History of Policies and Activities to 1940* [Cambridge, Mass.: Harvard University Press, 1957]).

22. World Bank, *World Development Report 1993* (New York: Oxford University Press, 1993), pp. 242–43; 256–57.

23. For a brief discussion of some basic democratic issues raised by science and technology in modern society, see Harold Lasswell, "Must Science Serve Political Power?" *American Psychologist* 25 (1971): 117–23. Some recent considerations of the problem include: Alvin M. Weinberg, "Technology and Democracy," *Minerva* 28 (1990): 81–90; David Guston, "The

Essential Tension in Science and Democracy," *Social Epistemology* 7 (1993): 3–23; *Technology for the Common Good*, ed. Michael Shuman and Julia Sweig (Washington, D.C.: Institute for Policy Studies, 1993); *Technology in the Western Political Tradition*, ed. Arthur M. Melzer, Jerry Weinberger, and M. Richard Zinman (Ithaca, N.Y.: Cornell University Press, 1993).

24. Carnegie Commission on Science, Technology, and Government, *Partnerships for Global Development* (New York: Carnegie Commission on Science, Technology, and Government, 1992), p. 57.

25. United Nations Development Programme, *Human Development Report 1994* (New York: Oxford University Press, 1994), pp. 129–218. There is, of course, no simple dividing line between "industrialized" and "developing" nations, and there is considerable heterogeneity within both groups (especially the latter). This report identifies thirty-three industrialized nations representing 22 percent of the world's population in 1990 and including much of the former Soviet Union.

26. Ibid.

27. World Bank, *World Development Report 1994*, pp. 224–25.

28. United Nations Development Programme, *Human Development Report 1994*. For an anecdotal account of the problem of concentration of technological capacity, see Mike Holderness, "Down and Out in the Global Village," *New Scientist* (May 8, 1993): 36–40.

29. Eugene B. Skolnikoff, *The Elusive Transformation: Science, Technology, and the Evolution of International Politics* (Princeton, N.J.: Princeton University Press, 1993), chapter 4; Enrique Martín del Campo, "Technology and the World Economy: The Case of the American Hemisphere," in *Globalization of Technology*, ed. Janet H. Muroyama and Guyford H. Stever (Washington, D.C.: National Academy Press, 1988), pp. 141–58; Francisco R. Sagasti, "Underdevelopment, Science and Technology: The Point of View of the Underdeveloped Countries," in *Views of Science, Technology and Development*, ed. Eugene Rabinowitch and Victor Rabinowitch (Oxford: Pergamon Press, 1975), pp. 41–53.

30. Dan L. Burk, Kenneth Barovsky, and Gladys Monroy, "Biodiversity and Biotechnology," *Science* 260 (June 25, 1993): 1900–1901.

31. Carnegie Commission on Science, Technology, and Government, *Science and Technology in U.S. International Affairs* (New York: Carnegie Commission on Science, Technology, and Government, 1992), p. 22.

32. United Nations Development Programme, *Human Development Report 1992* (New York: Oxford University Press, 1992), p. 40.

33. For example, see Nathan Rosenberg, *Exploring the Black Box: Technology, Economics, and History* (Cambridge: Cambridge University Press, 1994): 121–38.

34. For example, Robert Gilpin, *The Political Economy of International Relations* (Princeton, N.J.: Princeton University Press, 1987), pp. 303–5; John F. Devlin and Nonita T. Yap, "Sustainable Development and the NICs: Cautionary Tales for the South in the New World (Dis)Order," *Third World Quarterly* 15 (1994): 49–62.

35. United Nations Development Programme, *Human Development Report 1994*, pp. 142–43. This report adjusts per capita gross domestic product to reflect personal purchasing power, by compensating for "such factors as the degree of openness of an economy...and possible overvaluation of exchange rates" (United Nations Development Programme, *Human Development Report 1993*, p. 109). "Real per capita GDP" includes a further adjustment for inequitable distribution of wealth within a nation.

36. Fernand Braudel, *The Wheels of Commerce*, vol. 2 of *Civilization and Capitalism, 15th–18th Century*, trans. Sîan Reynolds (Berkeley: University of California Press, 1993), 507.

37. For a summary of the resource challenges facing the developing world, see World Commission on Environment and Development, *Our Common Future* (New York: Oxford University Press, 1987).

38. Thomas Malone, "Perspectives on Technology Transfer," the Harrelson Lecture, North Carolina State University, Raleigh, N.C., March 15, 1993, p. 3 (emphasis in original).

39. When useful products do emerge—such as satellite imagery—the costs of acquiring and analyzing the resulting data are often prohibitive.

40. United Nations Development Programme, *Human Development Report 1992*, p. 40.

41. For example, Skolnikoff, *Elusive Transformation*, chapter 4; Andrew Barnett, "Knowledge Transfer and Developing Countries: The Tasks for Science and Technology in the Global Perspective 2010, *Science and Public Policy* 21 (February 1994): 2–12; "While the Rich World Talks," *The Economist* (July 10, 1993): 11–12.

42. World Resources Institute, *World Resources 1994–95* (New York: Oxford University Press, 1994), chapter 4.

43. Charles Tilly, *Coercion, Capital, and European States, AD 990–1992* (Cambridge, Mass.: Basil Blackwell, 1992), p. 220.

44. For example, Ralf D. Hotchkiss, "Ground Swell on Wheels," *The Sciences* (July/August 1993): 14–18; Elizabeth Brubaker, "India's Greatest Planned Environmental Disaster," *Ecoforum* 14 (1989): 6–7.

45. Carnegie Commission on Science, Technology, and Government, *Partnerships for Global Development*, p. 95.

46. World Bank, *World Development Report 1993*, pp. 25, 152–53. "Global burden of disease" is a quantitative measure of public health encompassing "losses from premature death" and "loss of healthy life resulting from disability."

47. U.S. Congress, Office of Technology Assessment, *Status of Biomedical Research and Related Technology for Tropical Diseases*, OTA-H-258 (Washington, D.C.: U.S. Government Printing Office, September 1985); Tore Godal, "Fighting the Parasites of Poverty: Public Research, Private Industry, and Tropical Diseases," *Science* 264 (June 24, 1994): 1864–66.

48. Jon Cohen, "Bumps on the Vaccine Road," *Science* 265 (September 2, 1994): 1371–73.

18 "Amish Technology: Reinforcing Values and Building Community"

Jameson M. Wetmore

Even when we recognize that technologies are value-laden, it is difficult to intentionally choose technologies that promote specific values. Typically, a single technology represents a number of different values, making it impossible to choose a technology without making compromises. But one group that has done an impressive job of taking on this task is the Old Order Amish. In this piece, Jameson Wetmore explains how the Amish evaluate technologies. As a group they reflect on whether integrating a certain technology into their society will help to promote, preserve, or dissipate the values they hold most dear. They try to choose those technologies that they believe will ultimately benefit their society and avoid those they fear will undermine it. As they develop new needs, they do not simply take existing technologies off the shelf. They actively design their own artifacts, regulations, and systems of use in an effort to ensure that their values are not disturbed by the values inscribed into technology by others. The process may not be a perfect democratic way of directing technology, but it is an example of a conscious effort to reflect on the relationship between technology and values in order to build a more desirable society. This article complements Daniel Sarewitz's chapter since the Amish do not share the mainstream western belief in technological or economic progress. The Amish recognize the importance of technology in building a society and attempt to promote and solidify their religion and community by reflecting on technological change.

On late-night TV and in popular jokes, the Amish are usually portrayed as rural farmers who live in a bygone era.[1] They are supposed to be a people who would never set foot in automobiles, never study the workings of a diesel engine, and never admit change into their society. And yet, when a non-Amish person—or "English" person as the Amish call their English-speaking neighbors[2]—travels through an Amish community, he or she discovers something very different. An observer may see an Amish woman talking on a pay phone, an Amish carpenter using a drill press, or even an Amish teenager driving a car. This revelation is often startling, but scenes like these are in fact the norm. They are not examples of Amish straying from their faith, but evidence that stereotypes obscure the intricacies of Amish life.

The relationships the Amish have with the outside world and technology may at first seem arbitrary, but they are the result of careful consideration. The Amish are not fundamentally anti-technology; rather, they believe that change does not necessarily result in desirable ends. They have not banned all machines and methods invented in the past 150 years, but they do exercise extreme caution when dealing with new

From *IEEE's Technology & Society Magazine* 26, no. 2 (Summer 2007): 10–21. Reprinted with permission.

Figure 18.1
Even though power lines tower over an Amish farm, they choose not to connect to the grid.

technologies. The Amish are cautious because they fear the changes that can accompany new technology. What a modern observer might see as potentially undesirable effects—like pollution and injuries caused by heavy equipment—however, are not major concerns for the Amish. The foremost reason the Amish carefully regulate technology is to preserve their culture [2].

Like many scholars of technology, the Amish have rejected the idea that technologies are value-free tools. Instead, they recognize that technology and social order are constructed simultaneously and influence each other a great deal. Implicitly they agree with the argument that technology and the social world are co-produced, that technology, in Sheila Jasanoff's words, "both embeds and is embedded in social practices, identities, norms, conventions, discourses, instruments and institutions—in short, in all the building blocks of what we term the social" [3, p. 3]. The Amish believe that technologies can reinforce social norms, enable or constrain the ways that people interact with one another, and shape a culture's identity. But despite the fact the Amish believe technology is so powerful, they are not technological determinists [4]. They do not view technology as an autonomous force, but rather as a tool that can be actively used to construct and maintain social order. The Amish recognize both the power of technology to shape their world and their power to shape technology.

The Amish have not, however, developed these ideas out of some sort of theoretical or academic interest. (In fact, they do not believe in education past the eighth grade.[3])

Figure 18.2
A bulk tank and mechanized agitator used to meet grade "A" milk regulations.

Rather, they reflect on the relationship between technology and society because they believe it is crucial if they are to understand and strengthen their culture, religion, and community. Their belief that technology and society simultaneously influence each other has both inspired and informed Amish attempts to maintain their way of life. The Amish regulate which technologies are to be used, when they are to be used, how they are to be used, and why they are to be used because they believe that one of the most important ways they can promote and reinforce their values is by actively embedding these values in their relationships with technology.

This chapter explores the way the Amish actively try to shape their society through technological decision-making. It can be tempting to simply point to various technologies the Amish use and ask—why? But because the Amish do not view technology as entirely separate from their society, any faithful explanation of their technology cannot either. Thus in order to convey the full picture this chapter will examine numerous facets of Amish life including their codes of conduct, the process of becoming an adult member of the church, economic pressures, business needs, and family life.

One Amish person succinctly explained the Amish approach to technology in the following way: "Machinery is not wrong in itself, but if it doesn't help fellowship you shouldn't have it" [6].[4] This article argues that the Amish pursue this goal of fostering community through technological choice in at least two interrelated ways. They first seek to prohibit those technologies they believe are antithetical to their values and choose those they believe will reinforce and strengthen their values. This straightforward approach is very important to the Amish, but it cannot explain all of their decisions. The Amish also recognize that the technologies they use have become a crucial part of their identity and they use this link between technology and identity to strengthen their community. Thus when making decisions about technology, the Amish rely on a second criterion—they deliberately choose technologies that are different from those used by other Americans in order to maintain their unique culture. The Amish believe that their way of life depends as much on the technologies they choose as any of the other social institutions that govern their work, religion, and community. The Amish practice of reflecting on their own their relationship with machines and techniques makes Amish culture a window into the ways in which technologies, societies, and values are interwoven.

Figure 18.3
Amish buggies stand in stark contrast to the trucks and minivans driven by their Indiana neighbors.

Amish Community and Values

To begin to understand why the Amish make the decisions about technology that they do, one must first understand Amish values. This can be difficult for those raised with very different social norms, but there are a few basic ideas that can help one begin to appreciate why the Amish make the choices they do. The Amish are a sect of Christianity and, as such, share the same Bible and many basic theological beliefs with other Protestant churches.[5] There are a few important points on which they differ in both emphasis and approach, however. One of the church's fathers, Menno Simons, advised his people to "rent a farm, milk cows, learn a trade if possible, do manual labor as did Paul, and all that which you then fall short of will doubtlessly be given and provided you by pious brethren, by the grace of God" [9, p. 451].[6] The idea of honest work, living a simple life, relying on their fellow believers, and trusting in God has shaped the Amish way of life to this day. They place great importance on values like humility, equality, simplicity and community.

Community is especially important to the Amish. They have gone to great lengths to carry out the scripture passage that implores them to "be not conformed to the world" [11]. The Amish believe that the world is full of distractions that must be avoided if they are to live piously. To steer clear of these distractions and ensure that they rely on their "pious brethren," they have separated themselves from those that do not share their faith.[7] Today the Amish live in groups of between 30 and 50 families called districts. They go to school together, worship together, play together, work together, and make decisions about technology together. The Amish believe that these separate communities provide the fertile soil in which they can best understand their place in the world, pass on their values to the next generation, and live the humble lives they believe are so important.

Rules that Bind and Nurture

Community is so essential to their way of life that the Amish have very carefully shaped the way it is organized. The primary method by which they do this is known as the "Ordnung"—a code of conduct that varies slightly from district to district.[8] The Ordnung is comprised of the district's long established traditions, as well as more recently agreed upon norms, and governs every aspect of Amish life—including the format of church services, the color of clothing to be worn, and which technologies are acceptable and which are unacceptable. The Ordnung is not written down, but it is understood and adhered to by the adult members of the community because it is continually being conveyed by example and occasionally by instruction when someone breaks a rule or inquires about a rule.

The Ordnung structures the life of the Amish in two interconnected ways. First, it provides the members of an Amish district with a template for living that they believe will nurture their community, their religious beliefs, and their values. For example, the Ordnung emphasizes the Amish dedication to nonviolence by forbidding Amish people from becoming soldiers and it requires that church services be held at a different family's house each week so that members of the community are continually supporting and relying on each other.

A number of Amish rules are designed to aid them in their quest to remain humble. For instance, to ensure that no individual becomes prideful about the way they look, each district specifies the color and design of clothing its members are to wear. Many districts go as far as to even reject buttons as "unnecessary" or potentially "prideful" adornment and require Amish to use straight pins to fasten their clothing. The Ordnung is also designed to promote humility by encouraging Amish adults to avoid being photographed in such a way that a viewer can distinguish who particular individuals are. This helps to reinforce the idea that an Amish person should not stand out as an individual, but rather is part of a community.

Through measures like these, the Amish use the Ordnung to promote their values, instill responsibility, pass down traditions, and build strong ties with one another. One Amish minister described the effective use of an Ordnung when he stated: "a respected Ordnung generates peace, love, contentment, equality, and unity" [10, p. 115]. Because it lays out how their life should be lived, in a very real sense the Ordnung is what makes an Amish person Amish.

The second way the Ordnung structures Amish life is by defining what is not Amish. In a sense, the Ordnung is the line that separates the Amish from the non-Amish; it is what gives the Amish their distinctly separate identity. For instance, each of the rules that detail what an Amish person should wear not only ensures that they will look

Figure 18.4
A sophisticated sawdust collection system, powered by a diesel engine, services an Amish carpentry shop.

Amish, but also that they will be easily distinguished from outsiders. In an interview, one Amish man used a parable to describe how this aspect of the Ordnung can promote community [12]. He said that if you own a cow and your property is surrounded by green pastures, you need a good fence to keep it in. For the Amish, who are as human as anyone and are tempted by the outside world to abandon their faith and way of life, there need to be good fences as well. The Ordnung defines what the Amish cannot do and makes those who are not adhering to the faith readily visible. Because they believe the outside world is a distraction that must be mediated, the Ordnung provides the barriers that keep community members focused on their fellow Amish and their faith.

Ordnung and Amish Change

Although a district's Ordnung is meant to convey the traditions of the community, it can be—and occasionally is—changed. When individual members begin exploring new abilities and possibilities that raise some concerns, the district must decide whether or not such activities should be allowed. To facilitate this process, twice a year each Amish district holds a counsel meeting. The counsels are led by the district's bishop (its religious and secular leader) but all of the adult members of the church —men and women—vote on the practices in question. To ensure that the implications of new practices are carefully considered, the voting system is designed such that change is very difficult. If two or more people (out of a possible 60–100) reject the change, the Ordnung remains unaltered. Thus the Amish allow for change, but the emphasis on tradition is built into the mechanisms that allow this change.

At least one other factor also helps to ensure that these deliberations are conservative. When considering a modification to their Ordnung, the members of a district must consider the other districts around them. If they make a change that neighboring districts believe is too radical they may be shunned, i.e., the offended districts could break off all communications with them and no longer recognize them as fellow Amish. This threat is of particular concern not only for community reasons but also because there are often close family ties between districts. An Amish woman might, for instance, decide that voting for allowing electrical appliances in the home is not worth risking the very real possibility that she may never again get to talk to her daughters who married into other districts. There are often small differences in the Ordnung of neighboring districts. For instance one may allow rubber carriage tires or bicycles while others do not. But because of the threat of being shunned, change to a district's Ordnung is usually incremental and often done in concert with other districts.

While Amish counsel meetings address all aspects of Amish life, beginning in the late 19th and early 20th centuries, the conversations increasingly began to focus on modern technologies. Only a few years earlier, it might have been difficult to distinguish the Amish from many other rural American communities. Their dress may have been a bit different, and their buggies less flashy, but they farmed in largely the same way and used many of the same technologies. The development of powerful new technologies like electricity, the automobile, and the airplane, however, generated a significant amount of concern in Amish communities. There was a suspicion that technologies like these would cause a significant disruption in the Amish world. To limit the ways in which machines and techniques negatively impact their society, the Amish have developed rules to govern their use.

Regulating Technological Change

The precise reasons why specific technologies were—and continue to be—regulated is difficult to pin down. The Amish have left very few, if any written explanations; non-Amish are not allowed to attend the Amish counsels and most Amish are very hesitant to discuss the details of counsel meetings with outsiders [13], [14]. Despite these obstacles, conversations with and further study of the Amish can begin to shed some light on the decision making process. As with any democratic process, there were likely many factors taken into account and different people involved may have had very different ideas about why things happened the way they did. But there are a few general themes that can help begin to explain the rationale behind Amish decision-making.

An Amish minister described the decision making process in the following way: "We try to find out how new ideas, inventions, or trends will affect us as a people, as a community, as a church. If they affect us adversely, we are wary. Many things are not what they appear to be at first glance. It is not individual technologies that concern us, but the total chain" [15, p. 16]. The Amish believe that social change is often closely tied with technological change and therefore tend to be suspicious of new technologies. They are strikingly different from most English in that they do not see an inherent value in technological progress. They must be fully convinced that a given technology will benefit the things they do value—their ethics, their community, and their spiritual life—before they will accept it.

As with the Ordnung in general, the Amish formulate rules about technology with two interconnected goals in mind. First, when deciding whether or not to allow a certain practice or technology, the Amish first ask whether it is compatible with their values. If they fear that a particular technology might disrupt their religion, tradition, community, or families, they are likely to prohibit it. The Amish not only believe that the English world is distracting, but also that many English machines and methods are distracting. For instance, the Amish believe that the pride, sense of power, and convenience that can come from owning an automobile may cause a person to focus on him or herself as an individual and thereby neglect the group. The Amish believe that technologies in general must be mediated in order to avoid situations like this and help to ensure that their way of life is not compromised.

The second purpose of the Ordnung—to create a fence between the Amish and non-Amish—has also played an important role in the Amish decisions about technology. Today, the most visible differences between the Amish and English worlds are the technologies they use. Most Americans do not see the Amish as different because they

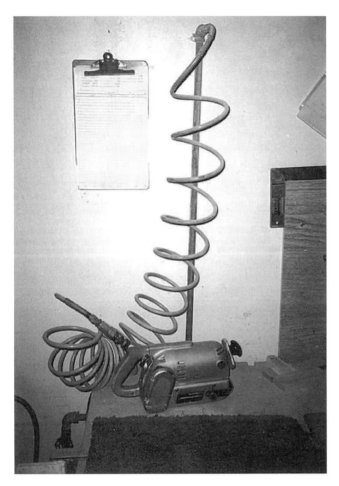

Figure 18.5
A pneumatically powered belt sander lies on a workbench in an Amish carpentry shop.

believe in adult baptism, but rather because they drive buggies, use horse drawn plows, etc. These differences were not accidental. The major technologies being developed in the non-Amish world at the beginning of the 20th century—like electricity, the automobile, and the airplane—very quickly became symbols of the modern world. The Amish rejected many of these technologies in part to retain their identity as separate from the modern world.

When asked today why they have rejected a specific technology, many members of the church will simply reply: "Because it's not Amish."[9] This argument is circular, but it emphasizes the way in which the Amish link their identity to the technologies they use. By banning these highly visible technologies, the Amish developed a new way of distinguishing themselves and strengthening the fence between themselves and the English world.

Regulating Electricity

The way in which the Amish make decisions about technology to promote both their values and their identity can be seen in an example where there is some historical record. The strict Amish regulation of electricity began in 1910 when Isaac Glick, an Amish farmer in Lancaster, Pennsylvania, hooked an electric light up to a generator [10, pp. 198–201]. His use of the new technology led to a counsel debate and the decision was made not to allow it.

Donald Kraybill, who recounts this story, argues that the reason was twofold. First of all, the Peachey Church, a group that broke off from the Mennonites as the Amish had, had just decided to allow electricity and the Amish were looking to prove that they were distinct from this new congregation. Secondly, they believed that physically hooking one's house up to the grid, a public utility owned by large corporations, did not help in the drive to be separate from the modern world. As one Amish farmer feared: "It seems to me that after people get everything hooked up to electricity, then it will all go on fire and the end of the world's going to come" [10, p. 200]. Instead of linking to the grid, the Amish continued to use the power sources they had been using—kerosene and natural gas—to cook their food and illuminate and heat their homes.

To this day, power lines bypass Amish houses. But the justification for this rule may have changed over the years. Many Amish today argue that the desire to avoid a physical connection to the English world is not the reason they reject getting power from electric companies [6], [14]. They point out that they have tapped into natural gas lines (or would if a utility provided them) rather than have to pick up canisters in town. The precise reasons why the Amish initially deemed connection to the grid as a threat to their community no longer matter, if they ever did matter. What is more important is that the Amish have defined electricity as the domain of the outside world, and thus any use of the technology must be very carefully considered. Even if the Amish link themselves to the outside world by piping gas into their homes from a public utility, they are still reaffirming their identity by forging a different relationship to power than their English neighbors.

Amish Transportation

Another area of technology that the Amish have carefully considered in order to ensure that it reflects their values and reinforces their identity is transportation. Traditionally the Amish have relied on horse drawn carriages to transport themselves, but in the early part of the 20th century they were faced with a new option. In 1907, an auto-

Figure 18.6
The interior of an Amish kitchen is nearly identical to a modern kitchen except that it has no electrical appliances and is lit with sunlight and gaslight.

mobile manufacturing company was formed in Lancaster, Pennsylvania, the heart of Amish country. This company advertised its product as "the king of sports and the queen of amusements," and immediately turned off the Amish, who saw it as an unnecessary luxury and dangerous source of pride [10, p. 214]. By the second decade of the 20th century, after a few Amish had purchased motorcars, every Amish district in the United States independently decided to prohibit the use of the automobile [16, pp. 37, 73]. Because most of the people an Amish family knows live relatively close to their home; because the Amish are not relegated to a strict schedule that demands speedy transportation; because horses have become practically family members; and because buggies are relatively inexpensive (costing today between two and three thousand dollars new), require little maintenance, and last for up to twenty years, the Amish saw no reason for changing their traditional way of life [17, p. 8].

But economics are not the only reason why the Amish have chosen to keep their buggies. Some argue that buggies are a social equalizer because they are uniform, free from excess bodywork and color, and because one buggy cannot be made significantly faster or slower than another. Automobiles, on the other hand are criticized for providing an abnormal sensation of power that can be used to not only show up one's neighbors, but to abandon them altogether. As one Amish man noted, "Young people can just jump in the car and go to town and have a good time in it.... It destroys the family life at home" [18]. Buggies are deemed better because they slow the pace of life to ten or twelve miles-per-hour, giving people a chance to interact with their environment rather than fly by it. The Amish believe the automobile is not very compatible with the values they hold dear.

Despite these criticisms, however, there are several situations today in which an Amish person would be allowed to make use of a motor vehicle. For instance, it is not uncommon for an Amish woman to be driven to the grocery store by an English friend; for an Amish family to travel from Indiana to Florida via bus; for an Amish business to lease a car indirectly through a non-Amish employee; or even for an Amish teenager to actually drive and own an automobile. While these at first may seem to contradict Amish principles, each case signifies an arrangement that the Amish believe can help strengthen their community, and is therefore allowed under the Ordnung.

In the first scenario, it is probably not a necessity that the Amish woman be driven to the store—it is likely that she could take her own horse and buggy—but because she is not the one driving the car, it is acceptable behavior. She does not have the freedom to roam as she pleases, but rather must depend on another person. Some English people have gotten so involved in transporting the Amish that they have started their own thriving taxicab companies. These services are welcomed by the Amish because they satisfy a need and still make it inconvenient for a person to tour about on a whim.

The second scenario is a response to the fact that the Amish are spread across the United States. Many young Amish move miles away from their families to find land and work. It would be extremely difficult to travel by buggy to visit family members that lived a thousand miles away. Thus the Amish allow the use of public transportation (other than airplanes) to visit family and even to take vacations. The Amish community is a highly structured environment, but it is not a prison. Such trips allow them to reinforce their family ties and their ties with other Amish communities.

The third scenario reflects a fairly recent change that will be explored later. To sum up quickly, this scenario is the result of the belief that many Amish businesses cannot survive without an automobile. For instance, Amish businesses that specialize in building fences would likely run out of work rather quickly if they did not accept jobs outside of the area easily traversed in a buggy. To make this possible, some districts grant businesses special permission to lease a car, but only if they agree to certain restrictions. Under no condition would an Amish person be able to drive it; he or she must instead hire and be dependent upon an English employee. A district may even prohibit parking the car near an Amish home to decrease the temptation to use the car for trivial things. Some districts allow Amish businesses to use motor vehicles, but take a number of precautions to limit the potential negative impacts they perceive.

"Running About"

The fourth scenario is the result of a deeply rooted Amish tradition that will require further explanation of how the Amish structure their society. The Amish understand that it is difficult to be Amish. It requires a significant amount of humility, patience, and dedication. They also understand that because their lives are so intertwined, members who do not accept these responsibilities can threaten the active and united nature of their community. Therefore, the Amish go out of their way to ensure that their members truly want to be Amish.

The primary technique they use is the church admission process itself. To curtail immature and uninformed decisions no one is allowed to enter into the church until they are in a position where they can readily think for themselves. The Amish contend that it takes not just age but also experience to develop such wisdom. Therefore they give their children the opportunity to explore alternatives to Amish life by turning a blind eye to those who violate the Ordnung and choose to adopt some English ways. The Amish term for this phase of life is "rumspringa," or "running about."[10]

Many Amish youths take the opportunity to experience what another life would be like. Amish adolescents may begin with relatively small violations such as curling the brim of their hat or driving the family carriage faster than their parents would. (It is often said that one can tell that a teenager is driving a buggy whenever it is going fifteen miles-per-hour, rather than the average of ten to twelve.) But the "running about" period also gives Amish youth the chance to experiment with modern technologies. Many of them are drawn to the outside world because they are fascinated by the devices they see English people using. Thus Amish teenagers may find ways to watch television, listen to music on the radio, operate their own ham radio, or even drive automobiles.

By their early 20s, most Amish children decide that they are not satisfied by English customs and technologies. Many of them begin to see more clearly the benefits of Amish culture and sincerely regret their actions [12]. Over eighty percent of children (and as many as ninety-five percent in some places) decide to become adult members of the Amish church [19]. The period of rumspringa helps to ensure that this is an informed commitment to community and church. Offering children the option to leave the rigorous and humble life of an Amish person and explore what the outside world has to offer—including its technologies—ensures that the people that make up the community truly want to be there and will henceforth work for the good of the Amish people.

Modern Pressures

While the questions of whether to adopt electricity and automobiles were important for the Amish to resolve, these were only the beginning of the difficulties their society encountered in the 20th century. Although they work hard to remain separate, many changes in American government, economics, society, and technology have had a significant effect on the Amish. In recent years the stability of the Amish has been put to rigorous tests. In their efforts to meet these challenges and stay focused on their values as much as possible, the Amish have chosen to alter some of their traditions and, in particular, the technologies they employ.

An example of this can be seen in the Amish response to new milk regulations imposed by a number of states in the 1950s and 1960s. These regulations required farmers to install electric powered bulk tanks with cooling systems if they wanted their milk to continue to be rated Grade "A" quality. The regulation clashed head on with the Ordnung of Amish communities.

This put the Amish in a bit of a dilemma. Much of their tradition is built upon an intimate relationship with the land. Many Amish view farming as the ideal way to earn a living. They have kept themselves separate and free from the outside world by working the land upon which they settle. The Amish did not want to significantly compromise one of the cornerstones of their culture.

Therefore, in 1968, a group of five Lancaster bishops and four milk inspectors from Pennsylvania met to iron out an agreement that would satisfy both parties [10, pp. 202–205]. The inspectors' primary concern was that the milk be kept refrigerated. They suggested simply installing normal electric refrigeration units. But the Amish refused to run electric lines into their barns. Instead they developed an "Amish solution." They agreed to install coolers, but chose to power them using diesel engines salvaged from old trucks.

The inspectors also required that the milk be automatically stirred five minutes every hour. This was a difficult request for the Amish to grant because the very word "automatic" bothered them, but they eventually consented to a newly devised system that used a 12-V battery, rather than 110-V electricity, to run an automatic starter. The fact that the Amish had traditionally used batteries to power a few devices like flashlights made this a bit more palatable.

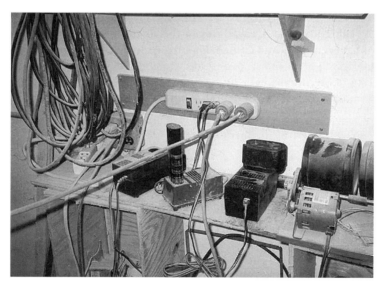

Figure 18.7
An "Amish power strip" draws power from a generator to supply energy to various batteries for a carpentry business.

Finally, the inspectors wanted the milk picked up every day to decrease spoilage. At this point, the Amish drew a line they would not cross for any reason. They would not allow anyone to interfere with Sunday, their day of rest and church services. Because the Amish were a major producer of milk in the area, the bulk milk industry agreed to readjust its practices slightly by picking up milk a second time on Saturday instead of Sunday morning. With this specially devised arrangement, the Amish won a minor battle in keeping their community economically sound and their culture relatively unchanged.

The resolution of the milk controversy is an instance in which the Amish accepted new technology, but they did it in a uniquely Amish way and for Amish reasons. The compromise was important because it protected the ability for the Amish to continue to earn a living doing the work they find most rewarding—farming. Yet while they introduced new technologies into their society, they made sure that the machines were different from those used by their English neighbors and that the electricity they generated could not be easily put to other uses. With this new—seemingly modern—technology, the Amish were able to meet an economic need while still retaining their identity and practice of being different from the outside world.

Amish Entrepreneurs

Despite compromises like these, the Amish have not been able to rely completely on farming to support themselves economically. For at least the last forty years, they have been in the middle of a land squeeze. Because married couples desire to have many children and the Ordnung prohibits contraceptives, an Amish family has an average of seven children [20]. Even though not every Amish child enters the church, this has resulted in a constant rise in Amish population. As of 2001, the Amish numbered over 180,000 children and adults [10, p. 336]. They have sought new farmland by gradually spreading into 25 American States and the province of Ontario. But the English population is also increasing and land prices are rising. There simply is not enough farmland to go around.

Young Amish adults increasingly have to look for employment other than farming. In the first half of the 20th century nearly all the Amish in the area surrounding Arthur, Illinois, were farmers. By 1989, that number was less than half [17, p. 9]. In Indiana the changes have been even more marked. While over fifty percent of Amish men under the age of 35 were farming in some Indiana areas in 1993, less than twenty-five percent of young Amish men were farming in 2001 [1, pp. 119–120].

Many young Amish who are not able to farm have found work in English factories, supermarkets, or stores in their area. Generally, they are treated well and receive a good wage. But being employed by the English can disrupt an Amish community. The hours and location of the business can restrict an Amish person's ability to participate in his or her culture and the exposure to the culture of the modern world can exert an influence as well. As one Amish woman noted, "The shops coming in were a good thing.

They gave our young people jobs among our own people. But now they've got money and they go to town" [21].

Because of their concern that working for outsiders will dilute their culture and traditions, Amish communities have begun developing their own entrepreneurial talents and have increased the number and variety of businesses they own and operate [22]. Amish people have explored business ventures as diverse as machinery assembly, log house construction, upholstering, engine repair, grocery stores, bookstores, and cabinetry building. Economic forces have made the Amish ideal of communities comprised primarily of farmers impossible. But by developing their own businesses, the Amish ensure that they can work relatively close to home, work with their fellow church members, be free to attend community events like weekday weddings, and help reinforce their separation from the outside world.

As the Amish have entered these new fields—many of which are dominated by large American corporations—they have chosen to make some compromises when it comes to technology. They believe that in order to produce and sell an affordable product in the modern age, some increase in technology is necessary. As an Amish bishop put it, "To make a living, we need to have some things we didn't have fifty years ago" [6].

An example of this can be seen in the issues faced by Amish carpenters. Because the Amish have traditionally been good at building and feel that it is admirable to work with one's hands, carpentry has become one of their key industries. However, it would have been very difficult to survive on the output one could create using hand powered tools. Therefore, the Amish struck another bargain. They still strongly disagreed with running electric lines into their shops, so they motorized hand tools in a different way. A number of carpentry shops purchased regular electric saws, routers, and sanders and retrofitted them with motors that could be powered with air pressure. They then installed large diesel engines just outside their shops and strung pneumatic lines to the various work stations.[11]

Why go to all the trouble and expense to create such an intricate power system when electricity does the same job? In part because it distinguishes the Amish as different from their neighbors. But also because, as an Amish minister explained, "so far no Amish person has ever figured out how to run a television with an air compressor" [17, p. 3]. Television is seen as a technology that is contradictory to Amish ideals because it brings the outside world into the home and can distract one from one's family and neighbors. It is often used as a barometer by the Amish to determine whether or not something is acceptable. The Amish allow certain forms of electricity, but choose those forms that make it difficult to power devices like kitchen appliances, radios, and televisions.

The Amish have also developed ways of gaining the business benefits of certain technologies while maintaining their distance from them.[12] One way they do this is by hiring English companies to take care of certain aspects of an industry that they do not want to do themselves. As was already mentioned, the Amish will often hire English drivers to transport them to work sites, etc. But the Amish may also rely on non-Amish businesses to help them attract and interact with customers in ways they cannot or prefer not to do themselves. For instance, the Amish have been able to tap into the market for remodeling kitchens in far away cities by contracting with companies who do the on-site work. It is also now possible to buy Amish-made furniture online through websites developed and maintained by English companies. These arrangements help the Amish economically and yet minimize the distraction and compromises that come with using particular technologies themselves.

Line Dividing Home and Work

Despite all of the detailed explanations given above, the fact that the Amish use such a wide array of modern technologies may still seem fairly surprising. It does not mesh with many English people's visions of what Amish life should be. Many Amish feel a similar unease. They believe that they must adopt some new practices to remain economically viable, but that does not mean that they are enthusiastic about such changes. To compensate for these distractions, the Amish have tried to protect the simplicity of the home. While they have adjusted the Ordnung to promote Amish businesses, they are much less likely to change rules that govern the life in the home.[13] A stark example of this demarcation is the fact that diesel generators and pneumatic equipment are not allowed in the Amish home; kitchens are empty of electric appliances and interiors are still lit by candles, gas lamps, and windows.

The desire to protect the home has also shaped the Amish rules concerning telephones [27], [28]. Traditionally the Amish have been opposed to owning telephones because they believe that phones disrupt the natural interactions between people. An Amish buggy maker contended that "if everyone had telephones, they wouldn't trouble to walk down the road or get in the buggy to go visiting anymore" [17, p. 3]. Telephones are seen as distracting; they give the outside world an easy entrance into Amish households and make them needlessly noisy.

But the English companies and customers that the Amish rely on have abandoned many of the forms of communication that the Amish prefer. Without a phone it is difficult for furniture shops to communicate with distant customers, for stores to order merchandise, or even for farmers to coordinate milk and produce pick-ups with dairy and grocery companies. To remedy this problem, these businesses began to use the phones of their non-Amish neighbors. But as businesses got bigger and were sometimes far away from English phones, this became increasingly difficult. Gradually many Amish districts have begun to allow telephones, but with certain qualifications that ensure they do not compromise their lives at home.

Most districts maintain the rule that telephones are not allowed inside buildings owned by Amish people. Instead they are usually placed in small structures, or "Amish phone booths," that are kept "a safe distance away" from Amish dwellings. Typically the telephones are purchased by either the community in general or by specific Amish businesses, but they are kept accessible to the entire community. They are outfitted with a log so that calls can be recorded and payments can easily be made by individual people.[14] This arrangement encourages cooperation, reduces the impact on traditional forms of communication, and allows Amish businesses to develop. But most of all it keeps telephones outside of the home. It helps keep the home free from the distractions of the modern world.

Where the Amish Stand Today

The Amish are continually debating whether or not to introduce new technologies into their society—a process which can be contentious at times. A young Amish farmer noted that he (and every other Amish dairy farmer) would love to install glass piping that would quickly transport the milk from the cows to the refrigerators and relieve him of a lot of work, if only it were allowed [29]. Yet despite his desire, this farmer is still firmly committed to his community. Like many other Amish, he struggles with the Ordnung, but has agreed to and recognizes the benefits of a society that does not accept rapid change.

These struggles will continue as changes in American government, business, farming, and technology exert increased pressure on the Amish way of life. In response to some of these stresses, the Amish have chosen to accept some somewhat marked changes in technology. Some Amish communities now allow battery-operated typewriters, electric cash registers, and fax machines [25]. These new machines have led to a vigorous debate because many of them require 110 volt electricity (easily done by coupling invertors to their existing diesel engines), which could also be used to power a television. But some districts have decided that their businesses cannot survive without them.

The Amish are not, however, about to relax their control over technology. Because they believe that technology call shape those things they value above all others—their culture, their community, and their values—they continue to closely monitor and regulate its use. One Amish man admitted, "We realize ... that the more modern equipment we have and the more mechanized we become, the more we are drawn into the swirl of the world, and away from the simplicity of Christ and our life in Him" [30, p. 95]. The Amish see technology as a potential disruption to their simplicity, humility, and separation and work to make sure that it disrupts their lives as little as possible. A bishop explained his difficult position in the following way: "Time will bring some changes; that's why our responsibility is so great.... We can prolong out' time. I'll do what I can" [6]. Why this dedication when the world around them is changing so quickly? One Amish farmer argued that "If it hasn't worked for the good of [English] families, why will it work for our society? It's not good community" [31]. The Amish exert control over technology in an effort to protect themselves from the values and distractions of the English world.

The Amish believe that their society and their technology are inextricably intertwined. In an effort to maintain and protect their community of believers, therefore, the Amish require that every technology they use not only conforms to, but reinforces their tradition, culture, and religion. They achieve these goals through two primary techniques. First they choose technologies that they believe will best promote the values they hold most dear—values like humility, equality, and simplicity. Thus they have rejected the speed, glamour and personal expression of automobiles in favor of modest, slow, and community-building horse-drawn buggies. Second, they deliberately choose tools that are different from those used by the outside world. This differentiation helps them maintain their unique identity, bonds their community, and ensures that they will continue to be able to accept technology on their own terms. The Amish view technologies as value-laden tools and use these tools to reinforce their values and build their community. While many scholars of technology have argued that this is the case, the Amish employ the idea in order to build the world they want to live in.

Acknowledgment

I would like to thank Louis B. Wetmore and Gordon Hoke both for helping me to get in touch with Amish communities and for enlightening conversations; Michael Crowe, Deborah Johnson, Shobita Parthasarathy and two anonymous reviewers for comments on various drafts; and the Menno-Hof Museum in Shipshewana, Indiana, and the Mennonite Historical Library at Goshen College for their assistance in locating resources. Most importantly, I would like to thank the Amish people who took the time to share their culture and their experiences with me.

Notes

1. This article primarily refers to the Old Order Amish. Because this is the largest and most recognizable group of Amish people, they are typically referred to as simply "Amish." For an explanation of the different types of Amish see [1, pp. 21–22].

2. When talking to one another, most Amish speak a derivative of German usually referred to as "Pennsylvania Dutch."

3. The Amish rejection of advanced education is based on their belief that "the wisdom of this world is foolishness with God" [5, p. 91].

4. This chapter is partly based on a handful of interviews conducted by the author in Amish communities in Indiana, Illinois, and New York. Because the Amish value their privacy their names will not be cited. For an interesting discussion on the difficulties of interviewing the Amish see [7].

5. For a detailed account of Amish history see [8].

6. Menno Simons was the founder of the Mennonites. Tile Amish church broke from the Mennonites in the late 17th century in part because they believed the Mennonites were straying from Menno Simons' teachings [10].

7. Although the Amish separate themselves for the good of their own people, they have not forgotten the outside world. Their desire to help others is often directed towards those outside their community. Should a non-Amish neighbor's barn burn down, the Amish will band together and help with the erection of a new one, just as they would for a fellow

Amish person. Above and beyond this, some Amish communities are known to participate actively in hunger and disaster relief projects across the world.

8. While each district has its own distinct Ordnung, they are similar on many points. As such this article will often refer to "the Ordnung" of the Amish in general for those issues on which there is almost universal agreement.

9. Nearly every Amish person interviewed for this article gave this answer at one point or another.

10. "Rumspringa" has recently been subject to a fair amount of media coverage in the United States because of the 2004 UPN television show "Amish in the City" and the 2002 feature-length documentary Devil's Playground. These programs can be a bit misleading as they focus on the most extreme examples of Amish rebelliousness. Most Amish teenagers do not live in Los Angeles, parade up and down the red carpet at movie premieres, or deal drugs.

11. These new systems proved to be so efficient that a few English companies now produce them for non-Amish shops [23]. The Amish are surprisingly inventive in other fields as well and have even been awarded patents in a few cases. For instance, they have developed a cook-stove that employs an airtight combustion compartment that some claim is the "only significant advance in wood-fire stoves in 300 years" [24, p. 30]. The Amish also have designed a horse-drawn plow fitted with a hydraulic lift so that rocks do not present as much of a problem to farming [25].

12. The Amish relationship with medicine follows a similar rule. While they rely on homeopathic remedies for many things, if they find an English doctor that they trust, the Ordnung does not prohibit them from receiving medical care that uses advanced technologies.

13. A number of scholars have criticized this stance as just one more method the male dominated society uses to repress women [26].

14. Whether and how the Amish can receive phone calls varies from district to district [27]. For instance, some do not allow incoming calls to be answered; some allow calls to be prearranged; and some use voice mail services provided by phone companies.

References

[1] T. J. Meyers and S. M. Nolt, *An Amish Patchwork: Indiana's Old Orders in the Modern World*. Bloomington, IN: Quarry, 2005.

[2] R. E. Sclove, "Spanish waters, Amish farming: Two parables of modernity?" in *Democracy and Technology*. New York, NY: Guilford, 1995, pp. 3–9.

[3] S. Jasanoff, Ed., *States of Knowledge: The Co-production of Science and Social Order*. New York, NY: Routledge, 2004.

[4] M. R. Smith and L. Marx, *Does Technology Drive History? The Dilemma of Technological Determinism*. Cambridge, MA: MIT Press, 1994.

[5] T. J. Meyers, "Education and schooling," in *The Amish and the State*, D. B. Kraybill, Ed. Baltimore, MD: Johns Hopkins Univ. Press, 1993, pp. 86–106.

[6] Interview with an Amish bishop, Shipshewana, IN, Feb. 3, 1996.

[7] D. Z. Umble, "Who are you? The identity of the outsider within," in *Strangers at Home: Amish and Mennonite Women in History*, K. D. Schmidt, D. Z. Umble, and S. D. Reschly Eds. Baltimore, MD: Johns Hopkins Univ. Press, 2002, pp. 39–52.

[8] J. A. Hostetler, *Amish Society*, 4th ed. Baltimore, MD: Johns Hopkins Univ. Press, 1993.

[9] M. Simons, "Brief and clear confession," (1544) in *The Complete Writings of Menno Simons*, J. C. Wenger Ed. Scottdale, PA: Herald, 1956, pp. 422–454.

[10] D. B. Kraybill, *The Riddle of Amish Culture*, rev. ed. Baltimore, MD: Johns Hopkins Univ. Press, 2001.

[11] Bible, Romans 12:2.

[12] Interview with an Amish carpenter, Shipshewana, IN, Jan. 27, 1996.

[13] Interview with a bishop, Seneca Falls, NY, Nov. 21, 1997.

[14] Interview with a businessman, Finger Lakes Region, NY.

[15] E. Stoll and M. Stoll, *The Pioneer Catalogue of Country Living*, Toronto, Canada: Personal Library, 1980.

[16] D. O. Pratt, *Shipshewana: An Indiana Amish Community*, Bloomington, IN: Quarry, 2004.

[17] R. Mabry, Be Ye Separate, *A Look at the Illinois Amish*, Champaign, IL: Champaign News-Gazette, 1989.

[18] V. Larimore, director, *The Amish: Not to Be Modern*, 1986.

[19] T. J. Meyers, "The Old Order Amish: To remain in the faith or to leave," *Mennonite Quart. Rev.*, vol. 68, pp. 378–395, July 1994.

[20] K. Pringle, "The Amish dilemma: The attraction of the outside world," *Champaign-Urbana News-Gazette*, p. E1, Aug. 30, 1987.

[21] Interview with an Amish housewife, Arthur, IL, Mar. 30, 1996.

[22] D. B. Kraybill and S. M. Nolt, *Amish Enterprise: From Plows to Profits*, 2nd ed. Baltimore, MD: Johns Hopkins Univ. Press, 2004.

[23] Interview with an Amish carpenter, Arthur, IL, Mar. 30, 1996.

[24] E. Brende, "Technology Amish style," *Technology Rev.*, vol. 99, no. 2, pp. 26–33, Feb./Mar. 1996.

[25] E. Tenner, "Plain technology: The Amish have something to teach us," *Technology Rev.*, vol. 108, no. 7, p. 75, July 2005.

[26] S. D. Reschly, "'The parents shall not go unpunished': Preservationist patriarchy and community," in *Strangers at Home: Amish and Mennonite Women in History*, K. D. Schmidt, D. Z. Umble, and S. D. Reschly, Eds. Baltimore, MD: Johns Hopkins Univ. Press, pp. 160–181, 2002.

[27] D. Z. Umble, *Holding the Line: The Telephone in Old Order Mennonite and Amish Life*. Baltimore, MD: Johns Hopkins Univ. Press, 1996.

[28] H. Rheingold, "Look who's talking," *Wired*, vol. 7, no. 1, Jan. 1999.

[29] Interview with a young Amish farmer, Arthur, IL, Mar. 30, 1996.

[30] M. Good, *Who Are the Amish*? Intercourse, PA: Good Books, 1985, p. 95.

[31] Interview with an Amish corn and dairy farmer, Arthur, IL.

IV THE COMPLEX NATURE OF SOCIOTECHNICAL SYSTEMS

The premise of this book is that if we want to build a better future, we must recognize the important role that technology plays and we must proactively direct technologies toward the future we want. To emphasize the importance of thinking about technology as we build our future, section II explored a variety of ways in which machines and societies influence each other and section III demonstrated numerous ways in which technologies can reflect, embody, reinforce, and undermine values. To build a better future, however, requires one more important step; it requires that we develop the ability to steer technological development.

While social constructivists tell us we have the capacity to shape and direct technology, putting this into practice is no small feat. It is difficult not because taking action is difficult, but because the effects of those actions are not always easy to predict. Sociotechnical systems are immensely complex. Understanding how they are sustained, how they change, and how our actions affect them is a daunting task.

Technological innovation is often presented as a simple step-by-step process. The traditional view suggests that it occurs in the following sequence: science discovers new aspects of nature; governments and corporations fund projects to further explore these discoveries and put them into systematic form; engineers apply the new science to create new products; and consumers buy and use the new products. In this account the process seems to be linear and controllable. We can speed up or slow down the process by intervening at various points, turn it in this or that direction, etc. But while this view may accurately depict how a few technologies have gotten to the marketplace, the sequence described is far from the norm and hides many of the forces that are typically at work.

This section focuses on the complexities and challenges in the development and functioning of sociotechnical systems. Sociotechnical systems are built, maintained, and modified through complex processes involving many actors, balancing and trading-off diverse factors, and dealing with a good deal of uncertainty. Complexity is the norm in all of the stages through which a technology is conceived, developed, and used (or rejected) as well as in the ways it ultimately affects the world. To begin to sort out and understand the complexities here, it is important to explore at least two kinds of complexity.

The first and most widely recognized category has to do with the multiplicity of social factors that may affect the development of sociotechnical systems. The complexities of human relationships and human interactions are a vital facet of sociotechnical systems since they are rarely ever conceived, designed, and produced by a single individual. Governments and corporations decide which projects to fund, engineers make important decisions about how a machine will be constructed, marketers influence who will decide to buy a product and users ultimately deploy products to achieve certain ends—including those not envisioned by the companies, governments, and engineers that originally designed it.

Because of the number and variety of people that have the ability to shape a sociotechnical system, it is difficult to precisely predict whether an invention will be adopted or how new technologies will be integrated into society. As Langdon Winner argued, something as seemingly innocuous as a bridge may allow certain people to cross a road while simultaneously restricting the movement of other social groups. Users can also be surprisingly inventive in the ways they take up technologies. Examples of unpredictable adoption and use of technology include the use of airplanes as attack weapons, the use of new chemical-glues to get high, and the use of cell phone cameras to take pictures of unaware people in public showers. Each person or group involved in technological development leaves an imprint and can knowingly and unknowingly shape the values affected by the technology.

The second category of complexity has to do with the fact that the technical aspects of a design may never be completely understood. Although engineers and scientists have a deep understanding of how things work, their knowledge is always limited; there are factors of which they may be unaware, situations they haven't considered, or aspects they haven't had time to test. Engineers and scientists often do things that have not been done before or done in quite the same way. Examples of the uncertainties of engineering are found throughout the history of engineering. There are many cases in which conscientious engineers who, because of incomplete knowledge, built something that did not work as intended. For example, the disasters associated with the Three-Mile Island nuclear facility, the Challenger Shuttle, and the Tacoma-Narrows Bridge were all the result of factors that were not clearly understood or anticipated before the failure occurred. While it is important for engineers to continually increase their understanding, they must also develop methods for working without a precise knowledge of all the factors involved. Non-engineers, in turn, must also approach technology with recognition that its precise behavior will never be absolutely certain.

The combination of social and technical complexity has an important impact on all who seek to influence the development and direction of sociotechnical systems. Just how a sociotechnical system will develop and what kind of impact it will have cannot be precisely predicted. Thus, efforts to direct technology will never be perfect. But if we abdicate our responsibility to try, we will ensure that we will have no control over our future. Recognizing the existence and effects of uncertainty can help us better understand, anticipate, and prepare for the risks inherent in the sociotechnical systems that we build and that constitute our lives. Acknowledging the complexities of sociotechnical systems is a key step in being able to act effectively.

Of course, this section cannot cover every possible complexity and unintended consequence involved in the creation and use of technology. For instance patent systems designed to inspire innovation may compel drug companies to focus on developing new chemical compounds rather than exploring potentially helpful (but unpatentable) natural remedies. Regulations intended to ensure that medical procedures are safe may convince university researchers to test them on disadvantaged communities. And the threat of product liability cases may dissuade corporations from doing pioneering work in potentially dangerous areas. But while these issues are not addressed here, the readings in this section begin to explore the wide variety of ways in which social influences and technical uncertainty complicate the development of new technological systems. Reflecting on historical examples can inform how we approach the challenges of the future. The readings in this section prepare us to think through new sociotechnical systems; navigate their complexities; and envision, and, where appropriate, circumvent possible effects.

Questions to consider while reading the selections in this section:

1. How and why are values hidden in the development of new technologies?
2. How do social complexities and technical complexities compound one another?
3. How can we begin to manage the uncertainties inherent in sociotechnical change?
4. What lessons can we learn about unanticipated consequences by studying past and current examples of sociotechnical change?
5. What important complexities are not explored in this section?

19 "Will Small Be Beautiful? Making Policies for Our Nanotech Future"

W. Patrick McCray

One of the earliest stages in the development of any technology is to obtain funding. Brilliant ideas remain only ideas unless there is money to increase knowledge, test hypotheses, and develop products. The process of securing funding is not a straightforward one. Because there are neither obvious solutions to critical needs nor even obvious needs that demand some sort of solution, those advocating new technological developments must develop a case for their vision. In essence, new technologies—from toothbrush bristle patterns to genetic engineering—are marketed before they are even built. One of the most powerful and frequently used methods to do this is to present a vision of the future that describes how certain technical achievements will create specific social goods. This article explores the ways in which nanotechnology was sold to the U.S. government as the next great frontier and the next great set of solutions. The U.S. Congress could have launched billion-dollar-a-year projects to eradicate malaria from the world; to build ore-mines on the bottom of the Atlantic Ocean; or to pursue technologies to enable human cloning. But while they passed on these possible initiatives, they were inspired by what McCray calls "nano-utopian visions." Both the goals and the technologies we pursue are not predetermined; lawmakers, engineers, corporations, and others must sift through numerous possibilities and actively decide what problems are most crucial and which solutions are most likely to be effective. The optimistic—and even at times deterministic—future that nanotechnology enthusiasts predicted would result from research in the field (including the ideas presented in "Nanotechnology: Building the World Atom by Atom" that is reprinted in section I) succeeded in convincing lawmakers to fund research in nanotechnology.

It was as if nanotechnology had gone through a phase transition; what had once been perceived as blue sky research...was now being seen as the key technology of the 21st century.[1]

On 24 October 2003, California politicians, academics, and civic leaders attended the groundbreaking ceremony for the California NanoSystems Institute in Santa Barbara. This US$55 million high-tech building, with its ultra-filtered clean rooms and modular laboratories, represented only a fraction of the burgeoning national investment in nanotechnology.[2] Since 2000, when President William J. Clinton announced a new national initiative to foster the tools of the "next industrial revolution," dozens of universities and corporations have initiated programs where scientists and engineers research phenomena and technologies at the nanoscale.[3]

From *History and Technology* 21, no. 2 (2005): 177–203. Reprinted with permission.

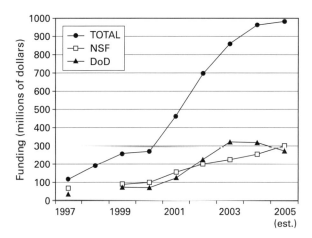

Figure 19.1
Funding profile for nanotechnology research in the U.S., 1997–2005. Total funding is shown along with monies allocated specifically to the National Science Foundation and the Department of Defense. 2005 figures are projected; breakdowns by agency not available for 1998. Source: Roco, "The U.S. National."

Following proclamations by Clinton and other leaders, government and commercial investment into nanotechnology soared. In 2004, federal agencies in the USA dispensed almost a billion dollars a year to eager researchers (figure 19.1 shows the federal funding profile for nanotechnology research).[4] Meanwhile, forecasts, not from intemperate prognosticators but sober-minded science managers, predicted that the international market for nano-goods would be US$1 trillion by 2015. The production of these goods, supporters said, would require a new high-tech (and highly paid) workforce of some two million people, potentially leading to major restructuring of the global workforce.[5]

To its advocates, nanotechnology is a "transcendent realm," a regime where research in the physical and biotechnological sciences may converge with information technologies and the cognitive sciences.[6] While definitions of what nanotechnology is vary—indeed, one of its rhetorical strengths would seem to be its interpretive flexibility—its basic feature is "smallness."[7] According to official government definitions in the USA, nanotechnology involves three key components; research and technology development at the length scale of 1–100 nm; creating structures and devices that have novel properties and applications because of their small to intermediate size; and the ability "to control or manipulate on the atomic scale."[8]

Nanotechnology, in other words, is engineering done at the molecular scale and represents the ability to deliberately shape new structures, atom by atom. Compared with the behavior of individual atoms or molecules (the latter being about 1 nm in size) or the properties of bulk materials, the behavior of structures in the size range of a few to a 100 nm is quite different. In this regime, the effects of surface area, interfaces, and quantum behavior can create new behaviors and properties. Because the nanoscale is where the basic structure of materials are established and their properties defined, advocates of nanotechnology predict the creation of revolutionary nanostructures from the "bottom up." If realized, the ability of nanotechnologists to shape materials and structures with atomic precision would be the culmination of humans' expanding ability to manipulate their material world.

Humans, with varying degrees of intent, have been nanotechnologists for millennia. Ancient artisans have consciously manipulated metal, glass, and ceramic materials to create certain properties. Consider the Lycurgus Cup, a 4th-century glass vessel in the collection of the British Museum. The object's unusual optical properties are caused by a haphazard dispersion of nanometer-sized particles of a gold–silver alloy in a glass matrix.[9] Nanotechnology's potential, its advocates claim, rests in its promise for *precise control*. In their view, it offers the ability to organize atoms and molecules into usable novel materials, structures, or systems with physical, chemical, and biological properties that are either known or predictable via sophisticated computer modeling.

The marshalling of Congressional support for the National Nanotechnology Initiative (NNI) occurred against the background of post–Cold War (and, later, post–September 11) science and technology policies. Proponents of nanotechnology offered seductive visions of a tomorrow that appealed to citizens, scientists, and policy makers, visions that this article examines. As one Nobel laureate said in a government brochure designed to make nanotechnology visible to the public, "Nanotechnology has given us the tools...to play with the ultimate toy box of nature....The possibilities to create new things appear endless."[10] In testimony to Congress and in books written for the public as well as science and policy communities, technological visionaries and science administrators alike predicted that nanotechnology could "change the nature of almost every human-made object in the next century" and, in doing so, not only alter our "comprehension of nature and life...[but] even influence societal and international relations."[11] These predictions, in turn, helped spur lawmakers to allocate billions of dollars for research and commercial development. Despite arguments from skeptics attempting to re-direct policy and commercial research and development efforts, government leaders eventually adopted and ratified these utopian predictions in national science policy when they approved the National Nanotechnology Initiative in 2000.

The sheer dollar amount, the rapidity of the funding increases, and the accompanying flood of books, websites, and pop culture references to what one author called the "nanocosm" are sufficient to command attention from a wide array of disciplines in the academy.[12] While scholars have made initial efforts to analyze the economic, social, ethical, and legal challenges posed by the predicted "next wave" of technological innovation, historians have paid comparatively little attention to nanotechnology.[13] While this lacuna is understandable as nanotechnology's potential benefits still reside largely in research laboratories (the major commercial products exploiting nanoscale research thus far are stain resistant pants and improved cosmetics), the relative dearth of historiography concerning nanotechnology with which one can engage makes the historian's task more difficult. It is too early to assess the economic or societal effects of nanotechnology; it is not premature, however, to examine the roots of and rationale for the United States' massive investment in the nanocosm. This article is therefore to be read as an exploratory reconnaissance of the funding and policy terrain of nanotechnology in the last decade and is written with the understanding that future documentation will add much more depth and detail to the story presented here.

What is required to create a successful new science initiative? Previous studies of policy-making for science have identified several variables for victory. These include the role of entrepreneurs who lobby for policies that support their goals as well as perceived national needs. Coordinated activity among agencies, interest groups, and researchers as well as budgetary and institutional contexts are other factors.[14] Howard McCurdy, in his discussion of how Americans imagined space exploration, rightly notes that public policy also has an important subjective dimension whose roots are located in imagination and visions of the future.[15] Why was the NNI successful in obtaining so much federal support and what was its "subjective dimension"?

This article addresses the area where, in the late 1990s, utopian imaginings and policy rationales met fiscal realities. National science policy is frequently analyzed through, if not indeed seen as commensurate with, the politics and annual changes of science budgets.[16] This article goes beyond the conflation of policies with budgets to explore how and why such monies were allocated. In describing how advocates for the NNI gradually orchestrated its funding, this paper offers a historical perspective on the promotion of a major national initiative to support nanotechnology in the USA. In looking at the rationales offered by lawmakers and science managers, I pay special attention to the depiction of future nano-utopias in support of policy goals.

Historians will recognize the marshalling of techno-utopian visions as a theme common to the history of many technologies. From household appliances to nuclear power and the Internet, the introduction of new technologies has often been accompanied, if not presaged, by wildly optimistic claims and unbridled enthusiasm.[17] More often than not, enthusiasts and supporters of new and "revolutionary" technologies have seen their predictions tempered by fiscal, social, or regulatory realities. The belief that the next industrial revolution will come from nanotechnology existed in conjunction with other more prosaic rationales, some relics from the Cold War, for funding the NNI. Besides considering how advocates promoted the NNI and how its funding reflects broader changes in national science and technology policy, it is important to ask why utopian techno-visions continue to flourish given their predilection to ultimately disappoint.

Before the Flood

Historians of science and technology have long recognized the existence of "creation stories," myth-like narratives for research communities that trace the development of a particular idea or invention back to a singularity—lone inventor or small teams who create a revolutionary breakthrough. For geneticists, there is Watson, Crick, and the double helix; for electrical engineers, Shockley, Brattain, and Bardeen's invention of the transistor looms large.

Nanotechnologists' creation story can be traced to a precise point in space and time—the evening of 29 December 1959 when Caltech physicist Richard P. Feynman gave an after-dinner speech in Pasadena to members of the American Physical Society. Feynman, one of the most original physicists of the 20th century, gave his address, whimsically entitled "There's Plenty of Room at the Bottom," at a time when the mass production of microelectronics was just beginning and computers still occupied entire rooms.[18]

Feynman described the potential benefits of "manipulating and controlling things on a small scale," which he believed could be accomplished while still obeying the basic laws of physics. Scientists, he predicted, would build microscopic machines that could manipulate matter at a very small scale while engineering done at the "bottom" would provide plenty of attractive research opportunities. Human control of matter at the atomic scale was, according to the Caltech physicist, a "development which I think cannot be avoided" and he backed up his assertion by offering a thousand dollars "to the first guy who can take the information on the page of a book and put it on an area 1/25,000 smaller in linear scale in such manner that it can be read by an electron microscope."[19] In his talk, Feynman drew upon utopian visions proffered by contemporary fiction writers, which suggested, ironically, how the scientific and visionary aspects of nanotechnology were interconnected from its very beginning.[20]

Feynman never did any research that could be construed as nanoscience.[21] Indeed the word "nanotechnology" did not appear until 1974 and then in an obscure proceedings of a Japanese industrial conference.[22] Nevertheless, Feynman's address became a crucial touchstone for future advocates of nanotechnology. In both popular treatments as well as policy documents, quotations from it are rampant while Feynman's authorship serves as a rhetorical anchor to scientific authority. This critical linkage was strengthened by the fact that Feynman later shared the Nobel Prize for his research on theoretical quantum electrodynamics. According to what some have appropriately called the "standard story" of nanotechnology, Feynman presented the crucial initial vision of the innovative nano-research that scientists could do, the achievement of which only required the proper tools.[23]

These tools—namely new and improved electron microscopes—were incrementally developed by scientists and engineers during the 1960s and 1970s. This led to the next milestone in the "standard story" that helped set the stage for the eventual promotion of a national nanotechnology initiative. In 1986, Gerd Binnig and Heinrich Rohrer, researchers at an IBM research laboratory in Zurich, received the Nobel Prize for their design of the scanning tunneling microscope. With a properly functioning instrument (the achievement of which is still something of a black art), researchers now had the ability to "see" individual atoms as their probe scanned a sample's surface and rendered an image of its topology on a computer screen.[24]

Now able to image atoms, the next milestone in realizing Feynman's now-mythic vision was the ability to move them. In a 1990 paper published in *Nature*, two researchers reported that they had manipulated individual xenon atoms on the surface of a single crystal of nickel cooled to almost absolute zero with "atomic precision."[25] In their demonstration of the ability to "fabricate rudimentary structures of our own design," Eigler and Schweizer honored their corporate sponsor by using 35 xenon atoms to spell "IBM." A major article in *The New York Times* followed in which a reporter described how he laboriously moved about a few atoms with Eigler's coaching.[26] Engineering at the nanoscale, as the *Times* described it, was not only potentially ground breaking but entertaining—as Eigler wrote in his 1989 lab notebook entry, "I am really having fun!"

Stunts like this made international headlines and provided the capstone to Feynman's roadmap—in principle, researchers could now organize matter as well as see at the atomic level. However, doing so was still a challenging enterprise that required a good deal of expensive equipment, tacit knowledge, and specialized skills.[27] While a new community of electron microscopists coalesced around the new tools and techniques in the late 1980s and early 1990s, much of this happened out of sight from the larger scientific community and the general public. Another element, one that Feynman did not account for, was needed: a messenger to boldly proselytize for nanotechnology.

In the 1980s, K. Eric Drexler emerged as the most visible and vocal proponent of nanotechnology and its transforming potential. Born in 1955, Drexler attended MIT and, while working at that school's Space Systems Laboratory, wrote a 1981 paper that outlined his early vision for nanotechnology.[28] He argued that the ability to design protein molecules could lead to the manufacturing of molecular machines that, in turn, could make "second-generation machines" and the "construction of devices and materials to complex atomic specifications."[29] Drexler went on to teach the first university course on nanotechnology, published an equation-laden textbook on the subject, and was a co-founder of the Foresight Institute, a non-profit group devoted to promoting the benefits of nanotechnology.

The first full articulation of Drexler's nano-utopia was his 1986 book *Engines of Creation*. Written for the general public, Drexler's book considered the potential benefits and risks arising from the wholesale production of nanomachines.[30] At the heart of his vision was what Drexler called "assemblers," hypothetical autonomous machines, protein-sized or smaller, that could produce virtually any material or object. Such ideas were incorporated by science fiction writers who imagined a future in which a person had access to a "utility fog" from which they could call forth the material necessities of life. While skeptical, one review of Drexler's book described it as a "hopeful forecast, remarkable for an unembarrassed faith in progress through technology."[31]

A full treatment of Drexler's activities and the attention they received is beyond the scope of this paper. It will suffice to say that a 1991 article in *Science* called Drexler "the apostle of nanotechnology" and noted that he was already a lightning rod for praise as well as vitriolic scorn from fellow scientists.[32] Criticisms came from both science journalists and other researchers, with the detractors claiming that Drexler's vision of self-assembling nanomachines not only drew attention (and possibly funding) from actual lab-based endeavors but also would violate basic laws of conservation.[33] Referring to Drexler's missionary zeal and the fervor he inspired among his supporters, one critic said, "You might as well call it nanoreligion."[34]

However, for the sake of promoting research and development of nanotechnology to the top of the national agenda, Drexler and other nano-advocates performed an important function. In promoting radical futurist scenarios, which his critics placed in the realm of fiction rather than science, Drexler and his colleagues brought nanotechnology to the attention of policy makers. As such, Drexler and company may be considered the vanguard of a wave that eventually brought potent government support for more conventional and staid research programs devoted to nanotechnology. This was not long in coming.

In June 1992, Senator Albert Gore, recently returned from the United Nations Earth Summit in Rio de Janiero, convened a congressional hearing devoted to "New Technologies for a Sustainable World."[35] Drexler was one of the witnesses invited to appear before Gore's subcommittee. He testified that "molecule-by-molecule control," which nanotechnology suggested, could become "the basis of a manufacturing technology that is cleaner and more efficient than anything we know today." Nanotechnology, Drexler argued, certainly met "the criteria for an environmentally critical technology."[36] Gore, for his part, displayed great enthusiasm for Drexler's ideas and his knowledgeable reference to Feynman's 1959 speech also confirmed that nanotechnology's creation myth still possessed great currency.[37]

Like many predictions of technological utopias, Drexler's congressional testimony says more about contemporary circumstances than it makes accurate statements about the future. Drexler's pronouncements resonated with concerns held by American politicians and the public about the environment, sustainable development, and American economic competitiveness. It depicts a future ameliorated and achieved through the adoption of a "technological fix," one where nanotechnology would raise "the material standard of living worldwide, while decreasing resource consumption and environmental impact."[38]

Drexler's appearance marked the most visible discussion yet on Capitol Hill of nanotechnology as a potentially valuable area for federal investment. Topics that politicians care most about—health, the economy, and national security—all looked to benefit from nanotechnology. While yet a minor blip, nanotechnology was now on the federal government's radar screen.

Planting the Flag

The utopian nano-futures portrayed by Drexler and others garnered a great deal of attention in the early 1990s, and the claims could be quite extravagant. In 1995, a nano-enthusiast claimed that nanotechnology could even produce immortality.[39] Consequently, many scientists and science writers were skeptical if not openly derisive and hostile toward such visionary claims. In his review of the book *Nano! Remaking the World Atom by Atom*, chemist David E. H. Jones, for example, concluded that until scientific questions surrounding the most optimistic claims of nano-enthusiasts were answered, nanotechnology itself would "remain just another exhibit in the freakshow that is the boundless optimism school of technical forecasting."[40] Perhaps more seriously, other critics noted that advocates who promised great benefits from nanotechnology and drew on Feynman's reputation for their scientific authority were squandering the political and social capital accumulated by real scientists who were doing actual laboratory work.[41]

Scientists were clearly expending a great deal of effort to establish distinct boundaries between what they saw as legitimate research and nano-visions that seemed to have more in common with science fiction than laboratory realities.[42] In order to gain access to the resources, i.e. funding from the federal government and industry they believed necessary to pursue a full-scale research program in nanotechnology, a compelling vision for nanotechnology needed to be articulated by people not on the fringe of the research establishment.

Mihail C. Roco was a key person who took the stage in the mid-1990s determined to strip nanotechnology of its exaggerated claims and to promote an ambitious, realistic nano-research program as a national priority. Born in 1947 in Bucharest, Roco was a mechanical engineer who came to the USA in 1980, where he carried out research at a number of institutions before joining the National Science Foundation (NSF) in 1990. "Captivated by the unity and coherence encountered in nature," Roco wanted to likewise promote the "coherence of science and technology."[43] Soon after joining the NSF, he helped create a new, modestly funded program that investigated the synthesis of nano-sized particles.[44] Gradually, Roco began to lay the foundation for a much more ambitious inter-agency program that would support a whole range of research in nanotechnology.[45]

Roco launched his entrepreneurial plans at a promising time. Following the end of the Cold War, there was much debate among scientists, policy makers, and government agencies about how to restructure US science policy. There was general agreement that that US science policy needed revision yet achieving consensus on how best to do this was difficult. National defense against a monolithic Communist threat no longer served as sufficient justification for funding basic science. Just as significantly, the end of the Cold War signaled to policy makers that the next arena of conflict would be in the global marketplace in the form of increased economic competition from European and Asian countries.[46] This was an especially salient fact to US lawmakers who read reports of how Japanese and European companies were taking the basic research done in American laboratories to the global marketplace and profiting handsomely in the process. In 1991, Japan's Ministry of International Trade and Industry pledged some US$225 million for nanotechnology, with research aimed at benefiting that country's electronics industries.[47] News like this fed existing fears among corporate and government leaders that the US was falling behind in a potentially key economic area and spurred more calls for a revamped national science policy in the wake of the Cold War.

Articulating an appropriate policy to deal with this competitive threat from abroad became a priority for lawmakers and policy analysts. More than half a dozen major reports appeared that suggested how to re-structure US science policy in response to the new global environment.[48] Solutions included improving America's publicly funded research infrastructure and investing public monies in technologies whose benefits were only on the horizon. While not named as such, advocates soon touted nanotechnology as one of these under-supported fields with great potential. Many of these suggested changes were embraced in the first official review of US science policy in the post–Cold War era. *Science in the National Interest*, released in August 1994, described basic research and technological development not as the keys to national security but as elements that "underpin the nation's economy." Gore stated, in words reminiscent of Vannevar Bush's half-century old vision, "Technology is the engine of economic growth; science fuels technology's engine."[49] US science policy, the report said, should enhance connections between fundamental research and national goals and one of these goals was, of course, long-term economic growth.

Compared with the structured plans put forth by the Japanese, the US effort in nanotechnology was under-funded and disorganized. For example, since 1978, the NSF had funded the National Nanofabrication Facility at Cornell University.[50] The US$40 million facility served as a place where researchers from around the country could come to do research and use equipment their home institutions lacked. The National Institute of Standards and Technology also began an initiative in nano-research. Given these disparate initiatives, which *in toto* were spread over 12 different agencies, Roco and others claimed that the US lagged behind other countries when it came to a major national investment in nanotechnology.[51]

As a first step in coordinating the people, infrastructure, and funding for nanotechnology research in the US, Roco, along with Richard Siegel (Rensselaer Polytechnic Institute) and Evelyn Hu (University of California at Santa Barbara), organized a workshop in Arlington, Virginia. This 1997 meeting was held under the auspices of the World Technology Evaluation Center (WTEC), formerly known as the Japanese Technology Evaluation Center. Charged with examining foreign research in selected technology areas and operated out of Loyola College in Baltimore in cooperation with the NSF, WTEC provided information for policy makers concerned about American industrial competitiveness.

The WTEC meeting marked the beginning of an effort to ascertain research trends in nanotechnology worldwide.[52] Roco claimed that nanotechnology was at "a level of development comparable to that of computer/information technology in the 1950s" yet was poised—with sufficient funding—to become a key emerging technology for the 21st century.[53] At the meeting, attendees described the current level of US spending on nanotechnology—in fiscal year 1997, this was US$115 million with more than half of the money coming from the NSF—and emphasized the need for interagency collaboration. The report Roco and his colleagues prepared also noted that governments in Japan and Europe were already funding comprehensive programs.[54] However, compared to the far-reaching and visionary claims by Drexler and other nano-utopians, the soberly written WTEC report served to clarify the current state of nano-research rather than promote nano-utopian visions.

After the May 1997 meeting, Roco helped form an Interagency Working Group on Nanoscience, Engineering, and Technology (IWGN) which he chaired. The IWGN answered to the cabinet-level National Science and Technology Council established by President Clinton in 1993. Throughout 1998, the IWGN met and worked out a vision for what ultimately became the National Nanotechnology Initiative. Like all major government initiatives, putting the pieces in place took a great deal of time and coalition building.

While Roco and his colleagues were formulating their proposal for the NNI, the US government reviewed its investment in scientific research yet again. In February 1997, soon after the second Clinton administration began, House Speaker Newton L. Gingrich asked the House Committee on Science to develop a long-range science and technology policy. The committee's report, released in September 1998, reaffirmed that, with the end of the Cold War, the principal contributions of science and technology would not be for national defense but economic competitiveness.[55] As Gingrich framed it, "the competitions we are engaged in now are less military and largely economic."[56]

The 1998 report also articulated aspects of America's new "mega-era" of science policy that advocates of the National Nanotechnology Initiative soon embraced.[57] Grants to individual scientists and engineers were the "primary channel by which the government stimulates knowledge-driven basic research," while the research done at universities and national laboratories must be continually "applied to the development of new products or processes," so as to further strengthen linkages between university laboratories and commercial ventures formed after the passage of the Bayh–Dole Act in 1980.[58] Just as significantly for the success of the NNI, the report acknowledged that private industry was unlikely to fund areas like nanotechnology that had no foreseeable practical application; this role would therefore devolve to the federal government.

As members of Congress re-examined national science policy, several prominent people joined the campaign for a national nanotechnology effort. At this point, as the push for the NNI's funding intensified, the first noticeable traces of nano-utopian visions crept into the rhetoric of scientists and administrators. In April 1998, Neal Lane, President Clinton's science advisor, appeared before the House Basic Research Subcommittee. Lane, whose specialty was theoretical atomic and molecular physics, made a statement that was to be widely quoted: "If I were asked for an area of science and engineering that will most likely produce the breakthroughs of tomorrow, I would point to nanoscale science and engineering."[59] Noting how NSF support had enabled nanotechnology to transform "from the realm of science fiction to science fact," Lane went on to describe ideas that drew as much from the former as the latter: nano-sized medical probes, new classes of materials, and—straight from Drexler's writings—"nanomanufactured objects that could change their properties automatically or repair themselves." As Lane put it, the "possibilities of nanotechnology are endless."

On 11 March 1999, Roco formally presented plans for the National Nanotechnology Initiative to Neal Lane and Thomas Kalil, Clinton's Deputy Assistant for Technology and Economic Policy. During their two-hour meeting, Roco described an initiative for fiscal year 2001 costing some US$500 million. At the end of the meeting, nanotechnology won out over competing initiatives and now had two well-placed advocates in the White House.

According to Roco, the next task was to educate both the public as well as Congress as to what nanotechnology was and what it could offer. A glossy brochure entitled *Nanotechnology: Shaping the World Atom by Atom* made the case to the public and drew upon nano-utopian rhetoric that was rapidly becoming *de rigueur*—the vision of Feynman's 1959 talk, the progress made in manipulating atoms via electron microscopy, the potential to fabricate nanoscale electronic devices, the now-famous IBM logo writ in xenon atoms, and the vast commercial and societal benefits that investment in the NNI could open up.[60]

Congress still required persuading, however. On 22 June 1999, the House of Representatives' Committee on Science met to review the federal government's support of nanotechnology and to discuss its potential economic implications. Chairman Nick Smith opened the meeting by noting that the United States did not dominate nanotechnology. "How much effort," he asked, "should the Federal Government be putting into tax-payer funded research in this area?"[61]

Several witnesses answered Chairman Smith; all drew in some capacity upon the utopian visions of nanotechnology's potential that had been gestating for over a decade. Of those who spoke that afternoon, chemist Richard E. Smalley of Rice University was perhaps most persuasive. Smalley was a co-discoverer of carbon-60 molecules, better known as "buckyballs," for which he had shared the 1996 Nobel Prize. Buckyballs were closely related to carbon nanotubes that Japanese scientists had discovered in 1991, and advocates of nanotechnology referred to these new structures as clear examples of how matter could be deliberately controlled at the atomic level.

"There is a growing sense in the scientific and technical community," Smalley told Congress, "that we are about to enter a golden new era."[62] Citing his own experience battling cancer, Smalley connected research at the nanoscale with an array of societal benefits that could result, including medical breakthroughs. The Nobel laureate shrewdly pointed out that research in nanotechnology was "intrinsically small science." Because research in nanoscience was not centralized in one or two national laboratories, funding the NNI would combat the trend towards "Big Science," offering a way to fund individual researchers. Funding "small science" (in both senses) would, Smalley argued, help attract students into science careers just as *Sputnik's* launch in 1957 had inspired his generation.

The NNI, Smalley avowed, would also support broader national policy goals. "There's this immediacy between the ivory tower pure scientist and the technologist" in nano-research, Smalley said, and the NNI would simultaneously strengthen both while creating valuable linkages between basic research and commercial applications.[63] Nanotechnology, Smalley concluded, presented a "tremendously promising new future." What was needed was someone bold enough to "put a flag in the ground and say: 'Nanotechnology, this is where we are going to go....'"[64]

In urging Congress to fund a major national initiative for nanotechnology, advocates of the NNI deployed an array of effective arguments, some new and some already familiar: the specter of economic competition with Europe and Japan; the opportunity to recruit a new generation of scientists and engineers; and support for traditional individual, peer-reviewed research rather than giving a handout to a few giant national laboratories. Persuasive utopian images and frontier metaphors were added to the rhetorical brew. One witness that afternoon, for example, compared nanotechnology with Kennedy's declaration to put a man on the moon.[65] What the USA needed to do, advocates insisted, was to be bold once again and plant a flag on a new frontier.

The Floodgates Open

Richard Smalley and the others who testified in June 1999 were largely pushing against an open door as they made their case for a major national investment in nanotechnology. Reflecting new ideas in circulation about national science policy, questions from congressional representatives were more praising of nanotechnology than they were probing. Such enthusiasm is understandable as the House committee that heard Smalley and others was convened to gather information, not appropriate tax-payers' money. Congressional questioning would become more intense once real dollars were at stake.

Roco and other NNI advocates focused their efforts next on federal agencies such as the NSF, NASA, and the Department of Defense, which would likely administer funding for a major new initiative. By November 1999, the Office of Management and Budget and the White House were studying plans for the NNI. That same month, the Presidential Council of Advisors in Science and Technology (PCAST) issued its formal review of the proposed NNI.[66] More circumspect in its espousal of the now-familiar utopian predictions for nanotechnology, PCAST nevertheless identified nanotechnology as a potential path to a "new industrial revolution," one that might lead to unforeseen advances in "materials and devices of all sorts... medicine and health, biotechnology and agriculture, as well as national security."[67] In an increasingly familiar comparison, nanotechnology was likened to the state of solid-state physics research in the 1950s but, unlike the lead that America enjoyed in transistors and microchips when their commercialization began, the USA lacked "an early lead in nanotechnology." PCAST argued that the USA could not afford to be in second place: "The country that leads in discovery and implementation of nanotechnology will have great advantage in the economic and military scene for many decades to come." The NNI, therefore, represented an opportunity to swiftly create an infrastructure for nano-research and perhaps avoid the precarious state the USA found itself in during the 1980s when fears of Japanese technological supremacy ran rampant. A subsequent letter from the PCAST to President Clinton offered firm and clear support of the NNI.[68]

With these endorsements in hand, President Clinton traveled to Pasadena, California where he addressed a standing-room only crowd at Caltech's Beckman Auditorium. On 21 January 2000, Clinton articulated the need to strengthen America's investment in science and technology. After paying homage to Caltech and Feynman (as well as an image suspended above him of the Western Hemisphere constructed with individual gold atoms), Clinton announced the inclusion of a US$2.8 billion "Twenty-First Century Research Fund" in his 2001 budget, which was an element in a larger strategy to double the NSF's budget over the next five years. A "top priority" was a major increase in the NSF's funding for nanotechnology—over 124% or US$227 million in the next year alone—which, when combined with funding for other agencies, would nearly double the federal investment. Clinton rhapsodized

Just imagine, materials with 10 times the strength of steel and only a fraction of the weight; shrinking all the information at the Library of Congress into a device the size of a sugar cube; detecting cancerous tumors that are only a few cells in size. Some of these research goals will take 20 or more years to achieve. But that is why—precisely why...there is such a critical role for the federal government.[69]

A major Presidential address now incorporated aspects of the utopian views suggested by Drexler and adopted by science managers. Returning to Washington, DC on Air Force One, Neal Lane and NSF Director Rita Colwell reiterated the promise of nanotechnology for reporters. According to Lane, "nanotechnology may not be something that everybody talks about every day, but they will, because it's really the whole next generation in manufacturing."[70] Not surprisingly, Colwell and Lane praised the speech and predicted wide bipartisan support of the President's plan. In fact, just a few months before Clinton's speech, an editorial in *The Washington Post* by former House Speaker Gingrich claimed nanotechnology would "have as big an impact on our lives as transistors and chips did in the past 40 years" and called for a substantial increase in federal support for science.[71]

Despite the utopian gloss advocates could put on the importance of nanotechnology to national needs, the annual appropriations process grounded the American nanotechnology initiative to more prosaic political and budgetary struggles. Initial congressional response to the NNI was positive. At a Senate roundtable discussion in April 2000, Senator Evan Bayh, [whose father, Birch Bayh,] twenty years earlier co-sponsored the Bayh–Dole Act that helped re-shape the relationship between universities and the federal government, called nanotechnology "extremely important to future rates of innovation" and the overall economy in the USA.[72] He and several of his colleagues heard presentations from Richard Smalley as well as Donald Eigler of IBM. Improvising on his earlier testimony, Smalley noted how basic sciences like physics and chemistry were at the core of nanotechnology. A vote for the NNI was, in effect, a vote for basic science research and education.[73]

As spring arrived in Washington, it became clear that the science advocates would not secure funding for the NNI without a struggle. Concerns over the NNI centered around several issues. Some Republican lawmakers warned that Roco's multi-agency plan would be hard to manage and result in a duplication of research. The NNI lacked, one congressman complained, "a clear strategy to ensure coordination within the government."[74] Meanwhile, a non-partisan report from the Congressional Research Service noted that, as "nanotechnology was still in its infancy," large-scale practical applications were at least a decade away.[75] Finally, those lobbying congress to double the entire NSF budget opposed any funding for the NNI that came from research programs already established.

In May 2000, the Senate subcommittee that controlled appropriations for both NASA and the NSF met. Senator Christopher Bond, the panel's chair, expressed doubts not about the NNI's value *per se* but about how the NSF would manage the rapid infusion of funding.[76] The Missouri Republican told NSF officials to "count me skeptical" about multi-agency initiatives like the NNI and asked the agency to clarify how it would allocate the requested increase in funds for nano-research. Bond's minority counterpart, Senator Barbara Mikulski of Maryland, expressed her own worries about allocating monies to the NNI without any kind of long-term plan. As she told Lane and Colwell, nanotechnology seems "like our little secret...we need visibility."

A few weeks before the Senate hearing, nanotechnology did indeed receive greater visibility, although not the kind its advocates wanted. In April 2000, the technology magazine *Wired* published an article by Bill Joy called "Why the Future Doesn't Need Us."[77] Joy, one of the co-founders of Sun Microsystems, could not be dismissed as a neo-Luddite. In warning about future technological dystopias, Joy singled out nanotechnology—especially the Drexlerian visions of autonomous and self-replicating nano-assemblers—as a potential threat to humanity. Joy's article became especially infamous for highlighting the "grey goo" problem, the idea that uncontrolled self-replication by nanomachines might destroy entire ecosystems and threaten what it meant to be human. In contrast to the utopian views of Drexler or the more attainable goals Roco and other NNI supporters proposed, Joy, writing as a "skeptical nanoist," sounded the tocsin for a decidedly dystopian vision of a nano-future and urged a moratorium on further research.[78]

Nano-advocates recognized the potential danger opponents like Joy posed. Just as utopian visions can generate public support and shape favorable policy, dystopian ones can derail them. At the May 2000 Senate hearing, Senator Bond remarked on the "hysteria and fear" already surrounding genetically modified foods and it was easy for nano-advocates to imagine a similar public reaction to nanotechnology.[79] There is no direct evidence that Joy's concerns influenced congressional debate of the NNI in the summer of 2000.[80] Furthermore, Joy did not represent the dominant view of industry. According to one participant in the establishment of the NNI, a "critical factor" in the selling of the NNI in the summer of 2000 was the vocal support industry representatives, especially those in the semi-conductor business, gave on the Hill.[81]

Over the summer of 2000, these lobbying efforts paid off as a swell of bipartisan support for the NNI swept away concerns, fiscal, dystopian, or otherwise. Maryland's Senator Mikulski, renowned in science circles as a staunch advocate of NASA, jumped on the utopian bandwagon. In a speech at NASA headquarters in June 2000, Mikulski told her audience that "we are on the verge of a new revolution—the nanotechnology revolution" and urged her colleagues to support the NNI, one of the "least noted and most important documents of the Clinton administration."[82] Across the senate aisle, Republican Senator Trent Lott praised nanotechnology in a letter to his colleagues on the Senate Appropriations Committee. Lott hailed the NNI not so much for its utopian potential but because the national research infrastructure it would help create might be based in regions like Lott's home state of Mississippi, which traditionally received few federal research dollars.[83]

At the end of the summer, Roco's committee formally unveiled its proposal for the NNI, now honed by months of fine-tuning. *The National Nanotechnology Initiative: The Initiative and its Implementation Plan* requested US$495 million—US$150 million of which the NSF would dispense—to expand research at the nanoscale.[84] At the heart of the interagency plan was an increase of some US$90 million for basic research to fund individual scientists and small teams. This was a major increase, up about 50% from the previous fiscal year to US$177 million. While defined as "research leading to new fundamental understanding and discoveries," the proposal clearly tied this research to the "capacity to create new affordable products with dramatically improved performance."[85]

The report also identified several "Grand Challenges," broad areas of research targeted for interagency funding.[86] These included creating new nanostructured materials, nanoelectronics, and bio-nanosensors and linked basic science and engineering to ambitious, long-term economic goals. Congress directed the remaining money to construct a national infrastructure with "centers of excellence" where nano-research would work with the stated goal of producing new discoveries to be "rapidly commercialized by industry."[87]

After last-minute political wrangling, Congress approved budgets for the NNI's various agencies in the fall of 2000. Some US$465 million of federal money was directed to fund the NNI, representing a 72% increase in the national investment in nanotechnology compared to the previous year.[88] Several states followed suit. California, with 12% of the US population and the world's fifth largest economy, has frequently been on the leading edge of new technological developments. In late 2000, then-Governor Gray Davis approved a US$100 million state initiative to foster nanotechnology in California: "It's my hope," Davis explained, "to replicate Silicon Valley."[89]

The overall funding profile for the national investment in nanotechnology is shown in figure 19.1. Three important facts stand out. In the span of little more than half a decade, nanotechnology advocates like Roco and two different administrations had increased national funding for nanoscale science and engineering by some 850%. Second, the NSF and the Department of Defense control the largest amounts of NNI funding and the relative amounts of money going to civilian versus defense-related research in nanotechnology reflect changing national priorities after the September 11 attacks. Finally, national enthusiasm for nanotechnology, at least in terms of budget allocations, only increased after George W. Bush was selected President in 2000. The new Republican administration generously supported nanotechnology, acting in part on the belief that the nascent field would contribute significantly to national security.[90] On 3 December 2003, Bush signed the "21st Century Nanotechnology Research and Development Act" which authorized US$3.7 billion more for nano-research over the next four years.[91] If one accepts the premise that budgets reflect policy, any doubts about nanotechnology's place as the first "boom science" of the 21st century or immediate concerns about environmental or societal consequences were swept away by the flood of money that began pouring into universities and national laboratories in 2001.[92]

Nano-Boom or Gloom?

It is too soon to judge the effects the NNI may have on society, industry, or the scientific community. Indeed, one might even hesitate to call nanotechnology a full-fledged "technology" as so very few of the devices that advocates had envisioned have materialized. Furthermore, nanotechnology has yet to infiltrate public awareness fully, as a recent poll indicated.[93] However, we can try to understand the reasons and rationales that prompted lawmakers to direct vast sums of taxpayer dollars to support a realm of technoscience that retains strong links to fiction as much as facts.

The NNI emerged at a salient point in US history. In the late 1980s and throughout the 1990s, economic competitiveness replaced the twilight military struggle of the Cold War, and science advocates could no longer claim national defense as the prime rationale for funding basic research. Lawmakers were still struggling to adapt national policies to the new global environment and to the implications of the increasing commercialization of science. Advocates of the NNI proposed it at a propitious time— between the end of the Cold War and a renewed preoccupation with national security after 11 September 2001—when lawmakers were trying to reshape national science policy.

One recent analysis correctly situates the NNI in the period of "post-academic" science.[94] This regime is characterized by an emphasis on the utility of science and the enlistment of academic research as a "wealth-creating technoscientific motor for the whole economy," views clearly expressed in the documents and testimony supporting the NNI.[95] While the end of the Cold War is certainly relevant, the changing nature of research funded by the federal government and conducted at universities is even more significant. Since the passage of the Bayh–Dole act in 1980, the "triple helix" of relations between the academy, industry, and government has been significantly altered and strengthened.[96] The borders between science and technology, as the NNI implementation plan shows, have blurred while the commercialization of academic science has become a key driver for its funding.

The favorable timing of the NNI only partly accounts for its success. It cannot explain fully the justifications deployed in support of a major new initiative that would benefit science, technology, and industry. As we have seen, nano-advocates presented an array of heterogeneous arguments to Congress and the public. Despite the new period of "post-academic science," many of these justifications were framed in familiar ways. They included the appeal to national competitiveness, introduced in the 1980s in the face of Japan's economic success, and the need for America to "plant a flag," as did the Apollo astronauts, on the cutting edge of a new technology. While American scientists claimed two-thirds of the 1996 Nobel Prize for their discovery of buckyballs, such honors did not ensure that the USA would reap commercial benefits from them. That Japan and Europe had or were planning major government programs for nanotechnology allowed advocates of the NNI to argue that it was imperative that the USA do likewise.

Congressional witnesses predicted that nanotechnology would draw American youth into science, a rationalization for science funding that dates to the *Sputnik* era at least. Chemist Richard Smalley reached back even further in time in his testimony, claiming that nano-research as "small science" was a way for the individual scientist to do more personal research commensurate with the days of sealing wax and string, and after September 11, national security, the major rationale for so much Cold War–era research and development, emerged once again as a major justification for funding nanotechnology.[97]

The NNI also offered lawmakers and science managers an opportunity to revitalize areas of research they saw as under funded during the 1990s. Clinton's announcement in 2000 of major federal support for nanotechnology, for example, was part of a larger strategy to significantly increase the NSF's budget and an opportunity for physics, chemistry, and engineering to regain lost ground due to stagnant budgets during the 1990s. The NNI, one editorial noted, "may prove to be one of the most brilliant coups in the marketing of basic research" since Nixon announced the "War on Cancer."[98] However, while the public could understand the necessity of research on cancer, the NNI relied on strategic and utopian marketing to enable its success.

Indeed, one can question whether the NNI is really even *science* policy *per se*. After leaving the NSF, Neal Lane, one of the advocates for the NNI inside the White House, admitted that the NNI was actually much more a *technology* policy and those who advocated the applications (as opposed to the fundamental knowledge) enabled by nano-research had won the day.[99] The NNI funded the basic research that advocates like Smalley called for, but much of this was directed toward specific commercial applications and national goals.[100]

Finally, one must acknowledge the energy that entrepreneurial science managers like Roco invested in crafting a policy and plan for the NNI. As Roco explained, the NNI was "prepared with the same rigor as a science project between 1997 and 2000: we developed a long-term vision for research and development."[101] These efforts succeeded in winning over lawmakers initially skeptical of a complex multi-agency effort as well as the NSF's ability to manage a major infusion of funds.

One can compare the success the NNI had with other major research initiatives. Opponents of the ill-fated Superconducting Supercollider, for example, criticized it as a mega-project that would benefit a small and elite research community as well as a single state (Texas).[102] In comparison, the NNI's funding was not earmarked for a single project, discipline, or region. Many parts of the country would presumably benefit from Congressional support for the NNI and the research monies would benefit a large number of researchers in both the life and physical sciences. In this sense, the organization and funding of the NNI was a decentralized form of federally supported Big Science similar to the Human Genome Project. Advocates for both claimed the programs would foster pluralism and decentralization in science as well as local investments and research initiatives.[103] The establishment of national policies for nanotechnology also resembles the rush to fund research in high-temperature superconductivity in the late 1980s. Descriptions of levitating trains, medical breakthroughs, and robust economic competitiveness all spurred federal and commercial investment in fundamental research conducted with an eye toward future applications from the outset.[104]

Overshadowing all of these rationales are two meta-reasons, both familiar and suspect to historians of technology, which also help account for the success of the NNI. While technological determinism may be out of fashion with scholars, nano-advocates regularly made arguments employing various shades of it. A typical line of reasoning invoked Moore's Law (which states that the computing power on a microchip doubles roughly every 18 months) and described the emergence of economic crises once engineers reached the physical limits of what was possible with microtechnologies.[105] Indeed, industry advocates used the seeming end of the technological improvements described by Moore's Law to push for the NNI.[106] This depiction of technological inevitability arose from nothing more than what experts claimed was the "natural" development of computing technology. Advocates presented nanotechnology as one way to avert these limitations.[107] Indeed, according to some enthusiasts, revolutionary developments at the nanoscale were themselves inevitable, yet another phase in human mastery over nature.[108]

Even more powerful as a motivating force were predictions of the utopian benefits nanotechnology could bring to American society and the economy. Visions of tomorrow that Roco, Smalley, and others successfully presented—which themselves tapped into Richard Feynman's mythic vision of nanotechnology's potential and the futuristic, sometimes outlandish, ideas put forth by people like Drexler—attracted and excited lawmakers. With the end of the Cold War and lawmakers' continued interest in economic competitiveness, nanotechnology offered to reshape, literally, how our world was constructed. As David Nye has pointed out, American narratives of technological utopias come in a variety of categories: *natural* where technologies are a natural outgrowth of society; *ameliorative* in which new machines improve everyday life; and *transformative* in which technologies reshape social reality.[109] Nanotechnology appears as a hybrid example as advocates drew upon elements from all three narratives in their lobbying efforts.

While the most extreme scenarios of a future made better through nanotechnology originated with people like Drexler, who operated at the vanguard (and sometimes the fringe) of nanotechnology, they were embraced and repackaged by mainstream science managers and renowned researchers. Mihail Roco and his colleagues described nanotechnology with an exuberance that matched the prognostications of Drexler and his allies.[110] Roco promoted the coming convergence of nanotechnology with biotechnology, information technology, and cognitive science ("Nano–Bio–Info–Cogno" in policy jargon), which could produce a "golden age...an epochal turning point in human history." As expressed, "If the Cognitive Scientists can think it, the Nano people can build it, the Bio people can implement it, and the IT people can monitor and control it."[111] Fully espousing an understanding of history predicated on an unshakable belief in technological progress, the funding of nanotechnology put humanity "at the threshold of a New Renaissance of science and technology."[112] Whether this convergence is indeed real or whether it represents another phase in the marketing of nanotechnology is yet to be seen.[113]

Visions of a future enabled and ennobled by nanotechnology share many common themes with scenarios presented by advocates of nuclear power, space exploration, and genetics technology. For example, supporters of nuclear power claimed that it would leave practically no part of society unimproved. Funded by the federal government and viewed with fear and fascination by the public, advocates presented the power of the nucleus as the solution to a host of economic, social, and national security problems. By eliminating hunger, unemployment, environmental degradation, and the drudgery of housework, nuclear power promised "dazzling possibilities of life, liberty and the pursuit of happiness."[114] Outer space, meanwhile, offered Americans new frontiers to explore as well as endless research opportunities for scientists.[115] Decades later, advocates of genetic engineering assured the possibility of disease prevention and therapy along with an improved understanding of what it meant to be human. Supporters of nanotechnology, deliberately or no, conflated these disparate utopian aspects. Nanotechnology could "offer the opportunity to understand life processes...cure and prevent disease, heal injured bodies, and protect society" all while providing more "endless frontiers" for exploitation and exploration by scientists and entrepreneurs.[116]

Dystopian warnings, put forth initially by Joy and others, did not abate after the NNI was funded. If anything, the surge of new money directed toward nano-research and development fueled anxieties.[117] Researchers spoke out publicly on the possible toxicity of newly designed nanoparticles and the insurance industry, still engaged in asbestos-related lawsuits, began to estimate litigation costs stemming from nanotechnologies.[118] Politicians were also aware of a potential backlash against the initiative they had approved. At an April 2003 hearing on nanotechnology's societal effects, Vicki L. Colvin, director of the Center for Biological and Environmental Nanotechnology at Rice University, warned Congress that nanotechnology needed to avoid the "wow-to-yuck" arc that genetically modified organisms had taken.[119] At the other end of the spectrum, Eric Drexler, arguably the most well-known and visionary advocate of nanotechnology's transformative potential, criticized the NNI for its timidity and business-as-usual funding strategy. Leaders of the NNI, Drexler charged, had neglected the original "Feynman vision" of nanomachines because "public concern regarding its dangers might interfere with research funding."[120] Despite the ample and growing funding directed toward nanotechnology, debate over the trajectory its research and development would take remained robust and unsettled.

In the 1920s, visionaries like Henry Dreyfuss and Norman Bel Geddes believed that utopia could literally be designed and society thereby perfected.[121] Advocates of nanotechnology expressed similar views; namely that the future could be crafted atom by atom. Whether future generations will look at predictions about nanotechnology's ability to re-shape the world with the same bemusement that the phrase "energy too cheap to meter" draws remains to be seen. Nanotechnology remains as much a futuristic vision as an active area of real-life research and development. With the inevitable disappointments that almost always accompany predictions of technological utopias, one may wonder why they remain so attractive to scientists, lawmakers, and the public. As the history of nanotechnology continues to unfold, perhaps historians will address this question again and arrive at a more nuanced and specific answer. Until that time, a simpler explanation remains—visions of technological utopias, despite their dubious claims to success and the darker visions that accompany them, still sell and compel.

Acknowledgments

The research for this article was supported by a grant from the California NanoSystems Institute and National Science Foundation Grant SES 0304727 (as part of a Nanoscale Interdisciplinary Research Team Project). Several references cited in this article are available in on-line format only. Where this is the case, a URL link is provided and a hard copy of the document is in the author's working files. He wishes to acknowledge the assistance and suggestions of Nicole Archambeau, Larry Badash, Justin Bengry, Evelyn Hu, John Krige, Kristen Shedd, and the anonymous reviewers.

Notes

1. "Interview with Mihail C. Roco," http://www.nano.gov/html/interviews/MRoco.htm.

2. A second facility is being built at the University of California at Los Angeles. Almost all publications, regardless of audience, tend to use "nanoscience" and "nanotechnology" interchangeably with a strong preference for the latter and presumably more all-encompassing term. Despite the obvious historiographical conundrum this poses for the historian of technology, I have adopted this style except where I am explicitly referring to the actual scientific research on nanoscale phenomena and techniques.

3. The White House, "National Nanotechnology Initiative."

4. M. C. Roco, "Broader Societal Issues," Office of Science and Technology Policy, "National Nanotechnology Initiative."

5. Roco, "International Strategy."

6. Roco, "Towards a US National," 435.

7. A nanometer (the prefix is the Greek word for "dwarf") is one-billionth of a meter and an oft-quoted unit of measurement cited in nanotech literature, the width of a human hair, is some 10,000 nanometers wide.

8. Definition adapted from www.nano.gov, the official website for the National Nanotechnology Initiative.

9. Barber and Freestone, "An Investigation."

10. Horst Stormer quoted in Amato, *Nanotechnology*, 1.

11. First quote from John Armstrong, formerly chief scientist at IBM, quoted on p. viii of Roco *et al.*, *Nanotechnology Research Directions*;" second quote from Roco, "The US National," 6.

12. Atkinson, *Nanocosm*.

13. See, for example, the papers collected in Baird *et al.*, *Discovering the Nanoscale*. The issue of patents is presented in Meyer, "Patent Citation Analysis" while a view of potential risks is articulated in Swiss-Re, "Nanotechnology" Risk Perception Series. See also Crow and Sarewitz, "Nanotechnology and Societal Transformation;" Fogelberg and Glimell, *Bringing Visibility to the Invisible*. An example of the few articles with a focus on the history of nanotechnology include Johnson, "The End of Pure Science?" which addresses policy issues and Mody, "From the Topografiner" which discusses instrumentation.

14. See, for example, Kingdon, *Agendas, Alternatives, and Public Policies* as well as papers in Sabatier, *Theories of the Policy Process*. See also Korsmo and Sfraga, "From Interwar to Cold War" and Cook-Deegan, "The Human Genome Project" for examples of how specific research programs were advocated.

15. McCurdy, *Space and the American Imagination*.

16. Sarewitz, "Does Science Policy Exist?"

17. See essays in Corn, *Imagining Tomorrow*, as well as the more recent collection by Sturken *et al.*, *Technological Visions*.

18. Feynman, "There's Plenty of Room at the Bottom."

19. Ibid., 26.

20. Feynman's 1959 talk appropriated ideas by science fiction author Robert Heinlein in his 1942 novella *Waldo* and, as Colin Milburn notes, nanotech's real origin can be argued to rest not with Feynman but in science fiction: Milburn, "Nanotechnology."

21. A search of the World Wide Web in 2004 for the most popular web pages dealing with Feynman returned as many references to his 1959 paper as they did his Nobel prize. Also

worth noting is the fact that the Foresight Institute, a non-profit organization devoted to promoting nanotechnology, annually awards a prize named after Feynman.

22. Taniguchi, "On the Basic Concept." Taniguchi's term, used in a much different fashion from today, described precision machining of devices with tolerances less than a micrometer.

23. Baird and Shew, "Probing the History."

24. Mody, "From the Topografiner." As Mody notes, scientists in the United States gradually adopted the STM in a variety of settings including corporate and national research labs as well as university departments. In 1986, the same year that Binnig and Rohrer won the Nobel Prize, Digital Instruments was founded in Santa Barbara, California and became the first company to market their version of an STM, the "Nanoscope."

25. Eigler and Schweize, "Positioning Single Atoms."

26. Siebert, "The Next Frontier."

27. Mody, "From the Topografiner."

28. Drexler, "Molecular Engineering."

29. Ibid., 5278.

30. Drexler, *Engines of Creation*.

31. Monmaney, "Nanomachines to Our Rescue."

32. Amato, "The Apostle of Nanotechnology."

33. Stix, "Waiting for Breakthroughs."

34. Amato, "The Apostle of Nanotechnology," 1311.

35. Science Committee on Commerce, Science, and Transportation, "New Technologies for a Sustainable World."

36. Ibid., 21.

37. Ibid., 30.

38. Corn, "Epilogue" and Drexler testimony in Science Committee on Commerce, Science, and Transportation, "New Technologies for a Sustainable World," 22.

39. Du Charme, *Becoming Immortal*.

40. Jones, "Technical Boundless Optimism" which was a review of Regis's *Nano!*.

41. Stix, "Waiting for Breakthroughs."

42. The permeable boundary between science and science fiction is considered in Milburn, "Nanotechnology." Moreover, the efforts of scientists to delineate between "real" science

and the long-term visions put forth by people by Drexler stands as a classic example of boundary work; see, for example, Gieryn, *Cultural Boundaries of Science.*

43. Roco, "The US National," 10.

44. This was funded at about US$3 million in 1991, a modest amount by NSF standards.

45. A note on the funding process: The President's budget is typically submitted in January each year. In the US, new actions require a two-step congressional process—authorization and appropriation, two interdependent yet distinct activities. The actual allocation of money requires the prior passage of an authorization statute and the authorization process is done by a pair of committees, one in the House and one in the Senate.

46. This was a point made in a number of studies including a report by the Council on Competitiveness called "Science in the National Interest."

47. Crawford, "Japan Starts a Big Push."

48. Boesman, "Analysis of Ten Selected."

49. Office of the Press Secretary, "White House Releases."

50. This was originally called the National Research for Submicron Structures; the name was changed in 1988. Six years later, the NSF funded the National Nanofabrication Users Network which expanded the Cornell model to four other schools.

51. Roco, "The US National," 2.

52. A report from the May 1997 workshop was later published; Siegel *et al.*, *WTEC Workshop Report on R&D Status and Trends*. According to one person involved in the process, some scientists were uncomfortable with the fact that Roco appeared to have already decided that the US presence in nanotechnology was not sufficient. Evelyn Hu, 4 January 2005 personal communication with the author.

53. Ibid., 1.

54. A follow-up report addressed the specific question of where the US stood in relation to Japan and Europe; Siegel *et al.*, "Nanostructure Science."

55. House of Representatives Committee on Science, 105th Congress. "Unlocking our Future."

56. Gingrich, "February 12, 1997 letter."

57. According to the 1998 report, the pre–World War Two period was the first "mega-era" of science policy while the US support of science and technology for national defense during the Cold War typified the second "mega-era."

58. The Bayh–Dole Act encouraged the utilization of inventions produced under federal funding and promoted the participation of universities in the commercialization of ideas derived from basic research.

59. Lane's 22 April 1998 testimony can be found on the web site for the House Committee on Science: http://www.house.gov/science/.

60. Amato, *Nanotechnology*. What is unknown is how effective this campaign was or how many people it reached.

61. House of Representatives Committee on Science, "Nanotechnology," 1–2.

62. Ibid., 7.

63. Ibid., 21.

64. Ibid., 12.

65. Ibid., 10.

66. Pres. George H. W. Bush established PCAST in 1990. It is a non-governmental advisory group formed without Congressional approval. Therefore, each President must renew the group's existence. President Clinton established his PCAST in November 1993 by executive order.

67. This and subsequent quotes from the PCAST review; PCAST's review available at: http://www.ostp.gov/PCAST/pcastnano2.html.

68. 14 December 1999: http://www.ostp.gov/cs/pcast/letter_to_the_president_121499.

69. William J. Clinton; 21 January 2000 speech. Text available at: http://pr.caltech.edu/events/presidential_speech/pspeechtxt.html.

70. Lane quote from Jones, "Aboard Air Force One."

71. Jones, "Former House Speaker Gingrich."

72. Leath, "Administration and Congress."

73. Leath, "Senate Science and Technology Caucus."

74. Southwick, "Nanotechnology."

75. Davey, "Manipulating Molecules."

76. Leath, "NSF Hearing."

77. Joy, "Why the Future Doesn't Need Us."

78. With apologies to Robert Boyle and his 1661 book *The Sceptical Chymist*.

79. For example, the popular science fiction show *Star Trek: Voyager* featured a main character who could assimilate other life forms by injecting them with "nano-probes."

80. Joy's article did encourage policy makers to stress the need to better understand (i.e., fund studies of) the environmental and societal risks of the nano-frontier. The article was also seen as a sufficient threat so as to warrant a response from John S. Brown, a researcher at Xerox PARC at a September 2000 conference the NSF convened to address the ethical and

societal issues of nanotechnology; see Brown, "Don't Count Society Out." Glimell, "Grand Visions and Lilliput Politics" discusses Joy's article and Roco's reaction—what Glimell calls "doing the 'Roco'-motion"—to it.

81. Evelyn Hu, 4 January 2005 personal communication with the author. More work clearly needs to be done to understand the complexities of the support industry provided to the NNI.

82. Quote from p. 272 of Roco and Bainbridge, *Societal Implications*. Mikulski's support of science, and NASA in particular, is detailed in Munson, *The Cardinals of Capitol Hill*, which provides an insightful look at the appropriations process.

83. Roco and Bainbridge, *Societal Implications*, 272.

84. National Science and Technology Council, "National Nanotechnology Initiative." This was submitted by the Subcommittee on Nanoscale Science, Engineering, and Technology (NSET), chaired by Roco, which replaced the earlier IWGN.

85. National Science and Technology Council, "National Nanotechnology Initiative," 42.

86. The "Grand Challenges" areas were to receive a total of some US$133 million, $62 million more than the previous year. Some of this was administered by the NSF but the Department of Defense was slated to be the single largest recipient of "Grand Challenge" funds.

87. Some US$28 million (about 6% of the NNI funding) was also set aside for workforce training and studying societal and ethical implications. National Science and Technology Council, "National Nanotechnology Initiative," 30.

88. Of this, the NSF received US$150 million; the Department of Defense US$125 million.

89. Markoff, "California Sets Up 3 Centers."

90. The effects of the terrorist attacks of 11 September 2001 on funding for nanotechnology is complex and cannot be more than noted here. A 3 August 2003 letter from John H. Marburger, Pres. Bush's science advisor noted that nanotechnology was "essential to achieving the President's top three priorities: winning the war on terrorism, securing the homeland, and strengthening the economy;" included in National Science and Technology Council, "National Nanotechnology Initiative: Research and Development" (Supplement to the President's FY 2004 Budget).

91. US$982 million was requested for fiscal year 2005 alone.

92. Sarewitz, "Does Science Policy Exist," notes the relation, for better or worse, between budgets and policy in the USA.

93. Fewer than 30% of people polled had heard of "nanotechnology" and less than 20% could offer a reasonable definition of it. The Royal Society, *Nanoscience and Nanotechnologies*," 11.

94. Johnson, "The End of Pure Science?" while the term post-academic science' is itself from Ziman, *Real Science*.

95. Ziman, *Real Science*, 73.

96. The literature on this phenomenon is vast; a classic paper on the topic is Etzkowitz, "Entrepreneurial Science in the Academy."

97. For example, for fiscal year 2003 the Bush Administration added another "Grand Challenge" to the NNI which would focus on homeland security via detection and protection against chemical, biological, explosive, and radiological (CBER) threats.

98. Editors, "Megabucks for Nanotech."

99. Baird and Shew, "Probing the History," 10.

100. The focus on applications is clear in Congressional debates over science budgets. As Jones, "House Appropriations Bill" notes, discussion of the fiscal 2005 budget referring to nanotechnology specifically asks the NSF to increase research support for practical applications such as improving semiconductor devices.

101. Roco, "The US National," 3.

102. Riordan, "The Demise."

103. Heilbron and Kevles offer a comparison between the SSC and the Human Genome Project in "Finding a Policy; also Cook-Deegan, "The Human Genome Project."

104. Nowotny and Felt, *After the Breakthrough*.

105. Roco *et al.*, *Nanotechnology Research Directions*, x.

106. Evelyn Hu, 4 January 2005 personal communication with the author.

107. Roco, "The US National," 6.

108. For example, an article by Drexler's wife and partner at the Foresight Institute draws a direct line from the Greek concept of the atom to the future benefits of nanotechnology, suggesting the inevitability of this path; Peterson, "Nanotechnology."

109. Nye, "Technological Prediction," 171.

110. Roco, "Coherence and Divergence;" Roco and Bainbridge, "Converging Technologies."

111. Quotes from Roco and Bainbridge, "Converging Technologies," 285 and 289.

112. Roco and Bainbridge, "Converging Technologies;" 281.

113. According to Evelyn Hu, Roco's emphasis on nano's convergence with other areas of research may be part of a "deliberate repackaging" of the NNI to ensure that interest and funding remain strong. Evelyn Hu, 4 January 2005 personal communication with the author.

114. Del Sesto, "Wasn't the Future," 61.

115. McCurdy, *Space and the American Imagination*.

116. Committee for the Review of the National Nanotechnology Initiative, *Small Wonders, Endless Frontiers*.

117. The 2003 publication of Michael Crichton's bestselling potboiler *Prey*, the plot of which centers around hostile swarms of nanomachines, undoubtedly tarnished the limited public perception of nanotechnology as well.

118. Monastersky, "The Dark Side of Small;" Swiss-Re, "Nanotechnology."

119. House of Representatives Committee on Science, "Societal Implications of Nanotechnology," 49.

120. Drexler, "Nanotechnology."

121. Seagal, *Technological Utopianism*.

References

Amato, Ivan. "The Apostle of Nanotechnology." *Science* 254, no. 6036 (1991): 1310–11.

———. *Nanotechnology: Shaping the World Atom by Atom*. Washington, DC: National Science and Technology Council, 1999.

Atkinson, William I. *Nanocosm: Nanotechnology and the Big Changes Coming from the Inconceivably Small*. New York City: Amacom, 2003.

Baird, Davis, et al., eds. *Discovering the Nanoscale: A Reader of Workshop Manuscripts; from an International Conference at Darmstadt Technical University, October 9–12, 2003*. Darmstadt: Darmstadt Technical University, 2003.

———, and Ashley Shew. "Probing the History of Scanning Tunneling Microscopy." In *Discovering the Nanoscale: A Reader of Workshop Manuscripts; from an International Conference at Darmstadt Technical University, October 9–12, 2003*, edited by Davis Baird et al. Darmstadt: Darmstadt Technical University, 2003: 3–19.

Barber, D. J., and I. C. Freestone. "An Investigation of the Origin of the Color of the Lycurgus Cup by Analytical Transmission Electron Microscopy." *Archaeometry* 32, no. 1 (1990): 33–45.

Boesman, William C. (coordinator). "Analysis of Ten Selected Science and Technology Policy Studies." Washington, DC: Congressional Research Service, 1997.

Brown, J. S. "Don't Count Society Out: A Response to Bill Joy." In *Societal Implications of Nanoscience and Nanotechnology*, edited by Mihail C. Roco and William Sims Bainbridge. Boston: Kluwer Academic Publishers, 2001.

Committee for the Review of the National Nanotechnology Initiative. *Small Wonders, Endless Frontiers: A Review of the National Nanotechnology Initiative*. Washington, D.C.: Committee for the Review of the National Nanotechnology Initiative, Division on Engineering and Physical Sciences, National Research Council, 2003.

Cook-Deegan, Robert Mullan. "The Human Genome Project: The Formation of Federal Policies in the United States, 1986–1990." In *Biomedical Politics*, edited by Kathi Hanna. Washington, DC: National Academy Press, 1991: 99–167.

Corn, Joseph, ed. *Imagining Tomorrow: History, Technology, and the American Future*. Cambridge, MA: MIT Press, 1986.

———. "Epilogue," In *Imagining Tomorrow: History, Technology, and the American Future*, edited by Joseph J. Corn. Cambridge, MA: The MIT Press, 1986: 219–29.

Crawford, Robert. "Japan Starts a Big Push Toward the Small Scale." *Science* 254, no. 5036 (1991): 1304–05.

Crow, Michael M. and Daniel Sarewitz. "Nanotechnology and Societal Transformation." In *AAAS Science and Technology Policy Yearbook*, edited by Albert H. Teich *et al*. Washington, DC: AAAS, 2001: 89–101.

Council on Competitiveness called "Science in the National Interest." Washington, DC: National Science and Technology Council, 1994.

Davey, Michael E. "Manipulating Molecules: The National Nanotechnology Initiative." CRS Report no. RS20589, 2000.

Del Sesto, Steven L. "Wasn't the Future of Nuclear Engineering Wonderful?" In *Imagining Tomorrow: History, Technology, and the American Future*, edited by Joseph J. Corn. Cambridge, MA: The MIT Press, 1986: 58–76.

Drexler, K. Eric. "Molecular Engineering: An Approach to the Development of General Capabilities for Molecular Manipulation." *Proceedings of the National Academy of Science* 78, no. 9 (1981): 5275–78.

———. *Engines of Creation: The Coming Era of Nanotechnology*. New York: Anchor Books, 1986.

———. "Nanotechnology: From Feynman to Funding." *Bulletin of Science, Technology, and Society* 24, no. 1 (2004): 21–27.

Du Charme, Wesley M. *Becoming Immortal: Nanotechnology, You, anal the Demise of Death*. Evergreen, CO: Blue Creek Ventures, 1995.

Eigler, D. M. and E. K. Schweizer. "Positioning Single Atoms with a Scanning Tunneling Microscope." *Nature* 344, no. 6266 (1990): 524–26.

Editors. "Megabucks for Nanotech." *Scientific American* 285, no. 3 (2001): 8.

Etzkowitz, Henry. "Entrepreneurial Science in the Academy: A Case of Transformation of Norms," *Social Problems* 36, no. 1 (1989): 14–29.

Feynman, Richard P. "There's Plenty of Room at the Bottom." *Engineering and Science* 23, no. 5 (1960): 22–26.

Fogelberg, Hans and Hans Glimell. *Bringing Visibility to the Invisible: Towards a Social Understanding of Nanotechnology*. Goteborgs: Goteborgs Universitet, 2003; URL link is http://www.sts.gu.se/publications/STS_report_6.pdf.

Gieryn, Thomas. *Cultural Boundaries of Science: Credibility on the Line*. Chicago, IL: The University of Chicago Press, 1999.

Gingrich, Newton L. "February 12, 1997 letter to House Committee on Science Chairman F. James Sensenbrenner, Jr.:" http://www.house.gov/science/science_policy_study.htm.

Glimell, Hans. "Grand Visions and Lilliput Politics: Staging the Exploration of the 'Endless Frontier.'" In *Discovering the Nanoscale: A Reader of Workshop Manuscripts; from an International Conference at Darmstadt Technical University, October 9–12, 2003*, edited by Davis Baird *et al*. Darmstadt: Darmstadt Technical University, 2003: 147–64.

Heilbron, John L. and Daniel Kevles. "Finding a Policy for Mapping and Sequencing the Human Genome: Lessons from the History of Particle Physics." *Minerva* 26, no. 3 (1988): 299–314.

House of Representatives Committee on Science. "Unlocking our Future: Toward a New National Science Policy." Washington, DC, US Government Printing Office, 1998: http://www.house.gov/science/science_policy_report.htm.

———. "Nanotechnology: The State of Nano-Science and its Prospects for the Next Decade." Washington, DC, US Government Printing Office, 1999.

———. "The Societal Implications of Nanotechnology: Hearing before the Committee on Science." Washington, DC, US Government Printing Office, 2003.

Johnson, Ann. "The End of Pure Science? Science Policy from Bayh-Dole to the National Nanotechnology Initiative." In *Discovering the Nanoscale: A Reader of Workshop Manuscripts; from an International Conference at Darmstadt Technical University, October 9–12, 2003*, edited by Davis Baird *et al*. Darmstadt: Darmstadt Technical University, 2003: 20–33.

Jones, David E. H. "Technical Boundless Optimism." *Nature* 374, no. 6525 (1995): 835–37.

Jones, Richard M. "Aboard Air Force One: Press Briefing by Neal Lane and Rita Colwell." *FYI: The AIP Bulletin of Science Policy News*, no. 10 (2000); available at: http://www.aip.org/fyi/2000/fyi00.010.htm.

———. "Former House Speaker Gingrich on Doubling Research Funding." *FYI: The AIP Bulletin of Science Policy News*, no. 154 (1999); available at: http://www.aip.org/fyi/1999/fyi99.154.htm.

———. "House Appropriations Bill Recommends Cut in FY 2005 NSF Funding." *FYI: The AIP Bulletin of Science Policy News*, no. 99 (2004); available at: http://www.aip.org/fyi/2004/099.html.

Joy, William N. "Why the Future Doesn't Need Us." *Wired* 8, no. 4 (2000): 238–62.

Kingdon, John W. *Agendas, Alternatives, and Public Policies*. New York: Addison-Wesley-Longman, 1995.

Korsmo, Fae L. and Michael P. Sfraga. "From Interwar to Cold War: Selling Field Science in the United States, 1920s through 1950s." *Earth Sciences History* 22, no. 1 (2003): 55–78.

Leath, Audrey. "Senate Science and Technology Caucus Holds Briefing on Nanotechnology." *FYI: The AIP Bulletin of Science Policy News*, no. 44 (2000); available at: http://www.aip.org/fyi/2000/fyi00.044.htm.

———. "Administration and Congress See Promise in Nanotechnology." *FYI: The AIP Bulletin of Science Policy News*, no. 46 (2000); available at: http://www.aip.org/fyi/2000/fyi00.046.htm.

———. "NSF Hearing: Good News, Bad News." *FYI: The AIP Bulletin of Science Policy News*, no. 51 (2000); available at: http://www.aip.org/fyi/2000/fyi00.051.htm.

Markoff, John. "California Sets Up 3 Centers for Basic Scientific Research." *The New York Times*, 20 December 2000: A30.

McCurdy, Howard E. *Space and the American Imagination.* Washington, DC: Smithsonian Institution Press, 1997.

Meyer, Martin. "Patent Citation Analysis in a Novel Field of Technology: An Exploration of Nano-Science and Nano-Technology." *Scientometrics* 51, no. 1 (2001): 163–83.

Milburn, Colin. "Nanotechnology in an Age of Posthuman Engineering: Science Fiction as Science." *Configurations* 10, no. 2 (2002): 261–95.

Mody, Cyrus. "From the Topografiner to the STM and the AFM: What the History of Probe Microscopy Might Tell Us About Nanoscience." In *Discovering the Nanoscale: A Reader of Workshop Manuscripts; from an International Conference at Darmstadt Technical University, October 9–12, 2003*, edited by Davis Baird *et al.* Darmstadt: Darmstadt Technical University, 2003: 59–77.

Monmaney, Terrance. "Nanomachines to Our Rescue." *The New York Times* 10 August 1986: BR 8.

Munson, Richard. *The Cardinals of Capitol Hill: The Men and Women who Control Government Spending*. New York: Grove Press, 1993.

Monastersky, Richard. "The Dark Side of Small: As Nanotechnology Takes Off, Researchers Scramble to Assess its Risks." *The Chronicle of Higher Education*, 10 September 2004, A12.

National Science and Technology Council. "National Nanotechnology Initiative: The Initiative and Its Implementation Plan." Washington, D.C., Subcommittee on Nanoscale Science, Engineering and Technology, 2000.

———. "National Nanotechnology Initiative: Research and Development Supporting the Next Industrial Revolution, Supplement to the President's FY 2004 Budget." Washington, D.C., Subcommittee on Nanoscale Science, Engineering, and Technology, 2003.

Nowotny, Helga and Ulrike Felt. *After the Breakthrough: The Emergence of High Temperature Superconductivity as a Research Field*. Cambridge: Cambridge University Press, 1997.

Nye, David E. "Technological Prediction: A Promethean Problem." In *Technological Visions: The Hopes and Fears that Shape New Technologies*, edited by Marita Sturken, Douglas Thomas and Sandra Ball-Rokeach. Philadelphia: Temple University Press, 2004: 159–76.

Office of the Press Secretary. "White House Releases National Science Policy Report." Washington, DC, The White House, August 3, 1994.

Office of Science and Technology Policy. "National Nanotechnology Initiative: Research and Development Funding in the President's 2005 Budget." Washington, DC: Office of Science and Technology Policy, 2004: http://www.ostp.gov/html/budget/2005/FY05NNI1-pager.pdf.

Peterson, Christine L. "Nanotechnology: Evolution of the Concept." In *Prospects in Nanotechnology: Toward Molecular Manufacturing*, edited by Markus Krummenacker and James Lewis. New York: Wiley, 1995: 173–86.

Regis, Edward. *Nano! Remaking the World Atom by Atom*. Boston: Little, Brown, 1995.

Riordan, Michael. "The Demise of the Superconducting Super Collider." *Physics in Perspective* 2, no. 4 (2000): 411–25.

Roco, M. C. "Towards a US National Nanotechnology Initiative." *Journal of Nanoparticle Research* 1, nos. 5–6 (2000): 435–38.

———. "International Strategy for Nanotechnology Research and Development." *Journal of Nanoparticle Research* 3, nos. 5–6 (2001): 353–60.

———. "Coherence and Divergence of Megatrends in Science and Engineering." *Journal of Nanoparticle Research* 4 (2002): 9–19.

———. "Broader Societal Issues of Nanotechnology." *Journal of Nanoparticle Research* 5, nos. 3–4 (2003): 181–89.

———. "The US National Nanotechnology Initiative after 3 years (2001–2003)," *Journal of Nanoparticle Research* 6 (2004): 1–10.

———, S. Williams, and P. Alivisatos, eds. *Nanotechnology Research Directions: IWGN Workshop Report*. Baltimore, MD: World Technology Evaluation Center, 1999.

———, and William S. Bainbridge, eds. *Societal Implications of Nanoscience and Nanotechnology*. Boston: Kluwer Academic Publishers, 2001.

———, and W. S. Bainbridge. "Converging Technologies for Improving Human Performance: Integrating from the Nanoscale." *Journal of Nanoparticle Research* 4 (2002): 281–295.

Sabatier, Paul A., ed. *Theories of the Policy Process*. Boulder, CO: Westview Press, 1999.

Sarewitz, Daniel. "Does Science Policy Exist, and If So, Does it Matter: Some Observations on the U.S. R&D Budget." Discussion Paper for Earth Science Institute; Science, Technology, and Global Development Seminar, 2003.

Science Committee on Commerce, Science, and Transportation (Subcommittee on Science, Technology, and Space), United States Senate. "New Technologies for a Sustainable World." Washington, DC: Subcommittee on Science, Technology, and Space of the Committee on Commerce, Science, and Transportation, United States Senate, 102nd Congress, Second Session, 1992.

Seagal, Howard P. *Technological Utopianism in American Culture*. Chicago, IL: The University of Chicago Press, 1985.

Siebert, Charles. "The Next Frontier: Invisible." *The New York Times Magazine*, 29 September 1996: 137–39.

Siegel, Richard W., Evelyn Hu, and M. C. Roco, eds. *WTEC Workshop Report on R&D Status and Trends in Nanoparticles, Nanostructured Materials, and Nanodevices in the United States*. Baltimore, MD: International Technology Research Institute, 1998.

———. "Nanostructure Science and Technology: A Worldwide Study." Baltimore, MD, World Technology Evaluation Center, 1999.

Southwick, Ron. "Nanotechnology, the Study of Minute Matter, Becomes a Big Priority in the Budget." *The Chronicle of Higher Education* 2000: A38.

Stix, Gary. "Waiting for Breakthroughs." *Scientific American* 274, no. 4 (1996): 94–99.

Sturken, Marita, Douglas Thomas, and Sandra J. Ball-Rokeach, eds. *Technological Visions: The Hopes and Fears that Shape New Technologies*. Philadelphia, PA: Temple University Press, 2004.

Swiss-Re. "Nanotechnology: Small Matter, Many Unknowns." Swiss Reinsurance Company, 2004.

Taniguchi, Norio. "On the Basic Concept of 'Nano-Technology.'" In *Proceedings of the International Conference of Production Engineering*. Vol. 2. Tokyo: Japan Society of Precision Engineering, 1974.

The Royal Society. *Nanoscience and Nanotechnologies: Opportunities and Uncertainties*. London: The Royal Society, 2004.

The White House: Office of the Press Secretary. "National Nanotechnology Initiative: Leading to the Next Industrial Revolution." 21 January 2000; this document is available at: www.whitehouse.gov/news/releases/2003/12/20031203-7.html.

Ziman, John. *Real Science: What it Is and What it Means*. New York: Cambridge University Press, 2000.

20 "Sociotechnical Complexity: Redesigning a Shielding Wall"

Dominique Vinck

Engineering is often portrayed as simply the application of scientific laws to practical problems. However, producing technology, even if one focuses only on what happens in the lab, is far more complicated than such a model would lead one to believe. In this chapter, Dominique Vinck tells the true story of a young engineer who discovers first hand that daily life as an engineer involves more than facts, numbers, predictable phenomena, and well-defined problems. His day-to-day work requires a great deal of negotiation with people from other disciplines, other labs, and other companies just to put together what many of his co-workers believe to be one of the most mundane elements of the entire project. Examples like this demonstrate that precisely how technologies are put together is not preordained. Not only do the outcomes depend on the goals set by management or the regulations established by government, they also depend on how well individuals are able to communicate and argue for their particular priorities. This chapter describes the complexities faced by just one person making decisions about one very small part of a technology. The complexity increases exponentially when we add other stages and other actors to the picture.

Before embarking on a placement[1] in a design and engineering office, the young engineer does not really understand the complexity of the work awaiting him. Of course, he is ready to do complicated operations that must be dealt with at a high level of abstraction. He has also been trained to handle fairly sophisticated models and tools. He knows that he is bound to run up against difficult technical problems. Nevertheless, he has a certain number of working methods and tools under his belt that will get him out of many a difficult situation. He has the capacity to analyze problems, break them down into essential parts, and then model them. This ability to simplify things is supposed to help him get through the most complicated challenges. At least this is what he has been taught.

Yet the young engineering student still has to learn exactly how complex ordinary technical work really is. A placement period lasting just a few months will prove to be a real eye opener. He may have thought that an engineer's work is mainly technical, but he will quickly realize that, in reality, things are much more complex than that. He will also find that, if he wants to be an efficient engineer and get technically satisfactory results, he will have to decode and take into account only what appears to be real.

The aim of this chapter is thus to map and document the changing vision of young engineers after their entry into the industrial world. To build up our account of what

From Dominique Vinck, ed., *Everyday Engineering: An Ethnography of Design and Innovation* (Cambridge, Mass.: The MIT Press, 2003), pp. 13–27. Reprinted with permission.

typically happens, we will use the experience of an engineering student as he learns the ropes during a placement. Although the placement period in question is only 6 months, it must not be forgotten that the time usually required is much longer, about 2 or 3 years.

We will follow the work of an engineering student during his placement in a CERN (European Organization for Nuclear Research) design office in Geneva. For this student, the difference between what he learned at school and the way things really are in the design office is accentuated by the fact that his assignment seems to be quite simple: define the shape and dimensions of an object, and the materials to be used, so as to meet the specifications of the order givers and the laws of science and nature. Furthermore, this assignment is a good opportunity for the student to apply some CAD (computer-aided design) tools to a real case. The problem does not look complicated at the outset; our student needs only the initial data (the specifications defined by the order givers), a computer console, and a methodical approach.

However, what the young engineer will discover during his placement is that his pre-evaluation of it was much too simple and limited. To be able to fulfill his mission, he will have to change his views and his approach little by little. He will have to rework his initial impression of the design work. He imagined himself sitting in front of his computer designing an object (a scene that is consistently reproduced in literature on design methods). In fact he discovers a social world of varying shapes and sizes. He thought he would have to implement certain methods and apply certain cognitive processes. In fact he finds himself having to negotiate and settle on compromises. He thought the procedure to be followed would be straightforward, starting with the specifications, but everything is complicated by new requirements defined by the order givers following the draft of a first solution. Indeed, the story we are about to tell concerns not only the design of a technical object but also the re-design of an apprentice engineer.

A Strange Supervisory Board

Many young engineers have probably discovered the same thing when starting out on their careers. Few of them, however, have had to deal with the same kind of supervisors as this student. What is more, the specific framework in which the work is done should be underlined. It provides the opportunity to discuss and analyze the trainee engineer's experience and find the terms to express what he sees and feels. The framework therefore has a lot to do with what the young engineer experiences.

His mechanical engineering studies are coming to an end when one of his lecturers, Jean-François Boujut, talks about the possibility of his pursuing a DEA[2] or even a doctorate. Involving research work, the DEA gives students an additional non-technical skill. It is also, the lecturer explains, an opportunity to step back from the operational work required by the PFE. But the most surprising announcement is that the proposed subject is to be co-supervised by a sociologist. The student is interested to discover that the mechanical engineering teams and the sociology teams in Grenoble are used to working together. However, since our student has only devoted himself to mechanical engineering throughout the course of his studies, he prefers to concentrate on this area and the technical work in hand at the beginning of his placement period. His tutors nevertheless ask him to take an observer's view of the project and closely follow the design process and the actions and interactions that it generates. To begin with, the trainee thinks that his observations are unrelated to his work as a designer. They involve a different part of his mission. This part is non-technical, and the trainee cannot really see what the aim of making such observations is.

The student, placed in a CERN design office in Geneva, is put under the direct responsibility of the head of the office, Bertrand Nicquevert. To the student's great surprise, his engineering school tutors and his "industrial" tutor seem to work hand in hand. They apparently get on really well and share the same opinion of the work he has been given. They say that it is an interesting opportunity to decode and analyze the design process. The student also discovers that Nicquevert holds a master's degree in philosophy. Not your usual mechanical engineer! And as if his supervisors were not an unlikely bunch as it was, Pascal Lécaille—an anthropologist writing a thesis on simulation tools—joins the group in one of the first supervision meetings.

The young engineer can only explain this strange group of supervisors by the interest they have in the other part of his mission, i.e., the social aspects and all the other factors surrounding the actual design work: the language barrier and the cultural differences of the people in the design office, the different age groups and the probable consequences of people retiring, and, finally, the behavioral and relational problems of the office personnel, especially the more senior designers with respect to their young manager. This set-up, in which the social factors are peripheral, external, or simply tacked onto the technical job in hand, is not to be called into question. However, as the design work progresses, we will discover a different way of looking at things, based on the people concerned, the way they react, and their different relationships. Indeed, the problem can only be defined, and a solution found, if these elements are taken into account. Hence, the sociologist's view of the mission does not fall entirely outside the scope of the technical work; it is up to him to try and understand the dynamics of the technical work.

A Simple Object in a Complex Environment

And so, fresh from his mechanical engineering school, the student settles down in the open-plan office. He is given a work surface and a computer console, like the other fifteen office members. With a mission to fulfill and a place to do it in, he thinks that he will be able to get along fine. At school he has learned to use the models, the catalogues of technical solutions, and the appropriate methods for each design phase. With all this learning under his belt, he should have no difficulty finding the right solution to the problem, making the calculations, and checking his work.

What is more, the technical part that he has to design is very simple. It is a wall, or more specifically a shielding disk, to separate two parts inside the ATLAS particle detector. On one side is the calorimeter (for measuring particle energy); on the other is a superconducting magnet. The shielding must prevent all particles other than muons from interfering with the measurement of the trajectographs called "muon chambers." The shielding must absorb photons and neutrons. For this, materials with high absorption rates for such particles (such as polyethylene or copper) are used.

To get to the bottom of the problem, the student starts by reading up about the entire system in which these shielding disks are to be placed. For a week, he concentrates on learning the technical terms relating to the detector. Using a document referred to as the Product Breakdown Structure, he identifies each of the parts, its name, its abbreviation, and its dimensions. He makes several sketches in his logbook for future reference. In doing this, he discovers just how complex the detector really is. He also discovers its impressive size and weight: 25 meters high, 40 meters long, 10,000 tonnes. The detector is to be used to determine the identities, energies, and directions of the particles produced during frontal collisions with proton beams. It is made up of detection and measuring instruments (a trajectograph, a calorimeter, a muon spectrometer), confinement and regulation parts (superconducting coils, cryostats), a range of electronic systems, and various supporting and structural parts.

Designing such a detector obviously involves a large number of people, institutions, and countries. In all, 1,700 physicists and engineers, some of whom can be considered "order givers," are taking part in the project. The "order givers" are the future users. Each element is being designed by a specific team. The CERN design office is one of these teams. As the head of the design office gives a quick overview of the ATLAS detector, he points to different parts of a technical drawing, saying "This is us here, and that's a team in England over there." Working in partnership with others is difficult even if it is routine. The whole thing requires complex coordination among senior managers within CERN and among various project steering committees. The technical complexity of the detector is thus matched by the organizational complexity of its coordination and that of the technical information system.

As the engineering student listens to the explanations offered by the head of the design office, he discovers that coordination among the designers and with the physicists is, in general, a central issue. Far from being a relatively closed space, the design office has a tight working relationship with numerous people from various institutions. The head of the design office talks especially about two categories of partners: the safety department (which is in charge of checking all the calculations for the sensitive parts) and the physicists (who are seen as dreamy idealists always wanting to go one step further without taking into account how feasible their ideas are, or at least that's what it looks like). The design office sometimes relies on the former to temper the wishes of the latter. Even within the office, the question of how to work together is often raised in connection with people's cultural differences and differences in age, and also in relation to where they are seated in the office. Indeed, in the middle of the office there is a row of large cabinets in which 30 years' drawings are kept. This row of cupboards physically divides the office in two. However, in the middle of the row there is a gap about half a meter wide. The head of the design office says: "See that? I've made a space between the cupboards. It was like bringing down the Berlin Wall." This witty remark goes some way toward explaining the reluctance of certain designers to cross the office.

Normally the student would not be concerned with all these coordination problems. The shielding disk that he has to design is just one small part of the whole assembly. He should be able to deliver a detailed draft design of the disk within 3 months. What is more, people hardly seem to be interested in this part. The physicists, for example, have turned their attention to other parts of the detector. The shielding is not seen as a "noble component," says the head of the design office, unlike the particle detectors. It is one of the "common components"—parts that go between instruments to accommodate fluids (e.g. for cooling), cables, and support structures. Indeed, if it were possible, the physicists would like these parts to disappear altogether. For them, cooling should be immediate and homogeneous without having to bother with all the tubing and extras. As for the framework supporting the instruments, this really takes up far too much room. These "common" parts are seen as cumbersome by the physicists who will be using the detector. The designers, however, covet these commons because they impose themselves as constraining boundaries. Of course, when a designer needs just a few extra millimeters, he takes them. The problem is that he is not the only one to do so. In fact, it is the head of the design office who is in charge of making sure all these parts fit together and who has to bring these coveted boundary areas of the system apparatus into existence.

The shielding disk on which the engineering student is working is one secondary part that nobody is really interested in. This is why he is under the impression that he need only analyze the problem and find the solution, without having to get into lengthy negotiations with the physicists. To him, the problem is purely technical. From the outset he knows that the space available for the disk's external dimensions is limited by the external dimensions of the surrounding parts. Using the drawings given to him by his colleagues in the design office, he studies these surrounding parts and their dimensions. He takes into account a few general rules relating to safety and ergonomics, so that the detector can be accessed for maintenance. Thus, his scope of action is limited by a multitude of specifications and requirements imposed by various people involved in the project.

Interactions between Objects

During the first days of his placement period, the head of the design office takes the young engineer around the various departments. He is introduced to many people, some of whom are working on issues directly linked to his own study, some of whom are not. He also takes part in technical coordination meetings that deal with project planning, fitting the various parts together, and the safety of their design. He feels that such meetings are just a matter of procedure.[3] Their aim is above all to check how the project is going and swap information. However, the people in the meetings argue about dates and about documents that haven't been handed over. This has nothing to do with the technical side of the project. It has to do with how projects and meetings are organized and managed.

However, as they go through the corridors from one department to the next, the trainee and his tutor come across various people with whom the tutor strikes up conversation about details regarding various projects that he is in charge of and which he needs to keep in mind. The student is astonished to see that part of the project's technical coordination takes place in the corridors. Like the canteen, the corridors are used as a forum for solving many problems.

The more people they meet, the bigger the trainee's list of contacts becomes, although so many names frighten him. The words of his tutor hardly reassure him. He thought he had come to carry out an engineering assignment first and foremost and, as a kind of sideline, act as an observer. But he discovers he actually has to communicate information, get answers to questions, and negotiate. He discovers then why he has been put in charge of designing the shielding. In fact his tutor had known that it would not be easy. He had even said as much right from the start but the young engineer, judging from the simplicity of the object to be designed, had put this to the back of his mind. When the head of the design office had agreed to take on the student along with his strange group of university supervisors, mechanical engineers rubbing shoulders with sociologists, it was because he thought that an outside view of the situation would help the head of the design office to understand what was in play.

And so the young engineer discovers that his poor shielding disk is the object of important stakes in terms of its functional definition. Indeed, it has to fulfill two functions: to stop particles and to bear the weight of the muon chambers. The problem is that the shielding is surrounded by various neighbors that have to be taken into account.

Moreover, the word "neighbor" is used both to talk about neighboring physical objects and to refer to the designers of such objects or the order givers. This is why people talk about negotiating with the neighbor when talking about the cryostat, for example. The number of neighbors involved is already quite considerable: several types of muon chambers, the vacuum chamber, the tile calorimeter, the cryostat, the toroid, the rails on which the system has to run, and the electronic data capture boxes. The trainee discovers, for example, that the electronics engineers in charge of designing the data capture systems have designed an enclosure that is too big and have thus reduced the shielding designer's room for maneuvering. In fact, he needs a clearance margin, as it is difficult to know the exact dimensions of the parts once they have been built. If he can't have the data capture box redesigned or moved, the trainee will have to plan a cutout in the shielding disk. And the physicists will probably not like this. Furthermore, it will reduce the disk's rigidity. As the shielding is at the center of a series of neighboring relations, it is an intermediate object; thus, the young engineer has to argue his case if he wants to get the amount of space he requires.

Ten or so neighboring objects mean ten or so teams or individuals to be contacted for data and information concerning their parts, along with all the associated constraints. On the other hand, the trainee discovers that he has to validate his own designs with these people. Some of them work in the design office, but not all of them. Sometimes the trainee has a CERN physicist with a listening ear to deal with; at other times he must deal with a renowned Italian physicist who is impossible to find, or with a Parisian team that does not answer his e-mail requests for information.

Technical or Strategic Work?

The trainee also learns that the work on shielding is strategic for the design office. Indeed, some of the technical parts, such as the muon chamber, are already the objects of dimensioning studies by work groups such as the Muon Layout Collaboration. In order to define the job at hand, the trainee bases his studies on a technical design report drafted by the Muon Collaboration. This document lists the technical features of the chambers and all the teams working on them. It defines the part of the enclosure that concerns him and in which the disk has to fit. Thus, for some of the parts, things have already been defined, and it would be difficult for the trainee to change them. This means that the design office has less room for maneuvering in the design of the structural parts of which it is in charge. The trainee realizes that his tutor has chosen this moment in the project to assign him to the shielding job so that they can have their say in the matter as early as possible. It is essential for the design office not to have to work with a part that is already joined to the rest. The office therefore has to fight to preserve a certain amount of freedom in its design work. Relations between the design office and its partners are as important as all the problems relating to borders, space, and margins.

It is only at this point that the trainee realizes how poorly prepared he was for this situation. He does not really know what kind of attitude he should adopt in this complex social world of hierarchies (related to the organization itself but also to the reputation of people), divisions, and territorial occupations. Therefore, for several weeks the trainee has put these facts on the back burner, preferring to concentrate on what he can do best: a technical job performed at a computer console. He has memorized the environment of the shielding disk from the drawings given to him by a neighbor at the office. He has redefined the enclosures so that he can accurately assess the space available and the margins for maneuvering. And he certainly needs these margins to be able to reinforce the shielding disk so that it can withstand the weight of the muon chambers. Finally, having concentrated only on the technical aspects, he has learned to know where he stands from a technical point of view; thus, he has developed a line of arguments to use with his neighbors if ever he should have to negotiate.

So he beavers away at his computer console, designing, imagining, and calculating. He checks the possibilities for adding reinforcements without disturbing the layout of the muon chambers. These reinforcements are necessary to prevent the disks from buckling under the weight of the chambers. After talking to his office neighbor, who is in charge of integrating the chambers, he drafts some ideas for fixing them to the disk. He discovers that this is the most delicate part of his own design work, as the loads to be borne are considerable and he has little space for adding the framework. Although two-dimensional design software would give him a good idea of the surrounding space available, it does not really give him an overall view.

Having worked with three-dimensional simulation and viewing tools at his engineering school, he decides to use his training period to put one of the software programs to the test. Using it, he is able to show how the muon chambers and shielding disks fit together. (His mechanical engineering tutors are interested but not entirely convinced. They prefer working with concrete analyses rather than such calculation tools.) Next, the trainee designs a framework able to fit into the space available. He simulates various calculations of the framework with different diameters and materials so that he can get a realistic idea of the mechanical stress. He discovers that the framework will not be rigid enough unless it is closed at both ends.

For several weeks, he concentrates on the design of this framework, working in an environment that seems increasingly restrictive and hostile owing to the dimensional constraints and the problems of accessibility: little space available, the need to leave clearance for the detectors to be opened, and the overall suitability of the assembly. There are so many geometric constraints that his first concern is to find a solution that is able to fit inside the space available. While devoting all his time and energy to this problem, he is also able to build up good professional relations with his colleagues in the office. He discovers that everyone there has had to forge a place for himself. The space-related problems of everyone he meets when working on its project are reflected in the design office itself.

Having discovered the importance of "neighbors" when working on a design issue, our student undertakes to list them all, both the technical parts and the people working on them, along with the questions that he would like to ask them. The list is long indeed, and it gets longer since the very notion of neighborhood has to be revised. Before this, it was defined in terms of spatial proximity. It referred to "elements" that may or may not be in (physical) contact with the disk, but are not separated by another element. Perhaps it is his mechanical engineering background that has so far restricted his field of vision. There are in fact several kinds of relations among objects; geographical proximity is not the only one. Indeed, radiation goes through various parts of the detector along with heat, magnetic fields, vibrations (e.g. earthquakes), and gravity. (The detector may be symmetrical, but gravity does not see it that way.) The toroids in the detector generate an intense magnetic field that tends to cause the elements to come together. The magnetic field exerts a force and then checks whether this force affects it or not (or rather whether it affects the shielding). Added to this is the question of maintenance access to the detector. All these forms of interaction can bring distant elements within the system in contact with one another. Drawing up a list of these elements along with the people working on them seems to be the only way forward. In doing this, the trainee is in fact trying to identify and target all the neighbors with the biggest influence on his design work. Defining each one's territorial position (who does what and up to what point) now seems increasingly important to the trainee, as it will allow him to define his own technical work. Moreover, the breakdown of roles played does not now seem as clear as when he started. And so, having worked hard on the technical side of things, the student discovers how essential it is to be able to socially decode relations if he wants to complete his mission successfully. In other words, he has to ask himself who does what and how far is it possible to negotiate. He finds out that certain elements cannot be negotiated, as changing them would put them back on the drawing board. Thus, the trainee comes to analyze the interactions between people, the recurrent nature of certain practices, the rules applied, and the possible interference of all this in his work.

Stabilizing What the Neighbors Want

The trainee also comes to realize that the demands of each person involved are not always clear and are far from stable. The shielding is supposed to fulfill two functions, but when he takes the various neighbors into account he realizes that things are much more complicated. Each neighbor has his own expectations and requirements, only some of which have been put down on paper. What is more, under the instigation of the head of the design office, the trainee has begun a functional analysis.[4] This requires listing and quantifying all the functional features of the disk, with the aim of discontinuing to work on assumptions alone. For example, the physicists say that the shielding should be 100 millimeters thick and made of iron, with a copper base. But why? Which physicist decided that? And on what basis? Where are the data that led these physicists to give such specifications 3 years earlier? Would they say the same thing today? In fact, all these so-called constraints have to be studied again, and the people who defined them must be found and asked why they said what they said. It is no longer possible, at this stage, to continue to rely on the available technical data. It would be better to find out the logical reasoning behind the orders given and whether it is possible to re-negotiate. For example, just how far can the basic functions of the detector be revised? What seemed to be finalized is perhaps not. And so, after 4 months, after the student has done his design work on the supports for the muon chamber, the physicists decide that the way the chambers are mounted does not satisfy them. After viewing the assembly, they realize the need for maintenance access. The support function thus becomes even more complex, requiring the addition of a new structure that is mobile in relation to the disk. All the design work on the direct support of the chambers has to be reviewed. The concept of the mobile structure and that of its supporting copper base have to be validated at the same time. And yet, the young engineer has already spent several weeks and much energy finding a solution. Bringing to light a new element has led to a whole array of fresh constraints calling into question the initial concept. The trainee begins to wonder what he can base his work on. On top of this, he discovers that certain neighbors have taken up more space than was planned simply because they were not aware that neighboring elements had to be taken into account. As far as they were concerned, their elements were surrounded by emptiness. It is easier for members of the design office with the job of integrating the different elements to understand these "neighborhood relations" than for a subcontractor of a distant part.

Now that the neighbors have been identified, the next problem is getting them to talk. Would it be possible, and enough, just to get them around a table? Some of them come to CERN only once a month. Of course, our trainee engineer has neither the power nor the authority to convene them in a meeting. The head of the design office does not have authority to do so either, in view of the number of people involved in the project.[5] So the young engineer decides that the best thing to do is define a certain number of elements himself and draw up the specifications that the physicists should have drawn up in the first place. These would then be submitted with the aim of getting them to react and thus define their needs. It is at this point that he discovers how foreign the "culture of a specifications sheet" (Bertrand Nicquevert's words) is to physicists. They shy away from the idea. They think that if they put their expectations down in writing they will no longer have the power to change them. For the head of the design office, on the other hand, the act of writing them down will force the physicists to express their needs, even if they will have to be modified later. If this is not done, they will be defining a solution to a problem whose terms are unknown.

To begin with, the design work consists in studying each problem one after another. Little by little, the idea that it is necessary to have an overall view, and not just a technical one, emerges. Different people work on each technical element, and it is essential to know exactly what they want and how far it is possible to negotiate with them.

The trainee thus submits his solutions to his neighbors. The drawing of the disk is sent to a physicist so that, through simulation, he can check whether it is acceptable in relation to his needs (i.e., particle absorption). A proposal for modifying the cryostat cover is faxed to the Orsay team in charge of its design. Within the design office, showing drafts of drawings to different colleagues during lunch or in an informal context produces some interesting reactions. It allows the young engineer to see that work with each partner is carried out differently and requires different approaches. At times, the design proposed is provisional, insofar as an unhurried colleague is expected to provide some data. At other times, the engineer has to wait for a reaction to the proposed modifications to a particular part. The assistant technical coordination manager is soon to leave for the United States for a meeting where he should have a chance to raise the question of the muon chamber. A file has to be prepared for him, and he has to be persuaded to bring up the matter. The problem here is that he comes to the office only once every 2 months. Negotiations depend on mediators whose logical approach is not always fully understood by the members of the design office. For some neighboring elements (such as the calorimeter), negotiation is easier, as the colleagues involved work at the Geneva site. For each neighboring element, and hence for each neighbor, there is a specific coordination procedure. In this way, the young engineer comes to understand the interest of the head of the design office in having an outside view of the situation.

Finally, it is interesting to see how simply working on an element such as the shielding draws attention. It has become a subject of interest, enabling questions to be raised sooner rather than later when later would in fact be too late. The person in charge of the muons (a physicist and a close guardian of the muon spectrometer specifications) did not want to get involved in things to do with "services" or "common parts." And yet the design office needs answers to a certain number of questions. Indeed, the more questions it asks the more seriously it will be taken. If no one bothered to do this, the shielding would just become a kind of black hole, a bin for all the neighbors to throw their unresolved problems into. After all, the designers are bound to find a solution later on. Having questions raised with the project actors was, in fact, one of the objectives of the head of the design office when he took on the trainee. He later explained his numerous expectations in relation to his various responsibilities as follows:

As project engineer for the traction system, I didn't have anybody to argue for the shielding disk. Doing the design within the design office was going to enable me to monitor its compatibility with the entire assembly.

As a member of the ATLAS Technical Coordination team, where I am responsible for mechanical integration, the shielding disk presents a number of unlikely neighbors (some of which were only discovered through Grégoire's work). It is at the center of numerous problems but there is nobody to deal with them and the initial design plan was far too succinct. It was essential to have somebody prepared to dig further into this design. But most of all, from a sociological point of view, it was the ideal opportunity to study the dynamics of the design process through a physical experiment.

As a mechanical engineer, and one that works at CERN, there were several small mechanical challenges: calculating the disk, the support, etc. But what I hadn't banked on was that this object would take on a new life, thanks to the initiative taken and the work put in by Grégoire. One of the amusing consequences of this situation is that today I am being offered the responsibility of the shielding disk as project engineer, which is something I wasn't expecting to begin with.

As a philosopher/epistemologist, my questioning centers around technical issues.... There are many scientists who like to dabble in philosophy, but there are significantly fewer engineers. The latter are "much more aware of the material aspects of a technical issue" (O. Lavoisy). Following Galison's example, I'm hoping to be able to go further into the question of relations between theoretical physicists and experimental physicists but on the triple basis of "theory/experiment/instrument" as opposed to the traditional epistemological dual basis. Furthermore, using an engineer studying for a DEA, supervised by a human sciences committee, was the ideal opportunity for understanding this area in a much more structured way. It is also an opportunity for me to understand my own work as an engineer and what is being done in the area of How Experiments Begin?[6]

The work of the design engineer turned out to be considerably different from what the young engineer had imagined. He had thought that it was just a matter of finding the right solution to the problem in hand by applying the models and methods he had learned in the course of his studies. He knew from the start that he wouldn't be able to base all his work on these existing methods and that he would have to invent new ones, but he certainly didn't think he would have to go so far.

Operational Summary

1. Design work is complex, even for a simple object   Designing a technical part, however simple, can quickly prove to be complex when the part in question lies at the center of a whole system and is linked to a certain number of other technical parts.
2. The design work builds up around a network of relations among technical parts   Designing an object involves taking into account a series of other objects, which are not always in direct contact with this object. These are related to one another; however, the way they are related is not always known at the start, and does not necessarily become clear during the design process. To define the specifications, the designer must describe this network of relations among technical elements and must go through it regularly to check on changes made.
3. Objects and their relations are linked to people and social groups   These objects can be taken into account only if the designer knows these people or groups (i.e. who orders, who designs, and who uses), their relations, and the logic behind what they decide and what they do. Of course, this demands precise attention and a decoding ability to prevent judgment from being based on simplistic analyses at the beginning of the placement (saying that problems are due to people or technical ideals, etc.). There are actors behind each technical element, and they act as spokespersons for these elements. The elements are at the center of these people's interest.
4. It is not always clear at the beginning what all the constraints are   They are gradually revealed as their relations with other elements are explored. It is not possible to have them at the start, notably because the actors themselves do not really know them. The process of designing solutions and making them viable through drawings leads the actors concerned (or order givers) to talk about the requirements that they would like to see fulfilled. Bringing the intermediate objects into existence is therefore an important step that will help the actors to express their needs.
5. It cannot be taken for granted that requirements and technical data are given objectively   In other words, final judgment must be withheld, as the information available may be misleading. It is better to understand how the data are put together (socially and technically) and then regularly test how stable they are.
6. Showing interest in an object gives it life   Working on it, drawing it, and circulating the drawing helps to awaken the interest of the various people involved, to position work in relation to it, and to demonstrate responsibility for it. It also helps those who draw up formal requirements. If no one is interested in an object, it cannot live.
7. To manage relations between technical elements, the designer has to take into account how the actors react and behave in relation to their specific element   Taking into account people and groups means first of all examining how they act, both socially and physically, especially when they have to interact with others.

8. *Doing technical work is just one strategy among others* Concentrating on "technical" work, such as entering and processing information using calculation software and CAD, is sometimes seen as a good strategy that can help the designer to report on the situation, his position, what he would reasonably like to obtain from his colleagues, and his margins for maneuvering.

9. *Industrial design stimulates discussion* Industrial design is not only a technical means of viewing objects during their design; it also stimulates discussion between designers and other project partners.

Notes

This chapter sums up work by Grégoire Pépiot, Jean-François Boujut, Pascal Lécaille, and Bertrand Nicquevert. Grégoire Pépiot is a mechanical engineer studying for a research diploma in industrial engineering under the supervision of Jean-François Boujut (a mechanical engineer) and Dominique Vinck (a sociologist). Bertrand Nicquevert heads a technical office in the Experimental Physics Division of CERN in Geneva.

1. As a part of their studies, young engineers are placed with companies. The idea is that this gives them an opportunity to put what they have learned into practice.

2. A DEA (Diplôme d'Etudes Approfondies) is a one-year postgraduate research diploma. It can be done at the same time as a PFE (Projet de Fin d'Études—a placement project carried out by engineers in their final year of study).

3. The head of the design office does not share this opinion. On the contrary, he says that there are very few procedures involved.

4. In fact, he didn't think it was up to him to define the various wishes of the physicists.

5. The head of the design office says that this is not really a problem since nobody has overall authority over the people involved in the project. Indeed, the way physical research is organized at the end of the 20th century is based on partnership, which means going through a long series of discussions to reach a consensus about what is possible (mechanically, geometrically, and perhaps financially).

6. This harks back to the title of Peter Galison's book *How Experiments End*.

21 "The Naked Launch: Assigning Blame for the Challenger Explosion"

Harry Collins and Trevor Pinch

In addition to the social complexities presented in the Vinck chapter, there are a number of technical uncertainties in engineering as well. Engineers and scientists have developed a wealth of information about how materials and machines behave in the world. But we do not—and perhaps cannot—understand every detail. When investigators and the media searched for the cause of the tragic failure of the 1986 space shuttle Challenger launch, many assumed that all the variables were well understood and a reckless decision had been made. This chapter—a summary of Diane Vaughan's account of the disaster in *The Challenger Launch Decision* (University of Chicago Press 1996)—offers a different explanation and a window into the technical complexities of large systems. It argues that engineers worked hard to reduce uncertainties, but could not possibly have eliminated them. Not every technology is so complicated that it can only be built and maintained by "rocket scientists," but because it is impossible to test every "real-world" situation, there will almost always remain some technical uncertainties in every technology. It is the responsibility of engineers to manage this uncertainty and make informed decisions to minimize risk—without forgetting that risk is always present. Collins and Pinch argue that perhaps the biggest mistake NASA made was to downplay the uncertainty involved in space flight itself.

We always remember where we were when we first heard about a momentous event. Those over forty-five years old know what they were doing when they heard that John F. Kennedy had been assassinated. Similarly, anyone who was watching television remembers where they were at 11:38 a.m. Eastern Standard Time on 28 January, 1986 when the Space Shuttle *Challenger* exploded. The billowing cloud of white smoke laced with twirling loops made by the careering Solid Rocket Boosters proclaimed the death of seven astronauts and the end of the space programmme's "can do" infallibility.

Unlike the inconclusive Warren Commission that inquired into Kennedy's death, the Presidential Commission chaired by William Rogers soon distributed blame. There was no ambivalence in their report. The cause of the accident was a circular seal made of rubber known as an O-ring. The *Challenger*'s Solid Rocket Boosters were made in segments, and the O-rings sealed the gap between them. A seal failed and the escaping exhaust gas became a blow torch which burned through a strut and started a sequence of events which led to the disaster.

From *The Golem at Large: What You Should Know about Technology* (New York: Cambridge University Press, 1998), pp. 30–56. © 2002 Cambridge University Press. Reprinted with the permission of Cambridge University Press.

The Commission also revealed that the shuttle had been launched at unprecedentedly low temperatures at the Cape. Richard Feynman the brilliant, homespun American physicist is often credited with the proof. At a press conference he used a piece of rubber O-ring and a glass of iced water to show the effect of cold on rubber. The rubber lost its resilience. Surely this obvious fact about rubber should have been known to NASA? Should they not have realized that cold, stiff rubber would not work properly as a seal? Worse, it emerged that engineers from the company responsible for building the Solid Rocket Boosters, Morton Thiokol, had given a warning. At an impromptu midnight teleconference the night before the launch, they argued that the O-rings would not work in the bitter cold of that Florida morning. The engineers, it transpired, had been overruled by their own managers. In turn the managers felt threatened by a NASA management who expected its contractor to maintain the launch schedule.

In 1986 the production pressure on the shuttle was huge. A record number of flights—fifteen—had been scheduled for the supposedly cheap, efficient, and reusable space vehicle. Among the cargoes were major scientific experiments including the Hubble Space Telescope. The launch preceding *Challenger* had been the most delayed in NASA's history with three schedule "slips" and four launch pad "scrubs." The ill-fated *Challenger* had already been delayed four times and the programme as a whole needed to achieve a better performance to fit the fiscal constraints of NASA in the 1980s. In the glory days of the Apollo moon landings the "Right Stuff" had been matched by the right money, but no more.

The conventional wisdom was this: NASA managers succumbed to production pressures, proceeding with a launch they knew was risky in order to keep on schedule. The *Challenger* accident is usually presented as a moral lesson. We learn about the banality of evil; how noble aspirations can be undermined by an uncaring bureaucracy. Skimp, save and cut corners, give too much decision-making power to reckless managers and uncaring bureaucrats, ignore the pleas of your best scientists and engineers, and you will be punished.

The *Challenger* story has its victims, its evil-doers, and two heroes. Aboard the shuttle was a school teacher, Christa McAuliffe, who represented ordinary people; she was one of us. The evil-doers were managers, both at NASA and at Morton Thiokol. One hero was Richard Feynman, who needed only a couple of minutes and a glass of cold water to establish what NASA had failed to learn in fifteen years. The other hero was the whistle-blowing, Morton Thiokol engineer, Roger Boisjoly. On the eve of the launch, fearing a catastrophe, Boisjoly tried desperately to get his company to reverse the launch decision. After the Rogers Commission vindicated Boisjoly, he took up the fight against Morton Thiokol and NASA in a billion-dollar law suit, alleging a cover-up. Boisjoly is the little guy taking on the Government and corporate America.

After the event it is easy to slot the heroes and villains into place. It is harder to imagine the pressures, dilemmas, and uncertainties facing the participants at the time that the fateful launch decision was made. We have to journey backwards in time to recapture just what was known about the O-rings and their risks before the launch.

Before we start, note how hard it is to sustain the image of amoral, calculating managers causing the crash. NASA managers and bureaucrats were under pressure to get the *Challenger* launched; yes, they knew as well as the next person that the endless delays put in jeopardy the image of the shuttle as a reusable and efficient space vehicle; and, yes, it would be a publicity coup for the hard-pressed space programme to get the Challenger launched in time for "school teacher in space," Christa McAuliffe, to link up live with Ronald Reagan's forthcoming State of the Union address; but, why risk the space programme as a whole and their own futures for a matter of a few hours of scheduling? As George Hardy, then NASA's Deputy Director, Science and Engineering Directorate, Marshall Space Flight Center, told the Presidential Commission:

I would hope that simple logic would suggest that no one in their right mind would knowingly accept increased flight risk for a few hours of schedule. (Quoted in Vaughan, p. 49)

Safety has to come high in any manager's calculation of risks and benefits however selfish and amoral their intentions.

New research on the *Challenger* launch decision by Diane Vaughan shows that the dangers of the O-rings were not ignored because of economic or political pressures. They were ignored because the consensus of the engineers and managers who took part in the fateful teleconference was that, based on the engineering data and their past safety practices, there was no clear reason not to launch that night. With twenty-twenty hindsight we can see they were wrong, but on the night in question the decision they reached was reasonable in the light of the available technical expertise.

O-Ring Joints

A Solid Rocket Booster (SRB) works by burning solid fuel and oxygen. The shuttle SRBs are one hundred and forty nine feet tall, just shorter than the Statue of Liberty. They burn ten tons of fuel per second, which helps "boost" the shuttle off the launch pad. Huge pressures of hot gas build up inside each booster, only relieved as the exhaust rushes out of the fireproof nozzle, providing lift. This exhaust gas jet is a potent force quite capable of melting metal. It is vital that it does not escape from anywhere except the place designed for it.

It is much easier to load solid fuel propellant into a booster which can be broken down into sections. Each booster on the shuttle consists of four large cylindrical sections plus nose and nozzle. The sections are built at Thiokol's base in Utah, transported separately, and joined together at Kennedy. The joints between each section have to be specially designed to withstand the enormous pressures inside. During the first fraction of a second of launch, when the boosters first explode into life, each cylindrical section barrels outward causing the joints to bend out of true—a phenomenon known as "joint rotation" (figure 21.1). The barrelling outwards occurs because the joints with their supporting collar are much stiffer than the metal rocket casing above and below them.

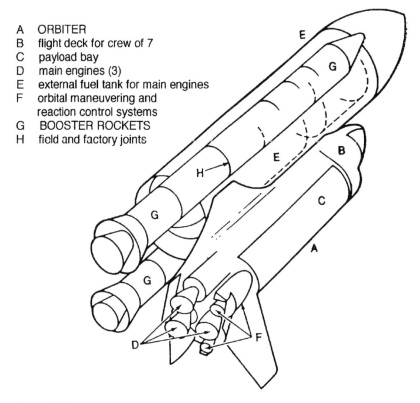

A ORBITER
B flight deck for crew of 7
C payload bay
D main engines (3)
E external fuel tank for main engines
F orbital maneuvering and reaction control systems
G BOOSTER ROCKETS
H field and factory joints

Figure 21.1
Space Shuttle components.

The joints used are known as "tang and clevis" joints. The cylindrical sections of the booster stack one on top of the other with the bottom lip of each section (the tang) fitting snugly inside the top of the section below in a special pocket just under four inches long (the clevis) (see figure 21.2). Each joint is wrapped by a steel collar and fastened by 177 steel pins. The joints are sealed by two rubber O-rings nestled in grooves specially cut on the inside of the clevis in order to prevent hot gases escaping during the fraction of a second of joint rotation.

An O-ring looks like thirty-eight feet of licorice, $\frac{1}{4}$ inch in diameter and formed into a loop. When the booster segments are stacked the O-rings are compressed (O-ring "squeeze") to seal the tiny gap. A special putty based on asbestos protects the O-rings from the hot propellant gases. To test the joints before launch a leak check port is located between the O-rings. It is vital to make sure that the rings have seated correctly after the booster sections have been stacked. Air is blown between the two rings and the pressure is measured to check for leaks. Ironically this pressure test was later found to

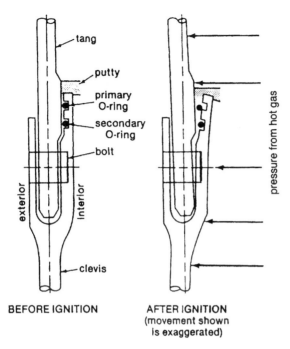

Figure 21.2
Joint rotation.

endanger safety because the air produced blow holes in the putty which the hot gases would seek out and penetrate, eventually reaching the O-ring seals and abrading them.

The joint between sections, like many components on the shuttle, is a lot more complicated than it looks. Even the question: "What is the exact size of the gap between each section during launch?" turns out to be difficult to ascertain. The gap when the booster is stationary is known to be about four thousandths of an inch. But during ignition this gap expands by an unknown amount because of joint rotation. The gap opens up for only about six-tenths of a second. This may not sound like a very long time, but it is, as one engineer stated "a lifetime in this business." (Quoted in Vaughan p. 40.) Finding out what exactly happens during this six-tenths of a second takes an extraordinary amount of engineering effort. At the time the *Challenger* was launched the best estimates for the maximum size of the gap varied between forty-two hundredths of an inch and sixty hundredths of an inch.

The Design and Testing of the SRB Joints

None of these difficulties were apparent in 1973 when Morton Thiokol won the contract to build the SRBs. The design of the joints was based upon that of the very dependable Titan rocket. The joint was believed by everyone to be highly reliable. The Titan had only one O-ring. For added safety on the shuttle a secondary O-ring was added as back-up for the primary seal. The secondary seal provided "redundancy." NASA aimed for redundancy with as many components as possible, and particularly with ones where a failure would lead to losing the whole mission. Obviously not all components can have redundancy; if the shuttle lost a wing this would be a catastrophic failure, but no-one suggested that the shuttle should have a back-up wing.

Both NASA and its contractor, Morton Thiokol, took responsibility for the design and testing of the boosters and joints and seals. The tests were often carried out in parallel at Utah, where Thiokol was located, and at NASA's rocket engineering centre—the Marshall Center, Huntsville, Alabama. Marshall was the jewel in NASA's crown. Its first director was the legendary German rocket scientist Wernher von Braun. He had established Marshall as the centre of technical excellence upon which the success of the Apollo programme was based. With this proud history, Marshall considered its expertise and facilities to be superior to anything Morton Thiokol could offer. The engineers at Marshall had a reputation for being conservative and rigorous; they saw it as their job to keep the contractor honest by trying to "shoot down" their data and their analyses. At Thiokol the Marshall people became known as the "bad news guys." Thiokol's engineers favoured more practical options and were known to get "defensive about their design," especially when under attack from Marshall. The differing attitudes of the two groups sometimes made for tense negotiations and led to some long-running disputes.

Both groups, of course, used the best science and engineering available to them. If there were technical conflicts they would try and resolve them until they both had the same results confirmed by multiple methods, multiple tests, and the most rigorous engineering analysis. But, as we shall see, rigorous engineering science, like golem science, is not always neat and tidy.

Early on the problem of joint rotation was recognized by both sets of engineers, but they differed in their estimates of its significance. Thiokol engineers calculated that, on ignition, the joint would close. Marshall engineers did not agree; their calculations showed that the joint would momentarily open. The consequences of the joint opening would be two-fold: (1) the compression or "squeeze" on the O-ring would be reduced and make it a less reliable seal; (2) an O-ring could become unseated. Ironically in *this* disagreement Thiokol and NASA engineers took exactly the opposite positions they would take on the night of the *Challenger* launch. In this case it was NASA who expressed reservations about the joint while Thiokol was confident the joint would perform as the design predicted.

The Hydroburst Test

To try and resolve matters a test was devised. The hydroburst test shoots pressurized water at the joint to simulate the pressures encountered at launch. It was conducted at Thiokol, in September 1977, where a joint was run through twenty pressure cycles—each cycle simulating the pressure at launch. The results indicated that NASA was right and that the joints could open for a short duration during ignition, blowing the seals out. Thiokol agreed with Marshall that the joint opened but they did not think that this threatened failure because the test was unrealistic. They had two grounds for challenging how well the test simulated an actual flight. Firstly, in a real flight, an O-ring would experience ignition pressure only once, not twenty times, and the data showed that the rings had worked perfectly for the first eight out of the twenty trials. Secondly, the test was carried out with the booster lying on its side. In use the booster would be vertical. They blamed the leaky joints on distortions in the horizontal rocket produced by gravity. Thiokol were so confident after this first test that they saw no need for any further tests. Marshall engineer Leon Ray, however, disagreed and insisted on more.

Similarity and Difference

The problem the engineers are facing here is a variant of one raised by the philosopher, Ludwig Wittgenstein. Whether two things are similar or different, Wittgenstein noted, always involves a human judgement. We make such judgements routinely in our everyday lives. Things appear similar or different depending on the context of use. When, for instance, we see the familiar face of a friend from varied angles and under varied lighting conditions we have no trouble in saying *in the context of everyday recognition* that the face we see is always the "same" face. In other words, we routinely treat this as a matter of *similarity*. If the context is, however, that of fashion photography, then crucial *differences* in the facial features emphasized in each photograph under varied lighting conditions or camera angles might become important. In this context we might routinely treat this as a matter of *difference*.

The same considerations apply to technological testing. Because most tests only simulate how the technology will be used in practice, the crucial question in judging test outcomes becomes: how similar is the test to the actual use? Morton Thiokol's interpretation of the hydroburst test was that the twenty ignition cycles test was sufficiently *different* to the actual use of the shuttle that it was not a good guide to what would happen during a real launch. NASA's position, on the other hand, was that the tests were *similar* to what might occur in a real launch. Therefore, for NASA, the tests were valid indications of problems with the seals.

As mentioned above, NASA–Marshall was well known for its conservative engineering philosophy. Thiokol engineers felt it had to be balanced with the need to produce a practical design:

You take the worst, worst, worst, worst, worst case, and that's what you have to design for. And that's not practical. There are a number of things that go into making up whether the joint seals or doesn't seal or how much O-ring squeeze you have, and they took the max and the min which would give you the worst particular case... all those worsts were all put together, and they said you've got to design so that you can withstand all of that on initial pressurization, and you just can't do that or else you couldn't put the part together. (Quoted in Vaughan, p. 99)

With these underlying differences in design philosophy, and the interpretative loophole provided by the need to make similarity or difference judgements, it is not surprising that tests alone could not settle the performance of the joint.

More Testing

More tests on the joints were carried out. These new tests, however, only made matters worse:

Continued tests to resolve the disagreement simply added to it. Marshall and Thiokol engineers disagreed about the size of the gap that resulted when the joint rotated. Although joint rotation was not necessarily a problem, the size of the gap mattered: if a gap were sufficiently large, it could affect the sealing capability of the rings. But tests to determine gap size continued to produce conflicting results.... Both sides ran test after test, but the disagreement went unresolved. (Vaughan, pp. 100–101)

One of these tests, known as the Structural Test Article, simulated the pressure loads expected on the cylindrical sections of the shuttle during launch. Electrical devices were used to measure the amount of joint rotation. The results indicated that the problem was even worse than Marshall had first suspected. The size of the gap was sufficient for both O-rings to be out of position during ignition. The primary would be blown out of its groove as the pressure impacted on it, but the secondary would be left "floating" in its groove, unable to seal should the primary fail later.

Thiokol again challenged Marshall's interpretation. In their view there was a problem not with joint rotation but rather with the electrical devices used to measure joint rotation. They maintained that these electrically generated measurements were "off the charts" compared to their own (for them, obviously better) physical measurements of rotation which they had carried out during the test. They concluded that the calibration of the electrical instruments must have been wrong. And, since their own physical measurements indicated a smaller gap, the secondary would actually be in a position to seal.

Once again we are encountering the ambiguity of test results. Now the issue is another aspect of the human practices at the heart of golem science. The issue is that of *experimenter's regress*—a term introduced in *The Golem*. This refers to the catch-22 in research frontier experiments where the outcome is contested. The 'correct' outcome can only be achieved if the experiments or tests in question have been performed competently, but a competent experiment can only be judged by its outcome.

The logic of the situation is like this: what is the correct outcome of the test? Is there a large or small gap? The correct outcome depends on whether there is a large or small gap to detect. To find this out we must design a good test and have a look. But we don't know what a good test is until we have tried it and found that it gives the correct result. But we don't know what the correct result is until we have built a good test... and so on, potentially *ad infinitum*.

For NASA, the good test is the one with electrical data which supports their view that the gap is large enough to cause problems with the seals. For Morton Thiokol the good test is the one based on the physical measurements which supports their view that the gap is small and the seals will function as expected. Confidence for both also came from belief that their measuring equipment was the "most scientific" and therefore produced the "most accurate" and "best" result.

Make Sure the Thing's Going to Work

Usually the experimenter's regress is soon broken by bringing in other considerations. If the two sets of engineers could not agree over the gap then perhaps another party could provide the necessary certainty. A third party with the requisite skills and familiarity with the O-rings was called in—the O-ring manufacturers. Unfortunately, in this case they could not resolve the issue either. They confirmed that the gap size was larger than industry standards and noted that the O-ring was being asked to perform beyond its intended design. But they then passed the buck back to the engineers, calling for yet more tests to be carried out which "more closely simulate actual conditions" (Vaughan, p. 103). As NASA's Leon Ray recounted:

I made the presentation to their technical folks, and we told them, the joints are opening up, and here's how much. What do we do? We've got the hardware built, we're going to be flying before long, what do you think we ought to do? And those guys said, well, that's a tough one, a real tough one. I feel sorry for you guys, I don't know what you're going to do, but we will study it over and give you an answer.... Later on they both wrote letters back to us and said, hey, you've got to go with what [hardware] you've got. Do enough tests to make sure the thing's going to work. Now that's their recommendation. Go on and do some tests, and go with what you've had. That's all we can tell you. So we did. (Vaughan, p. 103)

Although a redesign of the joint was contemplated by Ray, at this late stage it would probably have badly delayed the shuttle programme. Also, all participants, including Ray, felt that the joint they had could be made to work; they just weren't certain it would work in all circumstances.

It is wrong to set up standards of absolute certainty from which to criticize the engineers. The development of an unknown technology like the Space Shuttle is always going to be subject to risk and uncertainties. It was recognized by the working engineers that, in the end, the amount of risk was something which could not be known for sure. They would do their best to make sure it was "going to work." Marshall's Larry Wear puts the attitude well:

Any airplane designer, automobile designer, rocket designer would say that [O-ring] seals have to seal. They would all agree on that. But to what degree do they have to seal? There are no perfect, zero-leak seals. All seals leak some. It's a rare seal that doesn't leak at all. So then you get into the realm of, "What's a leaking seal?" From one technical industry to another, the severity of it and the degree that's permissible would change, you know, all within the same definition of seals.... How much is acceptable? Well, that gets to be very subjective, as well as empirical. You've got to have some experience with the things to see what you can really live with.... (Vaughan, p. 115)

And it should be borne in mind that the O-ring seal was just one of many components on the shuttle over which there was uncertainty.

Testing for Worst Scenarios

A new strategy to resolve the dispute over gap size now emerged. What mattered most was not the exact size of the gap but whether the O-rings would actually work. Attention was switched to testing the seals' efficiency when gap sizes were much larger than either NASA or Thiokol expected. This time agreement was reached and, despite violating industry guidelines, the primary was found to seal under conditions far more severe than anything expected during a launch. As a further test, the performance of the secondary seal was also examined. A primary seal was ground down to simulate seal erosion and was tested at ignition pressure. In these circumstances, the secondary was found to seal, ensuring that the joint had the desired redundancy.

The two groups next tried to make the joints as tight as possible, producing maximum O-ring "squeeze." They did this by making the O-rings larger, ensuring they were of the highest quality and by "shimming" the joints—wedging thin wafers of metal inside the joints to make them still tighter.

It was at this point, in July 1981, that Roger Boisjoly joined the Thiokol team. As was mentioned above, the amount of O-ring squeeze obtained was below the industry standard of 15 per cent (the minimum amount by which the rubber should contract under the pressure of the joint). Leon Ray for NASA and Roger Boisjoly for Thiokol worked closely together to reach a value acceptable to both sides. As Arnie Thomson, Boisjoly's supervisor, reported:

He [Roger] had a real good experience, hard, tough, head-knocking experience at first with Leon [Ray], because both happen to be vigorous people, and both of them, you know, stand for what they believe, and after, oh, gee, I guess it would be seven or eight months of difficult conversations, Roger and Leon came to an agreement on... about 7.5 percent squeeze [initial compression]... that was negotiated, you know, that we would never go below [that], and we would in fact select hardware [shims] so that that would not happen. (Quoted in Vaughan, p. 104)

Ray, on behalf of NASA, was reported to be happy with the joint. When the joint was tested he is quoted as saying:

We find that it works great, it works great. You can't tell any difference. You don't leak at $7\frac{1}{2}$ percent. (Quoted in Vaughan, p. 104)

Ray went on to summarize his views:

> We had faith in the tests. The data said the primary would always push into the joint and seal. The Titan had flown all those years with only one O-ring. And if we didn't have a primary seal in a worst case scenario, we had faith in the secondary. (Quoted in Vaughan, p. 105)

Both engineering communities acknowledged that the joint did not work as per design, but they felt that it worked sufficiently well that it was an acceptable risk. Both groups now agreed that the primary would seal, and both asserted that the joint had redundancy (i.e. the secondary would act as a backup). But they had rather different understandings of redundancy. NASA still believed there was a larger gap than Thiokol believed. NASA felt that there would be redundancy at initial ignition but not in the WOW (Worst on Worse) situation such as a primary O-ring failing late in ignition when the secondary O-ring might not be in position to seal. Thiokol, with their smaller gap measurements, felt that there would be redundancy at all times as the secondary would always be seated in the groove ready to seal. As Boisjoly remarked:

> In all honesty, the engineering people, namely Leon Ray...and myself, always had a running battle in the Flight Readiness Reviews because I would use the 0.042" [gap size] and they were telling me I was using a number too low, and I would retort back and say no, the horizontal number [Marshall's 0.060" gap size] doesn't apply because we really don't fly in a horizontal position. (Quoted in Vaughan, p. 106)

The Thiokol and Marshall designation of the joint as "an acceptable risk" was now officially endorsed by the NASA bureaucracy and the shuttle was made ready for its first flight. The joints had passed several formal bureaucratic risk evaluations as part of the process of certification.

First Flight of the Shuttle

Now the ultimate test of the SRB joints could take place—the first flight. Unlike in all previous tests, the gap between the *test* and the *actual technology in use* could be finally closed.

On 12 April 1981, the first Space Shuttle was launched. Two days later, after orbiting the earth thirty-six times, the shuttle landed safely at Edwards Air Force Base. The two SRBs were recovered from the ocean and disassembled. No anomalies were found in the joints; they had behaved exactly as the Marshall and Thiokol engineers had predicted.

1981–1985 Erosion and Blow-by Become Accepted and Expected

The second flight of the shuttle took place in November 1981. Thiokol sent a team of engineers and a photographer to Kennedy to inspect the recovered boosters. The manager of Thiokol's operations at Kennedy, Jack Buchanan, was one of the first to notice something had gone awry:

At first we didn't know what we were looking at, so we got it to the lab. They were very surprised at what we had found. (Quoted in Vaughan, p. 120)

What they were looking at was one eroded primary O-ring. Hot gases had burnt through the rubber by about five hundredths of an inch. This may not sound like very much, and only one of the sixteen joints had been affected, but O-ring erosion had never occurred on Titan, nor on any of the test firings of the shuttle engines, nor on the previous flight. It was the first time. The engineers immediately started investigating what had gone wrong.

An explanation was soon found. The putty protecting the eroded seal was found to have tiny holes in it. These allowed propellant gases to reach the seal itself. As a Marshall engineer explained:

The putty was creating a localized high temperature jet which was drilling a hole right into the O-ring. (Vaughan, p. 121)

Different compositions of putty and different ways of applying it were tried to prevent the problem repeating itself.

The eroded seal did bring some good news: there had been erosion, but the primary had still sealed. Also tests on the eroded seal enabled some specific safety margins to be drawn up. Small pieces were cut out of a primary O-ring to simulate almost twice as much erosion as had actually occurred. Even with this amount of erosion the joint was found to seal; this was tried at pressures three times greater than expected during ignition. As one Thiokol engineer commented:

We didn't like that erosion, but we still had a couple of mitigating circumstances. First of all, it occurred very early on in firing... and if the primary O-ring was burnt right through that early... the secondary should be in a good position to catch it. In addition... we found out that even with major portions of the O-ring missing, once it gets up into that gap in a sealing position it is perfectly capable of sealing at high pressures.... (Vaughan, p. 122)

With the new putty in place the next flight went according to plan. There was no erosion. This confirmed the belief amongst the engineers that they had the right explanation for the erosion.

Going Operational
On July 4 1982, after the fourth flight had landed safely, President Reagan publicly announced that the shuttle was now "operational." Despite such proclamations, R&D continued apace on many components and glitches often occurred. For example, on the very flight which led Reagan to make his announcement, the Solid Rocket Boosters

had mistakenly separated from their parachutes and had dropped into the ocean, never to be recovered. This deprived the engineers of an important set of data on the performance of the O-rings. The shuttle was certainly not an operational technology in the way that a commercial airplane is. Part of the shock caused by the *Challenger* accident comes from a mistaken image—an image NASA did nothing to discourage by flying US Congressmen and ordinary citizens in the vehicle. The shuttle always was, and will be for the foreseeable future, a high risk state-of-the art technology. Even today, with the joints redesigned after the *Challenger* accident and new safety procedures initiated at headquarters, the official risk is one catastrophic accident per hundred flights—astronomically greater than would be contemplated for any commercial vehicle.

More evidence of O-ring erosion was found in 1983 and 1984. The pattern was, however, frustratingly sporadic, usually only affecting one joint at a time, and the erosion could usually be explained away. In one case, the putty was again found to have had defects and in another case, the upping of the air pressure to check if the secondary was lodged in its groove before flight had induced blow holes in the putty. Tests and analysis continued and slowly the engineers thought they were getting a better understanding of how the joint behaved. It seemed that the hot gases only played on the rings for a very short period at the start of ignition and therefore only a limited amount of erosion occurred before the joint sealed. As the pressure across the joint equalized, the gas flow stopped and there was no more erosion. This led to the important notion that erosion was "a self limiting factor." The last two flights of 1984 had no erosion at all, seeming to confirm the engineers' belief that they were on top of the problem.

Blow-by

In 1985 the first cases of O-ring blow-by occurred. "Blow-by" refers to hot ignition gases that blow past the primary during the split second before it seals. Blow-by is more serious than primary erosion because the hot gases threaten the secondary O-ring, jeopardizing joint redundancy. The first launch of 1985 experienced three consecutive nights of record low Florida temperatures. Blow-by on this flight reached a secondary O-ring.

After this instance of blow-by, Roger Boisjoly in particular felt there might be a link between low temperature and the damage. He had inspected the joint after its recovery and found grease in it. This grease had been burnt black and came from between the two O-rings. It was for him an indication that the cold may have affected the ability of the primary O-ring to seal. Boisjoly immediately started looking for systematic data on cold and O-ring performance. Part of the problem he faced was that all previous research had considered the effect of excessive heat on the O-rings. (The temperature in the main combustion chamber was 6,000°F—above the melting point of iron.) Boisjoly had his "observations" of the seal, but what he did not yet have were "hard" data. The engineering culture at Marshall was such that for an argument to carry weight and be considered scientific, hard quantitative data were needed. Boisjoly set out to obtain these by initiating a test programme to investigate the effects of cold on the O-rings. This programme, however, was not urgent, ironically because the record low temperatures had been considered a fluke and unlikely to be repeated.

Although the engineers at Marshall and at Thiokol were alarmed about the first ever blow-by, they felt that they had a three-factor rationale for continuing to classify the joint as an acceptable risk. (1) The erosion was still within their basis of experience (it was less than the worst erosion experienced). (2) The amount of erosion that had occurred was within the safety margin established by testing a cut-away seal. (3) The phenomenon still seemed to be "self-limiting." Thus, despite the added worries about the temperature, Thiokol personnel (including Roger Boisjoly), concluded at the Flight Readiness Review for the next flight that "[it] could exhibit the same behavior. Condition is not desirable but is acceptable" (Vaughan, p. 156).

On an April 1985 flight (launched at seasonably high temperatures) there was more blow-by when a primary O-ring burned completely through and hot gases, for the first time, eroded a secondary. This caused great concern, but a detailed analysis found an idiosyncratic cause for the blow-by. The erosion on the primary was so bad that it must have occurred in the first milliseconds of launch and this meant the primary could not have been seated properly. Since none of the other seals were damaged, that seal alone must have been impaired—a piece of hair or a piece of lint trapped unnoticed in the joint would have been sufficient. To prevent this happening again the leak test pressure was raised (an improperly seated seal would leak more). Again, even though the primary had failed, the secondary had worked, thus confirming redundancy. That the engineers seemed to understand the joint's performance was confirmed when primary O-ring erosion occurred on the very next flight, only to be explained by blow holes in the putty produced by the now greater pressure of the leak test!

Certainly the events of 1985, with blow-by encountered for the first time, caused mounting concern among the engineers. All sorts of new reviews, analyses and tests were ordered, including, as mentioned above, tests on O-ring resiliency at low temperatures. Despite this, all the engineers directly involved, including Roger Boisjoly, still considered the joint to be an acceptable risk.

The *Challenger* Launch Decision

We are now ready to understand what many accident investigators found incredible, how despite all the previous warnings and with Thiokol engineers pointing out on the eve of the launch the possible link between O-ring damage and low temperatures, the *Challenger* was still launched.

As we have seen, the joint was not perfect, but neither were a lot of other components on the shuttle. Also the relevant communities of engineers had, over the years, gained what they thought was an understanding of the joint's peculiarities; that hard-won understanding was not going to be given up easily for an untried new design which might have even more problems. Much of the misunderstanding over the *Chal-*

lenger accident has come, not only from twenty-twenty hindsight, but also, and more importantly for this book, from the mistaken view that engineering knowledge is certain knowledge. The expert engineers' own views of the risks versus the inexpert outside view can be seen in the following excerpt from testimony given to the Presidential Commission, where an FAA attorney, Ms Trapnell, is interviewing Thiokol engineer, Mr Brinton:

Mr Brinton: Making a change on a working system is a very serious step.
Ms Trapnell: When you say working system, do you mean a system that works or do you mean a system that is required to function to meet the schedule?
Mr Brinton: What I was trying to say is the colloquialism, "If it ain't broke, don't fix it."
Ms Trapnell: Did you consider that system to be not broken?
Mr Brinton: It was certainly working well. Our analyses indicated that it was a self limiting system. It was performing very satisfactorily...
Ms Trapnell: Well, then, I guess I don't understand. You say on the one hand, that it was a self limiting situation.... But on the other hand you say that the engineers were aware of the potential catastrophic result of burn-through.
Mr Brinton: Well, let me put it this way. There are a number of things on any rocket motor, including the Space Shuttle, that can be catastrophic—a hole through the side, a lack of insulation. There are a number of things. One of those things is a leak through the O-rings. We had evaluated the damage that we had seen to the O-rings, and had ascertained to ours and I believe NASA's satisfaction that the damage we have seen was from a phenomenon that was self limiting and would not lead to a catastrophic failure.
Ms Trapnell: ...Does it surprise you to hear that one of Thiokol's own engineers [Boisjoly] believed that this O-ring situation could lead to a catastrophic failure and loss of life?
Mr Brinton: I am perfectly aware of the front tire going out on my car going down the road can lead to that. I'm willing to take that risk. I didn't think that the risk here was any stronger than that one. (Quoted in Vaughan, pp. 188–9)

The problems with the joint were not, as many accident investigators surmise, suppressed or ignored by NASA, the engineers were actually all too well aware of the problems and the risks. They had lived with this joint and its problems over the years and they thought they had a pretty good understanding of its peculiarities. They knew the joint entailed a risk and, furthermore, a risk to the lives of the astronauts, but so did countless other components on the shuttle.

It is important to understand that the engineers and managers going into the crucial teleconference were not in a state of ignorance. They went in to the conference with all their accumulated experience and knowledge of the joint—any new information they were given was bound to be assessed in the context of this experience and knowledge and judged by the criteria they had always used.

The Pre-launch Teleconference

Let us follow events at Utah as the Thiokol engineers prepared for the crucial teleconference.

The main concern for Thiokol was that cold weather would reduce the O-ring resiliency. After a lengthy discussion and analysis, Thiokol decided that they would recommend "no launch" unless the O-ring temperature was equal to or greater than 53°F which was the calculated O-ring temperature on the previous coldest launch.

Thiokol felt that they did not have a strong technical position either way, but decided to act conservatively. The projected temperature of the O-rings (29°F) at launch time was 24°F below their lowest experience base. The flight with the lowest O-ring temperature had the worst blow-by. The O-ring squeeze would be lower; the grease would be more viscous; the O-ring actuation time—the time for the O-ring to be extruded into the seal gap—would be longer: consequently there must be doubts whether the primary and secondary O-rings would seal.

It was noted by the Thiokol engineers just before they went on air that there was a flaw in their technical presentation. A chart of Roger Boisjoly's, comparing two instances of blow-by, made the point that the January cold-temperature launch had the worse damage of the two. This chart, however, did not contain any temperature data. But another Thiokol chart revealed that the temperature for the second flight was 75°F. By putting the two charts together, a huge gap in Thiokol's argument appeared. The two worst incidences of damage happened at both the highest and lowest temperatures!

The only way Thiokol could resolve the problem was to rely on Boisjoly's own observations that the blow-by really was a lot worse on the low-temperature launch.

The teleconference, with thirty four engineers and managers present, started at 8.15 p.m. (EST). The group did not divide neatly into engineers and managers since the structure of an engineering career means that everyone who was a manager had previously been a trained engineer. Thiokol presented all its charts and gave its rationale that the launch should be held back until the temperature reached 53°F. The flaw in Thiokol's argument that temperature was correlated to damage was soon noticed. Boisjoly, referring to his visual inspection, was repeatedly asked if he could quantify his concerns, but he was not able to do so. Other inconsistencies appeared in the Thiokol presentation.

Larry Mulloy, the head of NASA's SRB management, who was at Kennedy for the launch, led the attack. His main point was that Thiokol had originated and up until now had always supported the three-factor rationale as to why the joint was an acceptable risk. And this was so even after the damage on the previous low-temperature launch was reported. Now they wanted to introduce temperature as a new factor, but O-ring blow-by could not be correlated with temperature according to their own data. Even the decrease in O-ring squeeze with temperature was not decisive. Although the resiliency was diminished, he argued that even at 20°F the resiliency was positive and was greater than the minimum manufacturer's requirement. His conclusion was that the primary might be impaired in a worst case scenario, but that there was no evidence that the secondary would not seal.

All participants at the teleconference agreed that Thiokol had a weak engineering argument. The most controversial part was their recommendation of "no launch" below 53°F. This seemed to contradict their own data which had shown no blow-by at 30°F. Procedurally it seemed odd to create a new criterion at the last moment—one which Thiokol had itself breached in the past. For example, for nineteen days in December and fourteen so far in January the ambient temperature at the Cape had been below the 53°F limit. Indeed on the morning of the teleconference the ambient temperature was 37°F at 8.00 a.m. and Thiokol had raised no concerns. It appeared to many participants that Thiokol's choice of 53°F was somewhat arbitrary.

Mulloy made the point directly. Thiokol, in effect, were imposing a formal new criterion on the very eve of launch. This led to his infamous remark "My God, Thiokol, when do you want me to launch, next April?" This comment was somewhat unfortunate in the light of what happened and was used against Mulloy as evidence that he put the flight schedule ahead of safety. Although Mulloy's statement was exaggerated for rhetorical effect it made clear his concern, that Thiokol were basing very serious conclusions and recommendations that affected the whole future of the shuttle on a 53°F limit that he felt was not supported by the data.

Mulloy's views here were echoed by many of the participants at the teleconference. Marshall's Larry Wear said:

... it was certainly a good, valid point, because the vehicle was designed and intended to be launched year-round. There is nothing in the criteria that says that this thing is limited to launching only on warm days. And that would be a serious change if you made it ... (Vaughan, p. 311)

Another Marshall engineer, Bill Riehl, commented:

... the implications of trying to live with 53 were incredible. And coming in the night before a launch and recommending something like that, on such a weak basis was just—I couldn't understand. (Vaughan, p. 311)

After Mulloy's "April" comment the next to speak on the net was Mulloy's boss, George Hardy. Adding emphasis to Mulloy's argument he said that he was "appalled" at the Thiokol recommendation of a 53°F limit. Again he reiterated the weakness in Thiokol's engineering argument. He also added the well-remembered remark that, "I will not agree to launch against the contractor's recommendation." (p. 312)

Hardy's use of the word "appalled" does seem to have carried weight with some of the participants. As one Thiokol engineer remarked:

I have utmost respect for George Hardy. I absolutely do. I distinctly remember at that particular time, speaking purely for myself, that that surprised me ... And I do think that the very word itself [appalled] had a significant impact on the people in the room. Everybody caught it. It indicated to me that Mr. Hardy felt very strongly that our arguments were not valid and that we ought to proceed with the firing. (p. 312)

Other participants, familiar with Hardy, and with the cut and thrust of these sorts of debates, felt that there was nothing unusual about Marshall's response. As Thiokol's Bill Macbeth said:

No, it certainly wasn't out of character for George Hardy. George Hardy and Larry Mulloy had difference in language, but basically the same comment coming back, [they] were indicating to us that they didn't agree with our technical assessment because we had slanted it and had not been open to all the available information.... I felt that what they were telling us is that they had remembered some of the other behavior and presentations that we had made and they didn't feel that we had really considered it carefully, that we had slanted our presentation. And I felt embarrassed and uncomfortable by that coming from a customer. I felt that as a technical manager I should have been smart enough to think of that, and I hadn't. (p. 313)

The Marshall engineers were certainly vigorous in their rebuttal. But this was quite normal. The two groups of engineers had been going at it hammer and tongs for years. What was new, however, was that this was the first time that the contractor had made a no-launch recommendation.

Thiokol requested a five-minute off-line caucus. Diane Vaughan reports that all participants who were asked why the caucus was called responded that it was because Thiokol's engineering analysis was so weak. NASA was apparently expecting Thiokol to come back with a well-supported recommendation not to launch, but with a lower and more reasonable threshold temperature.

Back in Utah, the five minutes turned into a half-hour discussion. Senior Vice President, Jerry Mason, chaired the discussion and started by reiterating the points Mulloy had made. Boisjoly and Arnie Thompson vigorously defended their position, going over the same data they had presented earlier. Mason put the counter arguments. The other engineers were mainly silent.

Finally, Mason said if engineering could not produce any new information it was time to make a management decision. None of the engineers responded to Mason's request for new information. Alarmed that their recommendation was about to be overturned, Boisjoly and Thompson left their seats to make their arguments one last time. Boisjoly placed in front of Mason and his senior colleagues the two photographs of blow-by, which showed the difference in the amount of soot on the two launches. Sensing they were getting nowhere, the two engineers returned to their seats. Mason then polled his fellow senior managers: three voted to launch and one, Robert Lund, hesitated. To Lund, Mason said "It's time to take off your engineering hat and put on your management hat." Lund, too, voted to launch.

Subsequently, Mason's actions have been interpreted as replacing engineering concerns with a management rationale, where scheduling, the relationship between customer and client, and so on, were given priority. People who participated at the meeting, however, considered this to be a typical "engineering management" decision made where there was an engineering disagreement. As Thiokol's Joe Kilminster explained:

There was a perceived difference of opinion among the engineering people in the room, and when you have differences of opinion and you are asked for a single engineering opinion, then someone has to collect that information from both sides and made a judgement. (Vaughan, p. 317)

All the four senior Thiokol managers gave as their reason for changing their minds facts that they had not taken into account in their original technical recommendation. These were: no overwhelming correlation between blow-by and temperature; data showing that the O-rings had a large safety margin on erosion; and redundancy with the secondary.

The news of Thiokol's reversal was relayed as the teleconference went back on-line. After Thiokol's new recommendation and technical rationale had been read out, Hardy looked round the table at Marshall, and asked over the teleconference speakers if anyone had anything to add. People were either silent or said they had nothing to add. Finally the Shuttle Projects manager asked all parties on the teleconference whether there were any disagreements or any other comments concerning the Thiokol recommendation. No one said anything. The teleconference ended at 11.15 p.m. EST.

Conclusion

What has emerged from this re-examination of the shuttle launch is that the prevailing story of amoral managers, pressurized by launch schedules, overruling honest engineers, is too simple. There were long-running disagreements and uncertainties about the joint but the engineering consensus by the time of the teleconference was that it was an acceptable risk. Indeed, it was Thiokol's failure to meet the prevailing technical standards which led them to reverse their decision. They simply didn't have enough evidence to support a no-launch recommendation, particularly one that set a new low-temperature limit on the shuttle's operation that many considered unreasonable.

We are also now in a better position to evaluate another misconception—the one spread, perhaps inadvertently, by Richard Feynman: that NASA were ignorant of the effect of cold on O-rings. At the crucial teleconference this point was considered in some detail. NASA representatives had researched the problem extensively and talked about it directly with the O-ring manufacturers. They knew full-well that the O-ring resiliency would be impaired, but the effect was considered to be within their safety margins.

What the people who had to make the difficult decision about the shuttle launch faced was something they were rather familiar with, dissenting engineering opinions. One opinion won and another lost, they looked at all the evidence they could, used their best technical standards and came up with a recommendation.

Of course, with hindsight we now know that the decision they took was tragically wrong. But the certainty of hindsight should not be mistaken for the uncertainty of the golem at large. Without hindsight to help them the engineers were simply doing the best expert job possible in an uncertain world. We are reminded that a risk-free technology is impossible and that assessing the working of a technology and the risks attached to it are always inescapable matters of human judgement.

There is a lesson for NASA here. Historically it has chosen to shroud its space vehicle in a blanket of certainty. Why not reveal some of the spots and pimples, scars and wrinkles of untidy golem engineering? Maybe the public would come to value the shuttle as an extraordinary human achievement and also learn something about the inherent riskiness of all such ventures. Space exploration is thrilling enough without engineering mythologies.

Finally, the technical cause of the *Challenger* accident is to this day (1998) not absolutely certain. Cold temperature, erosion, and O-ring blow-by were a part, but other factors may have played a role: there were unprecedented and unpredicted wind shears on that tragic January day which may have shaken free the temporarily sealed joint. The current understanding of the size of the gap at rotation is that it is actually much smaller than either NASA's or Thiokol's best estimates. Ironically it is now believed that the excess squeeze from the much narrower gap was a contributory cause of the *Challenger* disaster. Golem technology continues its valiant progress.

All figures are reproductions of documents appearing in "Report to the President of the Presidential Commission on the Space Shuttle Accident" (Washington, D.C.: Government Printing Office, 1986).

22 "Bodies, Machines, and Male Power"

M. Carme Alemany Gomez

At first glance, one might think that the only value promoted and facilitated by a washing machine is cleanliness. This article argues that much more is at work in decisions about their development and use. Gomez shows the diverse ways in which gender and class relations affected the design, construction, and operation of a particular washing machine in Spain, and in turn how the design of the technology reinforced these relationships. Gomez explains the processes by which male engineers designed the machines. She argues that the design not only determines how clothes will be washed but also "predetermines the washing practices a woman must adopt down to the very movements she must make, over and over again, as she uses the machine." While Gomez seems to be a technological determinist, the more subtle idea is that if women were given a voice during the early stages of development, the technology would be designed very differently, and perhaps would save a great deal of backbreaking work. The people who design and build a technology—even a seemingly mundane one like a washing machine—make decisions about details that ultimately affect the user.

In this study from Barcelona we follow the trajectory of one technical artefact: the clothes washing machine.[1] The washing machine, it goes without saying, is not a new invention. The features today's models display have been evolving for over 40 years. We shall see the considerations that have gone into its design in the past, and the reasons behind a recent remodelling. We look at how it is manufactured and by whom, and discuss its design implications in use in the home.

The artefact's trajectory is analysed as a *social* process, one in which both gender and class relations can be seen to be constructed and activated. In following its course we can see something of the interrelation between productive work in the factory and reproductive work in the home, as part of gender and class relations.

This approach, expecting to see gender and class relations in the life story of a machine, is one of three theories with which we approached the study. Second, we draw on a tradition of thought that is by now strongly established in social studies of science and technology (Callon and Latour 1988; Callon 1989; Latour 1989). We suppose that technology and society are not two interconnected domains, which none the less overlap, but that they are superimposed on each other (Woolgar 1991). Technology, in this view, is not seen as "having social aspects," but as being social in its very constitution. It follows from this that in our study we do not understand technology as constructed

---

From Cynthia Cockburn and Ruža Fürst Dilić, eds., *Bringing Technology Home: Gender and Technology in a Changing Europe* (Buckingham: Open University Press, 1994), pp. 129–146. © Carme Alemany Gomez. Reprinted with permission.

outside gender and class relations. Rather we see technology, gender and class relations being a process in which all are a part, and all are simultaneously constructed.

Third, we see gender relations as a set of power relationships that not only involve the construction of values, behaviours and attitudes, but also constitute a process of bodily submission, as evidenced in postures, gestures and movements (Foucault 1979). Following the approach used in a different context by Michel Foucault, we extend it to the way in which gender and class relations are established by different social actors, placing particular emphasis on techniques and tactics of dominance.

The Spanish Domestic Appliance Sector

In 1990, the most recent year for which figures are available, the "white goods" industry in Spain had a turnover of 150,000 million pesetas and employed approximately 12,000 people. It was characterized by a heavy concentration of production and the dominant presence of large transnational groups. The sector's current form is the result of intense changes over the past decade, both in the sector itself and in Spain's industrial fabric as a whole, due partly to severe economic recession and partly to the processes culminating in Spain's entry to the European Economic Community (EEC) in 1986.

The Spanish white goods sector had grown fast throughout the 1960s and the first half of the 1970s, as a result of changes in both supply and demand. On the supply side, early in this period a technological transformation, led by Italian industry, began the sector's trend to mass production that eventually resulted in a significant reduction in prices (Owen 1983; Bianchi and Forlai 1988). By the end of the 1970s, white goods, especially refrigerators and washing machines, were no longer luxury items for a restricted market.

On the demand side, Spain saw a considerable demographic growth, rising incomes and rapid urbanization, and a development towards the consumer patterns of the more developed societies. These factors all favoured the purchase of domestic appliances by Spanish households. The tariff protection that operated in this period helped Spanish industry in particular to benefit from this growing market.

Since the manufacture of white goods calls only for relatively simple technology, the development tended to take place through medium-size enterprises. Spanish capital was clearly dominant in the process, even though the period did also see the expansion into the Spanish market of some transnationals—the American company Westinghouse and the Italian companies Ignis and Zanussi—with their own technology and brand-names.

A preferential treatment agreement signed by Spain and the EEC in 1970, anticipating Spain's accession to the Common Market, enabled Spanish white goods, among other industrial products, to enter the EEC countries under favourable conditions. Exports to non-EEC areas, particularly Portugal and North Africa, were also growing.

However, this period of economic expansion, or "soft growth," in the sector was curtailed at the end of the 1970s, due partly to economic crisis, but also partly to certain inherent problems besetting the domestic appliance sector. The recession came at a time when the first-purchase market was showing signs of saturation in the most important ranges, such as kitchen stoves, refrigerators, heaters—and washing machines. It was besides very clear that Spain's imminent entry into the EEC would mean the end of protection for the industry in its home market. It would be thrown into open competition with the European white goods industry at a time of considerable excess capacity.

All these factors impelled the white goods industry in Spain into far-reaching changes that were to persist well into the 1980s. The government's Industrial Reconversion Plan for the sector set about eliminating surplus capacity, adjusting supply to demand and effecting labour reductions. Its main instruments were a consolidation of production and a series of fiscal, financial and labour measures. Its implementation did indeed effect a drastic reduction in the number of firms. By 1988 four-fifths of production was concentrated in three large groups, two of which were transnationals, with a loss of almost half the sector's jobs.

Technological Innovation in Washing Machine Production

Turning now to washing machines in particular—we will see that the design of a new prototype to replace existing models was shaped more by these macroeconomic factors than by any intention to produce a machine that would ease the burden of domestic washing. What we found to be motivating the manufacturer was that the new model would make possible the introduction of a quite different manufacturing process that promised a substantial rise in productivity. How many customers, when purchasing a "technologically-improved" washing machine, understand that the improvements have been not to the machine's functioning but to the production line?

In Spain, 60 per cent of washing machine production is by two transnational groups, in both of which the central design and innovation unit is located at group headquarters. As a result, there is no research into new washing machine design in Spain itself. Technological innovation in domestic washing machine design is in any case limited. It is a technical appliance whose form has evolved over the decades to a point at which few improvements are now anticipated by its manufacturers. Engineers in the sector say that, to them, "innovation" now means mainly the progressive incorporation into models for the lower end of the market of technological features already in use in quality or high-performance machines.

The company in which the fieldwork was carried out produces washing machines, under a well-established brand name, for the middle-to-bottom end of the market. Its factory, where we interviewed personnel, is situated in a large industrial area in the province of Barcelona. The firm was badly affected by the crisis in the sector, as a result of which it joined the Industrial Reconversion Plan in 1986. Despite absorbing workers from another company, under the Plan the firm's overall workforce was reduced, by almost half, to 400 employees.

The aid received under the Plan did not bring about any modernization of the plant, nor the re-launch of the company. On the contrary, its only consequence was this shedding of labour and a virtual salary freeze for three years. The employees embarked on a long struggle to retain their jobs, and indeed this company has one of the highest levels of union activism.

In 1989, three years after the restructuring, the company was taken over, along with some others in the sector, by a large foreign group. Its new owners gave priority to increasing profitability by the following means. First, the entire management team was overhauled in a bid to put an end to the management policies of the 1980s. The manufacture of the various washing machine models was redistributed throughout the companies of the group in Spain and abroad, to enable longer production runs. Small-scale investments were made at various points in the production process with the aim of increasing manufacturing capacity and productivity. In particular, certain specific tasks that had been a source of conflict with the unions were now automated. Most important, however, was the introduction of the newly-designed washing machine mentioned above, so as to transform production methods and double output by 1993, without any increase in labour.

It will be evident, then, that it was not developments in the *washing process* but in the *manufacturing process* that were the main motive force behind this technological innovation. The quality of the machine as experienced by the user might be improved in the process but this would not have been its designers' prime concern.

The Design Process: Limited Input for Women

It is striking that the network of actors we found taking part in both commercial and technical decision-making about washing machine development was male to a man, while women only appeared as users. Technological design and development were the responsibility of a specialist department of the firm: Product Engineering. The only woman here was the secretary.

We learned from the engineer in charge that small-scale improvements to a model might arise from the initiative of any one of several social actors, including the "organization and methods" department, the production department, the marketing department and the department responsible for after-sales service. In the case of full-scale remodelling however the initiative would originate with Marketing. This department, working from market studies, from research carried out on their own products and from experience gathered through the commercial network, would draw up a detailed commercial specification for the new model, a blueprint or brief for Product Engineering, who would subsequently translate it into a design.

The interests of Marketing and Product Engineering do not necessarily coincide. The former is concerned primarily with the commercial success of the appliance. They must assure its popularity with the consumer, whose needs and motivations they must take into account. Product Engineering on the other hand introduces an additional objective: that of ensuring the new model is producible cheaply enough to be both competitive and profitable.

Women are explicitly considered only in the former perspective. Marketing make use of market surveys focused principally on women users. They also have women in mind in creating a public image for the washing machine and deciding an advertising strategy. Product Engineering, by contrast, with the excuse of working solely to technical objectives, seeks to ignore the fact that the principal users of the machine will be women. The engineer heading this department made this clear. "In the conception and design of the washing machine, engineering does not take into consideration the person who's going to use the appliance," he said. The concern of the department, he admitted, stopped with manufacturing. "This is a factory and we sell washing machines to our commercial representatives, and in some way it is at this point that there's a slight breakdown in communication with the outside world."

Management clearly prefer communication with the user to be limited to the commercial departments. None the less, like it or not, Product Engineering does in reality need to know the reactions and responses of users in a more direct form. Product Engineering, we found, arguing the deficiency of Marketing's research reports, had invented its own strategies.

In the first place, the engineers invoked something they called the "washing machine culture," informal knowledge accumulated over many years. In addition to beliefs about washing machine use generated within this culture, the image of the user was further shaped in the Test Laboratory where the technicians tried to "imagine the errors that might be made during use." This involved constructing a series of scenarios depicting women's "clumsiness" and "technical ignorance" that justified the incorporation into the machine of devices that, in the technicians' perception, would prevent it from malfunctioning however inept its use.

Images of the users so constructed are of course highly coloured by gender relations. The devices the *male* designer incorporates in the machine do not always lead to an improvement in the machine as actually put to work by *female* users. For example, in the prototype we studied at design stage, the drum was positioned higher in relation to the door than in the model it superseded. This had been done for a strictly techno-economic reason: reduction of assembly time and hence of production cost. The effect was that in use the water level in the machine would be higher than the bottom of the door. This in turn necessitated a safety catch (in fact incorporated in most modern washing machines), in case, as the engineers put it, "the woman were absent mindedly to open the door during the washing cycle," or "a child were to open the machine." Women and children were often associated in the engineers' thinking in a way that clearly reduced women's capabilities to those of a child. Product Engineering entirely failed to imagine, or perhaps having imagined chose to ignore, the clear disadvantage to the user of being unable to add extra items of clothing to the wash after the start of the cycle.

A further strategy of Product Engineering to get access to user knowledge is to build a limited number of prototypes of the new model and distribute them among members of the company's senior staff for their "wives" to test in use and give their opinion.

Such interaction between the firm and "the domestic sphere" is of course highly unusual, home life normally being rigorously distanced from working life. Its artificial importation in this way is explainable only as an effect of the sexual division of housework: company engineers (male) have no personal experience on which to judge the effectiveness of their designs. They justify using company wives by saying they are distant enough from the design process to be objective critics but close enough to the firm to be trusted to keep its secrets.

Finally, a prototype is also tested in the Test Laboratory, which communicates its findings back to Product Engineering. The laboratory is, not incidentally, staffed by women. Like the mobilization of company wives, this is a mechanism to introduce the user, in a strictly retrospective way, into a late stage of appliance design fully controlled by men.

Test Laboratory staff are not in a position to effect any fundamental change to design. Their concern is limited to the finer details of the machine in use, such as noise levels and the duration of washing cycles. They devote a good deal of attention to the design of controls, their size and ease of use, testing them against the strength of the hands of female employees, and they study difficulties encountered in loading clothes, assessing the optimum door size and fastening. By the time the prototype reaches the Test Laboratory, the basic conception of the machine is fixed, and it reflects men's interests rather than women's. Since the women laboratory technicians do not have sufficient technical training to query the technical aspects of the various devices incorporated, men's technical control of the appliance is not challenged.

Consider for instance the fundamental choice between top and front loading machines. Top loading is more convenient for the user, who avoids having to stoop or kneel, and can open the lid at whim at any point in the wash cycle. Yet, very few manufacturers in Spain today make top loading machines. They require more assembly time and are for that reason unpopular with production engineers. Instead, the firms' commercial departments use every means to promote their preferred front loading models, which as a result have become the norm in the Spanish market. At the same time, interior designers and manufacturers of kitchen fittings have been inflexible in designing washing machines into the kitchen, prioritizing worktop space and hence the front loader. This too has inhibited the demand for the top loading machine.

Housework is the "female domain" *par excellence*. These are the tasks considered to be the "natural" role of women. Yet what we see here, and what other research shows (Chabaud-Rychter 1987), is that *men model and control women's domestic work through the electrical appliances they design*. Furthermore, this control that men exercise extends even to women's *bodies*. The design of a washing machine not only specifies in general how clothes will be washed. It predetermines the washing practices a woman must adopt down to the very movements she must make, over and over again, as she uses the machine.

This control over women's bodies through technical design, even though it is not stated explicitly at any time, is detectable in the way the design engineers dwell on those gender differences that can be interpreted as physical inferiority in the female—

weaker hands for example. One can surmise too that the choice of the front loading solution in preference to top loading is a gendered choice. The stooped position, the bending motion required to put clothes in the front loader would be rejected out of hand were men the anticipated users: to be manly is to be upright. Bending and kneeling is an acceptable concomitant of the design only because the user is that already subordinated creature, a woman, and the designer is not. Had things been otherwise perhaps more energy would have been put into seeking "appropriate technical solutions" to avoid subjecting the user to this daily obeisance to the machine. We shall see some similarly gendered ergonomic processes occurring in the case of factory jobs.

Production: Myths of Strength and Skill

What we saw happening above, in "design for use," was totally gendered thinking in engineering. It is thinking in which in reality there is always a gender model, yet in which gender is never fully admitted in such a way as to be incorporated openly into professional knowledge and design skill. It has this semi-conscious, shadowy existence because to bring it fully to view would involve the engineers in admitting their own masculinity and partiality. For they are partial in two senses—biased and incomplete. An exactly parallel process visible in "design for manufacture" supports this interpretation.

Asked who they thought was going to do the various production jobs they were planning, the chief engineer said "when you design, you are automatically and logically thinking how the appliance can be made." It was this *how* he admitted as his concern, not a specific *who*. In this way the design and development appear to be asexual, or at least a process in which gender relations are peripheral. In practice, however, since the engineer is male and he does not bear a woman in mind, the process has reference to a male norm.

This is important, since the technological choices in the design of the appliance have, and are intended to have, direct implications for production, and hence for the work of the producers. Following the washing machine from design to production it became clear to us that those implications involved the reproduction of gender difference and inequality.

The engineers we spoke with, working in the design and manufacture of washing machines, were agreed that there were no longer any physical barriers to prevent women participating fully in the manufacturing process—with the unimportant exception of a few particular tasks that would in any case soon be redundant. Yet we found women to be only around 15 per cent of the workforce in this industrial sector and an even smaller minority (10 per cent) in the company in which our study was made.

An analysis of the distribution of tasks in the washing machine manufacturing process showed, as other gender studies of technology have shown (Cockburn 1985), that the technology factor in a job operates to exclude women. In some tasks, given differences in average male and female size and musculature, this is literally the case. Certain controls are designed to suit the size and strength of a male hand. The height of certain machines, such as sheet metal presses, is too great for the average woman's stature. The resulting jobs are considered by women and men alike as being more suitable for the latter. Women however are clear that this is due to the design of the tool or machine, not to the type of work, which they say is not essentially difficult.

It is not that tools and machines are designed or selected for the production line without conscious thought. On the contrary, size, cost and functioning are hotly debated within the development team. The neglect of gender, the failure to take professionally into consideration as one relevant design factor the ways in which women and men differ is, ironically, *gendered*. It has to be read as a specific and significant omission, an expression of gender power relations in the workplace.

As Foucault suggests, however, power is not simple repression but "a complex network of strategies" (Foucault 1979: 108). It achieves its very effect from the way we all, in one way or another, participate in it: even the oppressed gain something. Power operates in gender relations in such a way that women's oppression is sometimes expressed as, even experienced as, something positive. One male worker for instance declared "it's not right that a woman should work on the metal presses, because if a man loses a finger, that's one thing. But a woman!..." The exclusion of women from certain operations in the workplace may thus be legitimized as kindness—protecting them and their femininity from harm.

Weakness and Strength
Physical strength is an important factor in job design. Advances in technology, such as conveyor belts, handling equipment, automation and robotics, have gradually reduced the number of jobs calling for a great deal of physical strength. The jobs so transformed, incidentally, have often been the ones that were a source of labour conflict. In the manufacture of washing machines, however, certain jobs are still classified as "heavy." Press maintenance work, which calls for occasional removal of the extruders from the presses, was one such. The handling of the washing machine drum in the manufacturing process was another. Both these jobs were performed by men, who earned a special wage supplement entailed by law in the case of tasks involving the application of a force of more than 5 kg. In cases where the physical force required exceeded the legal limit of 25 kg, men had invoked the law and refused to do the work. Drum handling was, through such action, eventually lightened by introduction of mechanical aids.

It is important however to distinguish between the *absolute level* of strength required and the *frequency or continuity* with which effort has to be sustained. It is significant that the law only governs the former and has nothing to say about the continuous repetition of smaller amounts of exertion, as involved in many women's jobs. In washing machine manufacture we found some sections staffed exclusively by women—for example, bundling and attaching electrical cables for the programmers. These are jobs which seem "light," since they do not call for exertion of a 5 kg force at any one moment. Yet the movements involved impose repetitive strain on the hands, elbows, shoulders and back.

In these sections the typical health hazard is not the one-off sprain of a muscle or ligament, but a chronic condition caused by the rhythm of production that may in some cases be irreversible. For instance, a significant number of women in the cable section have had to have surgery for carpal tunnel syndrome due to repetitive strain injury to the wrists. The arduous nature of the work characteristically performed by women in the factory was not recognized by management however. They were unable to claim any wage supplement because their tasks were not heavy work as defined in law.

Some "light" repetitive jobs, for instance certain subassembly activities, as it happened involved both sexes. In these cases we found the behaviour of women and men in relation to the health hazard entirely different. Women had no hesitation in complaining of physical strain induced by their work. To do so in no way affronted their gender identity. The men, however, were inhibited by their sense of themselves as men. Men cannot, without diminishing their own virility, admit to fatigue when the job is "light" and calls for no visible level of physical strength.

In the mixed sections it was therefore the women's complaints that sometimes brought about an improvement in working conditions for both sexes. For instance, men, despite frequent cuts and other minor injuries, had refused to wear protective gloves on the production line. When women were introduced to the line, free from the constraints of masculine self-identity, they adopted the protective wear, which in due course became common practice for men too. This obviated one job at the end of the production line: wiping off the blood stains on the white enamel. It had been a woman's job, no doubt because it was felt to need a feminine hand to raise the product to the houseproud standard demanded by the market.

Finally, there are certain men's jobs in which automation is eliminating the demand for physical strength. Seen in this light, technological innovation benefits "the worker," but in practice the gains accrue mainly to men. Despite the fact that women could now do them, such jobs in the washing machine factory had seldom been opened up to women. The forklift truck had been introduced to take the strain out of "manhandling" heavy goods. Women however continued to be excluded from goods handling long after the objective justification had gone.

Because the forklift truck symbolizes strength, the job of driving it is masculine like the manual job it replaced. Implements, tools and machinery expand human potential to lift weights and manipulate materials. They could in theory empower women by enabling them to transcend their current physical limits. Given prevailing gender relations and gender symbolism however they serve instead to enhance male power.

Masculinity is associated not only with physical strength but also with physical mobility. We found women doing jobs that *required* less mobility than the jobs men did. For instance, the machines they operated called for less physical movement than men's machines. They were also in the main doing the kind of jobs that *permit* less mobility. We found women moved about less than men from one section, or one part of the factory, to another. Their exclusion from forklift driving was a case in point. Sections staffed only by women were often located in marginal areas of the factory from which the women rarely emerged. Male sections occupied more central areas, so men's presence was more visible and significant. Women performed more stable tasks that did not involve rotation nor permit of contact with many people. For example, there is one activity that offers just this kind of interaction. If a worker needs to leave an assembly line like this one for even a moment to visit the WC, she or he must call for a "pee relief." Although there were women who were well qualified, having performed every job on the line, we found not one filling this role.

Finally, women's work is often performed sitting down or in a bending position, postures that we have already seen are considered suitable or natural for women, but not for men. Men refuse such tasks whenever possible, opting where they can for an upright position. It may be more tiring, but at least it is not effeminate. This was brought home to us in interview by a young male operator. He mused on the women at their benches in the electrical cable section. "It looks as if they're *sewing*," he said. He showed what he meant by bowing over the bench, in an ugly gesture that spoke even more clearly than his words. "Having to work in a bending position like that isn't what you'd call *virile*."

We are used to analysing "women's jobs" in terms of the attitudes and aptitudes they call for. If we look, however, as we have done here, more closely at bodies and physique, we can see clearly just how men use technology to secure domination over and control of women.

Ignorance and Knowledge

We found the association of masculinity with technology in the factory to be profound and widespread. Men were given priority over women in the retraining schemes made necessary by the new manufacturing process. Men were taught about the technical features and functioning of the equipment, while women's training was limited to simple instruction in operations. Again, machine maintenance technicians in the factory were exclusively men, even though some had no specific training in the area. It was inconceivable that a woman with equally little training be offered such a job.

Women were assumed to be technologically ignorant. When a woman did take on a job that had formerly been done by a man, there was a general scepticism as to her ability to do it. She was obliged to prove her value to the group if she were to be recognized and accepted as an equal. Similarly, when women suggested modifications to tools or pointed to problems with the work methods imposed by the foreman, they were seldom taken seriously. Only when their suggestions were seen to work was their ability acknowledged. Getting their ideas implemented, however, involved such prolonged insistence that women were of course labelled "obstinate."

The exclusion of women from technical knowledge and technical work, as research in other countries has found (Kergoat 1978), blocked their possibilities of promotion. Women were stuck in the lower category jobs, while men climbed out of them. None of the positions of authority in the plant was held by a woman, even in all-female sections. Yet most of the male supervisors and foremen, as women well knew, were no more than workers promoted up from the production line or other manufacturing jobs. "It's not as if they have qualifications. We know them..."

Such supervisory jobs enable those who do them to accumulate technological knowledge. As men get these opportunities, and women do not, the gap between the sexes widens. Nor is women's exclusion from requisite knowledge unique to production. In the administrative departments too we found women had few possibilities for promotion. Women in both situations know full well what is happening. They seldom complain, however, because they consider promotion or professional advancement beyond their reach.

The difficulties encountered by women in getting their jobs reclassified or in getting promotion out of them were absent from the union's negotiating agenda. The union was simply ignoring women's problems. As one woman put it, "They look after 'working conditions,' not women's issues." We found very few women participating in union activities. Women agreed that if there were women on the committee, women's issues would be more likely to be put forward. Often however domestic responsibilities were a barrier to women organizing and making their own demands. One woman explained, "I'm very limited. I've got my children and my job...," whereas "if the men have a meeting, they think 'I'll see the children later,' and they leave them with their grandmother or someone." Inequalities at home and work reinforced each other.

Implications of the New Model Machine

As already mentioned, our research period coincided with the design stage of a new model of washing machine. Product Engineering's objective in this was "to double production and improve quality with an investment of about 1,000 million pesetas ($10.3 million or £5 million), without increasing the number of employees." The new model would introduce new materials and a different production process, including simplification of certain assembly sequences. In turn, these innovations would call into existence a new mode of work organization, most importantly the virtual elimination of existing assembly lines, each with their 30 work stations and quality control at the end.

Instead the machine was to be constructed in a series of "islands," in which each individual would carry out several work sequences and be responsible for quality control and self-rectifying of mistakes. The production job would thus become more varied and responsible. The assembly line as such would be virtually defunct, reduced to six unskilled functions at the end of the new island processes.

The "new technical solutions," as the engineers called them, were expected to reduce production time by 60 per cent and greatly improve product quality. By eliminating the conflict-prone jobs they would eradicate nearly all the sources of worker unrest that had been troubling the factory. They would also make less onerous some other jobs that, though not a source of strife, did have a record of low productivity and high absenteeism for reasons of ill health. Since these improvements would benefit the workers, we may perhaps acknowledge them as being, in part, won by class struggle.

The management clearly recognized that the new methods were going to call for a somewhat different quality in the worker. The engineers were talking about a new "manpower culture" and envisaged a need for conversion training and re-skilling. In principle the restructuring in the factory could have been used as the occasion for a new personnel policy, geared to reducing the sexual division of labour and lifting women out of the low job categories to which they were confined. There were no signs of this happening. Indeed, listening to the engineers in Product Engineering and Organization and Methods, we felt the innovations were likely to reinforce existing gender relations.

When, in our interviews, the discussion was on a general level, they used generic terms such as *gente* (people), *personal* (personnel) and *personas* (people of both sexes). However, whenever the conversation turned to specific jobs calling for degrees or other qualifications, they automatically slipped into using masculine nouns and pronouns. For example, one engineer explained the island system as follows:

It would be a job where the *person* [*la persona*, i.e. man or woman] would be more fulfilled. It wouldn't any longer be just tacking on your little piece and then the next *man* [*el siguiente*, masculine gender] fixes his on and passes it on to the one after...

The masculine form used here is a deliberate masculinizing of the statement, since its usage in fact breaks the rules of Spanish grammar. *La persona*, though a feminine noun, refers equally to woman and man. The correct correlative would be the sex-neutral *la siguiente*.

This was by no means an isolated slip. We heard, "The *person* will have responsibility. But if a part is defective, it's not *his* problem [*no es problema de el*]. *He* just has to reject it." And again,

In the assembly line, one *woman* [*una*] put on the water connection, another [*otra*] fixed on the electricity connection, and each was one job. But in the new system, the one *fellow* [*este senor*] will put on several things, the water connection, the electricity supply, the condenser and so on.

Despite the management's hope that the changes could be effected without increasing staff, it appeared that a small mount of recruitment would be unavoidable. Who would get the chances? One engineer foresaw "every attempt will be made to retrain the people [*la gente*, women and men] who already work here." Gripped by this pervasive inertia, however, he added "although we'll need some new *operarios* [grammatically, *working men*]." And he went on, "We are planning more complete jobs so that the *operario* will be more satisfied with *his* work."

Some would be good jobs calling for technological qualifications. The engineering departments foresaw taking on staff "who could be *delineantes* [draughts*men*] or *tecnicos* [male technicians], a *man* who also knows how to calculate." They were talking about employing a *maestro industrial* (graduate engineer) or an *ingeniero tecnico* (technical engineer). Alternative feminine forms for these terms do exist in Spanish, but clearly did not spring to mind. The engineers did not have the same difficulty in finding feminine terms, however, when the discussion turned to the need for additional office staff.

The engineers had given no thought to the possibility of adapting the new jobs to the physical capabilities of women, for instance by specifying appropriately sized tools and parts. The modifications considered had any and every reason except involving women more equally in technical production. Certain men's jobs, we knew, were going to be "lightened." When we asked one engineer specifically whether these would in future be unisex jobs, he prevaricated. "Yes, there's no reason why not, since there's no great strength needed—nor—no, there's no reason why not. Of course it depends on how many workers we have and the new ones we'll have to take on. I don't know. But in principal...," and so on, and so forth.

Jobs are thus conceptualized, evaluated and structured by those who have power over them, in line with masculine values. Women are left with residual tasks that are not "proper jobs" for men. An engineer in Product Engineering made it clear enough: "Women are far more effective than men for repetitive, routine tasks." Another in Organization and Methods echoed him: "Women are ideal for assembling wiring...they are more careful and delicate, much more skilful manually." They were in total harmony in agreeing that "it's better for women to be in the jobs where a certain delicacy is needed, because they're more refined and careful than men."

In all these subtle and not-so-subtle expressions of gendered thinking we see how men's power is perpetuated and expanded in the process of technological innovation, how always justifications are ready to hand for excluding women from all but the jobs men do not want.

The Washing Machine in Use

As with other researchers in this European project we particularly wanted to make the conceptual connections: workplace/home, production/consumption, design/use. We therefore went on to interview seven washing machine owning couples, the better to understand the social relations of washing machine use. They were heterosexual couples, married and unmarried, between 25 and 50 years of age, representing low, middle and upper-middle income brackets. In each instance the woman had paid work outside the home. Two of the couples were in fact employed by the washing machine manufacturer. Each individual was interviewed separately, some being seen at work, some at home.

Of course the numbers here are small and even our in-depth interviews cannot permit generalization. None the less, we found some interesting examples of how a household technology-in-use is shaped by the gender dynamics of the heterosexual relationship.

Ownership of a washing machine depends partly on the duration and stability of a marriage or partnership. Students and young unmarried couples not long living together seldom have a washing machine. We therefore found ourselves interviewing couples who had been together some years, and who had jointly chosen and purchased their machine.

Among the men of these seven couples we noted a clear age difference in attitudes to housework. In the oldest couple interviewed (45 and 50 years), the man had never approached the washing machine to put in a load, nor yet to take one out. Nevertheless, he was quite familiar with its operation since he and his wife were employed by the manufacturer. He, besides, had a second occupation repairing appliances.

His wife, it transpired, was in any case unwilling to abandon control of the laundry, justifying herself by saying she understood better the care of her husband's clothes and the household linen. In any case, as she said, "with two jobs, my husband is never at home." This woman, it must be noted, perhaps by virtue of her involvement in washing machine manufacture, was fully capable of repairing and adjusting her washing machine.

The elder of the two daughters in this family, though she did not participate in any family laundry chores, often used the washing machine for her blue jeans for which she had her own routine: wash in cold water and spin dry at top velocity. She would also occasionally throw in dirty underwear, unearthed when cleaning her room.

The youngest couple in the study had an approach not dissimilar to that of this young woman. Their laundry was not done systematically, but coincided more or less with room cleaning. Should one or other have need of a specific article of clean clothing, that person would take steps to wash it. The man would do his own shirts on an emergency basis, and look after the washing of his jeans without devoting any special care to them. His partner, by contrast, used the washing machine thoughtfully having evolved a sorting system for the wash. While her companion often spontaneously threw in a half-load, she was the one who took responsibility for the greater part of the household laundry.

Justifying his practice, this man explained that his partner did a good deal of careful hand-washing and he did not want to risk damaging her delicate things by mixing them in with his own. She, for her part, said his rough approach to washing would not do for her silk blouses. "He doesn't have to take care of his clothing the way I do."

In some homes we found men participating in washing in a selective way. They would take on the washing of, say, towels and the children's jeans but leave the rest to the woman for fear of "making mistakes." Such men preferred a sporadic involvement, avoiding commitment while giving an impression of being cooperative.

In general it was the younger men in our small sample who were to some extent—if spasmodically—participating in household laundry. Perhaps we were seeing a generational change in the way men relate to housework. If so, the shift was not one from non-participation to full sharing, but rather one from non-participation to "showing an interest" in learning a few skills to "help" the woman, whose sphere the household continued to be. At best men's participation was unreliable and they continued to give priority to activity outside family and home.

There was an age difference in the washing practices of the women in the sample too, cut across by other factors, such as engagement or otherwise in outside work, and the skill level of that work. The older women were unshaken in their belief that washing is women's work. "Men don't know how to." They were therefore not eager to ask their menfolk for more help. In answer to our question whether they would welcome more collaboration from their husbands, they might answer "yes," but it would be with a smile of disbelief that said "that'll be the day!" When women, and it was mainly the younger women, did demand of their partners, directly or subtly, more participation in housework, men were not unresponsive. They would "help" more, though irregularly, and not to the extent of equal sharing. For that they would require a willingness they do not show to learn the finer arts of laundry skill.

On the other hand, women themselves impede such a transformation in the gender relations of housework by their own inflexibility. Many women would like more involvement from their partners but are themselves unwilling to see the household laundry practices modified. They are unwilling to negotiate their sense of propriety, their standards of cleanliness and their preferences for a certain order in domestic work. For instance, many women will wash a man's underwear but not expect him to wash hers, apparently feeling it an intrusion into an intimate and personal feminine space. The woman in the young couple cited above, albeit abetted by her partner's carelessness, could be seen as actively impeding his involvement by her protectiveness over her silk blouses. Again, men usually arrive home later in the evening than women, even when both work. Women often use this as a reason to be the one to start the wash, when objectively it would not suffer from a small delay.

Willingness to negotiate seems to be lowest when the woman's external work is non-skilled, low paid or low in satisfaction. We found women in that situation to be the ones tending to take their housework more seriously and to be more possessive about it, as if they were compensating thereby for the minimal responsibility and self-respect accorded them in the world outside the home.

Men's Power and Women's Bodies

This sketch of an artefact pursuing its course from conception through to use has highlighted the class and gender relations in technological processes. The interrelation between class and technology is clear enough. Class relations involve a dynamic opposition between two forces with contrary intentions. What capital seeks in technological innovation is greater profitability. Women and men of the workforce look to it for an improvement in their working conditions. We saw evidence of both these motivations in the company described.

The interrelationship of technology and gender is more subtle and less obvious, but no less profound. Technological innovation occurs within and reinforces existing, traditional gender relationships. We saw several gendered phenomena in the workplace that add up to systematic forms of control and physical domination of women that can only be interpreted as manifestations of male power.

There was the dominant position of men as a sex in the design and development of the artefact. There was the actual display of physical strength by men, and an affirmation of their physical superiority through association with machines that symbolized strength. There was the size of controls and levers on certain machines and tools, explicable only by the masculinity of the intended user. Simply by virtue of being men, men occupied not only management and professional positions in the firm, but the better paid manual jobs evaluated and graded according to masculine criteria of what is arduous work and what is technical work. Technology presents itself as an instrument in the hands of men that dramatizes and augments masculine supremacy.

Then again, we saw women excluded from jobs that, though once they called for a man's strength, no longer did so. Instead, women were performing jobs that needed manual dexterity or care in handling parts. Skills and characteristics learned in their lives as girls made women, assumed to be "natural" attributes of the female sex (Kergoat 1990), were being used to confine them to some jobs and exclude them from others. The effect was again, in a workforce where women were entirely absent from supervisory, professional or management positions, a clear advantage for men in pay, status and control. To compound women's disadvantage, male solidarity and dominance in the trade union combined with that in the plant to prevent them from organizing in support of their own demands. These are not incidental phenomena, but active affirmations of male supremacy through technology and organization.

Male dominance and female subordination persist in both public and private spheres, and the gender relations of workplace and home are mutually reinforcing. We saw how women's workload in the home deprives them of the time they need to be involved in negotiating their rights at work. Besides, as other researchers have pointed out, women's domestic responsibilities, so often cited by employers as an explanation of their unequal treatment at work, might just as well be seen as equipping them (precisely by tying their hands) as a supremely desirable, super-exploitable labour force (Chabaud-Rychter *et al.* 1985). Conversely, dissatisfied with their paid work, women tend to overcompensate with domestic chores—which in turn holds them back from making cogent demands on men to share them.

We found too that women were active participants in patterns of relationship that perpetuate their oppression. In the circumstances in which they are employed, women themselves are often obliged to define their physical needs relative to the work load, which inevitably involves stressing their own physical *limitations*. Such demands cannot help but play into the image of women as relatively weak. They do not force men to acknowledge them as equal and valuable. They are therefore well enough received by the men who hold power—just so long as they do not involve them in much expenditure.

Women occupied relatively less space, and less prominent space, in the factory than men. This too reflected women's own reticence: we found the majority of women eating their meal at their work station, leaving the dining room to men. A cycle is generated in which women come to feel like intruders in spaces into which they seldom venture. It is likely that this withdrawal from space nominally shared by both sexes is partly due to women's avoidance of sexual harassment.

The tale of the social life of this innocent artefact, the washing machine, then, uncovers a painful reality in which men can be seen to control women, not just through symbolic manipulation of difference but also through quite material bodily positioning and inferiorizing. We might hypothesize that the continual reaffirmation of men's physical superiority serves to keep alive a fear in women of actual physical aggression by men—the ultimate weapon of male supremacy.

Note

1. This research was financed by the Women's Institute (Ministry of Social Affairs, Madrid) and CICYT (Inter-Ministry Commission of Science and Technology, Madrid, Reference SEC92-0465). It was carried out at the Centre for Studies on Women and Society, Barcelona.

References

Bianchi, Patrizio and Forlai, Luigi (1988) The European domestic appliance industry 1945–87, in de Jong, H. W. (ed.) *The Structure of European Industry*. Dordrecht: Kluwer Academic Publishers.

Callon, Michel (1989) *La Science et Ses Reseaux: Genèse et Circulation des Faits Scientifiques* [*Science and Its Networks*]. Paris: La Decouverte.

Callon, Michel and Latour, Bruno (1988) Recherche, Innovation, Industrie [Research, innovation, industry], special issue of *Culture Technique* edited by these authors, No. 18, March. Neuilly sur Seine: Centre de Recherche sur la Culture Technique.

Chabaud-Rychter, Danielle (1987) La division sexuelle des techniques [The sexual division of technique], in *Les Rapports Sociaux de Sexe* [*Gender Relations*], Cahiers de l'APRE, No. 7, Vol. 1. Paris: Centre National de la Recherche Scientifique.

Chabaud-Rychter, Danielle, Fougeyrollas-Schwebel, Dominique and Sonthonnax, Françoise (1985) *Espace et Temps du Travail Domestique* [*The Space and Time of Domestic Work*]. Réponses sociologiques [Sociological answers]. Paris: Librarie des Meridiens.

Cockburn, Cynthia (1985) *Machinery of Dominance: Women, Men and Technical Knowhow*. London: Pluto Press.

Foucault, Michel (1979) *Microfisica del Poder* (*Microphysics of Power*). Madrid: La Piqueta.

Kergoat, Danièle (1978) Ouvriers = ouvrières? [Male workers equal women workers?] in *Critiques de l'Economie Politique* [*Political Economy Review*], new series, No. 5, October–December, pp. 65–97.

Kergoat, Danièle (1990) *Qualification et rapports sociaux de sexe* (*Skill and Gender*). Paper presented at the World Congress of Sociology, Madrid.

Latour, Bruno (1989) *La Science en Action* (*Science in Action*). Paris: Editions La Decouverte.

Owen, Nicholas (1983) *Economies of Scale, Competitiveness and Trade Patterns within the European Community*. Oxford: Clarendon Press.

Woolgar, Steve (1991) *Abriendo la Caja Negra* (*Science: The Very Idea*). Barcelona: Editorial Anthropos.

23 "Crash!: Nuclear Fuel Flasks and Anti-Misting Kerosene on Trial"

Harry Collins and Trevor Pinch

A great deal of the effort required to build sociotechnical systems is placed in the design of artifacts. The process cannot end there, however. Artifacts do not automatically become part of society once they are designed and produced. Their relationships with other artifacts, systems, and people must be worked out. This chapter examines a crucial part of the postdesign phase of a new technology—the process of convincing policymakers, corporate executives, the public, and other constituents that they can trust the system. Without this trust, those building the system will not be able to gather enough support to actually put it into place. As a society we have already put our trust in countless sociotechnical systems. We trust traffic lights to modulate a safe flow of traffic, we trust refrigerators to keep our food from spoiling, and we trust airplanes to transport us from one place to another safely though we are miles above the earth. As sociotechnical systems like these change slowly over time and without incident, our trust in them is reinforced. But some technologies are red-flagged as cause for particular concern and in need of special scrutiny. Justification for placing our trust in them must be demonstrated. As Pinch and Collins show, the goal of demonstration is different from that of experimentation or testing. The cases they present are provocative because they show that a successful demonstration is a matter of convincing the public that the technology has value, and this may have only a distant relation to the technology's capacity to achieve its intended—or advertised—functionality. Since the public presentation of a technology can make a huge difference in whether it is accepted, who will use it, and how it will be used, technology builders have to carefully plan these demonstrations. The demonstrations that Pinch and Collins describe raise questions about the accuracy of demonstration and the responsibilities of those who construct demonstrations.

The general public made the point, "well that's all right, but we've got to take the word of you experts...for it—we're not going to believe that, we want to see you actually do it." So well, now we've done it.... They ought to be [convinced]. I mean, I can't think of anything else.—If you're not convinced by this,...they're not going to be convinced by anything.

These words were uttered in 1984 by the late Sir Walter Marshall, chairman of Britain's then Central Electricity Generating Board (CEGB). The CEGB used the rail system to transport spent nuclear waste from its generating plants to its reprocessing plants. In spite of the fact that the fuel was contained in strong containers, or flasks, the public

From *The Golem at Large: What You Should Know about Technology* (New York: Cambridge University Press, 2002), pp. 57–75. © 2002 Cambridge University Press. Reprinted with the permission of Cambridge University Press.

was not happy. The CEGB therefore arranged for a diesel train, travelling at a hundred miles per hour, to crash head-on into one of their flasks to show its integrity. Sir Walter's words were spoken to the cameras immediately following the spectacular crash, witnessed by millions of viewers either on live television or on the nation's televized news bulletins. Sir Walter was claiming that the test had shown that nuclear fuel flasks were safe. (The source from which Sir Walter's quotation was taken and of the basic details of the train crash is a video film produced by the CEGB Department of Information and Public Affairs entitled "Operation Smash Hit.")

In America, in the same year, a still more spectacular crash was staged. This time it was a full-sized passenger plane that was crashed into the desert. The plane was a radio-controlled Boeing 720 filled with dummy passengers. It was fuelled, not with ordinary kerosene but with a modified version, jellified to prevent it turning into an aerosol on impact. The jellification was brought about by mixing the kerosene with the same kind of additive as is used to thicken household paint. The fuel was meant to help prevent the horrendous fires that usually follow aircraft crashes and allow passengers who survived the impact itself to escape. The new fuel was known as Anti-Misting Kerosene, or AMK; it had been developed by the British firm ICI. The Federal Aviation Agency (FAA) arranged the demonstration. Unfortunately the impact was somewhat too spectacular; the plane caught fire and burned fiercely. A television documentary reported the event as follows.

Before the experimental crash the FAA believed it had identified 32 accidents in America since 1960 in which AMK could have saved lives. But the awful irony is, the immediate result of the California test has been to set the cause back—maybe even to kill it off altogether.... Those damning first impressions of the unexpected fireball in the California desert last December remain as strong as ever.

The two crashes seem to provide decisive answers to the questions asked of them. The fuel flask was not violated by the train crashing into it and the airplane burned up in spite of its innovative fuel. The contents of the fuel flask had been pressurized to 100 pounds per square inch before the impact, and it is reported that only 0.26 pounds per square inch pressure was lost as a result of the impact, demonstrating that it had remained intact. Photographs of the Boeing after the flames died left no doubt that no passenger could have survived the fire in the fuselage. Each of these experiments was witnessed indirectly—via television—by millions of people, and directly by hundreds or thousands. The train crash was watched "on the spot," according to one newspaper, by "1,500 people including representatives of local authorities and objectors to nuclear power." Though the laboratory experiments and astronomical observations we described in the first volume of the Golem series were untidy, ambiguous, and open to a variety of interpretations, and though there is some uncertainty about the effectiveness of the technologies described in the other chapters of this volume, these two demonstrations seem to have been decisive. Is it that the need for careful interpretation that we have described throughout the two volumes can be avoided if experiments are designed well and carried out before a sufficient number of eyewitnesses? Is it that the key to clarity is not the laboratory but the public demonstration?

Two Crashes: A Solution to Technological Ambivalence?

According to the CEGB, the flasks used to transport spent nuclear fuel are made of steel. Their walls are about 14 inches thick and the lid is secured with 16 bolts, each capable of standing a strain of 150 tons. The flasks weigh 47 tons. For the crash, a flask and its supporting British Rail "flatrol" wagon were overturned on a railway track so that the edge of the lid would be struck by a "Type 46" diesel locomotive and its three coaches. Under the headline "Fuel Flask Survives 100 mph Impact," the *Sunday Times* of 18 July 1984 reported Sir Walter Marshall, chairman of the CEGB, describing it as "the most pessimistic and horrendous crash we could arrange." Sir Walter summed up the events for the TV cameras as described above (p. 407) stressing that no one should now remain unconvinced of the safety of the flasks.

NASA arranged that the Boeing 720 fuelled with AMK would land, with wheels up, just short of a "runway" on which were mounted cutters—to rupture the fuel tanks and rip the wings—and various sources of ignition. "Hundreds of spectators" watched. The *Sunday Times* (2 December 1984) said, "A controlled crash to test an anti-flame additive went embarrassingly wrong yesterday when an empty Boeing 720 turned into a fireball...," and, "A fireball erupted on impact." Certainly a large fire started immediately, and the plane was eventually destroyed. In a television documentary called *The Real World* it was said that "when the radio-controlled plane crashes into the desert, the British development team saw 17 years of work go up in flames." The commentator described events as follows: "...the left wing of the aircraft hits the ground first and swings the aircraft to the left and a fire starts. The senior American politicians cut short their visit to Edwards and the popular newspapers write AMK off as a failure."

A nuclear fuel "flask" is a hollow cube of metal, perhaps 10 feet high, with what look like cooling fins to the outside. In this case the flask was painted bright yellow. The square lid is bolted to the top. The train approached from a distance. The locomotive was painted blue and it towed a string of brown-painted coaches. Seen on television, everything seemed to happen in slow motion, though this impression may be strengthened by the many subsequent slow motion replays. Because the senses are unprepared for such an impact, it is, in a strange way, quite difficult to see—one does not know what to anticipate and what details to look for. It is as though one is preparing to look into the sun—one almost screws up one's eyes in anticipation. The train hit the flask with a tremendous explosion of dust, smoke, and flame, pushing the flask before it. Parts of the engine flew in all directions. Once the dust had settled one could see the flask among the debris. A section of the fins had been gouged by the impact but, as the pressure test showed, apart from cosmetic damage, the flask was otherwise inviolate. A promotional video of the event, made by the CEGB, describes how fuel flasks are constructed and repeats the history of the test; it shows the crash, underscoring its significance.

Figure 23.1
Train crashes into nuclear fuel flask.

With the airplane crash the sense of witnessing the, literally, inconceivable was even more pronounced. Here was a full-scale passenger plane, something of obvious high financial and technological value, something that we are used to thinking of as a symbol of exaggerated care and maintenance, shining silver against the crystal-line blue sky, about to crash into the pale desert floor, made even more perilous by jagged cutters, designed to rip the wings apart, set into the mock runway. The airplane struck with a vicious tearing; parts flew off; slow motion showed a door ripped off and the immediate billowing up of a red-yellow flame which enveloped the aircraft. A distant shot showed a column of black smoke climbing high into the desert sky. Subsequent close-ups showed the burnt-out fuselage and the burnt remains of dummy passengers.

The author of this chapter (Collins) has tried his own experiment, showing videos, or sets of slides, of the train crash to various audiences, and asking them to point out any problems—any reason why we should refuse to believe the evidence of our senses and fail to accept Sir Walter Marshall's claim. While Sir Walter's presentation, with its used-car-salesman's brashness, tends to encourage cynicism, audiences have rarely been able to come up with any technical reason for drawing contrary conclusions. Seeing the film of the plane crash in its initial cut has the same effect—one cannot imagine anything other than that it shows that AMK failed. The contrast with the ambivalent scientific experiments and technological forays we have described elsewhere in these volumes seems stark. We need, however, a more sophisticated understanding: we need first to make some conceptual distinctions between experiments and

Figure 23.2
Boeing 720 about to crash into obstacles in "demonstration" of anti-misting kerosene.

demonstrations, and then we need to re-analyse these two incidents with the help of experts....

Experiments and Demonstrations

The Golem series deals with matters which are controversial. In so far as the terms have any meaning at all, making use of a commonplace technological artifact—driving in one's car, writing a letter on one's word-processor, storing wine or whiskey in a barrel—is a demonstration of the reliability and predictability of technology: these things work; about such things there is no controversy. And that is why they are politically uninteresting—the citizen need know nothing of how such things work to know how to act in a technological society. But each of these technologies may have a more politically significant controversial aspect: Are diesel cars more or less polluting than petrol cars? Is literacy being destroyed by the artificial aids within the modern word-processor? Are the trace chemicals leached out of smoked barrels harmful to health (we have not the slightest reason to think they are)? The treatment of the controversial aspects must be different to the uncontroversial aspects. The same is true of what we loosely refer to as experiments: one does not do *experiments* on the uncontroversial, one engages in *demonstrations*. Historically, as Steven Shapin argues, there has been a clear distinction between these two kinds of activity.

In mid to late seventeenth-century England there was a linguistic distinction...between "trying" an experiment [and] "showing" it....The trying of an experiment corresponds to research proper, getting the thing to work, possibly attended with uncertainty about what constitutes a working experiment. Showing is the display to others of a working experiment, what is commonly called demonstration....I want to say that trying was an activity which in practice occured within relatively private spaces, while showing...[was an event] in relatively public space.

To move to the nineteenth century, Michael Faraday's mode of operation at the Royal Institution shows the same distinction between experiment and demonstration. Faraday practised the art of the experimenter in the basement of the Royal Institution but only brought the demonstration upstairs to the public lecture theatre once he had perfected it for each new effect. As David Gooding puts it:

Faraday's *Diary* records the journeys of natural phenomena from their inception as personal, tentative results to their later objective status as demonstrable, natural facts. These journeys can be visualised as passages from the laboratory in the basement up to the lecture theatre on the first floor. This was a passage from a private space to a public space. (p. 108)

The hallmark of demonstrations is still preparation and rehearsal, whereas in the case of an experiment one may not even know what it means for it to work—the experiment has the capacity to surprise us. Demonstrations are designed to educate and convince once the exploration has been done and the discoveries have been made, confirmed, and universally agreed. Once we have reached this state, demonstrations have the power to convince because of the smoothness with which they can be performed. Indeed, the work of being a good demonstrator is not a matter of finding out unknown things, but of arranging a convincing performance.

Demonstrations are a little like "displays" in which enhancements of visual effects are allowed in ways that would be thought of as cheating in an experiment proper. For example, in a well-known lecture on explosives, the energy of different mixtures of gases is revealed. If the mixtures are ignited in milk bottles they make increasingly loud bangs as more energetic reactions take place. When I watched such a lecture, a milk bottle containing acetylene and oxygen, the most powerful mixture, was placed in a metal cylinder while a wooden stool was set over the top of the whole arrangement apparently to prevent broken glass spreading too far. When the mixture exploded the seat of the wooden stool split dramatically into two pieces. Subsequent inspection [by the author] revealed that the stool was broken before the test and lightly patched to give a spectacular effect. In an experiment, that would be cheating, but in a display, no one would complain. A demonstration lies somewhere in the middle of this scale. Classroom demonstrations, the first bits of science we see, are a good case. Teachers often know that this or that "experiment" will only work if the conditions are "just so," but this information is not vouchsafed to the students.

The presentation of science to the general public nearly always follows the conventions of the demonstration or the display. If we are not specialists, we learn about science firstly through stage-managed demonstrations at school, and secondly through demonstrations-cum-displays on television. This is one of the means by which we think we learn that the characteristic feature of scientific tests is that they always have clear outcomes. A demonstration or display is something that is properly set before the lay public precisely because its appearance is meant to convey an unambiguous message to the senses, the message that we are told to take from it. But the significance of an experiment can be assessed only be experts. It is this feature of experiment that the Golem series is trying to make clear; some more recent television treatments of science are also helping to set the matter right. It is very important that demonstration and display on the one hand, and experiment on the other are not mistaken for one another.

One final and very simple point is worth making before we return to the story of the two public "experiments." The simple point is that it is not true that "seeing is believing." Were this true there would be no such profession as stage magic. When we see a stage magician repeat an apparently paranormal feat, even though we cannot see the means by which it was accomplished, we disbelieve the evidence of our senses; we know we have been tricked. The same applies in the case of the "special effects" we see at the cinema, the case of some still photographs, the case of "optical illusions" on the printed page and the case of every form of representation that we do not read at face value.

Contrasting, once again, the demonstration with the stage magician's feats, whether we believe what we see is not solely a matter of what we see but what we are told, either literally—as in the case of Sir Walter Marshall's comments to the television—or more subtly, through the setting, the dress, and the body language of the performers, and who is doing the telling; it is these that tell us how to read what we see.

The Crashes Reanalysed

The Train Crash Revisited
The self-same manipulation of reality can be both an experiment and a demonstration. It can demonstrate one thing while being an experiment on another; it can fail in one aim while succeeding in the other. Both crashes were like this, but in different ways. The train crash was a superb demonstration of the strength and integrity of a nuclear fuel flask, but at the same time it was a mediocre experiment on the safety of rail as a means of moving radioactive materials. The plane crash was a failed demonstration of the safety of AMK but, as we will see, it was a very good experiment. There is nothing more that needs saying about the demonstration aspect of the train crash, but it is

worth pointing out that Sir Walter, and the subsequent promotional material, encouraged us to read it as a successful experiment on nuclear waste transport policy. To find out why it was not such an experiment, we need to turn to the experts.

The conservationist pressure group "Greenpeace" employs its own engineers. Their interpretation was somewhat different to that of Sir Walter Marshall. By their own admission, the CEGB were so sure that the flask would not break under the circumstances of the crash, that they could learn little of scientific value from it. They already understood quite a lot about the strength of flasks from previous private experiments and knew the circumstances under which they would not break. If Greenpeace are to be believed, CEGB also knew circumstances under which the flasks would break. Greenpeace did not question that this particular flask had survived the 100 m.p.h. impact intact, but they suggested that even as a demonstration, the test showed little about the transport of nuclear fuel in 1984. They claimed that the majority of the flasks used at the time were of different specification: the test specimen was a single forging of 14-inch thick steel, whereas the majority of the CEGB's flasks were much thinner with a lead lining. Flasks of this specification, they suggested, had been easily broken by the CEGB in various preparatory experiments done in less public settings.

Furthermore, Greenpeace did not accept that the test involved the most catastrophic accident that could occur during transport of nuclear fuel. Greenpeace held the crash to be a far less severe test than could have been arranged; they believed that the CEGB, through their consulting engineers Ove Arup, were aware of a number of crash scenarios that could result in damage even to one of the stronger, 14-inch thick, flasks. These included the flask smashing into a bridge abutment at high speed after a crash, or falling onto rock from a high bridge. Thirdly, they believed that the design of the test carefully avoided a range of possible problems. The CEGB, said Greenpeace, had the information to stage things in this way because, while developing the demonstration, they had done extensive work on different types of locomotive, different types of accident, and so forth.[1]

Specifically, Greenpeace claimed that the "Type 46" locomotive used in the test was of a softer-nosed variety than many other types in regular use. Thus, to some extent, the impact was absorbed in the nose of the engine. They claimed that the angle of impact of train to flask was not such as to maximise the damage. They claimed that the carriages of the train had been weighted in such a way as to minimize the likelihood that they would ride over the engine and cause a secondary impact on the flask. They claimed that because the wheels of the flatbed wagon used to carry the flask had been removed, there was no chance of them digging into the ground and holding the flask stationary in such a way as would increase the force of impact. (The film of the crash shows that the wagon and flask were lifted and flung through the air by the train, offering little resistance to its forward motion.) Finally they claimed that the wagon and flask were placed at the very end of the rail track so that, again, there was nothing in the way of sleepers (ties) or track to resist the movement of the flask as it was pushed along the ground—everything beyond the impact zone was smooth earth—there was nothing to hold the flask or penetrate.

Figure 23.3
Nuclear fuel flask being examined for damage after impact.

We do not know whether the commentary of Greenpeace's engineers is more accurate than that of the CEGB and this is not our business. We cannot judge whether Greenpeace or the CEGB had it right. Perhaps the flask would have survived even if the wheels had been left on the wagon and the rails had been left in place. We do not have the expertise to be sure.

Given all this, the crash test seen on TV may have been spectacular and entertaining, but there are many ways in which a real experiment could have been a more demanding test of the waste transport policy. The test demonstrated what the CEGB already knew about the strength of flasks, but it did not show that they could not be damaged under any circumstances. Sir Walter Marshall, as represented on TV, encouraged the public to read the outcome not as demonstrating the CEGB's mastery over a small part of the physical universe, but as demonstrating that nuclear flasks were safe. The public, then, were put in a position to read a demonstration of one small thing as a proof of something rather more general.

The Plane Crash Revisited
In the case of the plane crash, several experts were interviewed by the makers of a British television programme and it is this expertise that we tap.[2] This programme attempted to reinterpret the crash so as to support the idea that the test had demonstrated not the failure of AMK, but its effectiveness. The programme reassembled much of the film shot from the various angles, overlaying it with explanation and comment from representatives of the Federal Aviation Authority and Imperial

Figure 23.4
Two identical sled-driven planes, fuelled with ordinary kerosene (top) and anti-misting kerosene (bottom), are crashed in tests prior to the public demonstration.

Chemical Industries (ICI), the British firm that made the jellifying additive and developed AMK. We see, among other things, that television does not have to simplify.

In spite of the fire and the destruction of the airplane, the programme team claimed that the test was not such a failure for AMK as it appeared to be. We will use our demonstration/experiment dichotomy to describe the findings of the programme team, treating the demonstration aspects first and the experimental aspects second.

The demonstration should have resulted in no fire, or a small fire at most. The demonstration had been worked up by studying many collisions using old non-flight-worthy airplanes propelled at high speeds along test tracks. This series of tests had shown that AMK should work; it remained only to convince the public and the aviation pressure groups with a spectacular test. According to an ICI spokesman,[3] the Federal Aviation Agency, NASA and ICI were all convinced of the effectiveness of AMK before the test, and thought of the test as a demonstration. But the demonstration did not go exactly as planned; the ground-based pilot lost control of the radio-controlled plane. Instead of sliding along the runway with wheels down, the plane hit the ground early, one wing striking before the other. The plane slewed to one side, crashed into the obstacles and came to a rapid stop. A metal cutter designed to rip into the wings and fuel tanks entered one of the still spinning engines, stopping it instantly and causing a far greater release of energy than would normally be expected. This, in turn, caused the initial fireball which would not have happened if the demonstration had gone according to plan.

It might be argued that airplane crashes never go according to plan, but this is to miss the point that AMK had already been shown to be safer than ordinary kerosene under many crash circumstances. Most crashes would not involve an impact between a spinning turbine and a rigid metal object. If AMK could improve survival chances in most crashes that would be enough to merit its use; there was no need for it to be effective in every imaginable case. Therefore it was appropriate to demonstrate the effectiveness of AMK in a demonstration of a typical crash, rather than an extreme crash such as the one that actually happened. Bear in mind also that there is no point in exploring the effectiveness of AMK in crashes that are so violent that the passengers would not have survived the impact; this again makes it clear that it was a relatively mild crash that was intended, and appropriately intended. (Compare this with the nuclear fuel flask case, where even one accident involving the release of radioactivity would be unacceptable.) It is clear that what we have in the case of the plane crash is a demonstration that went wrong.

The very fact that the demonstration went wrong moved the event into unknown territory for which the scientists and technologists had not prepared; that is, the intended demonstration turned into an experiment. An ICI spokesman said, "in [accidentally] introducing a number of unknown effects we were able to observe things that we had not seen before and were able to draw a number of useful conclusions from them."[4] Appropriately interpreted, this experiment can be read as a success for AMK even under the most extreme of survivable conditions.

The initial fireball, horrendous though it appeared, was less severe than might have been expected given the nature of the crash. The fire swept over the fuselage, but did not penetrate into the passenger cabin or the cockpit. It is claimed that a proportion of the AMK which spilled as a result of the first impact did not burn; on the contrary, it splashed over the fuselage and cooled it.

This initial fire died out after a second or so. Because it did not penetrate the fuselage, this fire was survivable if terrifying. Passengers could have escaped after the initial fireball had died away. An airplane fuelled with ordinary kerosene involved in a crash of this sort would have burned out immediately, giving the passengers no chance.

Though the plane was eventually destroyed, the major damage was caused by fuel entering the plane through an open cargo door and through rents in the fuselage made by the cutters as the plane slewed to one side. This second fire, though it was not survivable, did not start until after the passengers, or at least some of them, would have had time to escape.

Furthermore, the test showed in another way that AMK is far less flammable than ordinary jet fuel. This can be seen because even after the second fire had destroyed much of the plane, there were still 9,000 gallons of unburnt fuel left in and around the plane. The programme showed film of this unburnt fuel being salvaged.

A spokesman for ICI says on the programme:

Subsequently, with the analysis of what happened, those who have been technically involved actually believe it was a success—that... the aircraft had a smaller fire than had been expected if there had been jet-A [ordinary aircraft kerosene]—a benefit that we hadn't even expected at all in that... the anti-misting fuel helped to cool the aircraft and provide conditions within the aircraft which would have allowed people—some people—to escape.

Here television had a role to play in defining the test as a failure. The critical period is between the initial fireball and the subsequent extensive fire. It is during this period that passengers still able to move could have escaped from the aircraft. And yet the natural way of cutting this episode for television was to move from the first fireball to the smoke cloud in the desert: there is nothing spectacular about the long period in-between when nothing happens. One must remember that the direct evidence of the senses is not the same as television. Anything seen on television is controlled by the lens, the director, the editor and the commentators. It is they who control the conclusions that seem to follow from the "direct evidence of the senses." A respondent who was present at the NASA test—the scientist who developed the fuel additive—said that the meaning of the test was by no means obvious to those who were actually there.

The observation points were about half a mile from the actual landing site and it took about an hour to assemble back at the technical centre. The instantaneous reaction of the people in my immediate vicinity and myself, when we saw that, after the fire-ball (larger than expected) died away (about 9 seconds), the fuselage appeared to be intact and unscathed, was success—a round of applause!

In the case of the plane crash we are extraordinarily lucky to have been able to watch a television reconstruction informed by experts who had an interest in revealing the ambiguities that are normally hidden.

Once more, we do not intend to press ICI's case against the interpretation of the local experts, but simply to show that as soon as the plane crash is seen as an experiment, rather than demonstration, enormous expertise is required to interpret it, and a variety of interpretations is possible. It is not our job to choose one interpretation over the others; it is not that we do not care—after all, human lives depend on the solution—but it is not our area of expertise.

Figure 23.5
What the television viewer saw: the Boeing 720 crash as photographed from the television screen.

Imagining What Might Have Been Done

We can reinforce the distinction between demonstration and experiment by imagining how these things might have been done differently. Imagine how Greenpeace might have staged the demonstration if they had been in control. One would imagine that they would first experiment in private in order to discover the conditions under which minimum force *would* damage a flask. An appropriate demonstration might involve firmly positioned rails which would penetrate one of the earlier thin-walled flasks when it was struck by a hard-nosed locomotive going at maximum speed. Perhaps the accident would take place in a tunnel where the flask had initially been heated by a fierce fire. About six months after the CEGB's experiment (20 December 1984), a train of petrol tankers caught fire inside a tunnel producing temperatures well in excess of anything required during statutory tests of fuel flasks. It seems possible that a flask caught in that fire would have been badly damaged. At least the event revealed the possibility of a hazard not mentioned during the CEGB's crash test.

A simpler, but effective, Greenpeace demonstration could reveal what would happen *if* a nuclear flask were damaged, without worrying too much exactly how the damage might be caused (or assuming that terrorists could gain access to the container). Thus, a demonstration using explosive to blow the lid off and spread a readily identifiable dye far and wide would serve to show just how much harm would result were the dye radioactive.

Either of the imagined demonstrations described above would lead to a conclusion exactly opposite to that advanced by Sir Walter Marshall. They would both suggest that it is dangerous to transport spent nuclear fuel by train. Yet neither of the imagined demonstrations would be *experiments* on the safety of rail transport for nuclear fuel, and they would no more prove decisively that it is unsafe than the CEGB's demonstration proved decisively that it is safe. Nevertheless, by imagining alternative kinds of demonstration, we are better able to see what these demonstrations involved; we can see what a demonstration proves and what it does not prove.

The same applies to the plane crash. We can easily imagine the demonstration going according to plan, the Boeing 720 landing wheels-up without bursting into flames, and the TV cameras entering the fuselage of the unburnt aircraft and revealing pictures of putatively smiling unscorched dummies. Yet that scene too hides some hidden questions. Could it be that the particular crash was unusually benign so that AMK has the potential to make a difference on only very few crashes? Could it be that the extra machinery needed to turn jelly-like fuel into liquid before it enters the engines of jet planes would itself be a cause of crashes? Could it be that the transition period, requiring two types of fuel to be available for two types of plane, would be so hazardous as to cost more lives than the new fuel would ever save? Could it be that the extra cost of re-equipping airlines, airports, and airplanes, might jeopardise safety in other ways at a time of ruthless competition in the airline market? Again, perhaps there was too little time for anyone to escape between first and second fires; perhaps cabin temperatures rose so much during the first fireball that everyone would have been killed anyway; perhaps there is always unburnt fuel left after a major crash; perhaps everyone would have died of fright. Both the analysis and the conclusions to these questions need more expert input than we can provide.

Conclusion

We started this chapter asking whether the solution to experimental ambivalence was to be found in great public spectacles. In the case of the experiments we described in *The Golem* [Cambridge University Press, 1994], and in the case of the other technologies described here, experts, given the most detailed access to the events, disagree with one another, and an outcome to a debate may take many years or even decades to come about. Yet in the case of these two crashes it seemed that the general public, given only fleeting glances of the experiments, were able to reach firm conclusions instantly. The reason was that they were not given access to the full range of interpretations available to different groups of experts. The public were not served well, not because they necessarily drew false conclusions, but because they did not have access to evidence needed to draw conclusions with the proper degree of provisionality. There is no short cut through the contested terrain which the golem must negotiate.

Notes

1. Greenpeace claim that their "leak" as regards the CEGB's earlier programme of experimental work came from inside the CEGB itself, and from an independent engineering source close to Ove Arup, the firm that ran the test for the CEGB. They say that when their claims were presented to an authoritative seminar organised by the Institute of Mechanical Engineers, their arguments were not proved unfounded except in one small respect. (At the time, Greenpeace claimed that the bolts securing the engine of the locomotive had been loosened by British Rail. They retracted this later.) I have not done any "detective work" to test Greenpeace's claims, so I present this information merely to suggest further that the counter-interpretations put forward by Greenpeace were technically informed and can be held by reasonable persons.

2. The programme was called *The Real World*. A remarkable edition of this entitled "Up in Flames" was broadcast on Independent Television.

"Up in Flames" is an interesting TV programme in the way it re-interprets scientific evidence. The handling of scientific evidence and doubt is very sophisticated yet still fascinating. Nevertheless, the programme draws back from forming a general judgement about the widespread flexibility of the conclusions that can be drawn from experiments and tests. The programme presses the FAA view on the viewer. We do not endorse the FAA view but use the programme as a means of revealing the alternative interpretations than can be provided by different sets of experts.

3. Private communication with the author. Interestingly, the ICI spokesman used the terms "demonstration" and "experiment" without bidding.

4. Private communication with the author.

24 "When Is a Work Around? Conflict and Negotiation in Computer Systems Development"

Neil Pollock

It is commonly accepted that engineers, politicians, and corporations have the ability to shape technology. What is sometimes underappreciated, however, is that users play an important role in the fate of sociotechnical systems as well. Users have the ability to affect the use, meaning, endurance, and impact of these systems. In this chapter, Pollock argues that the original designers of technologies build scripts into them—that is, artifacts are designed to carry out certain tasks by certain users in certain ways. These scripts can be conveyed through user manuals, warning labels, and the physical design of the material object. But users are not forced to adhere to these scripts. Even when certain values are inscribed into an artifact, users can circumvent the designer's conception to fulfill his or her own desires and advance his or her own values. Users can, in effect, redesign technologies, employ them in new ways, and achieve new purposes. The fact that so many people can direct the power of sociotechnical systems leads to complex questions about where responsibility lies when something goes wrong.

Introduction

What is a "work-around"? Typically, the concept is used to explain how one actor is able to adjust a technology to meet his or her particular needs or goals. Indeed, one of the most significant analyses of the practice of work-arounds appeared in the work of Les Gasser some years ago; Gasser describes work-arounds in relation to the ad-hoc methods deployed by users of administrative computer systems who were attempting to fix problems or glitches in their work. Gasser wrote that working-around means "...intentionally using computing in ways for which it was not designed" or avoiding a computer's use and "...relying on an alternative means of accomplishing work" (1986, 216).[1] Sociologists in general, and sociologists of technology in particular, continue to be fascinated by the practice and process of work-arounds. This, it might be suggested, reflects a wider interest in showing how users are not simply shaped by technologies but how they are also shapers of technology. Put another way, the term is often a useful trope to emphasize the differences between the "logics of a technology" and the "logics of human work" (Berg 1998), with the actual practice of work-arounds highlighting the effort necessary to bring these two factors into line. Gasser, in this sense, can be read as an account of how actors, through deploying some form of effort or skill, are able to overcome a difficulty or a constraint imposed by a technology. A stronger version of this argument is perhaps Bryan Pfaffenberger's (1992)

From *Science, Technology & Human Values* 30, no. 4 (Autumn 2005): 496–514. © 2005 Sage Publications Inc. Reprinted by permission of Sage Publications Inc.

description of the "technological adjustments" carried out by users when a new production process or artifact is introduced into their work setting. As Pfaffenberger sees it, the users, rather than accept the discipline of the new system, "...engage in strategies that try to compensate for the loss of self-esteem, social prestige, and social power that the technology has caused" (1992, 286). Typically, then, the common understanding of work-arounds is clear and unambiguous; they represent resistance on behalf of users and the means by which they attempt to wrest control back from a technology or an institution.

What motivates this article is the way in which work-arounds have become a much-used resource within the sociology of technology but, with a few notable exceptions,[2] as a topic they remain for the most part surprisingly underinvestigated and theorized. What is often missing in many discussions is any reference to their genesis or outcomes other than these general notions that users also shape technology or that work-arounds correct a misalignment between a technology and the desired goals of its users.[3] In contrast, I argue that a reappraisal of the term is both important and timely for two main reasons. First, as computer systems, the technology discussed in this article, spread ever more widely and into increasingly diverse and new domains, powerful incentives for increased standardization are brought with them (cf. Agre 2000). This leads to inevitable tensions about which elements in these settings should be standardized and which elements should not (cf. Star & Ruhleder 1996). An analysis of user work-arounds remains an essential part of understanding how such "misalignments" are reconciled.

Secondly, in fields such as management and administration computer systems development, there is a blurring of the once-clear distinction between users and producers of technologies. Increasingly, many systems are designed and built so that they are customizable by their users (Brady et al. 1992), meaning that users also engage in the construction of these technologies. The upshot is that it is now increasingly difficult to say exactly who has responsibility for the final shaping of systems and their implementation (cf. Suchman 1994). In this context of changing and less determinate technical divisions of labor and responsibilities, there is a need for analysis that puts the user and his or her modifications, as well as the ambiguity surrounding the process, at the center of its concerns.

The aim of this article, then, is to reawaken our interest in the topic of work-arounds in light of these new and more complicated technological practices. To try to do this I present the example of a group tasked with the job of customizing and implementing a "prebuilt" management information computer system (known as MAC) within the centralized administration departments of a university. Modifying technology is a routine and necessary aspect of this group's work, although they often find that some of their work-arounds promote tensions between them and the original designers of the system. Below I attempt to develop a basic understanding of some of the factors that lead to these work-arounds—what I am calling "networks in place"—as well as some of the tensions that lead from them. A backdrop to this study is work stemming from the actor-network approach, particularly Madeleine Akrich's (1992) important article on how technologies embody "scripts." In this first part, I review this article as well as make some suggestions about how we might adapt and deploy this form of thinking.

The Designer-Script-User Approach

In one of the most cited articles in the sociology of technology, Madeleine Akrich (1992) describes how designers, when building technologies, also build "scripts" into those technologies. Users, she argues, once they take up and use a technology, can then be seen to be enacting a script,[4] though she is careful to point out that scripts are never enacted straightforwardly, as users will often perform work-arounds, or what she calls "mechanisms of adjustment," to modify an artifact (and script) to more closely fit their particular circumstances. To work through this concept, she discusses the design and use of a photoelectric kit that was providing electricity to a village in French Polynesia. She outlines how the photoelectric kit suffered from one major problem: when electricity was most in demand, the kit was apt to break down. The power outage occurred because it was possible to damage the kit if it was allowed to run down, and engineers, assuming the users would be unable or unwilling to properly maintain the kit (i.e., the script), installed a control device that would make the kit inoperable. However, as the control device was continually breaking the circuit, residents would call upon the local electrician, who, tired of receiving calls late into the evenings, eventually installed a fused circuit in parallel with the control device. This meant that when the power was cut off, the users could bypass the problem themselves by using the new fuse.

The key issue for consideration here is that if we are to accept that users play out scripts when using technologies, how are we to understand modes of use that deviate from the script? Are they simply a result of other scripts, the agency or skill of people, or something else? One reading of the Akrich paper is to say that the problem is posed at the level of a choice or a dilemma. The electrician can either let the users live with the technology as designed and succumb to its prescription (i.e., have the inconvenience of constant power interruptions) or he or she can install a fuse, but this might be to risk straining relations with the designers of the technology (the electric company). In other words, the suggestion is that the electrician, through deploying his or her skill, is able to exercise some form of discretion.

As already suggested, the danger is that without unpacking a work-around and looking at what leads to and from work-arounds, it is possible to read the situation as the electrician having control over the situation and being able to decide on possible outcomes and bring these about. Moreover, it can also be read that all of this will happen at the expense of certain other entities and actors (i.e., the electrician is wresting some form of control back from the technology or the electric company).[5] In contrast, Mike Michael (1996) makes the appealing argument that just as we can describe a technology as prescribing one form of use, perhaps the same technology might also incorporate a script that enables its abuse. A technology does not embody simply one script, or order, but, according to Michael, they can embody multiple scripts.[6] Moreover, these multiple scripts can often be contradictory, meaning that just as a car, for instance, can demand a certain form of use (i.e., safe and careful driving), it can also enable the reverse (i.e., road rage, which can be used to intimidate other drivers). While an interesting argument, it raises some further questions: if a technology does embody multiple or contradictory scripts, then why are certain uses more likely than others? Why, in the main, do car drivers follow the "safe and careful" form of use? Is the user disciplined toward one role over the other? Seemingly, yes, or at least this is what Bruno Latour (1992) argues that engineers bet on when they attempt to anticipate the desires or goals of their users. Latour writes that this way of counting on earlier distribution of skills to help narrow the gap between "built-in users" or "users in the flesh" is like a "pre-inscription" (257 [178 in this volume]). In short, what he is suggesting here is that the tendency toward one form of use is already present in the wider network. Another method of describing such networks might be to talk about a "network in place."

In the following discussion of work-arounds, there are aspects from the above that I want to take up and develop: these are Michael's concern for multiplicity and what I am calling, after Latour, a network in place. The argument is that the tendency for work-arounds is already present in the networks that those implementing the computer system (MAC) inhabit—these networks in place. Having established this, I then analyze these networks by considering some of their contingencies—for example, the other actors and entities on which these networks in place depend and by which they are constituted. In particular, I examine their connections to the original designers of MAC and the computer system itself. Both are pivotal actors who simultaneously demand and promise the possibility of work-arounds and major obstacles, questioning and hindering the progress of the implementation.

One final clarification is needed before we turn to the empirical material. It will have become apparent that I have been discussing the language and concepts associated with the use of a technology and that I am attempting to apply this to an example that is normally thought to be one of implementation. I think there are good reasons for doing so. Namely, as has already been suggested, designers and users are not well bounded. Mackay et al. (2000) argue that the conventional distinctions between production and use cannot always be applied to information technologies, as users are becoming more like producers.[7] We might, perhaps, advance this argument in the other direction and suggest that, just as the notion of the user was found to be more complex than was traditionally assumed, producers also increasingly play contrasting roles. For instance, a technology like MAC is not reliant on one set of clearly defined producers delivering a system to a user but on an extended network of computer professionals working in and for different organizations. The group implementing MAC at the local site was made up of people with various levels of skill and expertise, ranging from those who had experience with similar implementations elsewhere to those who had been recently seconded in from nontechnical roles in other parts of the university. This group found that they were one element in this long chain and that they were tasked to work with the system in a certain way; this was linked to the efforts of the original designers of the computer system to ensure that MAC's code was modified only in the ways they deemed appropriate. In other words, the designers were attempting to configure the local programmers as their "users."[8] Indeed, as Friedman describes, hardware and software suppliers often think of the computer-system developers to which they sell products as their users. It is clear, then, that the meaning of user is shifting as the nature of computing changes (Friedman 1989; Mackay et al. 2000). Indeed, for Suchman, the key is to deconstruct such simple terms as designer and user and, at the same time, bring to the fore the relevant social relations that cross the boundaries between these two groups. In this sense, it might be suggested that workarounds represent one aspect of the relations between these groups, as well as the means by which the producer/user boundary is constituted. This leads us into an examination of the MAC system.

MAC and the Delphic Oracle

The oracle of Apollo at Delphi that gave answers held by the ancient Greeks to be of great authority but also noted for their ambiguity.[9]

The material produced here is from an ethnographic study carried out at one of the university sites where MAC was being implemented. Indeed, the MAC exercise involved most universities in the UK, as the system resulted from a decision by the centralized Universities Funding Council (UFC) in 1988 to "...take action to meet the increasing need for more and better management information systems in universities" (Goddard and Gayward 1994, 45). The idea was that the "...cost would be reduced substantially by universities working together to develop new systems common to all" (ibid., 45):

The UFC therefore established the Management and Administrative Computing (MAC) Initiative, a unique attempt to transform administrative computing across the whole university system. The initiative was placed under the control of a MAC Initiative Managing Team and all institutions were to be brought into cooperative groups (called Families) with the aim of all members of each Family eventually using the same administrative computing software and jointly developing and maintaining it. (ibid., 45)

The original designers and builders of the computer system (hereafter, the Designers), which was implemented at the site where I did my study, work for one of the world's largest software organizations, Oracle (hereafter, the Technology Vendor). In order to manage the implementation, the universities created a company called "Delphic" that was directly responsible for liaising between the Technology Vendor and each of the sites. While several of the people that I worked with, especially those who had spent time at the Technology Vendor on behalf of Delphic, pointed out that "frictions" existed between the Programmers employed by the universities and those working for the Technology Vendor, it might also be suggested that the word "delphic" is an accurate description of this relationship. Most of this ambiguity existed around the so-called "80/20" rule. By this it was meant that the system was something of a "grey box" (cf. Fujimura 1992): the design and building of the bulk of the system, the responsibility of the Technology Vendor, would then be delivered to each of the sites. Importantly, however, a small part of the systems was left to the discretion of computer Programmers working at each of the universities, who—working in close relation with the Designers—would attempt to tailor the system to the specifics of each of the sites. The boundary between the 80 and 20 and this tailoring work was the focus of my study.

The Delphic Support Desk
I spent several months working with one group of Programmers in an attempt to understand just how they managed to get their MAC system to work. One of the most intriguing things about studying this group of Programmers, and something that I had barely anticipated before I started the research, was that they would sit for hour after hour in front of their terminals, barely uttering a word. To ask them a question would be seemingly to break their concentration with the machine, to disturb the peace of the office. Even when sitting inches away from them, I was to learn nothing about the implementation. Nonetheless, the longer I was there, it seemed, the more they got used to me. After a while, I found that every now and then they would stop working to tell me something about what they were doing.[10]

Much later, however, I would realize that even while we had sat there in silence, they were in fact speaking, sometimes shouting. Their method of communication was electronic mail. It was this realization that they were in fact talking in the main via e-mail (sometimes even preferring to e-mail the person sitting across from them!) that led me to begin to sift through old, archived messages. One particularly interesting source was something called the "Software Problem Bulletin," which was a sort of on-line help desk or Problem Log run by the *Delphic Support Desk*.[11] The Programmers used the Problem Log as a type of last resort: if they were unable to resolve difficulties concerning MAC within their own local communities, then they reported the problem to the Support Desk, who either suggested a possible solution or passed the message as a possible bug to the Technology Vendor. The Technology Vendor would respond to each message by appending their comments (i.e., their answer to the problem). The Log was available to all the Programmers at the various sites, and they too would often post suggestions in reply to a message.

Comprising some several thousand e-mails, the Log reads like a working history of all the steps taken so far on the project. I had heard the term *work-around* continuously from the moment I first became involved with the Programmers, and the word appears in the Log. I ran the "Find" facility on my word processor to gain access to other discussions about work-arounds. The first message found was a description of a problem. Over the Easter period, Carole, one of the Programmers working at a site, had attempted to install the latest release of the MAC system, version 1.4, just released by the Technology Vendor. At the same time, she attempted to upgrade the software platform that MAC would run on, Oracle 7.1.3. However, there was a problem: MAC 1.4

cannot be loaded onto Oracle 7.1.3. According to the e-mail, a small program called BuildMAC, written by the Technology Vendor to assist in such upgrades, would not perform as it should. The message goes on to mention how a similar problem was reported at one of the other sites some months earlier. A Programmer named Liz had been attempting the same process and, as for Carole, the BuildMAC program had not carded out the upgrade.

Intrigued by the discussion being carried out on the Problem Log, I continued to search the postings, hoping to understand more about the genesis of this problem. Seemingly, it had begun when Liz had written to the Delphic Support Desk, describing her difficulties, and was told by one of the Programmers:

Liz, unfortunately Oracle 7.1.3 is unsupported against all MAC software currently released, so these problems cannot be reported as bugs to [the Technology Vendor] but they may like to have the problems passed on for "information purposes only" to help them prepare MAC for Oracle 7.1.3.

In responding to this message, Liz pointed out that when they first ordered Oracle 7.1.3, they did ask the Technology Vendor which version would be most suitable, and they were told that their choice would be fine. The matter was not mentioned again in the Problem Log, and despite the fact that MAC is not supported against her particular software platform, Liz attempted to modify BuildMAC by reworking its code. Moreover, once the work-around was complete and Liz had loaded the new version of MAC, she posted the rewrite to the Log as information for others. I will develop this discussion in a moment, but first I want to consider a different issue: what can be said about the mode of use of the Programmers—their attempts to modify the code?

The Programmers at the site where I carried out my study defined work-arounds as a necessary and important aspect of their routine work. Things would never quite fit or be the way they should be. Often, a feature of the system would be too complicated for the end users, or one aspect of the new system would not work with the existing software infrastructure. Such problems require innovative fixes or the rewriting of code. Consider the following diary extract of a conversation I had with someone named David:

There are a few problems with loading the data into the system and David says: "It's OK, I'll work around it." David continuously talks about work-arounds. I laugh and say to him, "Another work-around. It seems to be all work-arounds here." "That's life," he replies (a little dryly).

To program is to perform work-arounds, to bypass constraints, and to rewrite code. In other words, we might think of these Programmers—and, indeed, it would be in keeping with how they think of themselves—as *bricoleurs par excellence*.[12]

The image I want to develop here is of people drawing on past—or existing—knowledge, experience, or skill to confront their current situation and problems. Thus we might understand these constant attempts to work-around the code as the "networks in place" of these Programmers. This is partially in keeping with what was suggested earlier: that the tendency toward one form of use is already present in the wider networks of the user, and this is what engineers bet on when they attempt to anticipate the desires or goals of their users (Latour 1992). Of course, the crucial aspect in understanding these networks in place is to focus on their contingencies (i.e., the other networks on which they depend and are constituted).

When Is a Work-Around? Returning to the discussion of Liz and Carole, what is important to note, in terms of the argument being developed in this article, is that while Liz is attempting to rewrite BuildMAC, she receives help from her colleagues across the other sites and the Designers at the Technology Vendor.[13] In an earlier message, for example, Liz describes some of this collaboration:

Thus investigated with [the Technology Vendor] how to get BuildMAC to use ProC1.6 and pick up the include files from sqllib/public. [They] initially suggested renaming executables and using links, but wanted a proper way, so—amended [their] standard .mk files (sqlmenu5.mk srw.mk sqlforms30.mk) changing the default ProC make file variables from 2.0 to 1.6 as follows...

What is interesting about this is that Liz's work-around is seemingly legitimate. In fact, it is a necessity if her system is to ever work. Here, the work-around—changing the default ProC file variables from 2.0 to 1.6—is used, and we view the Programmers and Designers as colleagues discussing possible solutions. Work-arounds are very much part of the work of implementing a system—"that's life," as David from the office puts it.

Several weeks later, however, one of the Designers at the Technology Vendor appended the following statement to Liz's message, essentially rejecting the recoding work that she did:

...thank you for supplying this information. Unfortunately I am forced to close the bug as rejected as this is the only state applicable as this code was not released for that version of the PRO*C compiler.

Despite the fact that the Technology Vendor did not support Liz's rewrite, Carole went on to use this solution when she encountered the same problem some time later. Yet, Carole's work-around was not as straightforward: she was unable to get the "PRO*C compiler" to work, and she was forced to ask the Technology Vendor for help. Some days later, one of the Designers posted a message to the Log, describing Carole's problem:

I mailed Carole to ensure that it was the v1.6 PRO*C compiler that was being used. It was. On further investigation by our DBA [Database Administrator], and after some consultation with Carole, it would appear that a patch applied to the 1.6 PRO*C application is the cause of the problem.

Here, the Designer identifies the problem as being with Carole's use of Liz's work-around (a patch applied to the 1.6 PRO*C application). In a further message to the Log a few days later he summarizes the situation in the form of a final report to the Delphic Support Desk:

As you may know, [Oldcastle University] migrated from [MAC] 1.3 to 1.4 last week and encountered some problems which we helped with. We also advised them to migrate to 1.5, as 1.4 was no longer supported. This they did over the weekend and again had some problems, which I have mentioned in the Log. They contacted me on Monday morning and I have been looking at the problem(s) over the last day and a half. We have carried out a few checks and offered some advice on overcoming some of the problems, but it would appear that the problem lies in the data that they are working with and not a problem in any of our code.... Quite simply, I cannot justify any more time on this problem as it does not appear to be a problem with our software, rather a problem on site which may well require a great deal of time to identify.... Their current work-around is to use the basket 4 forms against the basket 5 database. I have expressed my concern over this and warned them that this is unsupported, but they appear to be confident that they have an adequate work-around.

Sometimes work-arounds are not considered normal working practices. If we were to think of an image of a network in place we would see how the Designers, with sleight of hand, begin to disrupt this network. The Designer is not performing the collegiality that we saw before, but is attempting to establish difference (i.e., to reconfigure the Programmers' relationship with MAC). To glance at the network now, we can catch sight of other networks coming into play, flexing and pulling to create real distance between the modes of use: now it is easy to see when the mode of use is a work-around and when it is something else.

To summarize this section, these practices are proscribed because as the Programmers carry out their modifications, they call into question the Designers' responsibility toward MAC and thus the distinction between who should be doing what. In other words, either they infringe on an important part of the code or they combine, or bricolage, in ways the Designers do not like. At the same time, however, work-arounds are demanded by the Designers in order to tailor the technology to the specifics of each of the sites (to work with existing software platforms). Importantly, it would seem the Designers of the system bet on the skills of the Programmers to carry out such modifications. So, one aspect of the contingency of these networks in place is that they are reliant on, and constituted by, this ambivalent situation where work-arounds are both problematized and supported by the Designers—what might be called the tension of work-arounds.[14]

Reconfiguring MAC: the Skills of the Programmers

A further aspect of these networks is that they are reliant on the efforts of the Programmers and their skill in working with the code. Such a relationship is not, as you will see, a straightforward one. Sima, for example, one of the other Programmers who worked in the office with David, sat frustrated for weeks attempting a work-around on a printer script. Sitting opposite her, I listened to her frustration as she talked to her computer, urging the program to compile. She was telling me how in her sleep at night she would even dream of the problem, constantly working through the code in her head, taking her thoughts down the different paths, following what was, to her, the essence of the code as it made its own way through the structure of her program. I listened also to her doubts (expressed privately to me and to the others who sat in the office) that she would ever be able to make the work-around work, and her fears of letting the others (who were relying on her finished code) down. I am particularly struck by Sima's continuous struggle with herself and her negotiation during her sleep of the routes the code would take and her effort to understand the way the code—if you like—flowed. Consider the following diary extract:

Sima has sat silent for several days now[,] only occasionally disturbed by Allison[,] who comes in periodically to check her progress. Sima asks her if she is worried that she will not get it done, and Allison says[,] "a bit." Sima tells David and me how [the Department Manager] is scared to come and talk to her at the moment. I take it that this is because he has given her such a horrible job to do. The programmers are in many ways heroic figures. They are the "ones" who make things work and whom others rely on to do things.

Thus, one aspect of Sima's skill, then, is her ability to immerse herself in, and relate to, the code. However, to do so is about grasping the work of others (many others). This can often be a difficult thing to do. Finally, after a couple weeks of struggling with the same piece of code, Sima relents and suggests to her manager that they should call in one of the Designers to help with the work-around.

Sima talked with the Designer (who was here for the day) about her problem of making things work and of how she is trying to change the code to print a "bankcheck" instead of a report. They talk about details of the code. He sits beside her and makes suggestions. She has spent a lot of time on this. He tells her to try something, and he goes away to talk to Allison. Later he comes back to Sima, and finally they get the code to work. Their talk had been calm and rational. She was telling him what she had done, and he was suggesting to her what to try next.

Skill of this sort is neither a given nor an object, but has to be continually worked at and tested. To be at one with a technology, to use the code effectively, takes effort. How are we to understand the work of these "wizards"—in particular, their choice to carry out work-arounds? To speak of wizards is not to make a disparaging comment, for the Programmers that I observed were well qualified, highly skilled, and very motivated; rather, it is to emphasize the contingency and indeterminacy of work-arounds, and to suggest that the skill to perform them—to be in a position to make this choice—emerges from, and depends upon, networks elsewhere.[15] Indeed, if you read some of the recent literature on computer system implementation you will see that these difficulties are increasingly common. Georgina Born, in her study of the work of coding in a French research institute, writes about some of the problems of working on systems originally developed by others. The people she studied often complained that when they looked back on collaboratively written programs "...the complexity of the codes made it extremely difficult to reconstruct afterward what was done, and how, in the bits of program authored by colleagues, without asking them" (1996, 109):

> To manipulate the system effectively requires knowledge of the specific coded universe of different layers of code. Naive and inexperienced users are powerless to enter lower levels of the code hierarchy in order to alter or improve a program's functioning. More surprising is the fact that the problem of the opacity of the hierarchy of codes—its resistance to meaningful decoding—also seriously affects senior...programmers. (ibid., 109)

What is being suggested here is that rather than reduce everything to one simple determinant—for example, "it all comes down to skill"—we might think of skill as both a connection to certain networks and an ability to perform the order embodied in those networks. This is, of course, the actor-network theory principle of treating actors as effects, and the view that technologies, among other things, have implications for us as agents (Law 1994). Thus a further aspect for understanding work-arounds is to consider how MAC itself provides for such modifications. Conventionally, we might think of MAC as a passive technology that is used by active agents who choose to use this tool in a number of different ways. Another way of imagining this would be to attempt to confuse this relationship between the Programmers and MAC. Actor-network theorists commonly speak of "hybrids"—that is, something different from just active humans and their passive technologies. It is to also emphasize that technologies are active, and that along with their users they perform together to produce "...the set of relations which give them their shape" (Law 2000, 5). Thus MAC, according to this way of thinking, is an actor in its own right. Moreover, if the skills of the Programmers are those of connecting and performing the order embodied in MAC, how do they perform together?

To explore this further I want to focus on a conversation I had with Maurice, another of the Programmers who worked in the office with David and Sima. Maurice characterized his experience of working with MAC in the following way: there is this constant need to make changes, as someone wants one part of the system to do something different, and he describes how MAC is "not built in concrete" and that "you *can* make changes to it." Maurice then goes on to acknowledge how the system also seems to work against his efforts to make changes:

> I don't know if it was designed to be changed, however. Some of the code is tricky. I mean, it is doing some clever stuff. They must have some really clever people there, doing code better than I could do. Some of the code really takes a while for you to get your head around.... The whole system is so constrained by the finance part of it. It is like a wheel with finance being the hub and the other parts being the spokes. You have to be careful when you make changes because you don't know what effect this will have on the other parts.

To clarify, Maurice seems to find himself in a position where the system is asking contrasting things of him: make changes/avoid making changes. It offers him the possibility of discretion in the sense that he is able to choose between different courses of action (Law 2000). MAC is not built in concrete and it can be changed. But, the way he decides to rewrite the code will affect others. For instance, changes he makes to the finance part of the system will, among other things, affect the work of his colleagues who, elsewhere, are relying on his rewrites to allow them to get their own work done. MAC is central here because it can be easily modified, and it allows Maurice to decide on and attempt work-arounds. Indeed, numerous authors have commented on the abstract and malleable nature of software: Shapiro and Woolgar, for instance, make the argument that software naturally lends itself to "...all manner of personalized idiosyncratic development approaches" (1995, 16). They also make the point that for some programmers, they "...will primarily see opportunity while some will mainly feel burdened" by such malleability. The example of Sima unable to get her rewrite to work after a couple weeks and being forced to call in a Designer, and Maurice's comment about having to be careful because of the finance hub, are both illustrations of where the possibility of discretion is closed down.[16] Here, MAC plays a part again because as it introduces its complex constraints—what Born (1996) earlier described as the "problem of the opacity of the hierarchy of codes"—there are very few possible courses of action.[17]

Conclusion

How do we account for a work-around? Often, the suggestion is that the user, when faced with a technology that is constraining in some form, is able to carry out a work-around and thus exercise some form of discretion or resistance.

This is always possible, especially if we understand the user and the technology to be each well bounded—that is, the role of the user is tightly defined as in a script, and the user attempts to work against this (as is suggested by Akrich). However, if we consider new forms of computer systems and the prominent role the user is beginning to play in the shaping and customization of such systems, things are increasingly less clear-cut. MAC, like many of the computer systems increasingly used by organizations, is a flexible technology, or something of a grey box, in which users have the capacity to shape and customize the final design. What this suggests is that we will continue to witness more ambiguous sets of user-producer relations where it is often not clear who has responsibility for what. Because of these complex divisions of labor, various groups come to rely on each other as an integral part of their day-to-day working practices, often as resources for the resolution of technical difficulties and problems.

The actors discussed are not simply users but neither are they simply producers who have been attempting to routinely negotiate relationships and identities with others within these increasingly confused networks. Work-arounds represent one part of that negotiation process. And as we have seen with the MAC example, these connections are not simple or straightforward, but they are full of tensions. What I have hoped to achieve in this article is to convince the reader that there is arguably a need to develop an improved understanding of the practice and process of work-arounds in relation to these less determinate technical divisions of labor and responsibilities. Where this article adds to our understanding is through the description of some of the processes that might lead to work-arounds. In particular, as I have described, MAC and its associated networks provide not simply for one mode of use but, to paraphrase Michael (1996), they allow for multiple modes of use. Moreover, sometimes these contrasting modes will operate in unison and sometimes they will be in conflict. First, one aspect of this is that the Designers attempt to link the successful implementation of MAC to the Programmers and their ability to tailor the system to fit in with the existing software infrastructure. Following Latour (1992), I have described the competencies that the Designers appear to bet on as the networks in place of the Programmers. Thus, at one level, it would seem that the Programmers actively reconfigure MAC and the Designers enlist them in doing so. Secondly, there are some obvious problems with this. MAC itself asks for contrasting things from the Programmers. While MAC can be easily modified, and it allows Maurice to attempt work-arounds, it also introduces complex constraints (i.e., it acts against the possibility of work-arounds). A further element of the tension is that while work-arounds are demanded by the Designers, these practices are also sometimes proscribed, because as the Programmers carry out their modifications, they call into question the Designers' responsibility toward MAC (i.e., the work-arounds infringe on the "80") and their role in the implementation. Hence, just when a work-around is a supported form of use, and when it is not, becomes a crucial question that has obvious resource implications, and this in itself makes it an important topic for the sociology of technology.[18]

Acknowledgment

I would like to thank all those at my research site, particularly the programmers in Administrative Computing Services (ACS). I would also like to acknowledge the support of the UK Economic & Social Research Council (ESRC), under which this research was funded. The actual research was initially carried out with an ESRC studentship and later developed while I was working under the ESRC's Virtual Society? Programme. Finally, I would like to thank Mike Michael, Chris Stokes, Chris Ivory, David Edge, Luciana D'Adderio, and James Cornford for their comments on earlier versions of this article.

Notes

1. Such forms of use ranged from users entering inaccurate data to bypass weaknesses in existing systems, to users simply manually carrying out the procedures the computer system is meant to do and inputting the job after the work has been completed.

2. See the work of Claudio Ciborra (2002).

3. Kathryn Henderson (1999), for example, uses but does not develop the term in her recent book on engineers and their use of CAD. See also the article by Marc Berg (1997), in which he describes how nurses work around the limitations of a medical record system. For an example of how the notion of a work-around is used in the loose body of thought that comes under the heading of Computer Supported Co-operative Work (CSCW), see the article by Luff and Heath (1993). More recently, Button and Sharrock (1998) use the term to describe how programmers circumvent an incompetent manager.

4. Scripts, argues Akrich, are often simply the outcome of decisions made by designers about future users—their skills and abilities and what the technology should do in relation to this user. Through the script: "...the designer expresses the scenario of the device in question—the script out of which the future history of the object will develop" (216).

5. See Berg (1997), who makes a similar point about writing within the social studies of technology.

6. This differs from Akrich, who describes a script as embedded within an artifact, whereas Michael is suggesting that scripts are both in the technology *and* in the wider networks attached to the technology. In other words, Michael's is a more dynamic notion of script where notions of use are the upshot of an interaction between the artifact and this larger network. A technology can hardly be thought as separate from, say, its instructions for use, as the artifact's working depends on these. In paraphrasing Pfaffenberger, he writes: "...technologies don't have instructions for their use inscribed in their design. Discourses are needed which guide users in their appropriate use" (1996, 3).

7. Friedman (1989), for instance, writing in the field of information systems, lists at least six user roles, which include not only those who simply input and retrieve data but also users who initiate systems and those who are involved in development and implementation as well as maintenance.

8. See the article by Button and Sharrock (1994) in which they also describe programmers as users.

9. *Collins Concise Dictionary*, Fourth Edition, HarperCollins, 1999. For an explanation of this quote see the discussion below.

10. See Janet Rachel (1994), who makes a similar point when referring to her own ethnography and the apparent inactivity of the programmers she witnessed, though she notes that the *activity* of these programmers was "...produced through the appearance of inactivity" (819, *her emphasis*). Behind these seemingly still bodies, however, they were furiously typing away on keyboards "...networked together in an effort to accomplish change on a grand scale in other parts of the organization" (819).

11. The apposite image of a true "Delphic" Support Desk is the one that I want you to keep in mind here.

12. See Ciborra (2002) for a detailed discussion of bricolage.

13. In their article, Star and Ruhleder (1996) ask *when* and not *what* is an infrastructure. Here, in their intriguing article, they are rehearsing the sociology of technology common-

place that technologies are not just things with particular properties "frozen in time" but emerge for people in the practice of technology use. Likewise, infrastructure, they argue, is also a fundamentally relational concept: "It becomes infrastructure in relation to organized practice. Within a given cultural context, the cook considers the water system a piece of working infrastructure integral to making dinner; for the city planner, it becomes a variable in a complex equation."

14. The key paper for this form of ambivalence within the approach advocated by Akrich is Singleton and Michael (1996). They argue that while actor-network theory has tended to story "successful" networks as those where the actors strictly play out their allotted roles, in practice actors often move between different positions (i.e., sometimes critical, sometimes supportive of the network). Indeed, as they argue, this crossover of roles often enables the very continuation of the network.

15. For a discussion of emergent skills, see Andrew Pickering's *Mangle of Practice* (1995).

16. Leigh Star (1995) has described this as the "myth of infinite flexibility," where in principle, software can be modified, but in practice it is very difficult to do so as changes will affect other parts of the system. This is especially true for integrated software systems (see Pollock and Cornford (2004) for a discussion of the difficulties of customizing Enterprise Resource Planning Systems).

17. For a discussion of software as a mediator, see also Born (1997).

18. The outcome of this negotiation will decide if the local programmers will receive further help in modifying that aspect of the system.

References

Akrich, M. 1992. The de-scription of technical objects. In *Shaping Technology/Building Society: Studies in Sociotechnical Change*, ed. W. Bijker, and J. Law. Cambridge, MA: MIT Press.

Agre, P. 2000. Infrastructure and institutional change in the networked university. *Information, Communication & Society*, 4:494–507.

Berg, M. 1997. Of forms, containers, and the electronic medical record: Some tools for a sociology of the formal. *Science, Technology, & Human Values* 22 (4): 403–433.

Berg, M. 1998. The politics of technology: On bringing social theory in to technological design. *Science, Technology, & Human Values* 23 (4): 456–490.

Born, G. 1996. (Im)materiality and sociality: The dynamics of intellectual property in a computer software research culture. *Social Anthropology* 4 (2): 109.

Born, G. 1997. Computer software as a medium: Textuality, orality and sociality in an artificial intelligence research culture. In *Rethinking Visual Anthropology*, eds. Banks, M. and H. Morphy. New Haven: Yale University Press.

Brady, T., M. Tierney, and R. Williams. 1992. The commodification of industry applications software. *Industrial and Corporate Change* 1 (3): 489–514.

Button, G. and W. Sharrock. 1994. Occasioned practices in the work of software engineers. In *Requirements Engineering*, eds. Jirotka, M. and J. Goguen. London: Academic Press Ltd.

Button, G. and W. Sharrock. 1998. The organization accountability of technological work. *Social Studies of Science* 28 (1): 73–102.

Ciborra, C. 2002. *The Labyrinths of Information: Challenging the Wisdom of Systems*. Oxford: Oxford University Press.

Friedman, A. 1989. *Computer Systems Development: History, Organisation and Implementation*. Chicester: John Wiley.

Fujimura, J. 1992. Crafting Science: Standardised Packages, Boundary Objects, and "Translation." In *Science as Practice and Culture*, ed. A. Pickering. Chicago: University of Chicago Press.

Gasser, L. 1986. The integration of computing and routine work. *ACM Transactions on Office Information Systems* 4:257–270.

Goddard, A., and P. Gayward. 1994. MAC and the Oracle family: Achievements and lessons learnt. *Axix* 1 (1): 45–50.

Henderson, K. 1999. *On Line and On Paper: Visual Representation, Visual Culture, and Computer Graphics in Design Engineering*. Cambridge, MA: MIT Press.

Latour, B. 1992. Where are the missing masses? Sociology of a few mundane artifacts. In *Shaping Technology/Building Society: Studies in Sociotechnical Change*, eds. W. Bijker and J. Law. Cambridge, MA: MIT Press [reprinted as ch. 10 in this volume].

Law, J. 1994. *Organizing Modernity*. Oxford: Blackwell.

Law, J. 2000. Economics as interference. Draft, published by the Centre for Science Studies and Department of Sociology, Lancaster University at http://www/comp.Lancaster.ac.uk/sociology/soc034jl.html.

Luff and Heath. 1993. System use and social organisation: Observations on human-computer interaction in an architectural practice. In *Technology in Working Order: Studies of Work, Interaction, and Technology*, ed. G. Button. London: Routledge.

Mackay, H., C. Carne, P. Benynon-Davies, and D. Tudhope. 2000. Reconfiguring the user: Using rapid application development. *Social Studies of Science*, 30 (5): 737–757.

Michael, M. 1996. Technologies and tantrums: Hybrids out of control in the case of road rage. Paper presented at the "Signatures of Knowledge Societies" Joint 4S/EASST conference at the University of Bielefeld, Germany, October 10–13.

Pfaffenberger, B. 1992. Technological dramas. *Science, Technology, & Human Values* 17 (3): 292–312.

Pickering, A. 1995. *The mangle of practice: Time, agency & science*. Chicago: University of Chicago Press.

Pollock, N. and J. Cornford. 2004. ERP systems and the university as a "unique organisation." *Information Technology & People* 17 (1): 31–52.

Rachel, J. 1994. Acting and passing, actants and passants, action and passion. *American Behavioral Scientist* 37 (6): 809–823.

Shapiro, D., and S. Woolgar. 1995. Balancing acts: Reconciling competing visions of the way software technologies work. Working paper, CRICT, Brunel University, UK, 16.

Singleton, V. and M. Michael. 1996. Actor networks and ambivalence: General practitioners in the cervical screening programme. *Social Studies of Science* 23:227–264.

Star, S. L., ed. 1995. *Ecologies of Knowledge: Work and Politics in Science and Technology*. New York: State University of New York Press.

Star, S. L. and K. Ruhleder. 1996. Steps toward an ecology of infrastructure: design and access for large information spaces. *Information Systems Research* 7 (1): 111–134.

Suchman, L. 1994. Working relations of technology production and use. *Computer Supported Cooperative Work (CSCW)* 2:21–39.

V   TWENTY-FIRST-CENTURY CHALLENGES

Sections I through IV make it clear that technological endeavors are not just machines, gadgets and structures. They comprise and shape society; they build the future. The material structures built and used today constrain and constitute tomorrow's social relationships, social institutions, and cultural and political notions. The first four sections provide a picture of how technology and society work in step, shaping and being shaped by one another. Understanding this relationship is the key to building a better world.

This section returns to the future. It takes what was presented about technology and society in the previous sections and uses it both to understand the direction in which the world is moving and also to see if there are ways to influence that direction. The section challenges the reader to reflect not just about how the world is changing or will change, but about the changes that would make for a better future.

The complexities described in the preceding sections present a daunting challenge for those seeking to change the world. The complex relationship between society and technology, coupled with the fact that a range of actors influence technology based on an incomplete knowledge of how it will behave and what its effects will be, mean that we confront a world that is difficult to understand and predict. How can all of this be managed? How can we steer sociotechnical development to solve problems and realize values that are essential to human wellbeing?

In the face of these complexities it may seem that we have little control. Hence, it is important to keep in mind that technology does not just happen; it is the product of people. Sometimes adoption of a technology leads to shifts in power that were intended by those who designed the technology. Other times it comes with unintended consequences, and people choose to tolerate these consequences. Or they choose to do something about the consequences such as changing the design of the technology, changing the policies that regulate the technology's uses, or even suing the corporation whose products ultimately led to the unintended consequences. Our sociotechnical world is the result of a series of decisions made by people (individuals and groups) and when we are aware of this, we are better equipped to make decisions that lead to better outcomes.

To be sure, it is easy to forget that people control technology. For instance, many of the technologies currently being developed are referred to as 'emerging technologies.' This label is effectively a hidden form of technological determinism; it implies that these technologies are in nature and simply need to be found and uncovered by scientists and engineers. However, what 'emerges' from the processes of research and development is largely a function of what engineers, government agencies, social critics, consumers, politicians, and many others *decide* to do. These technologies are not yet "made," they are "in the making" and can still be made into "this" or "that" particular kind of system. Human beings decide what these technologies look like, how they are integrated into society, and even whether they will exist at all. Only when the connection between human decision making and technology is made visible can individuals see the implications of their own decisions or the decisions of others; only then can individuals make conscientious decisions about the future.

The readings in the preceding sections made the relationship between technology, society, and values more visible. We turn now specifically to the context in which the future is unfolding. This section examines trends and issues that face us in the twenty-first century. The readings ask (and try to answer) such questions as: What will globalization mean? Can we address environmental injustice? Will world needs for energy be met? Do we really have to give up privacy in order to achieve security? The readings present trends that are predicted to continue and intensify, problems and issues that cannot or should not be ignored, and technologies that are generating concern as we move further into the twenty-first century. Continuation of some trends could mean a better world enriched by improved health care and greater convenience leading to longer, healthier, and less stressful lives. These trends could mean more peaceful, more democratic, and more comfortable lives. However, continuation of other trends could mean political unrest, degradation of the environment, or infringements on basic freedoms. How technology is developed and steered will make a significant difference in whether the promises or threats of the future are realized. Indeed, it is for this reason that it is important to take seriously the concerns that are expressed about emerging technologies and current trends as well as to listen to the potential benefits. An understanding of the possibilities for good and ill puts us in a better position to use whatever power and influence we have to influence the direction of development.

In effect, this section asks the question—what kind of world do we want in the future? Since the world is continually changing this is a question that must be repeatedly asked; answers must always be tentative. As the readings in this volume emphasize, the future is being created through the interplay of technical and social, through interactions among many actors including engineers, entrepreneurs, consumers, public agencies, and others. Scientists and engineers tell us what is possible and public discussion and debate about these new possibilities influence what scientists and engineers choose to do. The readings in this section provide an opportunity to both observe that interplay but also to engage in it by trying to figure out what we should think about possible directions for the future.

Of course, the readings included in this section are only a sample of current discussions of the future. New sociotechnical enterprises are constantly being envisioned and hence those who want to contribute to the future must keep abreast of trends and developments in technological prowess. This section provides a starting point for thinking about the future. It stresses that the decisions we make today will have an important effect on the world of tomorrow. It challenges us to face the latest issues, to envision the world we want, and to actively build a sociotechnical future that is better for all.

Introduction to Part V

Questions to consider while reading the selections in this section:

1. If the trends described in this section continue, who will be the winners and losers?
2. What forces are influencing the trends or creating the problems discussed in this section?
3. What values are being realized or threatened?
4. In what ways can social change be more effective in solving problems than technical change?
5. How are decisions being made to solve the challenges or steer future development?

25 "Shaping Technology for the 'Good Life': The Technological Imperative versus the Social Imperative"

Gary Chapman

In the early twenty-first century, many consider globalization to be an unstoppable force. New technologies (including computers, satellites, and the Internet) are making communications across the Earth instantaneous. This facilitates global corporations and integrated economies such that decisions in Tulsa can have profound effects on the lives of farmers in Sri Lanka. Arguably, the intensified interaction may reduce or even eliminate differences in culture and tradition. Globalization has generated a considerable number of detractors with criticism aimed both at the institutions promoting it and the technologies that make it possible. Gary Chapman, however, argues that the problem is not technology, but rather the power and influence of what he calls the "technological imperative." He critiques the idea that our primary goals should be to develop technologies and systems that increase efficiency and lower costs. If we make technology our primary goal, then many of the values we hold dear may be lost in the mix. In essence he argues that if we believe in technological determinism, it will come true. If, however, we choose to privilege social values over technical goals, we will be able to create a world that we want. Chapman maintains that this vision is not a pipe dream and explores the slow food movement as a case study in countering technological trends and changing the direction of technological development. If, for instance, our primary goal is to re-engineer produce so that it grows faster, is cheaper to harvest, and easier to ship, then local flavors, local economies, and local jobs might be lost. If, on the other hand, we are primarily motivated by specific values, people, and institutions, rather than technology, we can still employ technology, but we can employ it to local ends. Chapman urges us not to abandon technology, but rather to see it as a means to achieving our goals and not as the goal itself.

Since the collapse of the Soviet Union and the communist Eastern bloc at the beginning of the 1990s, technology and economic globalization have become the chief determinants of world culture. Indeed, these two omnipresent features of modern, civilized life in the postindustrial world are so intertwined that they may be indistinguishable—we may speak more or less coherently of a "technoglobalist" tide, one rapidly engulfing most of the world today.

A significant point of debate raised by this phenomenon is whether, and to what extent, the rapid spread of technologically based global capitalism is inevitable, unstoppable, and even in some vague sense autonomous. There are those who believe that there is in fact a strong "technological imperative" in human history, a kind of

From Douglas Schuler and Peter Day, eds., *Shaping the Network Society: The New Role of Civil Society in Cyberspace* (Cambridge, Mass.: MIT Press, 2004), pp. 43–53, 60–65. Reprinted with permission.

"technologic" represented both in macro-phenomena such as the market, and in individual technologies such as semiconductor circuits or bioengineered organisms. It is not hard to find evidence to support such an idea. The prosaic description of this concept would be that technological innovations carry the "seed," so to speak, of further innovations along a trajectory that reveals itself only in hindsight. Moreover, the aggregate of these incremental improvements in technology is an arrow that points forward in time, in a process that appears to be accelerating, piling more and more technologies on top of one another, accumulating over time to build an increasingly uniform and adaptive global civilization. There often appears to be no escape from this process. As the allegorical science fiction villains of the TV series "Star Trek," the Borg, say in their robotic, repetitive mantra, "Resistance is futile. You will be assimilated." Stewart Brand, the *Whole Earth Catalog* guru and author who turned into a high-tech evangelist, put it this way: "Once a new technology rolls over you, if you're not part of the steamroller, you're part of the road" (Brand 1987, 22).

The growing and vocal antiglobalization movement, on the other hand, is questioning such assumptions and challenging the idea of a necessary link between a "technological imperative" and human progress. The protesters at antiglobalization demonstrations typically represent a wide range of both grievances and desires, so it is difficult to neatly characterize a movement that is repeatedly drawing hundreds of thousands of protesters to each major demonstration. But the one principle that seems to unite them—along with many sympathizers who choose not to participate in public demonstrations—is that the future is not foreordained by a "technological imperative" expressed via global corporate capitalism. There is the hope, at least, among antiglobalization activists that human society might continue to represent a good deal of diversity, including, perhaps especially, diversity in the way people adopt, use, and refine technology.

Who wins this debate, if there is a winner, will be at the heart of "shaping the network society," the theme of this book. The outcome of this ongoing debate may determine whether there are ways to "shape" a global technological epoch at all. If the concept of a "technological imperative" wins out, over all obstacles, then human beings are essentially along for the ride, whether the end point is utopia or apocalypse or something in between. Human consciousness itself may even be shaped by surrender to the technological imperative.

If, on the other hand, technology can be shaped by human desires and intentions, then the critics of the "autonomous technology" idea must either explain or discover how technologies can be steered deliberately one way and not another. Neither side in this debate has a monopoly on either virtue or vice, of course. Letting technology unfold without detailed social control is likely to bring many benefits, both anticipated and unanticipated. Setting explicit goals for technological development could, on the other hand, help us avoid pitfalls or even some catastrophes. It will be the search for balance that is likely to characterize our global discussion for the foreseeable future, a balancing of the "technological imperative" with what might be called our "social imperative."

The Technological Imperative Full Blown—Moore's Law and Its Distortions

In 1965, Gordon Moore, the cofounder of the Intel Corporation and a pioneer in semiconductor electronics, publicly predicted that microprocessor computing power would double every eighteen months for the foreseeable future. He actually predicted that the number of transistors per integrated circuit would double every eighteen months, and he forecast that this trend would continue through 1975, only ten years into the future. But in fact this prediction has turned out to be remarkably accurate even until today, more than thirty five years later. The Intel Corporation itself maintains a chart of its own microprocessor transistor counts, which have increased from 2,250 in 1971 (the Intel 4004) to 42 million in 2000 (the Pentium 4) ⟨http://www.intel.com/technology/mooreslaw/index.htm⟩. The technology industry is nearly always preoccupied with when Moore's Law might come to an end, especially as we near the physical limits of moving electrons in a semiconductor circuit. But there always seems to be some promising new technology in development that will keep Moore's Law alive.

Moore's prediction has so amazed technologists, because of its accuracy and its longevity, that it has become something close to a natural law, as in Newtonian physics. Of course, there is bound to be an end to the trend that makes the prediction accurate, which means that Moore's Law will someday become a historical curiosity rather than a "law." But the prediction has taken on a life of its own anyway. It is no longer regarded as simply a prediction that happened to be fulfilled by a company owned and controlled by the predictor, a company that spent billions of dollars to make sure the prediction came true and profited immensely when it did come true. Moore's law is regarded by some technophiles as "proof" that computers and computer software will increase their power and capabilities forever, and some computer scientists use the thirty five-year accuracy of Moore's law as evidence that computers will eventually be as "smart" as human beings and perhaps even "smarter" (Kurzweil 1999; Moravec 1998).

There is no connection between transistor density on a semiconductor chip and whether or not a computer can compete with a human being in terms of "intelligence." It is not even clear what constitutes intelligence in a human being, let alone whether or not a computer might match or surpass it. Intelligence is an exceedingly vague term, steeped in controversy and dispute among experts. Computers are far better at some tasks than humans, typically tasks that involve staggering amounts of repetitive computation. But human beings are far better at many ordinary "human" tasks than computers—indeed, an infant human has more "common sense" than a supercomputer with billions of transistors. There is no evidence that human beings "compute," the way a computer processes binary information, in order to cogitate or think. Nor is there evidence that the von Neumann computer model of serial bit processing is even a simulacrum for human information processing.

This is not to say that Moore's law is insignificant or irrelevant—the advances in computer processing power over the past thirty five years have been astonishing and vitally important. And the fact that Gordon Moore was prescient enough to predict the increase in a way that has turned out to be amazingly accurate is fascinating and impressive. But as others have pointed out, the semiconductor industry has spent billions of dollars to make sure that Moore's prediction came true, and it is worth mentioning that Moore's own company, Intel, has led the industry for all of the thirty five years Moore's law has been tested. Nevertheless, if the prediction had failed we would not be talking about it the way we do now, as a cornerstone of the computer age, and for this Moore deserves credit. But there is nothing about Moore's law that points to the kinds of future scenarios that some authors and pundits and even engineers have attributed to it. There is nothing about Moore's law that makes it a true "law," nor is there any imperative that it be accurate indefinitely, except the industry's interests in increasing transistor density in order to sell successive generations of computer chips. In any event, there are trends now that suggest that this measure of progress, of increasing chip density, is gradually losing its significance (Markoff and Lohr 2002).

Moore's law is an example of how a thoughtful and interesting prediction has been turned into an argument for the technological imperative, that society must invest whatever it takes to improve a technology at its maximally feasible rate of improvement—and to invest in a specific technology, perhaps at the expense of a more balanced and generally beneficial mix of other technologies. Advocates of the semiconductor industry argue that semiconductor chips are the "seed corn" of the postindustrial economy, because their utility is so universal and significant to productivity. But so is renewable energy, or human learning, or sustainable agriculture, all things that have experienced a weakness in investment and attention, at least in comparison to semiconductors and computer hardware.

In April 2000, Bill Joy, vice president and cofounder of Sun Microsystems—a very large computer and software company in Silicon Valley—published a provocative essay in *Wired* magazine titled "Why the Future Doesn't Need Us" (Joy 2000). Joy raised some troubling questions for scientific and technological researchers, about whether we are busy building technologies that will make human beings redundant or inferior beings. Joy's article created a remarkable wave of public debate—there were public discussions about his thesis at Stanford University and the University of Washington in Seattle, he was invited to present his ideas before the National Academy of Sciences, and he appeared on National Public Radio. His warnings were featured in many other magazine articles.

Joy refers to the arguments of two other well-known technologists, Ray Kurzweil and Hans Moravec, who have both written extensively about how computers will one day become "smarter" than human beings. Humans will either evolve in a way that competes with machines, such as through machine implants in the human body, or else "disappear" by transferring their consciousness to machine receptacles, according to Kurzweil and Moravec. In his *Wired* article, Joy accepts this as a technological possibility, perhaps even an inevitability, unless we intervene and change course. "A technological approach to Eternity—near immortality through robotics," writes Joy, "may not be the most desirable utopia, and its pursuit brings clear dangers. Maybe we should rethink our utopian choices" (Joy 2000).

But the question then becomes, Can we rethink our utopian choices if we believe rather thoroughly in a technological imperative that propels us inexorably in a particular direction? Can Moore's law coexist with a free ethical choice for technological ends? Joy says that perhaps he may reach a point where he might have to stop working on his favorite problems. "I have always believed that making software more reliable, given its many uses, will make the world a safer and better place," he notes, adding that "if I were to come to believe the opposite, then I would be morally obligated to stop this work. I can now imagine such a day may come." He sees progress as "bitter sweet": "This all leaves me not angry but at least a bit melancholic. Henceforth, for me, progress will be somewhat bittersweet" (Joy 2000).

There is a great deal to admire in such emotions; these are the musings of a thoughtful and concerned person, and, given Joy's stature and reputation in his field, such qualities are welcome precisely because they seem so rare among technologists.

But behind such ideas is an unquestioned faith in the technological imperative. Joy suggests that unless we simply stop our research dead in its tracks, on ethical grounds, we may create dangers for which we will be eternally guilty. This is certainly possible, but it is distinctly one dimensional. Could we not redirect our technological aims to serve *other* goals, goals that would help create a life worth living rather than a life shadowed by guilt and dread? Joy begins to sound like Theodore Kaczynski, the "Unabomber," whom he quotes with some interest and intrigue (as does Kurzweil). Kaczynski also believed the technological imperative is leading us to our doom, hence his radical Luddite prescriptions and his lifestyle, not to mention his deadly attacks on technologists for which he was eventually sent to jail for life. Kurzweil, Joy, and Kaczynski all portray technology as an all-encompassing, universal system—Joy repeatedly uses the phrase "complex system"—that envelops all human existence. It is the "totalizing" nature of such a system that raises troubling ethical problems for Joy, dark fantasies of doom for Kaczynski, and dreams of eternity, immortality, and transcendence for Kurzweil. But might there not be another way to adapt technology to human-scale needs and interests?

The Slow Food Movement in Italy

The Italian cultural movement known as "slow food"—not a translation, it is called this in Italy—was launched by Roman food critic and gourmand Carlo Petrini in 1986, just after a McDonald's hamburger restaurant opened in Rome's magnificently beautiful Piazza di Spagna (Slow Food, 2002). Petrini hoped to start a movement that would help people "rediscover the richness and aromas of local cuisines to fight the standardization of Fast Food." Slow food spread very rapidly across Italy. It is now headquartered, as a social movement, in the northern Piedmont city of Bra.

Today there are 65,000 official members of the slow food movement, almost all in Western Europe—35,000 of these members are in Italy. These people are organized into local groups with the wonderfully appropriate name *convivia*. There are 560 *convivia* worldwide, 340 of them in Italy (Slow Food, 2002). Restaurants that serve as "evangelists" of the slow food movement display the group's logo, a cartoon snail, on the front door or window. If one sees this logo displayed by a restaurant, one is almost guaranteed to enjoy a memorable dining experience.

Slow food is not just about sustaining the southern European customs of three-hour lunches and dinners that last late into the night, although that certainly is part of the message. Petrini very cleverly introduced the idea of a "Noah's ark" of food preservation, meaning a concerted effort to preserve nearly extinct natural foodstuffs, recipes, and, most of all, the old techniques of preparing handmade foods. From this idea, the slow food movement has broadly linked gastronomy, ecology, history, and economics into a benign but powerful ideology that nearly every southern European citizen can understand. Slow food, in addition to being an obvious countermovement to American "fast food," has developed into a movement that promotes organic farming and responsible animal husbandry, community-based skills for the preservation of regional cuisines, and celebrations of convivial, ceremonial activities such as food festivals and ecotourism. The Slow Food organization has even sponsored a film festival, featuring films with prominent scenes about food, and plans to award an annual "Golden Snail" trophy, the slow food equivalent of an Oscar.

Slow food is thus one of the more interesting and well-developed critiques of several facets of globalization and modern technology. Specifically, it is a response to the spread of globally standardized and technology-intensive corporate agriculture, genetic engineering, high-tech food preparation and distribution, and the quintessentially American lifestyle that makes "fast food" popular and, for some, even imperative. Italian proponents of slow food and their allies in other countries—*convivia* are now found in other European countries as well as in the United States—view this as a struggle for the soul of life and for the preservation of life's most basic pleasures amidst a global trend pointing to increased competition, consumerism, stress, and "hurriedness":

We are enslaved by speed and have all succumbed to the same insidious virus: Fast Life, which disrupts our habits, pervades the privacy of our homes and forces us to eat Fast Foods....

Many suitable doses of guaranteed sensual pleasure and slow, long-lasting enjoyment preserve us from the contagion of the multitude who mistake frenzy for efficiency. (Slow Food, 2002)

In 1999, the slow food movement spun off a new variation of itself: the slow cities movement. In October of that year, in Orvieto, Italy, a League of Slow Cities was formed, a charter was adopted, and the first members of this league elected as their "coordinator" Signor Paolo Saturnini, the mayor of the town of Greve in Chianti (Città Slow, 2002).

The Charter of Association for the slow cities movement (which in Italian does have the Italian name "città slow") has a rather sophisticated and subtle view of globalization.

The development of local communities is based, among other things (sic), on their ability to share and acknowledge specific qualities, to create an identity of their own that is visible outside and profoundly felt inside.

The phenomenon of globalization offers, among other things, a great opportunity for exchange and diffusion, but it does tend to level out differences and conceal the peculiar characteristics of single realities. In short, it proposes median models which belong to no one and inevitably generate mediocrity.

Nonetheless, a burgeoning new demand exists for alternative solutions which tend to pursue and disseminate excellence, seen not necessarily as an elite phenomenon, but rather as a cultural, hence universal fact of life (Slow Food, 2002).

Thus, slow cities are those that

- Implement an environmental policy designed to maintain and develop the characteristics of their surrounding area and urban fabric, placing the onus on recovery and re-use techniques
- Implement an infrastructural policy that is functional for the improvement, not the occupation, of the land
- Promote the use of technologies to improve the quality of the environment and the urban fabric
- Encourage the production and use of foodstuffs produced using natural, ecocompatible techniques, excluding transgenic products, and setting up, where necessary, presidia to safeguard and develop typical products currently in difficulty, in close collaboration with the Slow Food Ark project and wine and food Presidia
- Safeguard autochthonous production, rooted in culture and tradition, which contributes to the typification of an area, maintaining its modes and mores and promoting preferential occasions and spaces for direct contacts between consumers and quality producers and purveyors
- Promote the quality of hospitality as a real bond with the local community and its specific features, removing the physical and cultural obstacles that may jeopardize the complete, widespread use of a city's resources
- Promote awareness among all citizens, and not only among inside operators, that they live in a Slow City, with special attention to the world of young people and schools through the systematic introduction of taste education. (Slow Food 2002).

Slow food and slow cities are thus *not* neo-Luddite movements; both acknowledge the importance of technology, and they specifically mention the benefits of communications technologies that allow a global sharing of ideas. But both movements are committed to applying technology for specific purposes derived from the values of the "slow" movement as a whole: leisure, taste, ecological harmony, the preservation and enhancement of skills and local identities, and ongoing "taste education." The subtleties of this worldview are typically lost on most Americans, who have no idea where their food comes from, nor would they care if they did know (this applies to most, but not all, Americans, of course). Many Italians and French are concerned that this attitude might spread to their countries and wipe out centuries of refinement in local cuisines, culinary skills, agricultural specialties, and other forms of highly specific cultural identity. As the Slow Cities Charter of Association straightforwardly asserts, "universal" culture typically means mediocre culture, with refinement and excellence reserved in a special category for people with abundant wealth. The "slow" alternative is to make excellence, identity, and "luxury" available "as a cultural, and hence universal fact of life," something not reserved for elites but embedded in daily life for everyone. This will require "taste education," something meant to offset the damage of mass marketing and the advertising of mass products.

Slow food might be dismissed as a fad among the bourgeoisie, a cult of hedonists and effete epicureans. It is certainly a phenomenon of the middle class in southern Europe, but it may serve as a kind of ideological bridge between more radical anti-globalization activists and older, more moderate globalization skeptics. The slow food movement is not thoroughly opposed to globalization, in fact. It is concerned with the negative effects of globalization and technology, which have prompted the movement's appearance and its eloquence—with respect to the leveling of taste, the accelerating pace of life, and the disenchantment with some of life's basic pleasures, such as cooking and eating.

In this way, the "slow" movement is an intriguing and perhaps potent critique of modern technology, which is otherwise widely viewed as propelling us toward the very things the "slow" movement opposes. In the United States, for example, bioengineering is typically regarded as inevitable, or even promoted as the "next big thing" for the high-tech economy. In Europe, by contrast, genetically engineered crops and foodstuffs are very unpopular, or, at best, greeted by deep skepticism, even among apolitical consumers. In the United States, there is widespread resignation in the face of the gradually merging uniformity of urban and especially suburban spaces. Shopping malls all look alike, and even feature the same stores; many Americans look for familiar "brand" restaurants and attractions such as those associated with Disney or chains like Planet Hollywood, indistinguishable no matter where they are found; a "successful" community is one where the labor market is so identical to other successful communities that skilled workers can live where they choose; suburban tracts of new homes are increasingly impossible to tell apart. Critics of this universal trend in the United States bemoan the appearance of "Anywhere, U.S.A.," typified by the suburban communities of the West, the Southwest, and the Southeast. These communities are often characterized by a sort of pseudo-excellence in technology—they are often the sites of high-tech industries—but mediocrity in most other amenities of life. Such trends are not unknown in Western Europe—Europeans are starting to worry that these kinds of communities are becoming more and more common there too—but there is at least a vocal and sophisticated opposition in the slow food movement and its various fellow travelers.

At bottom, the slow food and slow cities movements are about the dimensions of what Italians call "il buon vivere"—the good life. For Italians and many other Europeans (as well as many Americans and people in other countries, of course), the "good life" is one characterized by basic pleasures like good food and drink, convivial company and plenty of time devoid of stress, dull work, or frenzy. The advocates of slow food see the encroaching American lifestyle as corrosive to all these pleasures. The American preference is for convenience and technology that replaces many time-intensive activities, rather than for quality and hard-won skills. Plus, the American style is one of mass production aimed at appealing to as many people (or customers) as possible, leaving high-quality, customized, and individualized products and services reserved for the wealthy. In the United States, this trend has now taken over architecture, music, films, books, and many other things that were once the main forms of artistic and aesthetic expression in a civilized society. Many fear this American trend toward mass appeal, mediocrity, and profit at the expense of quality, excellence, and unique identity is now taking over the Internet as well.

It must be mentioned that the orientation toward local communities, the preservation of skills, skepticism about globalization and technology, and so on is not without its own pitfalls. At the extreme ends of this perspective are dangerous and noxious political cauldrons, either of nationalism and fascism on the one hand or neo-Luddite, left-wing anarchy on the other. There are already discussions in Italy about the similarities in antiglobalization sentiments shared by the extreme right and the extreme left, which in Italy represent true extremes of neofascism and full-blown anarchism. At the same time that these two forces battled each other in the immense antiglobalization protests in Genoa in the summer of 2001—when neofascists on the police force allegedly beat protesters to their knees and then forced them to shout "Viva II Ducel," a signal of the fashionable rehabilitation of Mussolini among the Italian right wing—intellectuals on the right of Italian politics were musing in newspapers about how much of the antiglobalization rhetoric of the protesters matched positions of the neofascist parties.

The slow food movement has nothing to do with this dispute—the movement is capable of encompassing the entire political spectrum, except perhaps the extreme left wing. Nevertheless, there is the possibility that an emphasis on local and historically specific cultural heritage, even limited to cuisines, could become a facade for anti-immigrant or even xenophobic political opinions, which are discouragingly common in European politics today. Moreover, there is a long history of skepticism and outright opposition to technology among right-wing extremists, who often try to protect nationalist and conservative traditions threatened by technologies such as media, telecommunications, the Internet, and reproductive technologies, to name a few.

So far, it appears that the slow food and slow cities movements are admirably free of such pathologies, at least in their public pronouncements and in the character of their concerns, such as environmental, agricultural, and water quality in developing nations. Both movements argue that technology has a place in their worldview, especially as a tool to share ideas globally, and as a means of protecting the natural environment. Far from being tarred with the neofascist surge in Italy, or with the anarchist tactics of French activist Jose Bove—who has smashed a McDonald's in Provence and burned seeds of genetically altered grain in Brazil—slow food and slow cities have been linked to the concept of "neohumanism," a phrase used by Italian poet Salvatore Quasimodo in his acceptance speech when he received the Nobel Prize for Literature in 1959. "Culture," said Quasimodo, "has always repulsed the recurrent threat of barbarism, even when the latter was heavily armed and seething with confused ideologies" (Quasimodo 1959).

This seems to me the idea of culture that the slow food and slow cities movements represent, using cuisine and urban planning as vehicles, like Quasimodo's poetry, to both repulse barbarism and strengthen "il buon vivere." It seems no mere accident that slow food has been born in the same country that gave us humanism and the greatest outpouring of art and aesthetic beauty the world has ever seen, roughly 500 years ago. Now, in an age of high technology, "neohumanism" seems like our best course for the future....

Tying It All Together

All of the elements described above require *action*, both individual and collective, as opposed to the passive behavior that allows an (alleged) technological imperative to take its course. These elements require understanding and consideration, and in some cases active seeking of information, connections, and ideas, all activities that go against the grain of current trends on the Internet. A basic premise is that technological development is not in fact inevitable, or that a particular outcome of a technology's trajectory is not foreordained. Making appropriate choices thus becomes not only possible but of paramount importance, and it helps a lot if multiple choices work together to enhance and promote values worth defending.

Slow food, for example, has admirably linked culinary enjoyment to environmental protection and responsibility as well as to a critique of economic forces that push people into a harried life. The Slow Cities League has extended this idea to the cultural identity of towns, cities, or regions, and then in turn to the desire for excellence and distinctiveness among individuals. The Open Source and free-software movement is clearly linked to grassroots struggles over intellectual property laws and to the challenge of balancing hard-won skills and ease of use. So far, technology activists attracted to the Open Source and free-software movement have had less sympathy with addressing the digital divide, although there are many activists who have made this bridge. This is simply an area where more work needs to be done.

It is obvious that there is not yet a completely realized "consciousness"—for lack of a better word—about the new paradigm described above, a way of thinking about the world that is not only opposed to the negative, dehumanizing aspects of globalization but that recognizes globalization's positive features and incorporates them into a worldview that both embraces technology and shapes it to different ends. Many young antiglobalization activists are—unfortunately—opposed to all aspects of globalization, and some are even opposed to modern technological development in general (even though antiglobalization organizations are very effective users of the Internet, by and large). Some young anarchists mistakenly want to preserve cultures and forms of social organization that are guaranteed to be stuck in poverty and ignorance indefinitely. Among the most visible antiglobalization activists—for example, those who are willing to resort to violence and overt confrontation with the police—there is hardly more intellectual content than an inchoate rage or the thrill of rebellion, which are useless for building alternative visions of good living.

The antiglobalization movement, on the other hand, has tapped into some powerful feelings, and not only among young rebels. Thoughtful people who simply see a world with intensified forms of current trends—including industry concentration and gigantism; a magnetic pull toward blandness, sameness, and mediocrity; a life chock-full of advertisements; and the replacement of authentic and free culture with ersatz and "payper" "experiences"—have reasons to be both alarmed and angry. People whose Internet experience stretches back to the years before the dot-com frenzy of the 1990s are often appalled at what the Internet has become. There is a sense that when commerce touches anything these days, it either dies or is transformed into something artificial and alien, a poor substitute for the "real." This may continue indefinitely, but not without resistance and friction.

Conclusion: A New Bipolarity?

During the decades of the Cold War, the "intellectual frame" of the world was organized around a competitive bipolarity between the capitalist West and the communist East, with other parts of the world judged on their orientation to these two poles. For historical and political reasons, the bipolarity of the Cold War was both ideological and geographic, symbolized by the phrase—and the reality of—the "Iron Curtain." This experience predisposed many of us to think of cultural and political bipolarities in geopolitical terms.

Globalization has not only ended this way of thinking about the world, but it has introduced the potential, at least, of thinking about competing worldviews that have no geographic "map"—that is, there are different social movements representing different goals for culture and history that transcend nation-states or regions. Since the tragedy of September 11 there have been speculations about a colossal and terrifying "clash of civilizations" between the secular West and the Muslim crescent that stretches from the west coast of North Africa to Indonesia. Such scenarios typically leave out the fact that there are millions of Muslims in Western countries, largely an epiphenomenon of globalization. The September 11 terrorists themselves spent years outside of Islamic countries. The old framework of geopolitical rivalries representing alternative views of how society should be organized and governed is no longer as relevant as it once was. People with any particular political orientation that claims to be relevant to the entire world can cheaply and easily find allies and surrogates around the globe by using the Internet and other communications technologies. The impact and significance of "memes"[1]—or germs of cultural ideas—have been increasing in a dramatic fashion because of these developments. Even the Zapatistas, the political movement of the impoverished and largely illiterate population of southern Mexico, have used the Internet in such effective ways that their conflict in Mexico has been described as a form of "information warfare" (Ronfeldt and Martinez 1997). The use of the Internet by antiglobalization protesters around the world has become a key component of their success in organizing massive and repeated demonstrations at the meetings of world leaders (Tanner 2001).

But it is the slowly emerging threads of a different vision of how to live life—and how to think about technology—that are most interesting, especially as they begin to interweave and create a positive, attractive, and feasible alternative to large-scale corporate organization and mass consumerism. Slow food and slow cities are two examples, and the Open Source and free-software movement is another. The growing opinion that biotechnology is a distraction from addressing the world's crisis in biodiversity is yet another example. The tumultuous politics in the Middle East are supplying new urgency to questions about postindustrial societies' dependence on fossil fuels for energy. There are other examples. So far, these "alternative" views, or challenges to the status quo, have yet to cohere into a complete ideological framework that might appeal to large numbers of people, enough people to transform political agendas. But this could happen. The antiglobalization protests are large enough and frequent enough—particularly in Western Europe—that they may begin to have sweeping effects on political discourse and elections. In the United States, there is growing evidence of a widening gulf between the two main political parties. The activists of each party, the Democratic Party and the Republican Party, are increasingly polarized; even the leadership of the Democratic Party is often viewed by its own supporters as too timid and compromised by deals with wealthy donors and corporations. Moreover, the two ends of the political spectrum are coming to represent not just political choices but entirely separate "lifestyles," or two distinct cultures, as intellectuals on both the right and the left now admit.

The familiar strategy of reigning in capitalism when its excesses need to be checked, typically through state institutions with regulatory powers, is losing its appeal among many people who are looking for completely new ways of reconfiguring economic sectors such as agriculture, communications, entertainment, and health care. We seem to be in the process of creating two cultures that will coexist but with constant friction, especially because of the economic fragility of small-scale enterprises in comparison to their more powerful global counterparts. If French farmers are put out of business because of world-trade agreements and the French lose their cherished traditions of regional culinary specialties, a lot of French people will be looking for ways to fight back. If the Internet turns completely into a pipeline for junk e-mail, pop-up web ads, online scams, garish "infotainment" sites, and pornography, with little material of quality and excellence as compensation, there will be a backlash. If all these things happen at more or less the same time—say, within the next twenty years—and enough people see this trend as folly and tragedy, a new, nongeographic bipolarity, may develop, one between people who care about such things and those who do not, or between people who have one vision of the good life and a competing group who see their vision as being fulfilled by contemporary capitalism. This is already happening in one form or another in most modern countries. Global capitalist culture is already uniting the interests of billions of consumers. What we are only starting to see is the merging of interests of people who have reservations about that particular fate for the world. Pieces of an alternative global culture are beginning to appear, but are not yet meshed into a coherent picture.

The world has many problems and not all of them will be solved by "alternative" ways of thinking. We are likely to see more wars, more anguish among entire peoples, perhaps even more terrorism, and there will be responses to all of these things that will shape history in important ways. There has certainly been a dramatic reinforcement of conventional security institutions in the United States since September 11, 2001, for example. The big global problems like poverty, illness, and violence will still need to be addressed by institutions with consensual powers of authority.

Nor will we see complete harmony and a convergence of ideas and goals among people who deliberately oppose the status quo. So many degrees of difference among such people may persist that they will continue to look like a chaotic, cacophonic mob rather than a historic force of change. It may only be over the course of many years that we come to recognize an emergent, new way of thinking, which is likely to take different but related forms all over the world. The common thread that may unite many disparate but like-minded efforts, from food activists to digital rights activists, is thinking about technology as malleable, as capable of serving human-determined ends, and as an essential component of "il buon vivere," the good life. It is only by working with that premise that the idea of shaping the network society makes sense.

Note

1. The word *meme* is defined by *The Merriam-Webster Dictionary* as "an idea, behaviour, style, or usage that spreads from person to person within a culture."

References

Brand, S. 1987. *The Media Lab: Inventing the Future at MIT*. New York: Viking Press.

Città Slow. 2002. Slow cities website. http://www.cittaslow.stratos.it/, accessed September 12, 2007.

Joy, B. 2000, April. "Why the Future Doesn't Need Us." *Wired* 8.04, 238–262 [reprinted as ch. 6 in this volume].

Kurzweil, R. 1999. *The Age of Spiritual Machines: When Computers Exceed Human Intelligence*. New York: Viking Press.

Markoff, J., and Lohr, S. 2002, September 29. "Intel's Big Bet Turns Iffy." New York Times. Http://www.nytimes.com/2002/09/29/technology/circuits/29CHIP.html, accessed September 12, 2007.

Moravec, H. 1998. *Robot: Mere Machine to Transcendent Mind.* Oxford: Oxford University Press.

Quasimodo, S. 1959. Quasimodo's Nobel Prize acceptance speech can be found at http://nobelprize.org/nobel_prizes/literature/laureates/1959/quasimodo-speech.html, accessed September 12, 2007.

Ronfeldt, D. and Martinez, A. 1997. "A comment on the Zapatista 'Netwar.'" In J. Arquilla and D. Ronfeldt, eds., *In Athena's Camp: Preparing for Conflict in the Information Age*, 369–391. Santa Monica, CA: RAND.

Slow Food. 2002. http://www.slowfood.com, accessed September 12, 2007. See also the website for Slow Food USA, http://www.slowfoodusa.org/, accessed September 12, 2007.

Tanner, A. 2001, July 13. "Activists Embrace Web in Anti-Globalization Drive." Reuters, http://www.globalexchange.org/economy/rulemakers/reuters071301.html, accessed September 12, 2007.

26 "The Feminization of Work in the Information Age"

Judy Wajcman

In the 1980s and 1990s, innovations in ICTs (information and communications technologies) and the shift from a manufacturing to service economy in the Western world were hailed by a number of scholars as advances that would help equalize the playing field in business for men and women. The idea was that as the need for employees in manufacturing (traditionally a masculine endeavor) was replaced by a need for service employees (traditionally a feminine endeavor), women would play a much more prominent role in the economy. Judy Wajcman's study demonstrates that such deterministic predictions didn't come true. Women have increased in the workforce, but don't hold the leadership roles that were predicted. The development and implementation of ICTs provides an opportunity to transform gender roles, but it can also be used to reinforce prevailing arrangements. Those who currently wield power—that is, managers and CEOs—will have a far greater impact on which positions women will have in corporations than technological change. Wajcman's work reminds us that technologies open up new opportunities, but they do not necessarily lead to specific social changes. The onus is on people, institutions, and corporations to decide what types of social change they would like to achieve, take advantage of the opportunities that new technologies make possible, and work to ensure that new technologies are interpreted and used in ways that help us progress toward such goals.

The future of work and the transformation of family life are key issues in contemporary social science. Many believe that the invention and diffusion of digital technologies are factors at the heart of these transformations. Much emphasis is placed on major new clusters of scientific and technological innovations, particularly the widespread use of information and communication technologies (ICTs), and the convergence of ways of life around the globe. The increased automation of production and the intensified use of the computer are said to be revolutionizing the economy and the character of employment. In the "information society" or "knowledge economy," the dominant form of work becomes information and knowledge-based. At the same time, leisure, education, family relationships, and personal identities are seen as molded by the pressures exerted and opportunities arising from the new technical forces.

Theorists of the information society have, for the most part, shown little interest in the question of changing gender relations. This, despite the fact that the feminization of the labor force has been heralded as one of the most important social changes of the twentieth century. Their oversight is particularly disappointing given the contribution

From Mary Frank Fox, Deborah G. Johnson, and Sue V. Rosser, eds., *Women, Gender and Technology* (Urbana: University of Illinois Press, 2006), pp. 80–97. Reprinted with permission of the University of Illinois Press.

that feminist scholarship has made to our understanding of the gendered nature of both work and technology. The sexual division of labor is central to technological development and the organization of work. Indeed, the literature demonstrates that the relationship between gender relations and ICTs is a complex mixture of interactive processes, a key site of which is the workplace (Wajcman, 2004).

The challenge, then, is to untangle the relationship between these dynamic processes and assess the implications for gender equality. The key underlying question is: to what extent are the older hierarchies of the gender order being destabilized in the digital economy? We begin by considering the meaning of the term "information society" and examine what the shift to a service economy means for women's position and experiences in the labor market. The chapter then looks at the implications of organizational restructuring for women's careers and power relations in management. Finally, we will look at how ICTs affect the spatial organization of economic activity.

The Information Society

The notion of the information society has been absorbed into everyday use, and it is also widely criticized (Castells, 1996; Loader, 1998; Webster, 1995). I use the term here because it captures the evolutionary and determinist frameworks employed by most theorists of the "information age." These frameworks are not only gender blind, but they also fail to recognize the operation of wider power relations. Many social commentators have written with confidence about the technological, economic, social, and cultural changes over the last two decades (Kumar, 1995; Held, Goldblatt, & Perraton, 1999). Several different schools of thought exist about postindustrial society, but the recurring theme is a claim that theoretical knowledge and information have taken on a qualitatively new role. Prominence is given to the intensity, extent, and velocity of global flows, interactions, and networks embracing all social domains.

One of the best known commentators of such change is Manuel Castells (1996), who argues that the revolution in information technology is creating a global economy, the product of an interaction between the rise in information networks and the process of capitalist restructuring. In the "informational mode of development," labor and capital, the central variables of the industrial society, are replaced by information and knowledge. In the resulting "Network Society," the compression of space and time made possible by the new communication technology alters the speed and scope of decisions. Organizations can decentralize and disperse, with high-level decision making remaining in "world cities" while lower level operations, linked to the center by communication networks, can take place virtually anywhere. "Information" is the key ingredient of social organization, and flows of messages and images between networks constitute the basic thread of social structure (Castells, 1996, p. 477). For Castells, the information age, organized around "the space of flows and timeless time," marks a whole new epoch in the human experience.

While optimistic and pessimistic visions of the information age exist, they all focus on the assumed outcomes for employment. The optimists see the expansion of the information-intensive service sector as producing a society based on lifelong learning and a knowledge economy. This implies that a central characteristic of work will be the use of expertise, knowledge, judgment, and discretion in the course of producing a product or service, requiring employees with high levels of skills and knowledge. The pessimistic neo-Braverman approach, by contrast, sees growing technology-induced unemployment and increased vulnerability to global capital (see review by Burris, 1998). It argues that automation standardizes worktasks and diminishes the need to exercise analytical skills and theoretical knowledge.

What is common to most of these understandings of the new social order is their tendency to adopt a technologically determinist stance. Castells explicitly builds on theories of postindustrialism, moving beyond a teleological model and giving the analysis a global reach (Bell, 1973). However, while he explicitly attempts to distance himself from technological determinism, he does not entirely succeed. The idea that technology, specifically information and communication technology, is the most important cause of social change permeates his analysis of Network Society. Similarly, there is a tendency to conceptualize these technologies in terms of technical properties and to construct the relation to the social world as one of implications and impacts. The result is a rather simplistic view of the role of technology in society. In this, Castells is typical of most scholars of the information society who fail to engage with the burgeoning literature in the social studies of science and technology (STS) that has developed over the last two decades (Jasanoff et al., 1995; MacKenzie & Wajcman, 1999).

While space is not available here to engage in any depth with this literature, the point of the STS literature is not to deny the transformative potential of technology. Rather, STS emphasizes that technological change is itself shaped by the social circumstances within which it takes place. STS studies show that the generation and implementation of new technologies involve many choices between technical options. A range of social factors affects which of the technical options are selected. These choices shape technologies and, thereby, their social implications. In this way, technology is a socio-technical product, patterned by the conditions of its creation and use. Understanding the place of these new technologies from such a perspective requires avoiding a purely technological interpretation and recognizing the embeddedness and the variable outcomes of these technologies for different social groups. Technology and society are bound together inextricably; this means that power, contestation, inequality, and hierarchy inscribe new technologies. ICTs can indeed be constitutive of new social dynamics, but they can also be derivative or merely reproduce older conditions. Moreover, it is increasingly recognized that the same technologies can have contradictory effects.

From Manufacturing to Services

Given the complexity of the relationship between technology and social change, it is apparent that the effects of ICTs specifically on women's work are extremely hard to isolate and assess. Many of the employment dynamics discussed in this chapter are functions of restructuring and work reorganization within industries. As Webster (2000, p. 123) notes, "The introduction and application of ICTs are a part of, as much as the consequences of, technological change." It is clear, however, that ICTs are profoundly implicated in changes in work reorganization and work location in particular industries that are central to the information society. What, then, are some of the key changes to women's work in the information society?

One of the most striking features of advanced capitalist economies is the feminization of the labor force. The sharp separation between work and home has been eroded by this development, at least for some social strata. The most important change to women's access to paid employment has been the sharp increase in the labor market participation of married women and women with young children. The causes of feminization are complex, but clearly they link to the substantial growth in service-sector activity and employment. In most advanced countries, the manufacturing sector has declined, with most new jobs being created in services. In one sense, this advantages women, since they have long been associated with service work, especially jobs involving caring for and catering to the needs of clients. Women have predominated numerically in clerical work, retail, catering, and the health and education professions, all of which are important providers of jobs in the modern service-based economy (Bradley et al., 2000; Webster, 1999).

Accompanying the feminization of the labor force has been a dramatic growth in economic inequality between different groups of women. This phenomenon is especially striking in the United States, where feminist demands for gender equality have been more potent than elsewhere, and where inequality in wealth and income has increased sharply in recent years (Jacobs, 1995). Although this trend to feminization has not brought about a major breakdown of gender segregation, there has been a significant movement of women into traditionally male professions such as law, medicine, and management. This elite of women has unprecedented access to well-paid, high-status occupations, while at the bottom of the occupational hierarchy, women have expanded their share of already femininized lower-skilled or lower-paid occupations. Indeed, in many traditionally female "semi-professions" like nursing and secretarial work, restructuring is creating a small minority of highly paid, highly skilled workers alongside a much larger number of poorly paid, minimally trained support workers. In these latter cases, polarization is occurring within a virtually all-female occupational category (Milkman, 1995). At the other end of the spectrum, advanced capitalist countries have seen an enormous growth among "contingent" workers, the majority of whom are women. Women make up the majority of part-time workers, temporary workers, and at-home "independent contractors" (Rubery et al., 1998).

The increase in "flexible" or "atypical" work that characterizes this era of economic globalization could not have occurred without the proliferation of ICTs that support it. For example, the financial service and telecommunications industries rely heavily on information technology (IT) for service and sales delivery. Business transactions ranging from personal retail banking to transnational financial market deals are increasingly mediated by IT. Researchers generally note that IT has three distinct features that can change dramatically the way service work is organized: the "automate, informate and networking capabilities" (Yeuk-mui, 2001, p. 178). Automation means the replacement of manual labor by the IT system to accomplish menial worktasks. The informating ability refers to the capability of IT to generate detailed information about the work process. The networking ability of IT refers to the use of company-wide intranets to coordinate worktasks, disseminate information, and exchange opinions between employees in different ranks and functional departments. These three distinct capabilities of IT have been widely acknowledged (see Zuboff, 1988). However, little consensus exists regarding their impact on work organization and how they affect the experience of women in employment.

One issue that has occupied feminist scholars has been the growing concern in contemporary firms with quality and customer service. Because of the complex interplay of technological and organizational developments, service work increasingly relies on "front-line" work that is people oriented. It has become common to argue that service work challenges our usual conceptions of work because the quality of the service is so intimately related to the personal qualities and social skills of the service providers (Leidner, 1993; Du Gay, 1996). Service work is being feminized in more than simply numerical terms, in that to an increasing extent these jobs require the supposedly feminine qualities of serving and caring. These new forms of labor have been variously theorized as "emotional labor" (Hochschild, 1983), "sexual labor" (Adkins, 1995), and "aesthetic labor" or "body work" (Tyler & Abbott, 1998).

The feminization of service work has specific implications for women because their physical appearance and "personality" become an implicit part of the employment contract. An illustration is the requirement imposed on female, but not male, flight attendants to weigh in periodically during routine grooming checks in order to maintain a strict weight-height ratio (Tyler & Abbott, 1998). The enforcement of weight standards by the airline industry leads to "enforced" dieting in pursuit of the thin, ideal body. This aesthetic labor or bodywork, like emotional and sexual labor, is an integral part of the "effort bargain" (Forrest, 1993). Yet it is not recognized or remunerated because it is seen as what women *are* rather than what women *do*.

Overall then, the shift from manufacturing to services must be viewed as a mixed blessing for women. On the one hand, women are more fully integrated into the paid labor force and are unlikely to be relegated to the domestic sphere of yesteryear. On the other hand, many of the jobs created in this sector are temporary and part-time. While these jobs offer women more flexibility, the use of IT by employers to fine tune their labor requirements can cost women dearly in terms of pay, conditions, and training opportunities. The skill requirements for much service sector employment tend to be social and contextual, making them less amenable to formal measurement. The issue of how skills or competencies are perceived, labeled, accredited, and rewarded is critical for women's ability to participate in and benefit from the "knowledge-based economy."

The failure to regard women's social and communicative skills as knowledge-based and reward them accordingly has strong echoes with the way in which women have been traditionally defined as technically unskilled, thus excluding them from well-paid work. The association between technology, masculinity, and the very notion of what constitutes skilled work was and is still fundamental to the way in which the gender division of labor is reproduced. Further, men's traditional monopoly of technology has been identified as key to maintaining the definition of skilled work as men's work. Machine-related skills and physical strength are basic measures of masculine status and self-esteem, and by implication, the least technical jobs are suitable for women. Technological innovation in the print and newspaper industry provides a clear illustration (Cockburn, 1983). For compositors, the move to computerized typesetting technologies was experienced not only as a threat to their status as skilled workers but also as an affront to their masculinity, and they resisted vigorously. The result is that machinery is literally designed by men with men in mind—the masculinity of the technology becomes embedded in the technology itself (see also Cockburn & Ormrod, 1993; Wajcman, 1991).

A recent report about women in the information technology, electronics, and communications (ITEC) sector confirms that women are making few inroads into technology-related courses and careers (Millar & Jagger, 2001). The report, which covers six countries, including the United States, found that women are generally underrepresented among graduates in ITEC-related subjects, despite the fact that they form the majority of university graduates overall. In the United States, for example, women were particularly underrepresented among graduates in computer and information science (27 percent) and engineering (16 percent). At the doctoral level, in computer and information science, women represented but 16 percent; in engineering, only 10 percent were women in 1995 (Fox 1999, 2001). Indeed, the number of women graduating in computer and information science declined from 14,966 in 1985–86 to 6,731 in 1996–97.

This bias in women's and girls' educational choices has major repercussions because ITEC employment is graduate intensive. It is reflected in women's low participation in ITEC occupations across the U.S. economy, which declined from 37 percent in 1993 to 28 percent in 2000. Women are relatively well represented in the lower-status ITEC occupations, such as telephone operators, data processing equipment installers and repairers, and communications equipment operators. By contrast, male graduates are heavily concentrated among computer system analysts and scientists, computer science teachers, computer programmers, operations and systems researchers and analysts, and broadcast equipment operators. In all the countries surveyed, women face considerable barriers when they attempt to pursue a professional or managerial career in ITEC. The result is that "women are chronically underrepresented in ITEC jobs that are key to the creation and design of technical systems" (Millar & Jagger, 2001, p. 16). For all the rhetoric about women prospering in the emerging digital economy, the gender gap is only being slowly eroded.

The multiple causes are all too familiar. Schooling, youth cultures, the family, and the mass media all transmit meanings and values that identify masculinity with machines and technological competence. An extensive literature now exists on sex stereotyping in general in schools, particularly on the processes by which girls and boys are channeled into different subjects in secondary and higher education, and the link between education and gender divisions in the labor market (see Light & Littleton, 1999). Furthermore, many children's computer games and educational software (such as the ubiquitous Game Boy by Nintendo) are clearly coded as "toys for boys." Many of the most popular games are simply programmed versions of traditionally male non-computer games, involving shooting, blowing up, speeding, or zapping in some way or another. They often have militaristic titles such as "Destroy All Subs" and "Space Wars," highlighting their themes of adventure and violence. No wonder, then, that these games often frustrate or bore the nonmacho players exposed to them. As a result, macho males often have a positive first experience with the computer; other males and most females have a negative initial experience.

The dominant image of the ICT worker is young, white, male "nerds" or "hackers" who work sixteen-hour days and who neither seek nor have access to family-friendly work practices such as part-time and flexible work. Indeed, it is rare to see a female face among the new dot-com millionaires. The "cyber-brat pack" for the new millennium—those wealthy and entrepreneurial young guns of the Internet—consists almost entirely of men. The masculine workplace culture of passionate virtuosity, typified by the hacker-style work, has been well described by Turkle (1984, p. 216) in a chapter entitled "Loving the Machine for Itself." Based on ethnographic research at MIT, Turkle describes the world of computer hackers as epitomizing the male culture of "mastery, individualism, nonsensuality." Being in an intimate relationship with a computer is also a substitute for, and refuge from, the much more uncertain and complex relationships that characterize social life. Turkle's account resonates with Hacker's (1981) studies of engineering where mastery over technology is a source of both pleasure and power for the predominantly male profession.

This is not to imply that there is a single form of masculinity or one form of technology. Rather, it is to note that in contemporary Western society, the culturally dominant form of masculinity is still strongly associated with technical prowess and power. Feminine identity, on the other hand, involves being ill-suited to technological pursuits. Indeed, the construction of women as different from men is a key mechanism whereby male power in the workplace is maintained. A successful career in ICT requires navigation of multiple male cultures associated not only with technological work but also, as we shall see below, with managerial positions.

Feminization of Management?

Compared with their generally unsteady progress in ICT employment, women have made great strides into traditionally male managerial careers, as indicated earlier

(Jacobs, 1995). This has given rise to a new orthodoxy that management culture, structures, and practices, which used to be deeply masculine and hostile to women, are being feminized. A growing number of organizational theorists and management consultants assert that, in the new service economy, there is a premium on less hierarchical, more empathetic and cooperative styles of management. According to the new orthodoxy, effective management needs a softer edge, a more qualitative approach (Applebaum & Batt, 1994; Handy, 2001; Kanter, 1989). Successful firms are described as people-oriented and decentralized, uncluttered by bureaucratic layers of management. Leadership is now concerned with fostering shared visions, shared values, shared directions, and shared responsibility. It is suggested that women have a more consensual style of management and thus are ideally suited to postindustrial corporations. The conclusion drawn in this literature is that the norm of effective management will be based on the way women do things.

As in the literature on service work, the management literature focuses on the advantages women have because of communication and social skills, considered to be natural attributes. Reflecting on the copious literature on leadership style, I am struck by the way in which it is permeated by stereotypical dualisms, such as that between hard and soft, reason and intuition (Fagenson, 1993; Helgesen, 1990; Powell, 1993). Instead of challenging the gendered nature of these dichotomies, they are simply inverted. Leadership traits that correspond with male traits, like dominance, aggressiveness, and rationality, are now presented negatively, while formerly devalued feminine qualities, like the soft and emotional, are presented positively.

My own research on senior managers in five high-technology multinational companies sets out to explore these claims; in particular, it explores whether there really is a difference in management style between men and women and, second, how women and men managers are faring in the new management cultures. The study combined a large-scale survey with detailed interviews of male and female managers, and featured a leading American computer corporation. (For a full description of the study, see Wajcman, 1998.)

Initially, I found that a high proportion of both women and men expressed the view that sex differences in management style do exist. On the whole they described women's difference in positive terms. Typical descriptions by both men and women of the male style include: "directive," "self-centred/self-interested," "decisive," "aggressive," and "task oriented." Adjectives used to describe the female style are: "participative," "collaborative," "cooperative," "people-oriented," and "caring." However, when respondents described *their own* management style, either as "participative style, people-handling skills, developing subordinates" or as "leading from the front, drive, decisively directing subordinates," there was no significant difference between the men's and women's responses.

These findings confirm the extent to which people characterize themselves in terms of dominant cultural values. Research shows that men and women tend to stereotype their own behavior according to learned ideas of gender-appropriate behavior, just as they stereotype the behavior of other groups (Epstein, 1988). An integral part of the identity of men and women is the perception that they possess, respectively, masculine and feminine qualities. So it is not surprising that women and men respondents subscribe to gender stereotypes of management styles.

It is only when asked to discuss more specifically their own work practices that the gap between beliefs and behavior emerges. The evidence from the qualitative case-study material reveals a major discrepancy between the current discourse of "soft," people-centered management and the "hard" reality of practice. Many of the interviewees commented that with the almost continuous downsizing of companies, management is returning to a more traditional hierarchical structure. Macho management is again in the ascendancy. Coping with uncertainty within the organization was a constant theme. All respondents found the process of making people redundant difficult, and no obvious gender differences in managerial style emerged in how they accomplished the task. Likewise, when people talked about how they handled other sensitive or conflictual issues, no gender differences emerged. Rather, both men and women told similar stories about how they dealt with such issues as pay, performance, and retrenchment.

The business context of continuous restructuring and job losses has greatly intensified pressures for senior managers. The traditional career-for-life model, based on employment security and promotion prospects, has been replaced by the logic of survival resulting in heightened individualistic competition for a dwindling number of career opportunities. In today's harsh economic climate, *both* men and women feel the need to conform to the "hard" macho stereotype of management because it is still, in practice, the only one regarded as effective.

Both the men and women in my study inhabit a male-dominated, high-tech, corporate world. Managers of both sexes must present themselves so as to project an image of the authoritative manager, adopting or adapting the hegemonic masculine mode. But while they may have the heart for it, women do not have the body for it. The employment contract grants employers command over the bodies of their employees, but these bodies are sexually differentiated. As several organization theorists have noted, male sexuality underpins the patriarchal culture of professional and organizational life (Cockburn, 1991; Gherardi, 1995; Hearn et al., 1989). The sexualization of women's bodies presents a particular problem for women managers.

Whereas writers on the service sector have emphasized the extent to which the sexual skills or services of employees are incorporated into their organizational role, in management, it is men's bodies that are inscribed in the managerial function and women's bodies that are excluded. As Acker (1990, p. 152) has emphasized, women's bodies are often ruled out of order: "women's bodies—female sexuality, their ability to procreate and their pregnancy, breastfeeding, and child care, menstruation, and mythic 'emotionality'—are suspect, stigmatized, and used as grounds for control and exclusion." Without constant vigilance regarding gender self-presentation at work, women run the risk of not being treated seriously as managers. I found that the women in my study had to abandon aspects of their femininity and develop attributes that resemble those of male executives. Far from witnessing a feminization of senior managerial work, my study concludes that women must generally "manage like a man" to succeed in their careers.

Within men's studies, too, there is a growing interest in the masculinity of managers (Cheng, 1996; Collinson & Hearn, 1996; Weiss, 1990) and whether it is being reshaped in the digital economy. Kerfoot and Knights (1996), for example, purport a shift away from the standard form of masculinity associated with modern management practice that is abstract, rational, highly instrumental, future-oriented, strategic, and wholly disembodied. This traditional and paternalistic masculinity is allegedly being displaced by a form of masculinity that displays entrepreneurialism and risk taking. A more egalitarian form of masculinity, it is consistent with the new management approach's emphasis on social relationships. Further feminization of managerial employment is promised, given the association of femininity with interpersonal skills. This claim is contested by Calas and Smircich (1993) who predict that more junior managerial positions, confined to national-level concerns, will continue to be feminized and downgraded while men colonize the more powerful and prestigious globalized functions. Multinational corporations are still run by men and the appointment of women to senior positions is still seen as a risky proposition.

Because of the enormous power held by senior managers and executives, gender inequalities in access to authority constitutes a key mechanism sustaining gender inequalities in the economy at large. For men, the transition from technical roles to managerial jobs is a relatively straightforward step in their career trajectory, whereas women's careers typically hit the proverbial "glass ceiling." An interesting finding of my research was that significantly more women than men reached senior positions via professional credentials. This is consistent with Savage's study, which also found that "women have moved into positions of high *expertise*, but not positions of high *authority*" (Savage, 1992, p. 124; see also Wright & Baxter, 1995). It appears that managerial power and authority are even more intrinsically gendered than technical expertise. While women have benefited enormously from the growth of professional occupations in the information society, men continue to monopolize the elite levels of corporate careers.

ICTs and Changing Location

All visions of the information society place great emphasis on the way ICTs allow for an increasing disassociation between spatial proximity and the performance of paid work (Castells, 1996; Sassen, 1996). The idea is that with the advent of technological innovations, production no longer requires personnel to be concentrated at the place of work. In this scenario, the home as workspace liberates people from the discipline and alienation of industrial production. Computer-based homework or telework offers the freedom of self-regulated work and a reintegration of work and personal life. Moreover, an expansion of teleworking will allegedly lead to much greater sharing of paid work and housework, as men and women spend more time at home together. Mothers are particularly seen as the beneficiaries of this development as working from home allows much greater flexibility to combine employment with childcare.

Futurologists commonly assume a dramatic increase in teleworking, but the term itself suffers from a lack of clarity. If one adopts a narrow definition of teleworkers, as those employed regularly to work online at home, the figures are rather small, perhaps 1–2 percent of the total U.S. labor force (Castells, 1996, p. 395). As many commentators like to joke, more people are researching teleworking than are actually doing teleworking. Nevertheless, teleworking has important implications for the way women's work is understood. We need to distinguish between skilled or professional workers who work from home and the more traditional "homeworkers" who tend to be semi-skilled or unskilled low-paid workers. The former certainly do have more choices about how they schedule their work to fit in with the rest of their everyday lives. However, these teleworkers, who tend to work in occupations like computing and consultancy, are typically men. Women who telework are mainly secretarial and administrative workers. So a rather conventional pattern of occupational sex segregation is being reproduced in this new form of work (Huws et al., 1996).

Indeed, women and men are propelled into teleworking for very different reasons. While women's main motivations are childcare and domestic responsibilities, men express strong preferences for the flexibility, enhanced productivity, convenience, and autonomy of such working patterns. The common media image of a woman working while the baby happily crawls across a computer is misleading. There is an important difference between being at home and being available for childcare. Women continue to carry the bulk of responsibility for domestic work and childcare and, for them, telework does not eliminate their double burden. Even among the minority of professional women who work from home, few are able to separate the demands of motherhood and domesticity from paid work (Adam & Green, 1998). For men, who can more easily set up child-free dedicated "offices" at home, telework often leads to very long and unsocial hours of work. These long hours tend to militate against a more egalitarian and child-centered way of life (Felstead & Jewson, 2000).

More significant than "pure" telework (although it has received much less attention) is the capacity of ICTs to facilitate and encourage people to bring work home from the office. Not only has there been a general increase in the number of hours worked at the workplace for managers and professionals, but also the expectation of availability has been greatly extended with the advent of mobile phones, e-mail, and fax machines. In this sense, the boundaries between the public world of work and the private home have become blurred. However, this is almost always in such a way as to facilitate the transfer of work into the home rather than the transfer of home concerns into the workplace. This has made the balance between work and home at senior levels even more difficult. ICTs may have raised productivity, but they have certainly not reduced working time (Schor, 1991).

Interestingly, here we have echoes of the earlier debate about how domestic technologies would save housewives time. Vacuum cleaners, refrigerators, electric ranges, and washing machines—to most Americans these are mundane features of everyday life. Yet, within the memory of many people living today, they were the leading edge of a revolutionary new laborsaving technology that promised to liberate women from a life of domestic drudgery. Although few would dispute that household technologies have made life easier for millions of women (and men), the average woman who is not employed outside of the home devotes just about the same amount of time to taking care of the household as did her mother and grandmother. Even for women who are employed full-time outside the home, increasing evidence casts doubt on there being a direct relationship between the ownership of household appliances and a reduction in the amount of labor required to accomplish routine housework. Rather than simply saving time, technology changes the nature and meaning of the tasks themselves, which can result in even "more work for mother" (Cowan, 1983; Bittman & Wajcman, 2000).

Any discussion of the geographical relocation of worksites must address the use of female labor in the developing world. Although the international division of labor is not a new phenomenon, innovations in ICTs allow a spatial flexibility for a growing range of tasks. The use of third-world female labor by multinational manufacturing industries offering poorly paid assembly jobs is well documented (Mitter & Rowbotham, 1995; Horton, 1996). Garment assembly and seamstress work is subcontracted to small, offshore companies in the South, while the process of design and cutting is carried out in the North.

However, with increasingly automated means of coordinating marketing, production, and customer demand on a daily basis, garment companies have begun to reverse their reliance on third-world labor. Western and Japanese companies alike are increasingly intent upon "close-to-market" strategies that involve subcontracting work to smaller companies in the West. In this move back to their host countries, companies mainly employ women from immigrant and ethnic minority groups, ensuring a captive, regional labor force, compelled to accept low wages and exploitative working conditions. Despite the high levels of capital investment and advanced ICTs used by these firms, there is little transfer of technical skills and expertise to the women who work in these manufacturing jobs.

White-collar, professional, and clerical jobs are also moving to developing countries, underpinned by the diffusion of ICTs. The emergence of the Indian information technology industry is a notable case. Skilled software design and development projects are sited in the West, whereas the programmers employed are located in offshore companies, maintaining consistently high production rates for very low wages. Countries like India have a ready pool of female labor available for software work—women who are well educated, English speaking, and technically proficient. It is estimated that, in India, women constitute over 20 percent of the total IT workforce, which is higher than women's participation in the Indian economy as a whole (Kelkar & Nathan, 2002). Despite the repetitive nature of the work and the lack of job security, women's income, authority in household matters, and social mobility have improved as a result.

Alongside the export of software jobs to countries such as India, Mexico, and China, a proliferation of cybercafes and computers can provide the means for women's groups to organize networks and campaigns to improve their conditions. Indeed, the combination of information technology with telecommunications, particularly satellite communications, provides new opportunities and outlets for women. In country after country, although women still account for a lower proportion of Internet users, and may use the Internet for different purposes than men do, their share is rapidly rising (Sassen, 2002). The fear that the globalization of communications would lead to homogenization and reduce sociability and engagement with one's community was ill conceived. On the contrary, new electronic media have helped build local communities and project them globally. Cyberspace makes it possible for even small and poorly resourced nongovernmental organizations to connect with each other and engage in global social efforts. These political activities are an enormous advance for women who were formerly isolated from larger public spheres and cross-national social initiatives. "We see here the potential transformation of women, 'confined' to domestic roles, who can emerge as key actors in global networks without having to leave their work and roles in their communities" (Sassen, 2002, p. 380). This is not to endorse utopian ideas of cyberspace being gender-free and the key to women's liberation. Rather, it is to stress that the Internet, like other technologies, contains contradictory possibilities and can be a powerful tool for feminist politics.

Conclusion

We need to keep a skeptical eye on purely technological interpretations of the effects of digital technology. Moreover, we need to recognize the embeddedness of social relations and the variable, sometimes contradictory, outcomes of these technologies for different groups of women. ICTs can be constitutive of new gender power dynamics, but they can also be derivative of or reproduce preexisting conditions of gender inequality at work. While the rise of the service economy has led to a feminization of the labor force, these new forms of work to some extent replicate old patterns of sex segregation. The skills that are exercised in predominantly female jobs are still undervalued, and women are slowly making inroads into the upper echelons of ICT occupations. This is directly related to the enduring problem of women's exclusion from senior levels of management. While the flexibility and spatial mobility afforded by ICTs have expanded opportunities for women, the male cultures associated with technical and managerial expertise still serve as a brake on progress toward gender equality at work.

References

Acker, J. (1990). Hierarchies, jobs, bodies: A theory of gendered organizations. *Gender and Society*, *4*, 139–158.

Adam, A., & Green, E. (1998). Gender, agency and location and the new information society. In B. Loader (Ed.), *Cyberspace Divide* (pp. 83–97). London: Routledge.

Adkins, L. (1995). *Gendered work: Sexuality, family and the labor market*. Buckingham, Eng.: Open University Press.

Applebaum, E., & Batt, R. (1994). *The new American workplace*. New York: ILR Press.

Bell, D. (1973). *The coming of post-industrial society*. New York: Basic Books.

Bittman, M., & Wajcman, J. (2000). The rush hour: The character of leisure time and gender equity. *Social Forces, 79*(1), 165–189.

Bradley, H., et al. (2000). *Myths at work*. Cambridge: Polity Press.

Burris, B. H. (1998). Computerisation of the workplace. *Annual Review of Sociology, 24,* 141–157.

Calas, M., & Smircich, L. (1993). Dangerous liaisons: The "feminine-in-management" meets "globalization." *Business Horizons,* March/April, 71–81.

Castells, M. (1996). *The rise of the network society*. Oxford: Blackwell.

Cheng, C. (Ed.). (1996). *Masculinities in organizations*. London: Sage Publications.

Cockburn, C. (1983). *Brothers: Male dominance and technological change*. London: Pluto Press.

———. (1991). *In the way of women: Men's resistance to sex equality in organizations*. London: Macmillan.

Cockburn, C., & Ormrod, S. (1993). *Gender and technology in the making*. London: Sage Publications.

Collinson, D., & Hearn, J. (Eds.) (1996). *Men as managers, managers as men*. London: Sage Publications.

Cowan, R. S. (1983). *More work for mother*. New York: Basic Books.

Du Gay, P. (1996). *Consumption and identity at work*. London: Sage Publications.

Epstein, C. Fuchs (1988). *Deceptive distinctions: Sex, gender, and the social order*. New Haven, Conn.: Yale University Press.

Fagenson, E. (Ed.). (1993). *Women in management: Trends, issues, and challenges in managerial diversity*. Newbury Park: Sage Publications.

Felstead, A., & Jewson, N. (2000). *In work, at home: Towards an understanding of homeworking*. London: Routledge.

Forrest, A. (1993). Women and industrial relations theory. *Relations Industrielles, 48,* 409–440.

Fox, Mary Frank. (1999). Gender, hierarchy, and science. In J. S. Chafetz (Ed.), *Handbook of the sociology of gender* (pp. 441–457). New York: Kluwer Academic.

———. (2001). Women, men, and engineering. In D. Vannoy (Ed.), *Gender mosaics* (pp. 249–257). California: Roxbury.

Gherardi, S. (1995). *Gender, symbolism and organizational cultures*. London: Sage Publications.

Hacker, S. (1981). The culture of engineering. *Women's Studies International Quarterly, 4,* 341–353.

Handy, C. (2001). *The elephant and the flea: Looking backwards to the future of work.* London: Hutchinson.

Hearn, J., et al. (Eds.). (1989). *The sexuality of organization.* London: Sage Publications.

Held, D., McGrew, A., Goldblatt, D., & Perraton, J. (1999). *Global transformations: Politics, economics and culture.* Cambridge: Polity Press.

Helgesen, S. (1990). *The female advantage: Women's ways of leadership.* New York: Doubleday.

Hochschild, A. (1983). *The managed heart: Commercialization of human feeling.* Berkeley: University of California Press.

Horton, S. (Ed.). (1996). *Women and industrialization in Asia.* London: Routledge.

Huws, U., et al. (1996). *Teleworking and gender* (Report 317). Brighton, U.K.: Institute for Employment Studies.

Jacobs, J. (Ed.). (1995). *Gender inequality at work.* Thousand Oaks, Calif.: Sage Publications.

Jasanoff, S., et al. (Eds.). (1995). *Handbook of science and technology studies.* Thousand Oaks, Calif.: Sage Publications.

Kanter, R. M. (1989). *When giants learn to dance: Mastering the challenge of strategy, management and careers in the 1990s.* New York: Simon & Schuster.

Kelkar, G., & Nathan, D. (2002). Gender relations and technological change in Asia. *Current Sociology, 50*(3), 427–441.

Kerfoot, D., & Knights, D. (1996). "The best is yet to come?" The quest for embodiment in managerial work. In D. Collinson & J. Hearn (Eds.), *Men as managers, managers as men* (pp. 78–98). London: Sage Publications.

Kumar, K. (1995). *From post-industrial to post-modern society: New theories of the contemporary world.* Oxford: Blackwell.

Leidner, R. L. (1993). *Fast food, fast talk: Service work and the routinization of everyday life.* Berkeley: University of California Press.

Light, P., & Littleton, K. (1999). *Social processes in children's learning.* Cambridge, U.K.: CUP.

Loader, B. (Ed.). (1998). *Cyberspace divide.* London: Routledge.

Mackenzie, D., & Wajcman, J. (1999). *The social shaping of technology* (2nd ed.). Milton Keynes, U.K.: Open University Press.

Milkman, R. (1995). Economic inequality among women. *British Journal of Industrial Relations, 33*(4), 679–683.

Millar, J., & Jagger, N. (2001). *Women in ITEC courses and careers.* London: Women and Equality Unit, U.K. Department of Trade and Industry.

Mitter, S., & Rowbotham, S. (Eds.). (1995). *Women encounter technology.* London: Routledge.

Powell, G. (1993). *Women and men in management*. Newbury Park, Calif.: Sage Publications.

Rubery, J., et al. (1998). *Women and European employment*. London: Routledge.

Sassen, S. (1996). *Losing control? Sovereignty in an age of globalization*. New York: Columbia University Press.

———. (2002). Towards a sociology of information technology. *Current Sociology, 50*(3), 365–388.

Savage, M. (1992). Women's expertise, men's authority: Gendered organisations and the contemporary middle classes. In M. Savage & A. Witz (Eds.), *Gender and bureaucracy*. Oxford, U.K.: Blackwells.

Schor, J. (1991). *The overworked American: The unexpected decline of leisure*. New York: Basic Books.

Turkle, S. (1984). *The second self: Computers and the human spirit*. London: Granada.

Tyler, M., & Abbott, P. (1998). Chocs away: Weight watching in the contemporary airline industry. *Sociology, 32*, 433–450.

Wajcman, J. (1991). *Feminism confronts technology*. University Park: Pennsylvania State University Press.

———. (1998). *Managing like a man: Women and men in corporate management*. University Park: Pennsylvania State University Press.

———. (2004). *TechnoFeminism*. Cambridge, U.K.: Polity Press.

Webster, F. (1995). *Theories of the information society*. London: Routledge.

Webster, J. (1999). Technological work and women's prospects in the knowledge economy. *Information, Communication & Society, 2*(2), 201–221.

———. (2000). Today's second sex and tomorrow's first? Women and work in the European information society. In K. Ducatel, J. Webster, & W. Herrmann (Eds.), *The information society in Europe* (pp. 119–140). Lanham, Md.: Rowman & Littlefield.

Weiss, R. S. (1990). *Staying the course: The emotional and social lives of men who do well at work*. New York: Free Press.

Wright, E. O., & Baxter, J. (1995). The gender gap in workplace authority: A cross-national study. *American Sociological Review, 60*, 407–435.

Yeuk-Mui Tam, M. (2001). Information technology in frontline service work organization. *Journal of Sociology, 37*(2), 177–206.

Zuboff, S. (1988). *In the age of the smart machine*. New York: Basic Books.

27 "Nanotechnology and the Developing World"

Fabio Salamanca-Buentello, Deepa L. Persad, Erin B. Court, Douglas K. Martin, Abdallah S. Daar, and Peter A. Singer

Nanotechnology has been touted as a technological fix for everything from cancer to the need for better armor plating. Many of the applications, however, seem to privilege the rich in the western world. The authors of this article contend that it need not be this way. They believe that this new array of technologies can also benefit the five billion people living in the developing world. To help spark this process, they contacted a number of nanotechnology experts from around the globe and encouraged them to reflect on the ways that the technology could meet some of the needs of the developing world. The result of this research is a list of "Top Ten Applications of Nanotechnology for Developing Countries." The authors argue that funding agencies—both government and private—must help ensure that nanotechnology is used responsibly by promoting the idea that it meet the needs of the entire world's population and not just the needs of the wealthy. In many ways the authors are advocates of the idea that globalization can be a great boon to the entire world's population, as long as the diversity of needs is actively addressed.

Nanotechnology can be harnessed to address some of the world's most critical development problems. However, to our knowledge, there has been no systematic prioritization of applications of nanotechnology targeted toward these challenges faced by the 5 billion people living in the developing world.

In this article, we aim to convey three key messages. First, we show that developing countries are already harnessing nanotechnology to address some of their most pressing needs. Second, we identify and rank the ten applications of nanotechnology most likely to benefit developing countries, and demonstrate that these applications can contribute to the attainment of the United Nations Millennium Development Goals (MDGs). Third, we propose a way for the international community to accelerate the use of these top nanotechnologies by less industrialized countries to meet critical sustainable development challenges.

Developing Countries Innovate in Nanotechnology

Several developing countries have launched nanotechnology initiatives in order to strengthen their capacity and sustain economic growth [1]. India's Department of Science and Technology will invest $20 million over the next five years (2004–2009) for their Nanomaterials Science and Technology Initiative [2]. Panacea Biotec (http://www.panacea-biotec.com/products/products.htm) (New Delhi, India) is conducting

---

From Nanotechnology and the Developing World, *PLoS Medicine* 2, no. 5 (May 2005): 0383–0386. Reprinted with permission.

Figure 27.1
Quantum dots may be used for cheap, efficient handheld diagnostic devices available at point-of-care institutions in developing countries. DOI: 10.1371/journal.pmed.0020097.g001.

novel drug delivery research using mucoadhesive nanoparticles, and Dabur Research Foundation (Ghaziabad, India) is participating in Phase-1 clinical trials of nanoparticle delivery of the anti-cancer drug paclitaxel [3]. The number of nanotechnology patent applications from China ranks third in the world behind the United States and Japan [4]. In Brazil, the projected budget for nanoscience during the 2004–2007 period is about $25 million, and three institutes, four networks, and approximately 300 scientists are working in nanotechnology [5]. The South African Nanotechnology Initiative (http://www.sani.org.za) is a national network of academic researchers involved in areas such as nanophase catalysts, nanofiltration, nanowires, nanotubes, and quantum dots (figure 27.1). Other developing countries, such as Thailand, the Philippines, Chile, Argentina, and Mexico, are also pursuing nanotechnology [1].

Science and technology alone are not the answer to sustainable development challenges. Like any other science and technology waves, nanoscience and nanotechnology are not "silver bullets" that will magically solve all the problems of developing countries; the social context of these countries must always be considered. Nevertheless, science and technology are a critical component of development [6]. The 2001 Human Development Report [7] of the UN Development Program clearly illustrates the important roles of science and technology in reducing mortality rates and improving life expectancy in the period 1960–1990, but it did not emphasize nanotechnology specifically. In a report released in early 2005 [8], the UN Task Force on Science, Technology and Innovation (part of the process designed to assist UN agencies in achieving the UN MDGs) addresses the potential of nanotechnology for sustainable development.

Top Ten Nanotechnologies Contributing to the MDGs

In order to provide a systematic approach with which to address sustainable development issues in the developing world, we have identified and ranked the ten applications of nanotechnology most likely to benefit developing countries. We used a modified Delphi Method, as described in our Top Ten Biotechnologies report [9] to identify and prioritize the applications and to achieve consensus among the panelists.

We recruited an international panel of 85 experts in nanotechnology who could provide the informed judgments that this study required, of which 63 completed the project.[1] We selected the panelists based on contacts identified in our previous study on nanotechnology in developing countries [1]. A conscious effort was made to balance the panel with respect to gender, specialty areas within nanotechnology, and geographic distribution. Of the panelists, 38 (60%) were from developing countries and 25 (40%) from developed countries; 51 panelists (81%) were male and 12 (19%) were female.

We posed the following open-ended question: "Which do you think are the nanotechnologies most likely to benefit developing countries in the areas of water, agriculture, nutrition, health, energy, and the environment in the next 10 years?" These areas were identified in the 2002 UN Johannesburg Summit on Sustainable Development [10]. We asked the panelists to answer this question using the following criteria derived from our previous Top Ten Biotechnologies study.

Impact How much difference will the technology make in improving water, agriculture, nutrition, health, energy, and the environment in developing countries?
Burden Will it address the most pressing needs?
Appropriateness Will it be affordable, robust, and adjustable to settings in developing countries, and will it be socially, culturally, and politically acceptable?
Feasibility Can it realistically be developed and deployed in a time frame of ten years?
Knowledge gap Does the technology advance quality of life by creating new knowledge?
Indirect benefits Does it address issues such as capacity building and income generation that have indirect, positive effects on developing countries?

Three Delphi rounds were conducted using e-mail messages, faxes, and phone calls. In the first round, the panelists proposed examples of nanotechnologies in response to our study question. We analyzed and organized their answers according to common themes and generated a list of twenty distinct nanotechnology applications. This list was reviewed for face and content validity by two nanotechnologists external to the panel. In the second Delphi round, the panelists ranked their top ten choices from

Table 27.1
Correlation between the Top Ten Applications of Nanotechnology for Developing Countries and the UN Millennium Development Goals

Ranking (score)	Applications of nanotechnology	Examples	Comparison with the MDGs
1 (766)[a]	Energy storage, production, and conversion	Novel hydrogen storage systems based on carbon nanotubes and other lightweight nanomaterials Photovoltaic cells and organic light-emitting devices based on quantum dots Carbon nanotubes in composite film coatings for solar cells Nanocatalysts for hydrogen generation Hybrid protein-polymer biomimetic membranes	VII
2 (706)	Agricultural productivity enhancement	Nanoporous zeolites for slow-release and efficient dosage of water and fertilizers for plants, and of nutrients and drugs for livestock Nanocapsules for herbicide delivery Nanosensors for soil quality and for plant health monitoring Nanomagnets for removal of soil contaminants	I,IV,V,VII
3 (682)	Water treatment and remediation	Nanomembranes for water purification, desalination, and detoxification Nanosensors for the detection of contaminants and pathogens Nanoporous zeolites, nanoporous polymers, and attapulgite clays for water purification Magnetic nanoparticles for water treatment and remediation TiO_2 nanoparticles for the catalytic degradation of water pollutants	I,IV,V,VII
4 (606)	Disease diagnosis and screening	Nanoliter systems (Lab-on-a-chip) Nanosensor arrays based on carbon nanotubes Quantum dots for disease diagnosis Magnetic nanoparticles as nanosensors Antibody-dendrimer conjugates for diagnosis of HIV-1 and cancer Nanowire and nanobelt nanosensors for disease diagnosis Nanoparticles as medical image enhancers	IV,V,VI
5 (558)	Drug delivery systems	Nanocapsules, liposomes, dendrimers, buckyballs, nanobiomagnets, and attapulgite clays for slow and sustained drug release systems	IV,V,VI

Table 27.1
(continued)

Ranking (score)	Applications of nanotechnology	Examples	Comparison with the MDGs
6 (472)	Food processing and storage	Nanocomposites for plastic film coatings used in food packaging Antimicrobial nanoemulsions for applications in decontamination of food equipment, packaging, or food Nanotechnology-based antigen detecting biosensors for identification of pathogen contamination	I,IV,V
7 (410)	Air pollution and remediation	TiO_2 nanoparticle-based photocatalytic degradation of air pollutants in self-cleaning systems Nanocatalysts for more efficient, cheaper, and better-controlled catalytic converters Nanosensors for detection of toxic materials and leaks Gas separation nanodevices	IV,V,VII
8 (366)	Construction	Nanomolecular structures to make asphalt and concrete more robust to water seepage Heat-resistant nanomaterials to block ultraviolet and infrared radiation Nanomaterials for cheaper and durable housing, surfaces, coatings, glues, concrete, and heat and light exclusion Self-cleaning surfaces (e.g., windows, mirrors, toilets) with bioactive coatings	VII
9 (321)	Health monitoring	Nanotubes and nanoparticles for glucose, CO_2, and cholesterol sensors and for in-situ monitoring of homeostasis	IV,V,VI
10 (258)	Vector and pest detection and control	Nanosensors for pest detection Nanoparticles for new pesticides, insecticides, and insect repellents	IV,V,VI

[a] The maximum total score an application could receive was 819.
DOI:10.1371/journal.pmed.0020097.t001

the 20 applications provided and gave reasons for their choices. To analyze the data, we produced a summative point score for each application, ranked the list, and summarized the panelists' reasons. Then we redistributed the top 13 applications, instead of the top ten, to generate a greater number of choices for increased accuracy in the last round. Thus, the highest score possible for an application was 819 (63 × 13). The final Delphi round was devoted to consolidating consensus by re-ranking the top ten of the 13 choices obtained in the previous round and to gathering concrete examples of each application from the panelists.

Our results, shown in table 27.1, were compiled from January to July 2004. They display a high degree of consensus with regard to the top four applications: all of the panelists cited at least one of the top four applications in their personal top four rankings, with the majority citing at least three.

To further assess the impact of nanotechnology on sustainable development, we have compared the top ten applications with the UN Millennium Development Goals (table 27.1 and figure 27.2). The MDGs are eight goals that aim to promote human development and encourage social and economic sustainability [11]. In 2000, all 189 member states of the UN committed to achieve the MDGs by 2015. The MDGs are: (i) Eradicate extreme poverty and hunger; (ii) Achieve universal primary education; (iii) Promote gender equality and empower women; (iv) Reduce child mortality; (v) Improve maternal health; (vi) Combat HIV/AIDS, malaria, and other diseases; (vii) Ensure

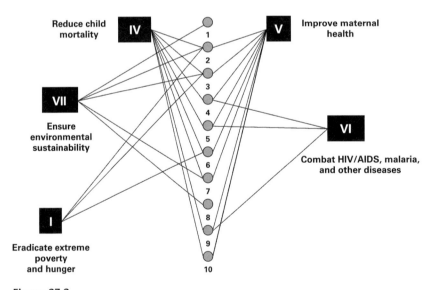

Figure 27.2
Comparison between the millennium development goals and the nanotechnologies most likely to benefit developing countries in the 2004–2014 period. DOI: 10.1371/journal.pmed.0020097.g002.

environmental sustainability; and (viii) Develop a global partnership for development. As shown in table 27.1 and figure 27.2, the top ten nanotechnology applications can contribute to achieving the UN MDGs.

Addressing Global Challenges Using Nanotechnology

What can the international community do to support the application of nanotechnology in developing countries? In 2002, the National Institutes of Health (NIH) conceptualized a roadmap for medical research to identify major opportunities and gaps in biomedical investigations. Nanomedicine is one of the areas of implementation that has been outlined to address this concern. Several of the applications of nanotechnology that we have identified in our study can aid the NIH in this process by targeting the areas of research that need to be addressed in order to combat some of the serious medical issues facing the developing world.

To expand on this idea, we propose an initiative, called "Addressing Global Challenges Using Nanotechnology," to accelerate the use of nanotechnology to address critical sustainable development challenges. We model this proposal on the Foundation for the NIH/Bill and Melinda Gates Foundation's Grand Challenges in Global Health [12], which itself was based on Hilbert's Grand Challenges in Mathematics.

A grand challenge is meant to direct investigators to seek a specific scientific or technological breakthrough that would overcome one or more bottlenecks in an imagined path to solving a significant development problem (or preferably, several) [12]. A scientific board similar to the one created for the Grand Challenges in Global Health, with strong representation of developing countries, will need to be established to provide guidance and oversee the program. The top ten nanotechnology applications identified in table 27.1 are a good starting point for defining the grand challenges.

The funding to address global challenges using nanotechnology could come from various sources, including national and international foundations, and from collaboration among nanotechnology initiatives in industrialized and developing countries. These funds could be significantly increased if industrialized nations adopted the target set in February 2004 by Paul Martin, Prime Minister of Canada: that 5% of Canada's research and development investment be used to address developing world challenges [13]. In parallel to the allocation of public funds, policies should provide incentives for the private sector to direct a portion of their research and development toward funding our initiative. The UN Commission on Private Sector and Development report *Unleashing Entrepreneurship: Making Business Work for the Poor* [14] underscores the importance of partnerships with the private sector, especially the domestic private sectors in developing countries, in working to achieve the MDGs.

Perhaps most importantly, our results can provide guidance to the developing countries themselves to help target their growing initiatives in nanotechnology [15]. The goal is to use nanotechnology responsibly [16] to generate real benefits for the 5 billion people in the developing world.

Acknowledgments

We are grateful to our panelists for providing their expertise, and to W. C. W. Chan and A. Shik for help with our analysis of the nanotechnologies. Grant support was provided by the Canadian Program on Genomics and Global Health (supported by the Ontario Research and Development Challenge Fund, and by Genome Canada through the Ontario Genomics Institute (Toronto, Canada); matching partners can be found at www.geneticsethics.net). EBC is supported by the Ontario Genomics Institute; DKM is supported by a Career Scientist award from the Ontario Ministry of Health and Long-Term Care; ASD is supported by the McLaughlin Centre for Molecular Medicine; PAS is supported by a Distinguished Investigator award from the Canadian Institutes of Health Research. The University of Toronto Joint Centre for Bioethics (Toronto, Canada) is a PAHO/WHO Collaborating Center for Bioethics.

Note

List of panel members found at DOI: 10.1371/journal.pmed.0020097.st001 (43 KB DOC).

References

1. Court E., Daar A. S., Martin E., Acharya T., Singer P. A. (2004) Will Prince Charles et al. diminish the opportunities of developing countries in nanotechnology? Available: http://www.nanotechweb.org/articles/society/3/1/1/1. Accessed 21 February 2005.

2. U.S., Indian high technology will benefit through cooperation (2003) Available: http://newdelhi.usembassy.gov/wwwhpr0812a.html. Accessed 27 January 2005.

3. Bapsy P. P., Raghunadharao D., Majumdar A., Ganguly S., Roy A., et al. (2004) DO/NDR/02 a novel polymeric nanoparticle paclitaxel: Results of a phase I dose escalation study. J Clin Oncol 22, 14S: 2026.

4. [Anonymous] (2003) China's nanotechnology patent applications rank third in world. Available: http://www.investorideas.com/Companies/Nanotechnology/Articles/China'sNanotechnology1003,03.asp. Accessed 27 January 2005.

5. Meridian Institute (2004) Report of the international dialogue on responsible research and development of nanotechnology. Attachment F. Available: http://www.nanoandthepoor.org/Attachment_F_Responses_and_Background_Info_040812.pdf. Accessed 21 February 2005.

6. Sachs J. (2002) The essential ingredient. New Sci 2356: 175.

7. UN Development Programme (2001) Human development report. Available: http://hdr.undp.org/en/reports/global/hdr2001. Accessed 12 May 2008.

8. UN Millennium Project 2005 (2005) Innovation: Applying knowledge in development. Task force on science, technology and innovation. Available: http://www.unmillenniumproject.org/documents/Science-complete.pdf. Accessed 12 May 2008.

9. Daar A. S., Thorsteinsdóttir H., Martin D., Smith A. C., Nast S., et al. (2002) Top ten biotechnologies for improving health in developing countries. Nat Genet 23: 229–232.

10. World Summit on Sustainable Development (2002) Available: http://www.un.org/jsummit/html/documents/wehab_papers.html. Accessed 30 June 2008.

11. United Nations (2000) UN millennium development goals. Available: http://www.un.org/millenniumgoals/. Accessed 27 January 2005.

12. Varmus H., Klausner R., Zerhouni E., Acharya T., Daar A. S., et al. (2003) Grand challenges in global health. Science 302: 398–399.

13. Government of Canada, Office of the Prime Minister (2004) Complete text and videos of the Prime Minister's reply to the speech from the throne. Available: http://www.pm.gc.ca/eng/news.asp?id=277. Accessed 27 January 2005.

14. UN Development Programme, Commission on the Private Sector and Development (2004) Unleashing entrepreneurship: Making business work for the poor. Available: http://www.undp.org/cpsd/indexF.html. Accessed 21 February 2005.

15. Meridian Institute (2005) Nanotechnology and the poor: Opportunities and risks. Available: http://www.nanoandthepoor.org/gdnp.php. Accessed 21 February 2005.

16. The Royal Society and The Royal Academy of Engineering (2004) Nanoscience and nanotechnologies: Opportunities and uncertainties. Available: http://www.royalsoc.ac.uk/policy. Accessed 27 January 2005.

28 "Nanotechnology and the Developing World: Will Nanotechnology Overcome Poverty or Widen Disparities?"

Noela Invernizzi and Guillermo Foladori

Invernizzi and Foladori wrote this article in direct response to the arguments posed by Salamanca-Buentello, et al. and others. They argue that it is conceivable that nanotechnology could be developed to meet the needs of those living in developing countries, but that drafting lists of possibilities will do little to make such development a reality. They point out that over the past several decades a number of grand visions of how technological advances will help build a better future for people in developing countries have been developed. However, the implementation of these technologies has not created the better world envisioned. Invernizzi and Foladori are concerned that when it comes to nanotechnology, once again visions of a better future for developing countries are being used to justify the enterprise, but the technologies developed will ultimately only benefit the rich. The fact that the nanotechnology products being sold in 2005 were for the most part luxury items like golf clubs, cosmetics, and stain resistant pants certainly supports their point. In order to accurately evaluate whether and how a particular technology can benefit a developing country, they implore us to carefully consider the social context of technologies and the complexities involved in building new sociotechnical systems. Invernizzi and Foladori point out that patent systems designed to encourage innovation can have the negative effect of giving corporations the power to decide which values new technologies will promote. They argue that we already have the technology to solve many of the developing world's problems—we just have not developed the systems to make sure they reach the people who need them. Rather than postulate the ways that new technologies can help the developing world, Invernizzi and Foladori encourage us to ask developing countries what they need.

Introduction

One of the most hotly debated issues, and one of the most difficult to discern in advance in the growing discussion on nanotechnology, is its possible effects on poorer countries and less fortunate segments of the population. There are optimistic stances, wherein nanotechnology is considered to be a panacea, and there are pessimistic viewpoints which suppose that the gap between rich and poor will widen as a result of the diffusion of this kind of technology. The debate on these different stances, supported by theoretical arguments and empirical data, is fundamental for arriving at a balanced viewpoint of the situation.

From *Nanotechnology Law & Business* 2, no. 3 (September/October 2005): 294–303. Reprinted with permission of Nanotechnology Law and Business. © 2005. All rights reserved. Please visit www.nanolabweb.com for more information.

In early 2005, several influential articles claimed that nanotechnology is a viable alternative for resolving most of the Millennium Development Goals of the United Nations.[1] Some scholars have even attempted to list the top ten nanotechnology applications that will most benefit those in poorer nations. While it is undoubtedly useful to determine which aspects of nanotechnology promise the most to poorer nations in terms of application, the international community needs to question this optimism, placing these new technologies in their social context.

I Nanotechnologies As Disruptive Technologies

Nanotechnology manipulates atoms and molecules to make or build things (or living beings). One may imagine a laboratory which, by combining suitable molecules in quality and quantity, could create electric drills. Although this is theoretically possible, it will take some time, at least until there are nanobots that could do the job by themselves. But this will be at a later stage, if ever. At the moment, what can be made are nanoproducts, in which scientists manipulate matter at the nanoscale to give them special or more efficient uses. Witness the introduction of nanotextiles in recent times.

According to the Nanotech Report, among the first products commercialized in 2004 with nanotechnological content were the following: thermal shoes (Aspem Aerogels); dust and sweat-repelling mattresses (Simmons Bedding Co.); more flexible and resistant golf clubs (Maruman & Co.); personalized cosmetics for different ages, races, genders, types of skin and physical activities (Bionova); dressings for cuts and burns that prevent infections (Westaim Corporation); disinfectants and cleaning products for planes, boats, submarines, etc. (EnviroSystems); spray that repels water and dirt that is used in the building industry (BASF); treatment for glass to repel water, snow, insects etc. (Nanofilm); cream that combats muscular pain (CNBC, Flex-Power, Inc.); and dental adhesives that set the tooth crown better (3M ESPE).[2] Lux Research, a company dedicated to the study of nanotechnology and its business, estimates that the sale of articles with nanoparticles will surpass the mark of $500 billion in 2010.[3]

At least four aspects make nanotechnology a great new development. First, it revolutionizes the manufacturing process. Nanotechnology builds from the smallest number of atoms and molecules to make the biggest final product—the bottom-up process. Nanotechnology can also reverse the process—instead of starting with physical matter as it is found in nature, according to its own structures, by reducing it to the size of the objects to use as has been done until now—the top-down process. Despite this road being familiar in chemical processes, the novelty is that, now, atoms and molecules can be *directly manipulated* to manufacture products.[4] This constitutes a novelty in the history of humanity and a new way of thinking in the world. Its consequences are unlimited. It is even conceivable to think that a process of production that manufactures by summing up molecules will, in theory, generate no waste.

Second, at this nanoscale level, there are few differences between biotic matter and abiotic matter in that it is potentially possible to apply biological procedures to material processes or interfere with materials in living bodies, adapting the latter to certain purposes or offering certain advantages. It may also be possible to manipulate biological matter or procedures to perform specific tasks. One example would be a way of allowing the body to rest without sleep, which would be very useful in war and other activities that are very physically or mentally demanding.

Third, nanoparticles may have physical and chemical properties (conductive, electric, mechanical, optical, etc.) which differ from the same elements on a macroscopic scale. By changing the physical properties of the matter, possibilities arise that surprise and excite scientists who are dedicated to this study. Many nanomaterials that are on sale offer great advantages in this way. Carbon nanotubes, for instance, are harder than diamonds and can be fifty to a hundred times stronger than steel.

Finally, nanotechnology combines several kinds of technologies and sciences such as information technology, biotechnology and materials technology. The latter is not a lesser element if we consider that the true development of nanotechnology will require a totally new professional education which will require rethinking schooling, maybe from the primary level.

The potential benefits of nanotechnology are impossible to calculate. Here we can mention a few of the more probable. In the field of health, it could increase the quality and length of life. Nanosensors, incorporated into the organism, could travel through the bloodstream similar to the way a virus does and detect illnesses before they spread to the rest of the body and combat them efficiently. In the future, drugs may no longer be generic for all people, but may be specifically designed according to the genetic make-up of the individual and his/her gender, age and diet. Ageing mechanisms could be retarded and even reversed, with the human lifespan's being lengthened significantly. With these artificial sensors, a person could become a bionic being, improving her biological capacities and developing others. Some even envision nanotechnology applications that will improve human perception and ability at fundamental levels. The field of prostheses is also among the most promising.

In the materials field, one novelty will be intelligent nanoparticles. Your wardrobe, for example, could be reduced to one single article. The item of clothing you have will react to changes in temperature, rainfall, snow and sun, among other elements, keeping the body always at the programmed temperature. Furthermore, it will repel sweat and dust, which will mean that it will not require washing. As if this were not enough, it would stop bacteria or viruses from penetrating it, protecting it even from possible bioterrorist attacks. In the case of an accident, your clothes would have healing effects, offering first aid. The same that applies to clothing could be adapted to certain dwellings and modes of transport. Another novelty is that carbon nanotubes are stronger than steel and only 1/6 of its weight. This will have a special impact on the aerospace, construction, automobile industries and many others.

The field of computer science will be one of the earliest industries affected and will enjoy the most revolutionary change. Computers can be a hundred times faster and much smaller and lighter, and can be custom built according to the tastes of the buyer in terms of design, size, shape, color, smell and resistance. Prototypes with built-in sensors will speed up designs, adapting to flexible production processes in different parts of the world, overcoming many of the barriers that distance now imposes. The old "just-in-time" production mode will become obsolete and may very well become the "as-you-need" mode of production. The possibilities for monopolistic concentration of production (global business enterprises) will multiply.

The combination of computerized systems, chemical laboratories, miniature sensors and living beings adapted to specific functions will revolutionize medicine (e.g., lab-on-a-chip) and also provide rapid solutions to the historical problems of contamination. Small bacteria with sensors may be able to consume bodies of water that have been contaminated by heavy metals, or decontaminate the atmosphere in record time. Nanocapsules with combined systems of sensors and additives will revolutionize the industries such as lubricants, pharmaceuticals and filters, to make no mention of others.

Nanotechnology may become a disruptive technology that will make obsolete the current competitive technologies, once established and entrenched in economies around the world. The social and economic effects on the international and national levels are difficult to foresee, but an effort must be made at this critical juncture in order to reduce the possible negative or unwanted consequences that have historically accompanied such dramatic transformations.

II Nanotechnology As a Solution for the Poor?

Despite the voices that warn of the possible negative consequences or risks of nanotechnology, there are others that suppose that the new technology will be beneficial to everyone, including the poor. In this light, the recent U.N. Millennium Project Report, *Task Force on Science, Technology and Innovation*, puts forward the idea that nanotechnology will be important to the developing world because it harbors the potential to transform minimal work, land and maintenance inputs into highly productive and cheap outputs; and it requires only modest quantities of material and energy to do so.[5] However, these same qualities could be seen as harmful because poor countries have abundant labor, and, in many cases, land and natural resources. In this way, nanotechnology may cause displacements and disruptions in the economies of poorer nations.

Reasoning in a purely technical and linear fashion, any country could theoretically join the nanotechnology wave. An effort for public funding may create the bases to establish specific nanocomponent industries to meet determined needs; or it is possible that businesses with a scientific tradition might justify a technological leap at a relatively low cost. This seems to be the opinion of the authors of at least one article[6] that has received a great deal of attention from the international scientific press.[7] Fabio Salamanca-Buentello and several of his colleagues from the Joint Center for Bioethics at the University of Toronto introduce nanotechnology as the solution to many problems in developing countries.[8] They understand the effort for harnessing nanotechnology in developing countries as a demonstration of the willingness of such countries to overcome poverty: "... we show that developing countries are already harnessing nanotechnology to address some of their most pressing needs."[9] After interviewing sixty-three experts in nanotechnology from several developed and developing countries, the authors identified the ten main nanotechnologies that could provide a solution to such problems as water, agriculture, nutrition, health, energy and the environment.[10] The technologies range from energy production and conservation systems, with sensors that will increase agricultural productivity and the treatment of water, to the diagnosis of diseases. In the article, the creation of a Global Fund is proposed for the development of these technologies for all developing countries. Overflowing with good intentions, the proposal reflects the mechanical idea that if a problem can be identified correctly, then all that has to be done is to apply a suitable technology, and it will be solved. Most of the examples used do not take into account the reality that the relationship between science and society is much more complex than identifying a technology and its potential benefits.[11] Let us put some of the examples in their social context.

1   The Experience of Poorer Nations with HIV/AIDS Pharmaceuticals

Salamanca-Buentello and colleagues suggest that quantum dots could detect HIV/AIDS molecules in the early stages, thereby facilitating the treatment of AIDS and reducing the number of new cases. While quantum dots may, in fact, provide a useful solution to the HIV/AIDS crisis in developing countries, Salamanca-Buentello's article does little to place this novel technology into the historical experience of poorer nations with advances in the medical field more generally.

The authors seem to forget the story of the last several years, which has been one of seemingly open war between multinational pharmaceutical corporations and the governments of countries that intended to manufacture antiretrovirals against AIDS. In this conflict, the World Trade Organization ("WTO") and the Commercial representative of the United States have systematically played the role of front-line soldier for these corporations. The rigors of the patenting system have used monopolist economics to drive medicine pricing for the last twenty years. This makes it impossible for poor people to buy medicine from companies that hold patents. Experience over the last several years has shown that when an epidemic occurs, some countries cannot afford to cover the cost of remedies very much needed by people within these poorer countries.

One of the most alarming historical cases, illustrating the behavior of the multinational pharmaceutical corporations that tends to undermine public health, was the action brought in 2001 by thirty-nine of the major pharmaceutical corporations against the government of South Africa. In that case, several of the corporations prevented the South African government from producing generic medicine for AIDS treatment. A lawsuit over the matter was soon filed. The lawsuit, which the South African government won, showed total insensibility to human rights on the part of the pharmaceutical corporations. The statements of one of the pharmaceutical company's representatives bore out this insensitivity. According to the reasoning of certain pharmaceutical companies, the court's ruling allowing the government of South Africa to produce affordable generic medicine could have a precedential effect allowing other governments to cheaply develop generic medicines: "while South Africa may represent less than 1% of world drug sales, the precedent of allowing a government to step on drug companies' patent rights would have far-reaching effects, beyond the questions of cost and crises."[12]

Nanotechnology products are already being patented, typically by the most important and largest corporations in the world. A patent in the U.S. costs $30,000 in legal bureaucracy, and a worldwide patent may be as much as a quarter of a million dollars.[13] For an underdeveloped country, it is very difficult to develop any medicine for which there is an important market (as is the case of AIDS) if we take into account the economic and legal medicine market's international "war," as well as the bureaucratic restrictions which drive up costs and reduce availability. This story has a simple moral: technology is produced in a given social context, and the efficiency and implications of its application depend on that social context.

2 The Experience of Poorer Nations with Biotechnology

Salamanca-Buentello and his colleagues identify nanotechnology as the solution to five of the eight Millennium Development Goals of the United Nations. Among these supposed solutions are nanosensors and nanocomponents to improve the dosage of water and fertilization of plants. With this technology, it would be possible to reduce poverty and hunger in the world. Simply identifying a potentially useful application, however, overlooks the clear historical experience of poorer countries. Not so long ago, in the 1980s, genetically modified organisms were hailed as the solution that would put an end to hunger and poverty. However, genetically modified organisms ended up being used mainly in developed countries; and three out of four patents are today in the hands of four large multinational companies. There has been no improvement for Third World countries; quite the contrary, transgenics turned up where they were not wanted or expected, as was the case of the contamination of corn in Oaxaca, Mexico. In the case of genetically modified organisms, commercial and technological dependence was increased, not reduced.[14] This historical example could well foreshadow the path that nanotechnology takes in worsening existing gaps between the developed and less developed world unless steps are taken now to avert a repeat of history.

It is far from a foregone conclusion to assume that agricultural nanotechnology will follow the controversial road taken by genetically modified organisms. However, avoiding such a situation requires a healthy debate concerning the possible social, economic and political implications in real time.[15] Michael Mehta highlights three lessons for nanotechnology that should be learned from the experience with biotechnology: (1) to provide legislation on nanotechnology products in such a way that public participation will not be undermined by science-based assessment; (2) to label products with nanocomponents in order to gain acceptance with the corresponding empowerment of the consumer; and (3) to use the precautionary principle in a way that could prevent serious risks without limiting the possible development of these sciences.[16]

One limitation with the above analysis is that nanoproducts are already facing political and economic pressure, in part responsible for building the nanotechnology revolution; and past experience, so far, plays too limited a role in this process. And the difficulty presented by efforts to categorize and regulate nanotechnology frustrates ready-made solutions for the industry to avert the problems encountered by the biotechnology industry with genetically modified organisms. Take the following two statements as an example: nanotechnology products face the paradox that they are (1) elementary particles of known chemical elements; and (2) manipulated in a way that is not natural. As for the first statement, nanoproducts do not always need to go through drug trials and registration. Regulations seem not to accompany the speed of technical improvements. A document by the Woodrow Wilson International Center for Scholars is explicit on the contradiction between the reality of nanoparticles and the ambiguity of the American regulatory standards, and it concludes on the need to reform the *Toxic Substance Control Act* ("TSCA").[17] Considering that first statement above, then, nanoproducts are nothing new; rather, they are part of nature.

The clear implication of the second statement is that nanoproducts are being patented as new elements which are not found naturally in that state. The following quote exemplifies this paradox: "[i]t is true that you cannot patent an element in its natural form as found in nature. However, if a purified form of this element is created with industrial uses—for example, [ ] neon—the new [forms] have a secure patent."[18]

The conclusion at first blush is that it seems as if "business as usual" characterizes the current debate about the implications of nanotechnology and the poor, rather than the old adage that learning from past experience prevents future mistakes.

The moral of the story is that the choice of a technology is not a neutral process. Choosing a technology depends on political and economic forces. It is not necessarily true that the technology which best meets our needs will be the one to survive.

III Without a Voice? The Poor and Their Inclusion in the Debate on Nanotechnology

Salamanca-Buentello and his colleagues also presuppose that interviewing thirty-eight scientists from developing countries and twenty-five from developed countries permits them to speak of the interests of the developing countries as if they were, in fact, spokespeople for those within developing countries. In a prior article,[19] three of the same authors maintained that the position adopted by Prince Charles[20] (arguing that nanotechnology will widen the gap between rich and poor countries) and by the ETC Group[21] (requesting a moratorium on public funding for nanotechnology) "ignores the voices of the people in developing countries."[22] Surely, Salamanca-Buentello and his colleagues intended to give voice to the people of developing countries on the issue of nanotechnology by conducting research interviews with nanotechnology scientists from the developed and developing world. Their genuine concern for those in the developing world is certainly not doubted here. Unfortunately, the opinion of scientists involved in nanotechnology does not necessarily fall within the most appropriate of pathways for satisfying the needs of the poor. The relationship between scientists and sociopolitical pressures are replete with examples of doubtful practices. In the biomedical arena, for example, we can find cases of independent determination of standards in biomedical trials compromised or auto-censored by the influence of pharmaceutical corporations;[23] and there are examples of funds given by pharmaceutical corporations to universities in order to have influence on decisions pertaining to research and development ("R&D") and to gain the right for subsequent licenses. Even still, there are examples of pharmaceutical companies' bankrolling academic studies that downplay their interests.[24] Some have made claims of fraudulent or doubtful laboratory trials conducted by some large pharmaceutical companies.[25] Still others describe the pharmaceutical corporations' inciting physicians to use governmental forms fraudulently in order to obtain reimbursements for medicine obtained for free from pharmaceutical companies.[26] Pharmaceutical corporations have also been accused of putting pressure on researchers to impede the flow of detrimental information into public forums.[27] And the list can go on and on. Academic opinions, therefore, can hardly be said to represent completely the voices of the poor.[28]

Technology is simply a part of a puzzle. Scholars may concur, for example, that infectious diseases constitute one of the main problems that the developing world is facing, but they may differ radically on how a solution to this problem should be attained. Prevention is not the equivalent of a cure. Nanotechnology is not necessary to reduce malaria radically, for example, as is suggested by Salamanca-Buentello and colleagues. There is no doubt that nanosensors could help to clean water, nor that nanocapsules could make drugs more efficient. Nevertheless, in the Hunan Province of China, malaria was reduced by 99% between 1965 and 1990 as a result of social mobilization backed up by fumigation, the use of mosquito nets and traditional medicine.[29] Vietnam reduced the number of malaria-related deaths by 97% between 1992 and 1997 with similar mechanisms.[30] The moral of this story is twofold: (1) scientists are not always the best spokespeople for the poor, even when they come from poor countries; and (2) there are many means to an end; and technology is not always the solution. Organizing people—which some refer to as *social technology*—can be just as important. In this way, identifying potentially useful scientific technologies for the developing world must become part of a much larger and inclusive social technology if gains are to be actualized in poorer countries.

IV Opening the Debate and Placing Nanotechnology in Its Social Context

The history of science and technology is full of examples of technologies that have not always helped the poor. In order to serve the needs of the poor, technology has to be used in a favorable socioeconomic context. Furthermore, the building characteristics of the technology, and the technological path, usually impede it from being freely used for the benefit of the masses in developing countries.[31]

Despite the optimistic assessments recently offered, experience suggests that nanotechnology could follow the mainstream economic trends that increase inequality.

First, the development of nanotechnology faces many of the same problems faced by prior technological developments because large multinational corporations are patenting the majority of the nanotechnology products. Patents are monopolistic guarantees of earnings for twenty years—something that certainly works against the rapid diffusion of the beneficial potentials of this technology for the poor.[32]

Second, nanotechnology's novel solutions and potentially laudable achievements may never come to fruition in developing countries because the main problem for a developing country is not so much the fixed costs of a laboratory of average sophistication, but the social context that is necessary for really incorporating nanotechnologies into the economy. Without fluid mechanisms of vertical integration between the sectors that produce nanoparticles and the companies that are potential buyers, the nanoparticles will never get out of the laboratory. From many accounts, this seems to be happening, nowadays, in developed countries. Wildson affirms, based on his conducting interviews of individuals with English companies that produce nanoparticles, that "nanoparticles are a solution in search of a problem."[33] Despite their numerous potential applications, the English producers say that they have a shortage of clients. This is confirmed by the cover story in a recent edition of *Business Week*, which, based on information from Lux Research, tells us that despite a promising future, many companies that sell nanotechnology products faced financial difficulties in 2004.[34] The linkages between the science and technology system and the productive sectors are very tenuous in most of the developing countries.

Third, nanotechnology's development in much of the world will do little to help the developing world due to the difficulty in finding qualified workers. A country's ability to foster and support technological careers requires a social context that supplies the necessary equipment and human capital in the long term. It will be difficult for many Third World countries to find the staff necessary to work interdisciplinarily in nanotechnology. Mexico, for instance, the thirteenth largest exporting power in the world, only has eleven research teams in three universities and two research centers in nanotechnology, with a total of ninety researchers and no official support program for field research.[35] Brazil, which launched a pioneer program for research and development in nanotechnology in Latin America (considering that it was in the same year as the U.S. initiative—2000) had between fifty and one hundred researchers in 2002 and probably around 300 in 2004.[36] Despite these seemingly impressive numbers in Brazil, challenging barriers remain which will continue to plague the ability of nanotechnology scientists in developing countries to produce benefits for the poor. Many nanotechnologists in developing countries may be enticed by higher wages out of poorer countries and into richer ones. The reason that this potentiality must be addressed now is as follows. Some estimate that nanotechnology will mean restructuring all learning to break down the traditional disciplinary frontiers, which, in practice, nanotechnology has already overcome. It is possible that changes in study plans would have to take place starting at primary education.[37] This means that multi-sector efforts are gambled on these changes, and elevated social demands are required. In many instances, poorer nations lack the resources, infrastructure and facilities for such interdisciplinary efforts as nanotechnology—particularly, where transformations must take place at so fundamental a level. Given the higher stakes and more interdisciplinary nature of nanotechnology, therefore, it is possible that the race for qualified scientists will heat up and increase the brain drain from the Third World into more advanced countries. This polarization of the labor market will punish poorer countries with less qualified labor. It is unlikely that the vast majority of developing countries will have the wherewithal, infrastructure and labor force to be able to join the nanotechology wave and capitalize on its potentials to transform society and industry.

Finally, even if large developing countries that could join the nanotechnology wave (such as China, India and Brazil, for example) can produce nanoproducts that could eventually result in clean and cheap energy options, in clean drinking water or in greater agricultural yields, this does not mean that the poor majority will benefit. For them socio-economic structure is a much more difficult barrier than technological innovation. Nanotechnology, even where fully integrated in developing countries, does nothing to change these socio-economic structures; instead, it could serve to exacerbate existing gaps and further the technological and socio-economic isolation of the poor.

V Conclusions

Nanotechnology is still in its early stages, but the later we choose to address its social and economic implications, the less chance there will be for the technology to help the poor before nanotechnology begins to put down roots within the mainstream hegemonic socioeconomic structure, characterized by worldwide inequality.

Notes

1. *See, e.g.*, Fabio Salamanca-Buentello et al., *Nanotechnology and the Developing World*, 2 PLoS Med. E97 (2005), *at* http://www.pubmedcentral.nih.gov/articlerender.fcgi?tool=pubmed&pubmedid=15807631 (last visited June 4, 2005).

2. *Top 10 Nanotech Products of 2004*, 3 Nanotech Report 1 (2004).

3. Stephen Baker & Adam Aston, *The Business of Nanotech*, Bus. Wk., Feb. 14, 2005, at 64. As a reference, all exports from Latin America and the Caribbean in 2004 totaled $461 billion; foreign debt in Latin America and the Caribbean in 2004 totaled $721 billion. Econ. Comm'n. of Latin Am. & Caribbean ("ECLAC"), Statistical Yearbook for Latin America and the Caribbean (2004), *at* http://www.eclac.org/estadisticas (last visited May 12, 2008).

4. Royal Soc'y & Royal Acad. of Eng'g, Nanoscience and Nanotechnologies: Opportunities and Uncertainties, July 29, 2004, *at* http://www.nanotec.org.uk/finalReport.htm (last visited June 27, 2005).

5. Calestous Juma & Lee Yee-Chong, U.N. Millenium Project Task Force on Sci., Tech. & Innovation, Innovation: Applying Knowledge in Development (2005), *at* http://www.unmillenniumproject.org/documents/Science-complete.pdf (last visited June 27, 2005).

See also David Dickson, *Scientific Advice "Essential" to Meet Development Goals*, SciDev.net, Jan. 10, 2005, *at* http://www.scidev.net/gateways/index.cfm?fuseaction=readitem&rgwid=4&item=News&itemid=1835&language=1 (last visited June 4, 2005).

6. Salamanca-Buentello et al., *supra* note 1.

7. Catherine Brahic, *Developing World "Needs Nanotech Network,"* SciDev.net (June 4, 2005), *at* http://www.scidev.net/News/index.cfm?fuseaction=printarticle&itemid=1923&language=1 (last visited June 4, 2005); Charles Q. Choi, *Top 10 for Developing World*, United Press Int'l (Apr. 18, 2005), *at* http://www.upi.com/view.cfm?StoryID=20050415-114140-8159r (last visited June 4, 2005); *Taking Nano to the Needy: A* Small Times *Q&A with Fabio Salamanca-Buentello*, Small Times, June 15, 2005, *at* http://www.smalltimes.com/Article_Display.cfm?ARTICLE_ID=270094&p=109 (last visited May 12, 2008).

8. It must be noted that these authors explicitly recognize that science and technology are not enough: "Like any other science and technology waves, nanoscience and nanotechnology are not 'silver bullets' that will magically solve all the problems of developing countries; the social context of these countries must always be considered." Salamanca-Buentello *et al., supra* note 1, at 2. The authors do, however, visualize science and technology investments and scientists involved in developing countries as an indicator of willingness to overcome poverty. *See id.*

9. Salamanca-Buentello *et al., supra* note 1, at 1.

10. "In order to provide a systematic approach with which to address sustainable development issues in the developing world, we have identified and ranked the ten applications of nanotechnology most likely to benefit developing countries. We used a modified Delphi Method ... to identify and prioritize the applications and to achieve consensus among the panelists." Salamanca-Buentello *et al., supra* note 1, at 3.

11. D. Sarewitz et al., *Science Policy in its Social Context*, Philosophy Today, 2004 Supplement, at 67.

12. Robert Block, *AIDS Activists Win Legal Skirmish in South Africa*, Wall St. J., Mar. 17, 2001, at A17.

13. A. Regalado, *Nanotechnology Patents Surge*, Wall St. J., June 18, 2004, at A1.

14. M. Schapiro, Blowback in Genetic Engineering *in* Alan Lightman, Daniel Sarewitz & Christina Desser, Living with the Genie (2003).

15. David H. Guston & Daniel Sarewitz, *Real-Time Technology Assessment*, 24 Tech. in Soc'y 93 (2002).

16. Michael Mehta, *From Biotechnology to Nanotechnology: What Can We Learn from Earlier Technologies?* 24 Bull. of Sci., Tech. & Soc'y 34 (2004).

17. Woodrow Wilson Int'l Ctr. for Scholars, Nanotechnology & Regulation: A Case Study Using the Toxic Substance Control Act (TSCA), Discussion Paper No. 2003-6 (2003).

18. Lila Feisee, *Anything under the Sun Made by Man*, Address at Biotechnology Industry Organization (April 11, 2001), *available at* http://www.bio.org/speeches/speeches/041101.asp (last visited June 27, 2005).

19. E. Court, A. S. Daar, E. Martin, T. Acharya & P. A. Singer, *Will Prince Charles et al. Diminish the Opportunities of Developing Countries in Nanotechnology?*, Nanotechweb.org, Jan. 28, 2004, *at* http://nanotechweb.org/articles/society/3/1/1/1 (last visited June 4, 2005).

20. *See* Geoffrey Lean, *One Will Not Be Silenced: Charles Rides into Battle to Fight a New Campaign*, Independent, July 11, 2004, *available at* http://news.independent.co.uk/uk/this_britain/story.jsp?story=540022 (last visited June 4, 2005) (explaining the position of Prince Charles on nanotechnology).

21. ETC Group, The Big Down (2003), *available at* http://www.etcgroup.org/documents/TheBigDown.pdf (last visited June 4, 2005).

22. Court et al., *supra* note 19.

23. Annabel Ferriman, *WHO Accused of Stifling Debate about Infant Feeding*, 320 British Med. J. 1362 (2000), *available at* http://bmj.bmjjournals.com/cgi/content/full/320/7246/1362?ijkey=334d739b3ac3456846aa637addf43d5bad31bbf0&keytype2=tf_ipsecsha (last visited June 7, 2005); *see also* Richard Woodman, *Open Letter Disputes WHO Hypertension Guidelines*, 318 British Med. J. 893 (1999), *available at* http://bmj.bmjjournals.com/cgi/content/full/318/7188/893/b (last visited June 7, 2005).

24. Richard Smith, *Medical Journals are an Extension of the Marketing Arm of Pharmaceutical Companies*, 2 PLoS Med. E138 (2005), *available at* http://medicine.plosjournals.org/perlserv/?request=get-document&doi=10.1371/journal.pmed.0020138 (last visited June 7, 2005); J. Montaner, M. O'Shaughnessy & M. Schechter, *Industry-Sponsored Clinical Research: A Double-Edged Sword*, 358 Lancet 1893 (2001); Eyal Press & Jennifer Washburn, *The Kept University*, 285 Atlantic Monthly 39 (2000).

25. S. Shah, *Globalization of Clinical Research by the Pharmaceutical Industry*, 33 Int'l J. of Health Servs. 29 (2003); T. Bodenheimer, *Uneasy Alliance—Clinical Investigations and the Pharmaceutical Industry*, 342 New Eng. J. Med 1539 (2000); Campaign Against Fraudulent Med. Res. ("CAFMR"), The Pharmaceutical Drug Racket—Part One (1995), *available at* http://www.pnc.com.au/~cafmr/online/medical/drug1a.html (last visited June 7, 2005); CAFMR, The Pharmaceutical Drug Racket—Part Two (1995), *available at* http://www.pnc.com.au/~cafmr/online/research/drug2a.html (last visited June 7, 2005); John Braithwaite, Corporate Crime in the Pharmaceutical Industry (1984).

26. Scott Hensley, *Pharmacia Nears Generics Deal on AIDS Drug for Poor Nations*, Wall St. J., Jan. 24, 2003, at A1.

27. J. Collier & I. Ilheanacho, *The Pharmaceutical Industry as an Informant*, 360 Lancet 1405 (2002).

28. This argument is not meant to dogmatically equate academicians or scientists with pharmaceutical companies. Rather, it is meant to expose the deep and entrenched connections between pharmaceutical corporations and scientific/academic research to illustrate

potentially how a new technology or medical application can be manipulated to prevent poorer nations from mechanically applying a useful technology to a problem which is identified by scientists/academicians, even if they come from poorer countries.

29. Sukhan Jackson, Adrian C. Sleigh & Xi-Li Liu, Economics of Malaria Control in China: Cost Performance and Effectiveness of Hunan's Consolidation Programme, Wld. Health Org. ("WHO") Social, Econ. & Behavioral Res. Report Series No. 1 (2002), *available at* http://www.who.int/tdr/publications/publications/sebrp1.htm (last visited June 27, 2005).

30. WHO, *Vietnam Reduces Malaria Death Toll by 97% within Five Years*, 2002, *at* http://www.who.int/inf-new/mala1.htm (last visited June 7, 2005).

31. D. Sarewitz et al., *supra* note 11.

32. Corporate intellectual property ("IP") departments, which have increasingly sought to turn patents into major revenue streams, are stoking the trend. According to data compiled by the National Science Foundation, IBM won the most nanotech-related patents in 2003. Also among the top ten: computer-memory giant, Micron Technology Inc. of Boise, Idaho; manufacturer, 3M Corp. of St. Paul, Minnesota; the University of California; and Japan's Canon Inc. See Regalado, *supra* note 13, at A1.

33. James Wildson, *The Politics of Small Things: Nanotechnology, Risk, and Uncertainty*, 23 IEEE Tech. & Soc'y Magazine 16 (2004).

34. Baker & Aston, *supra* note 3. ("A 2004 study by Lux Research found that many of the 200 global suppliers of basic nanomaterials failed to deliver what they promised.").

35. Ineke Malsch & Volker Lieffering, *Nanotechnology in Mexico*, Nov. 5, 2004, *at* http://www.voyle.net/Guest%20Writers/Drs.%20Ineke%20Malsch/Malsch%202004-0001.htm (last visited June 7, 2005).

36. Laura Knapp, *Brasil Ganha Centro de Pesquisa de Nanotecnologia*, O Estado de S. Paulo em linea, Jan. 20, 2002, *at* http://busca.estadao.com.br/ciencia/noticias/2002/jan/20/138.htm (last visited June 7, 2005).

37. C. L. Alpert, *Introducing Nanotechnology to Public and School Audiences*, NanoScience & Tech. Inst., *at* http://www.nsti.org/Nanotech2004/showabstract.html?absno=581 (last visited June 7, 2005) ("The NSF envisions a revolution in science education from elementary school through the post-graduate level; a systemic change that recognizes the convergence of research in physics, chemistry, biology, materials science, and engineering...").

29 "People's Science in Action: The Politics of Protest and Knowledge Brokering in India"

Roopali Phadke

Technological decisions often affect an enormous number of people. Sometimes the effects are small—they may simply mean that consumers have to decide between purchasing a razor with four blades or five. But the effects can also be profound; they can change the very fabric of people's lives. Because technologies can be so powerful, there has been a call by a number of academics, politicians, and NGOs to give ordinary citizens a voice in technological decisions. Lay citizens may exert some pressure on technological development through their purchasing habits, but such power is limited. It may spell the death knell for a razor with five blades, but it is unlikely to convince companies to develop new systems of shaving, let alone prevent nuclear waste from being buried in close proximity to a new housing development. A number of projects have been developed to allow members of the public to have a more active role in the development of new technologies. For instance, the Netherlands and several Scandinavian countries have developed what they call "science shops." Science shops bring together members of the lay public and panels of scientific experts to discuss concerns about technological issues like stem cell research, alternative forms of energy, and genetically modified foods. In a number of cases the reports generated by science shops have had a direct effect on legislation passed by these countries to govern, regulate, and direct technological development. In this article, Roopali Phadke tells the story of how a group of citizens working closely with NGOs were able to radically change a proposal for a new system of dams. In this case, as in many cases, local citizens can play a very constructive role in technological decision-making. Not only do they better understand the values that may be undermined or promoted by certain decisions, they have local knowledge about an area that outside experts may overlook. Citizens should not simply be seen as adversaries of technological change, but as a valuable resource that can help technology developers think through the variety of issues involved—including social, technical, and value issues—in order to make more informed and beneficial decisions.

Among international aid donors and nongovernmental organizations (NGOs), reforming state water infrastructure development has become an important platform for reenvisioning how natural resource agencies can promote local participation. Transnational NGO and social movement activism against "big dams" over the last two decades has demonstrably shifted public opinion away from top-down, technocratic approaches toward people-centered, bottom-up development (Khagram 2004; McCully 1996). This activist response has developed alongside, and often in association with,

From *Society and Natural Resources* 18 (2005): 363–375. Reprinted with permission of Taylor & Francis. http://www.informaworld.com.

important efforts by social scientists to document the deleterious environmental and social impacts of irrigation projects (Cernea 1999; Singh 1998).

The reform of state irrigation development is particularly interesting to examine in the context of India. Despite heavy public mobilization and mounting empirical evidence supporting the need for broad reforms in water sector development, Indian irrigation agencies continue to proceed with conventional models and tools for project development (Vaidyanathan 1999). The official model of technocratic expertise often excludes local communities from direct participation in project design. Rather than embracing their experiential and contextual knowledge base, farmers are often viewed by state engineers as passive recipients of privileged bureaucratic expertise. As a result of this technical approach, irrigation projects in India more often reinforce, rather than relieve, existing patterns of inequity, inefficiency and environmental degradation (Rangachari et al. 2000).

In light of this administrative inertia, it is particularly important to better understand the instances when farmer participation has impacted the design of government irrigation projects in India. This article explores the factors that are necessary to spur state technical bureaucracies into using locally generated data in technical planning. Through a case study of the Uchangi dam in Maharashtra, the article examines the vital role played by technical NGOs in catalyzing, facilitating, and translating local research into bureaucratic planning.

This article describes how and why the government irrigation agency was compelled to redesign the Uchangi dam as a result of both collective resistance and negotiation of an alternative technical design that legitimated the value of local knowledge. While illustrating the standard government approach to project planning, this case also provides a specific example of how local knowledge and expert science can come together toward the participatory design of technology. In this case, participatory research demonstrates that valuing the "localness" of community knowledge does not necessarily mean embracing an ontologically different, romantic or mythical understanding of the natural world. Rather, it can be a process of rendering visible a set of processes, discourses, and narratives that are absent from conventional state technical planning. By investing in participatory research, agencies can better meet local needs and reduce the political conflicts that impede the completion of infrastructure projects.

Employing an ethnographic approach, this research is based on field visits and semi-structured interviews with the governmental officials, nongovernmental engineers, activists, and village residents involved in the Uchangi dam case.[1] The first section of this article builds the theoretical connections between local knowledge and the implementation of participatory research. The second section, the case study description, explains why and how local communities organized their opposition and renegotiation of the Uchangi dam. Drawing lessons from this case, the final section argues that NGO engineers play an essential role in the process of legitimating participatory research by bridging local knowledge and bureaucratic technical expertise.

Facilitating Community Based Research

Participation in development projects is inherently a dynamic process that relies on exercises ranging from consultation at one end of the spectrum to decision-making at the other (Narayan 1995). With theoretical approaches to participatory research having significantly matured over the last 20 years, there is now a wide toolbox of techniques and institutional resources that are available to communities interested in conducting participatory research. Examples of participatory research techniques include rapid rural appraisal and community mapping (Rocheleau 1994; Chambers 1997).

Though the practice of conducting participatory research has gained popularity, the results of such research have rarely been valued in actual policymaking and project design. While participatory research often yields rich data and reveals locally relevant needs, the process is debunked by bureaucrats because it is highly variable, reflexive, and hard to replicate (van de Fliert and Braun 2002). Yet it is these very characteristics of participatory research that render it a radical and necessary departure from conventional development administration. Participatory research suggests shifting the initiative, responsibility, and action to rural people themselves with an emphasis on development as "process" (Mosse, Farrington, and Rew 1998). By recognizing the importance of "localness," this alternative approach views development projects as flexible, iterative, and contingent exercises.

Mainstreaming community-based research is difficult in practicality. Natural resource agencies, such as irrigation bureaucracies, may not be in a position to perform participatory research given their lack of sociologically trained staff and the general sense of mistrust between government officers and local residents (Kolavalli and Kerr 2002, 227). In contrast, intermediate NGOs have an important role to play in catalyzing and facilitating participatory research. Recognizing this need, the Indian government issued watershed development guidelines in 1994 that devolved a significant level of authority over project implementation to NGOs (Government of India 1994). Yet this process of decentralization has produced its own pitfalls. Within the NGO sector, there is great disparity in professional skills, development priorities, and financial capacities (Farrington, Turton, and James 1999, 3). While some NGOs have a sound record of public accountability and well-trained technical core staffs, others are shoestring operations that are as incapable of participatory research as government agencies. In addition, recent research has indicated that even NGOs known for "best practices" in the water sector are unable to scale up their development efforts beyond one or two village sites (Kolavalli and Kerr, 2002).[2] When a single NGO is charged with overseeing the implementation of multiple projects, its approach can become as model-driven as that of government agencies by losing focus on the importance of context and demand driven development.

While NGOs have an important role to play in facilitating local participation, the onus must remain on the state to implement infrastructure development. NGOs can play an effective intermediary role in the process by translating local knowledge into bureaucratic planning. Herein, the goal is not to make government irrigation expertise obsolete, but rather to create dynamic and iterative opportunities for exchange between farmers and engineering experts. The following case study provides an example of how farmers, NGOs engineers and government officials negotiated the boundaries between local knowledge and technical expertise.

Building the Uchangi Dam

Amidst a landscape of cashew nut trees and lush sugarcane crops, the Uchangi dam site is located in the Kohlapur district in southwestern Maharashtra. While Maharashtra is one of the most economically advanced states in India with a comparatively high per capita income, 37% of the population was listed below the poverty line in a 1993–1994 national government statistical survey (Kurian 2000). The rural poverty that exists in Kolhapur district is inextricably linked to a lack of access to water. Kohlapur district actually receives a large volume of water, averaging 1138 mm/year (IMD 1999, 409). The problem is that this rain falls in the form of monsoonal deluges over just 3 months. With only 15% of total cultivable area irrigated, the majority of villages in the district face water scarcity 6 months out of the year (Census of India, 1991).

To address the lack of irrigation infrastructure in Kohlapur district, the regional irrigation agency, the Maharashtra Krishna Valley Development Corporation (MKVDC), designed the Uchangi dam near the small city of Ajra (see figure 29.1).[3] The Uchangi dam site is located on the banks of the Taor stream, near the villages of Chafawade and Jeur. The project was initially proposed to store 617 million cubic feet (mcft) of water to irrigate 1797 ha of land (MKVDC 2000).

According to the dam design, 222 ha of residential and farm land from the villages of Chafawade, Jeur, Chitale, and Bhavevadi were to be submerged to build the main reservoir. The two main affected villages, Chafawade and Jeur, have a combined population of roughly 2000 people. Water from the Uchangi dam was proposed to benefit 10 small villages downstream of the dam.[4] When completed, the dam was intended to produce electricity using one 500-kW generator.

Though it was first approved by the government in 1985, due to fierce local opposition construction of the Uchangi dam was stalled for over 14 years. The main opponents of the project have been the villagers who were to have been displaced by construction of the reservoir. These project-affected peoples (PAPs) argued that the irrigation agency failed to involve them in the planning process.

While villagers accused the agency of such neglect, in fact, the MKVDC followed standard agency protocol in designing the Uchangi dam. Under agency norms, PAPs are not required to be informed about project details until the final stages of dam implementation. An MKVDC dam is designed by a team of civil engineers based on topographical, geological, and hydrological assessments. There is no requirement that local people be systematically involved or consulted in gathering data, or producing accounts, about water needs, availability, or use patterns. After a project has been

Figure 29.1
The Uchangi dam site in Maharashtra state.

designed by the engineers and contractors have been hired, the executive engineer responsible for the project approaches the district revenue department to begin acquiring land to construct the reservoir. It is during this final step, the land acquisition process, that farmers in the submergence zone officially receive their first notification about the project. Once PAPs are given this notice, it is illegal for them to sell or subdivide their lands. The main reason for providing this official notice is to prevent farmers from subdividing their properties so that they can claim greater compensation.[5]

Political Protest and Alternative Designs
Major opposition to the Uchangi project began in 1997, when irrigation officials began plotting markers for the submergence zone of the reservoir. In this same year, leaders from PAP villages approached a local political organization known as the *Shramik Mukti Dal* (SMD: Workers' Liberation League) for organizing support.[6] With the help of the SMD, the villagers of Jeur and Chafawade waged a long battle against the dam. During

the height of their struggle, opponents occupied the dam site for 50 days and nights, preventing even a single government vehicle from entering the project area. In June 1998, government officials tried to inaugurate dam construction at Uchangi. One thousand people, including women and children with their cattle, arrived in peaceful protest at the dam site to block the opening ceremony. While protests had effectively halted all work on the Uchangi project, SMD activists encouraged villagers to submit an integrated plan for watershed development to the agency. Given their perception of the agency's bias against local participation, SMD activists approached prominent regional technical NGOs to help villagers design a hydrologically sophisticated alternative. In 1999, two NGOs with watershed development experience, the Society for Promoting Participative Ecosystem Management (SOPPECOM) and the *Bharat Gyan Vigyan Samithi* (BGVS: India Science Knowledge Association), came to the assistance of Chafawade and Jeur residents. SOPPECOM is a small NGO based in the city of Pune with a core staff comprised of social workers, economists, and a cadre of retired high-ranking government irrigation engineers. Its funding comes principally from membership fees and research-related project grants from international and national agencies. In contrast, the BGVS is an out growth of the All India People's Science Network. This national organization, which receives government support, works at the grass roots through local chapters at the state and district levels.

Prior to their involvement in the Uchangi case, activists from SOPPECOM and BGVS had published the *Watershed Sourcebook* to guide activist organizations in conducting participatory watershed mapping (Paranjape et al. 1998). Participatory resource mapping (PRM) has been widely used throughout the developing world as a tool for community-based research (Chambers 1997). The PRM process involves training a group of community volunteers to conduct a plot-by-plot mapping of an area to take inventory of resource endowments, list resource constraints, and propose development projects. Though there are a range of participatory research tools available for assessing local development needs, Datta Desai of the BGVS argues that PRM is one of the most flexible research methods because it brings together experts and publics toward capacity building.[7]

To work out an alternative proposal in the Uchangi case, SOPPECOM and the BGVS orchestrated a PRM exercise in the villages of Chafawade and Jeur to gather missing information about local ecosystem characteristics and land use patterns. Fifty village youth were trained by SOPPECOM and BGVS staff to conduct the mapping exercise. In April 1999, these volunteers surveyed 150 plots in Chafawade and 100 plots in Jeur over a month period.

The PRM exercise was conducted over four phases. In the first phase, a household survey and land and water mapping exercise were performed. The socioeconomic survey assessed household income, property, access to health care, sanitation, and electricity. During the land and water mapping exercise, volunteer surveyors produced hand-drawn maps and data sheets that identified terrain features and land use patterns. Each farmer was asked about the direction of water flows on their plots, drainage problems and soil depth, color, and water holding capacity. Surveyors also noted the location of canals and lift-irrigation facilities. Last, they carefully studied groundwater use by identifying where wells were located, their size, and water fluctuation levels.

After generating the maps and compiling and classifying the data, this information was sent to SOPPECOM in Pune for secondary analysis. In the second phase of the PRM, SOPPECOM engineers used the information from the hand-drawn maps to determine macro soil, land, and water use patterns for the two villages. They then created derivative maps that located micro watersheds and indicated potential sites for irrigation and storage structures. The cost of these analyses was borne by SOPPECOM.

In the third stage of the PRM exercise, SOPPECOM officials sponsored village meetings in Chafawade and Jeur to discuss the picture that emerged from the data that had been collected by the communities. In particular, the micro-watershed derivative maps were used to discuss irrigation options. After these meetings, villagers and SMD activists worked with SOPPECOM engineers to develop a set of technical alternatives that could be submitted to the MKVDC to redesign the Uchangi dam.

In addition to helping generate specific alternatives, the PRM activity was important for building local technical capacity. In Jeur, college students were trained for the PRM. One of these students stated that through the PRM process "we learned that locals can do better surveys than the government surveyors. We get more information out of it." Farmers whose plots had been mapped suggested that through the PRM effort they were also instructed in basic soil conservation strategies, including learning about which crops were best adapted to specific soil types.[8]

In the fourth, and final, phase of the PRM, SOPPECOM engineers combined the locally generated data with the limited information they were given by the government agency to propose an alternative plan.[9] SOPPECOM's goals were to determine water storage sites that would limit land loss while increasing the beneficiary area. Local data about soil conditions, water flow, and water availability were used to gauge the best water storage sites in the valley. In the "Alternative Proposal to the Uchangi Dam" that was submitted to the agency, SOPPECOM argued that the standard cost-benefit approach does not study alternatives that limit land and homes lost. Furthermore, the authors stated that the "value of villagers' emotional bond with their lands and houses cannot be quantified in rupees" (Maharashtra Rajya Dharangrasth et al. 1999, 3).

The "Alternative Proposal" recommended constructing three supplementary storage dams instead of one large reservoir at Uchangi. The SOPPECOM plan also suggested building 11 additional weirs. SOPPECOM engineers drafted two different scenarios. Option 1, listed in table 29.1, provided 85% of the storage proposed for the Uchangi reservoir and significantly reduced the amount of land to be submerged (148 ha vs. 222 ha). The land that was to be submerged was either inferior rocky land or wasteland. While Option 1 was the preferred alternative, Option 2 provided for as much water storage as the proposed Uchangi dam with much less land displacement than the original plan (164 ha vs. 222 ha) (Maharashtra Rajya Dharangrasth et al. 1999, 6). SOPPECOM specified that the project should have phased construction and rehabilitation to allow affected and beneficiary groups to better participate in the process.

Table 29.1
Comparing the Alternatives

Name of site	Height (m)	Storage (million cubic feet)	Submergence area (ha)
Original design Uchangi	38	617	222
SOPPECOM option 1			
Khetoba	23.5	180.08	51.78
Dhamanshet	40.1	119.15	23.14
Cherlakatta	36.3	227.32	74.82
Total		526.55	148.74
SOPPECOM option 2			
Khetoba	23.5	180.08	50.78
Dhamanshet	40.1	119.15	23.14
Cherlakatta	39.3	314.74	90.10
Total		613.97	164.02

Note: The data in this table were extracted from the 1999 report by the Maharashtra Rajya Dharangrasth va Prakalpagrasth Shetkari Parishad et al. (1999, 6).

Negotiating an Alternative

After submitting the alternative proposal, there were a few rounds of negotiations with the agency. In addition to SOPPECOM, representatives from SMD and farmers from Chafawade and Jeur were present at these meetings. According to K. J. Joy, it was important for SOPPECOM officials to be present at these meetings because they were able to buttress the technical claims made by villagers through their irrigation expertise.[10]

In their negotiations with NGO and village leaders, MKVDC officials argued that they could not fully accept the alternatives that had been offered. The Deputy Engineer of the Uchangi project also argued that one large dam could not equal three smaller dams because they would lose electricity generation potential. However, the agency did concede to lower the height of the dam by 2 m. This reduction in scale meant that significantly less land would be submerged and not a single home would be lost. According to the final plan, the people of Jeur and Chafawade will have to shift some of their inferior farm lands. Farmers are currently in the process of selecting these new sites. By March 2001, 50% of these lands had been acquired.[11] Because some storage potential was lost by lowering the height of the dam, the agency did agree to investigate the storage site suggested in the "Alternative Proposal" at Khetoba. This site is located 8 km upstream from the Uchangi dam. The small Khetoba dam will submerge an additional 47 ha of land, of which 20 ha are forest area.

As table 29.2 indicates, as a result of the project re-design, villagers from Chafawade and Jeur are now included as project beneficiaries. The MKVDC has agreed to subsidize a lift-irrigation scheme for these two villages at 85% of the total cost. The provision of

Table 29.2
The Final Design

Project specifics	Original design	Final design
Water storage	617 mcft	624 mcft
Dam sites	Uchangi (38 m height)	New Uchangi (at 36 m height) and Khetoba
Submergence	222 ha (with full submergence of all homes in two villages)	75–100 ha (only farm land lost)
Beneficiaries	10 villages	12 villages (including Chafawade and Jeur)

lift-irrigation for PAPs marks a precedent for the MKVDC. Under this new policy, the agency has pledged that 6% of the total irrigation potential of any dam it constructs will go to PAPs upstream of the storage. In the Uchangi case, this translates to 98 ha of irrigation.

The provision of lift-irrigation for these two villages has been the most profound outcome of the re-design of the Uchangi dam. In India, it is exceptional to see PAPs shift their status from development victims, those who are ousted from their land, to project beneficiaries. After 15 years of protest, it is an extraordinary result that the majority of Chafawade and Jeur villagers will now be able to apply water to their fields from the same dam that had threatened to erase their material and cultural ties to their traditional lands.

Drawing Lessons in Participatory Technology Development

For over a decade, the role of local participation in government irrigation development has been actively researched and debated in international fora. Through field research and pilot programs, institutions like the World Bank, the International Water Management Institute, and the International Network for Participatory Irrigation Management have demonstrated that farmer participation in project design and management is central to improving the economic efficiency and social equity of water delivery (Narayan 1995).[12] Despite these long-standing efforts, irrigation agencies in India continue to pursue a standard top-down model of technocratic expertise. This is particularly the case in Maharashtra, where irrigation agencies are currently constructing hundreds of irrigation projects without the direct participation of affected communities.

This article has described how one rural social movement, which emerged in response to the construction of the Uchangi dam, induced the regional irrigation agency to renegotiate their approach to project design through both steadfast political opposition and articulation of alternative technical options. In light of the divisive water politics that dominate infrastructure development in India, it is important to understand why this social movement was successful in meeting its goals. In particular, what can this case teach us about the factors that compel agencies toward making use of locally generated data in the process of technical design?

The Uchangi case illustrates that agencies like the MKVDC are driven by a "blueprint" approach to project design. Rather than depending on locally determined needs or generations of farmer-based experience in water storage and distribution, the location and scale of the Uchangi dam were based on an idealized landscape represented by contours on a topographical map. By relying on a design process where technical decisions are made in the absence of local knowledge and participation, MKVDC projects have routinely lacked support from local stakeholders and have been mired in years of delays. According to SOPPECOM activist K. J. Joy, "no MKVDC project has been able to complete construction and water distribution on schedule because of protests from affected peoples."[13] This groundswell of organized opposition has drawn attention to the conceptual inadequacy of the MKVDC's modeling and technical design approach.

The successful redesign of the Uchangi project was the result of both collective resistance and community articulation of a technically sophisticated alternative plan. At the village level, SMD activists reported that where the dam displacees' oppositional efforts have been strong, they have been able to force agency officers to become more cooperative. SMD activist Sampath Desai stated that "Projects run smoothly when there is public cooperation. In places where there is opposition, officers can't even enter to do their surveys."[14] When asked about this suggestion, the MKVDC engineer responsible for the project reported in agreement. He stated that when villagers are well organized and can clearly articulate their demands, the agency is more amenable to negotiation.[15] In the Uchangi case, the agency had sunk significant funds into the construction of the dam. This project was also linked to overall district-level water storage targets. It was clear that the agency needed to complete the Uchangi project and was vulnerable to social movement activism.

While political opposition played a pivotal role in compelling the agency to negotiate, the manner in which the alternative was framed was perhaps as important. Toward producing and promoting an alternative design, villagers made an important decision to ally themselves with NGO technical experts. As intermediary support organizations, SOPPECOM and BGVS played key roles in catalyzing participatory research and translating local concerns into technical options. The retired government engineers from SOPPECOM who prepared the alternative design not only spoke a common language with government experts, but they also shared a professional culture. SOPPECOM activist K. J. Joy believes that the MKVDC was in the position to negotiate with villagers because the alternative proposal was written to fit neatly within agency analytical frameworks for water storage.

The ability of NGO engineers to translate the data that emerged from a local mapping project into a technical alternative that complied with agency norms was critical. Other alternatives could have been premised on radically different lines. For example, rather than focus on designing irrigation dams, a ban on water-intensive cropping could have been proposed in the valley. This would also have resulted in the need for a lesser volume of water storage. Yet it was a strategic choice to design an alternative that stored a comparable supply of water using medium-sized dams. The MKVDC was more likely to accept this alternative because of its institutional commitment and expertise in building large-scale infrastructure.

The use of technical norms as tools for negotiation did not undermine the value of local knowledge. In the Uchangi case, local knowledge was not ontologically distinct from technoscience. Rather, working with communities to understand their resource constraints and priorities was a way of privileging a different set of epistemological resources and material interests. The community mapping exercise asked a set of research questions that were very distinct from the agency's approach. Rather than simply investigating where water was located and how much of it could be stored, the PRM project explored how water was locally used, by whom, and at what social and ecological costs. The concerns and knowledge base of local farmers became the data sets from which NGO engineers drafted technical options.

The Uchangi case also sheds light on how local knowledge and engineering expertise can be hybridized. This includes distinguishing between the kinds of technical decisions that can be made by local residents and those that require expert intervention. While the demand for development projects should emanate from communities, there are specific aspects of technical design that must be handled by trained engineers. R. K. Patil, an agricultural economist with SOPPECOM, has suggested that local farmers are often the best people to consult about where a dam or canal should be sited because they intimately understand the vagaries of water flows and the nature of the local geological system. Yet he also argues that while small-scale water conservation structures can be built by volunteers, a "major dam cannot be designed by villagers."[16] The exact size and scale of a large dam, in terms of its volumetric storage capacity, is dependent on water availability calculations that are best made by irrigation engineers. During the technical design process, engineers will ideally consult villagers and create iterative opportunities for feedback.

The Uchangi case suggests that participatory resource mapping can be a valuable research tool. Yet government agencies may not be the best institutions to carry out such initiatives. In addition to lacking staff trained in sociological methods, villagers often view government "participation" programs with suspicion. While in theory participatory development involves cooperative and shared decision-making authority between research professionals and villagers, government efforts at implementing "participation" have entailed superficial, and at times even coercive, consultations with villagers (Pretty and Shah 1997; Cooke and Kothari 2002). In contrast, intermediary technical NGOs, like SOPPECOM and BGVS, are staffed by professionally trained social workers and engineers who can help carry out PRM-like exercises. Agencies such as the MKVDC can mainstream participatory research by using technical NGOs to help catalyze the collection and analysis of primary local data. By embracing more "contextualized" knowledge and design approaches, the MKVDC could improve its efficiency at delivering water by reducing the delays and cost-overruns that are caused by local political opposition. In addition, the incorporation of local priorities into dam design significantly improves the long-term sustainability of the project.

Conclusion

This article has sought to examine the factors that induce government agencies to include local knowledge and participation in the design of technical projects. The Uchangi case illustrates the importance of both organized political opposition and the construction of development alternatives. In the process of facilitating local research, intermediary NGOs played an important role in translating local needs and priorities into the technical language and cultural norms of bureaucratic agencies. Employing NGOs in this capacity does not relieve government of its responsibility to provide infrastructure development. Rather, it broadens the conceptual models that planners use to investigate technical options. By making the boundaries between local knowledge and expert science more porous, NGO-led community research can value add project design so that local priorities are reflected in project development. By reducing the need for locals to routinely oppose and delay the construction of water projects, this form of collaborative technology development can go a long way toward actually delivering water to where and when it is needed.

Beyond the villages described in this case study, the outcome of the Uchangi project has helped spur grass roots struggles throughout the state of Maharashtra. In August 2002, over 10,000 farmers gathered at a *Pani Parishad* (Water Meeting) in the city of Atpadi to challenge the MKVDC's approach to irrigation development. Calling on farmers to blend political opposition with the reconstruction of technical alternatives, this broader social movement has demanded that the agency invest in farmer participation and promote more equitable access to water. Within Kohlapur district, efforts are currently underway to redesign an MKVDC dam project that impacts over 50 villages in the Chikotra Valley.

As members of this broader movement, SOPPECOM and BGVS engineers and activists have championed an approach to agricultural knowledge and information sharing where multiple stakeholders mutually engage in a "collective learning process" toward problem identification and resolution (Roling and Wagemakers 1998, 17). Such approaches neutralize the expert/lay knowledge binary and demystify the role of local and traditional knowledge. What becomes relevant, instead, is how "individuals, groups, or institutions are able to make their particular definitions or situations authoritative, and in doing this to redirect material benefits in their direction" (Mosse et al. 1998, 27).

Notes

This research was conducted with generous support from the American Institute for India Studies, the International Water Management Institute, and the National Science Foundation (SES number 9984233). I would like to thank Katrina Smith Korfmacher, Lynne Fessenden, and four anonymous reviewers from *SNR* for their suggestions on this article. In India, I am indebted to Datta Desai, Sampath Desai, K. J. Joy, Ravi Kagalkar, Anant Phadke, and the villagers from Chafawade and Jeur for sharing their experiences and insights with me.

1. Fieldwork was performed between November 2000 and May 2001 in the cities of Pune, Kohlapur, and Ajra. Semistructured interviews with residents of Chafawade and Jeur villages and government project officers were conducted in March 2001. One focus group with the farmers, youth and NGO activists involved in the community mapping project was held in Chafawade village center in March 2001. In addition to interviews and site visits, content analysis included government project documents, newspaper articles, NGO pamphlets, and press releases.

2. "Scaling up" refers to extending the cumulative impact of institutional change so that it moves beyond the micro project level toward regional and national policy dimensions.

3. The MKVDC was established in 1996 to develop the water resources of the Krishna River where it flows through the state of Maharashtra. The agency is financed through both private commercial bonds and government contribution.

4. These villages are Shringarwadi, Yemekond, Shirsangi, Watangi (the major village), Kine, Posvatrwadi, Handewadi, Kolingdre, Uchangi, and Reddiwadi.

5. Interview with MKVDC official A. Surve, February 27, 2001.

6. The Shramik Mukti Dal (SMD) has been a major player in organizing the *Maharashtra Rajya Dharangrasth va Prakalpagrasth Shetkari Parishad* (Maharashtra State Dam and Project Oustees Association). As part of this association, SMD activists fight for the rights of dam displacees throughout Maharashtra. Recently, they have succeeded not only in halting construction of key dam projects but also in getting monetary compensation for displaced families, as well as irrigated land in command areas (Phadke 2000).

7. Interview with BGVS activist Datta Desai, March 27, 2001.

8. Focus group at Chafawade village center, March 9, 2001.

9. The agency did not provide SOPPECOM some important details about the main dam, spillway, irrigation sluices or dam foundation. For these reasons, SOPPECOM argued that they could not complete cost comparisons and that this should be done by the agency when reviewing the alternatives.

10. Interview with K. J. Joy, February 20, 2001.

11. Interview with MKVDC official R. Kagalkar, March 6, 2001.

12. In the state of Maharashtra, important contributions to this area of research have been made by NGOs, including the Hind Swaraj Trust, Gram Guarav Pratistan, and Watershed Organization Trust (WOTR).

13. Interview with SOPPECOM activist K. J. Joy, February 20, 2001.

14. Interview with SMD activist S. Desai, March 8, 2001.

15. Interview with MKVDC official R. Kagalkar, March 15, 2001.

16. Interview with SOPPECOM economist R. K. Patil, April 7, 2001.

References

Census of India. 1991. *Maharashtra State: Kohlapur District*. New Delhi, India: Office of the Controller.

Cernea, M., ed. 1999. *The economics of involuntary resettlement: Questions and challenges*. Washington, DC: World Bank.

Chambers, R. 1997. *Whose reality counts: Putting the first last*. London: Intermediate Technology Publications.

Cooke, B. and U. Kothari. 2002. *Participation: The new tyranny?* London, UK: Zed Books.

Farrington, J., C. Turton, and A. J. James, eds. 1999. *Participatory watershed development: Challenges for the twenty-first century*. New Delhi, India: Oxford University Press.

Gokhale, M. 1999. MKVDC defers Uchangi dam construction plan. *Indian Express*, June 6.

Government of India. 1994. *Guidelines for watershed development*. New Delhi, India: Ministry of Rural Development.

India Meteorological Department. 1999. *Climatological Tables: 1951–1980*. New Delhi, India: Director General of Meteorology.

Khagram, S. 2004. *Dams and Development: Transnational struggles for water and power*. Ithaca: Cornell University Press.

Kolavalli, S. and J. Kerr. 2002. Scaling-up participatory watershed development in India. *Dev. Change* 33(2):213–235.

Kurian, N. J. 2000. Widening regional disparities in India. *Econ. Polit. Weekly* February 12:538–550.

Maharashtra Krishna Valley Development Corporation. 2000. *Uchangi minor irrigation tank report*. Pune, India: MKVDC, Kohlapur Irrigation Circle.

Maharashtra Rajya Dharangrasth va Prakalpagrasth Shetkari Parishad, Shramik Mukti Dal, Western Regional Resource Center, and SOPPECOM. 1999. *Alternative proposal to the proposed Uchangi dam*. Pune, India: SOPPECOM.

McCully, P. 1996. *Silenced rivers: The ecology and politics of large dams*. London, UK: Zed Books.

Mosse, D., J. Farrington, and A. Rew, eds. 1998. *Development as process: Concepts and methods for working with complexity*. New York: Routledge Press.

Narayan, D. 1995. *The contribution of people's participation: Evidence from 121 rural water supply projects*. Washington, DC: World Bank.

Paranjape, S., K. J. Joy, T. Machado, A. Varma, and S. Swaminathan. 1998. *Watershed based development: A source book*. Calcutta, India: BGVS Watershed Programme.

Phadke, A. 2000. Dam-oustees movement in South Maharashtra—Path breaking achievements. *Econ. Polit. Weekly* November 18:4084–4086.

Pretty, J. N. and P. Shah. 1997. Making soil and water conservation sustainable: From coercion and control to partnerships and participation. *Land Degradation Dev.* 8(1):39–58.

Rangachari, R., N. Sengupta, R. Iyer, P. Baneri, and S. Singh. 2000. *Large dams: India's experience*. Capetown, South Africa: World Commission on Dams.

Rocheleau, D. 1994. Participatory research and the race to save the planet: Questions, critique, and lessons from the field. *Agric. Hum. Values* 11:4–25.

Roling, N. G. and M. A. E. Wagemakers, eds. 1998. *Facilitating sustainable agriculture*. Cambridge, UK: Cambridge University Press.

Singh, S. 1998. *Taming the waters: The political economy of large dams in India*. New Delhi, India: Oxford University Press.

Vaidyanathan, A. 1999. *Water resource management: Institutions and irrigation development in India*. New Delhi, India: Oxford University Press.

van de Fliert, E. and A. R. Braun. 2002. Conceptualizing integrative, farmer participatory research for sustainable agriculture: From opportunities to impact. *Agric. Hum. Values* 19:25–38.

30 "Security Trade-Offs Are Subjective" and "Technology Creates Security Imbalances"

Bruce Schneier

Since September 11, 2001, an enormous amount of attention has been given to various kinds of security in the United States. In this excerpt from *Beyond Fear*, Bruce Schneier explores both the expert and lay understanding of risk and argues that fears generated by the threat of terrorism have caused American society to make irrational trade-offs. One thing he finds especially troublesome is the idea that technology can be a "silver bullet" such that—if we get it just right—it will make us safe and secure. Schneier believes that relying too much on technology may make us less, not more secure. Our use of technology involves adopting systems that while making us safer in some ways also creates new risks and vulnerabilities on a larger scale. If Schneier is right, then the world of today is less safe than it was, say, 50 years ago, not just because of political change but because of the technologies we have adopted. Schneier implores us to consider the big picture; he does not deny the importance of technology in addressing security, but urges us to think about solutions in terms of sociotechnical systems. In some cases, for instance, political treaties might be far easier to achieve and more effective than implementing a new technological system. In other cases, a simple technological change might result in a significant increase in security. In still other cases, we might develop international agreements on technology that make the world safer for everyone. When problems are as big and pervasive as terrorism in the modern world, simple technological fixes are unlikely to succeed. Schneier's analysis makes clear that developing security strategies, weighing risks, and accepting trade-offs are not objective or obvious processes. We must combine our knowledge of technology and society wisely, and consider what values we hold most important, in order to arrive at solutions that are effective while at the same time not undermining the basic principles of our society.

Since 9/11, we've grown accustomed to ID checks when we visit government and corporate buildings. We've stood in long security lines at airports and had ourselves and our baggage searched. In February 2003, we were told to buy duct tape when the U.S. color-coded threat level was raised to Orange. Arrests have been made; foreigners have been deported. Unfortunately, most of these changes have not made us more secure. Many of them may actually have made us *less* secure.

. . .

From *Beyond Fear: Thinking Sensibly about Security in an Uncertain World* (New York: Copernicus Books, 2003), pp. 5, 13–14, 17–28, 87–101. © 2003 Bruce Schneier. Reprinted with kind permission of Springer Science + Business Media.

Unfortunately, many countermeasures are ineffective. Either they do not prevent adverse consequences from the intentional and unwarranted actions of people, or the trade-offs simply aren't worth it. Those who design and implement bad security don't seem to understand how security works or how to make security trade-offs. They spend too much money on the wrong kinds of security. They make the same mistakes over and over. And they're constantly surprised when things don't work out as they'd intended.

One problem is caused by an unwillingness on the part of the engineers, law enforcement agencies, and civic leaders involved in security to face the realities of security. They're unwilling to tell the public to accept that there are no easy answers and no free rides. They have to be seen as doing something, and often they are seduced by the promise of technology. They believe that because technology can solve a multitude of problems and improve our lives in countless ways, it can solve security problems in a similar manner. But it's not the same thing. Technology is generally an enabler, allowing people to do things. Security is the opposite: It tries to prevent something from happening, or prevent people from doing something, in the face of someone actively trying to defeat it. That's why technology doesn't work in security the way it does elsewhere, and why an overreliance on technology often leads to bad security, or even to the opposite of security.

The sad truth is that bad security can be worse than no security; that is, by trying and failing to make ourselves more secure, we make ourselves less secure. We spend time, money, and energy creating systems that can themselves be attacked easily and, in some cases, that don't even address the real threats. We make poor trade-offs, giving up much in exchange for very little security. We surround ourselves with security countermeasures that give us a feeling of security rather than the reality of security. We deceive ourselves by believing in security that doesn't work.

. . .

Security Trade-offs Are Subjective

There's no such thing as absolute security. It's human nature to wish there were, and it's human nature to give in to wishful thinking. But most of us know, at least intuitively, that perfect, impregnable, completely foolproof security is a pipe dream, the stuff of fairy tales, like living happily ever after. We have to look no further than the front page of the newspaper to learn about a string of sniper attacks, a shoe bomber on a transatlantic flight, or a dictatorial regime threatening to bring down a rain of fire on its enemies.

... [S]ecurity always involves trade-offs. A government weighs the trade-offs before deciding whether to close a major bridge because of a terrorist threat; is the added security worth the inconvenience and anxiety? A homeowner weighs the trade-offs before deciding whether to buy a home alarm system; is the added security worth the bother and the cost? People make these sorts of security trade-offs constantly, and what they end up with isn't absolute security but something else. A few weeks after 9/11, a reporter asked me whether it would be possible to prevent a repetition of the terrorist attacks. "Sure," I replied, "simply ground all the aircraft." A totally effective solution, certainly, but completely impractical in a modern consumer-oriented society. And what's more, it would do nothing to address the general problem of terrorism, only certain specific airborne attacks.

Still, this extreme and narrowly effective option is worth considering. Ground all aircraft, or at least all commercial aircraft? That's exactly what did happen in the hours after the 9/11 attacks. The FAA ordered the skies cleared. Within two and a half hours, all of the 4,500 planes in the skies above this country were pulled down. Transcontinental flights were diverted to airports in Canada and elsewhere. Domestic flights landed at the closest possible airport. For the first time in history, commercial flight was banned in the U.S. In retrospect, this was a perfectly reasonable security response to an unprecedented attack. We had no idea if there were any other hijacked planes, or if there were plans to hijack any other planes. There was no quick and efficient way to ensure that every plane in the air was under the control of legitimate airline pilots. We didn't know what kind of airline security failures precipitated the attacks. We saw the worst happen, and it was a reasonable assumption that there was more to follow. Clearing the skies of commercial airplanes—even for only a few days—was an extreme trade-off, one that the nation would never have made in anything but an extreme situation.

Extreme trade-offs are easy. Want to protect yourself from credit card fraud? Don't own a credit card. Want to ensure that you are never mugged? Live on a deserted island. Want to secure yourself against mad cow disease? Never eat meat products. But if you want to eat prime rib in a crowded restaurant and pay with your credit card, you're going to have to accept imperfect security. Choosing to stay away from dark alleys may seem like a simple decision, but complex trade-offs lurk under the surface. And just as the U.S. cleared the skies in the wake of 9/11, someone who never worries about where he walks might change his mind if there's a crime wave in his neighborhood. All countermeasures, whether they're to defend against credit card fraud, mugging, or terrorism, come down to a simple trade-off: What are we getting, and what are we giving up in order to get it?

Most countermeasures that increase the security of air travel certainly cost us significantly more money, and will also cost us in terms of time and convenience, in almost any way that they might be implemented. A new security procedure for airplanes, one that adds a minute to the check-in time for each passenger, could add hours to the total preparation time for a fully booked 747. The government has not seriously considered barring all carry-on luggage on airplanes, simply because the flying public would not tolerate this restriction. Positive bag matching of passengers with their luggage was largely accomplished with information systems, but a thorough check of every passenger's carry-on bag would be another matter entirely; already many travelers have decided to drive instead of flying, where possible, because airline security has become so intrusive.

Much of our sense of the adverse, then, comes down to economics, but it also derives from some nebulous cultural sense of just how much disorder, or larceny, or invasion of privacy, or even mere rudeness we'll tolerate. For example, Swiss attitudes toward security are very different from American ones. The public toilet stalls in the Geneva airport are walled cubicles with full-length doors and better locks than on the doors of many U.S. homes. The cubicles take up a lot of space and must be a real nuisance to clean, but you can go inside with your luggage and nobody can possibly steal it.

The Swiss are willing to pay higher prices for better security in myriad ways. Swiss door locks are amazing; a standard apartment lock is hard to pick and requires a key that can't be duplicated with readily available equipment. A key can be duplicated only by the lock manufacturer at the written request of the property owner. The trade-offs of this are pretty significant. One of my colleagues had a minor lock problem, and his whole family was locked out for hours before it could be fixed. Many Swiss families have only one or two house keys, regardless of the number of people living together.

The security a retail store puts in place will depend on trade-offs. The store's insurance company might mandate certain security systems. Local laws might prohibit certain security systems or require others. Different security systems cost different amounts and disrupt normal store operations to different degrees. Some decisions about security systems are optional, and some are not.

Because security always involves a trade-off, more security isn't always better. You'd be more secure if you never left your home, but what kind of life is that? An airline could improve security by strip-searching all passengers, but an airline that did this would be out of business. Studies show that most shoplifting at department stores occurs in fitting rooms. A store can improve security by removing the fitting rooms, but many stores believe that the resulting decrease in profits from sales would be greater than the cost of shoplifting. It's the trade-off that's important, not the absolute security. The question is: How do you evaluate and make sensible trade-offs about the security in your own life?

In many instances, people deliberately choose less security because the trade-offs for more security are too large. A grocery store might decide not to check expiration dates on coupons; it's not worth holding up the checkout line and inconveniencing good customers to catch the occasional fifty-cent fraud. A movie theater might not worry about people trying to sneak in while others are exiting; the minor losses aren't worth paying the extra guards necessary to forestall them. A retail shop might ignore the possibility that someone would lie about his age to get the senior citizen discount; preventing the fraud isn't worth asking everyone for proof of age.

In the professional lingo of those who work in security, there is an important distinction drawn between the words "threat" and "risk." A *threat* is a potential way an attacker can attack a system. Car burglary, car theft, and carjacking are all threats—in order from least serious to most serious (because an occupant is involved). When security professionals talk about *risk*, they take into consideration both the likelihood of the threat and the seriousness of a successful attack. In the U.S., car theft is a more serious risk than carjacking because it is much more likely to occur.

Most people don't give any thought to securing their lunch in the company refrigerator. Even though there's a threat of theft, it's not a significant risk because attacks are rare and the potential loss just isn't a big deal. A rampant lunch thief in the company changes the equation; the threat remains the same, but the risk of theft increases. In response, people might start keeping their lunches in their desk drawers and accept the increased risk of getting sick from the food spoiling. The million-dollar Matisse painting in the company's boardroom is a more valuable target than a bagged lunch, so while the threat of theft is the same, the risk is much greater. A nuclear power plant might be under constant threat both of someone pasting a protest sign to its front door and of someone planting a bomb near the reactor. Even though the latter threat is much less likely, the risk is far greater because the effects of such damage are much more serious.

Risk management is about playing the odds. It's figuring out which attacks are worth worrying about and which ones can be ignored. It's spending more resources on the serious attacks and less on the frivolous ones. It's taking a finite security budget and making the best use of it. We do this by looking at the risks, not the threats. Serious risks—either because the attacks are so likely or their effects are so devastating—get defended against while trivial risks get ignored. The goal isn't to eliminate the risks, but to reduce them to manageable levels. We know that we can't eliminate the risk of burglaries, but a good door lock combined with effective police can reduce the risk substantially. And while the threat of a paramilitary group attacking your home is a serious one, the risk is so remote that you don't even think about defending your house against it. The owner of a skyscraper will look at both attacks differently because the risk is different, but even she will ignore some risks. When in a million years would a 767 loaded with jet fuel crash into the seventy-seventh floor of a New York skyscraper? Threats determine the risks, and the risks determine the countermeasures.

Managing risks is fundamental to business. A business constantly manages all sorts of risks: the risks of buying and selling in a foreign currency, the risk of producing merchandise in a color or style the public doesn't want, the risk of assuming that a certain economic climate will continue. Threats to security are just another risk that a business has to manage. The methods a business employs in a particular situation depend on the details of that situation. Think, for example, of the different risks faced by a large grocery chain, a music shop, or a jewelry store.

There are many ways to deal with "shrinkage," as shoplifting is referred to in parts of the retail industry. Most chain grocery stores, for example, simply accept the risk as a cost of doing business; it's cheaper than installing countermeasures. A music shop might sell its CDs in bulky packaging, making them harder to shoplift. A jewelry store might go further and lock up all the merchandise, not allowing customers to handle any object unattended.

Then there's the matter of insurance. Of the three retail operations, probably only the jewelry store will carry theft insurance. Insurance is an interesting risk management tool. It allows a store to take its risk and, for a fee, pass it off to someone else. It allows the store to convert a variable-cost risk into a fixed-cost expense. Insurance reduces risk to the store owner because it makes part of the risk someone else's problem.

Like the U.S. government clearing the skies after 9/11, all of these solutions are situational. What a store does depends on what products it's selling, what its building is like, how large a store it is, what neighborhood it's in, what hours it's open, what kinds of customers it gets, whether its customers pay cash or buy using traceable checks or credit cards, whether it tends to attract one-time customers or repeat customers, et cetera.

These factors also affect the risk of holdups. A liquor store might decide to invest in bulletproof glass between the customer and the sales clerk and merchandise, with a specially designed window to pass bottles and money back and forth. (Many retail stores in south-central Los Angeles are like this.) An all-night convenience store might invest in closed-circuit cameras and a panic button to alert police of a holdup in progress. In Israel, many restaurants tack a customer surcharge onto their bills to cover the cost of armed guards. A bank might have bulletproof glass, cameras, and guards, and also install a vault.

Most of this is common sense. We don't use the same security countermeasures to protect diamonds that we use to protect donuts; the value and hence the risks are completely different.

And the security must balance the risk. What all of these businesses are looking for is to maximize profits, and that means adequate security at a reasonable cost. Governments are trying to minimize both loss of life and expenditures, and that also means adequate security at a reasonable cost. The decision to use guards is based not only on risk, but on the trade-off between the additional security and the cost of the guards. When I visited Guatemala City in the 1980s, all major stores and banks had armed guards. Guards cost less, per hour, in Guatemala than in the U.S., so they're more likely to be used. In Israel, on the other hand, guards are expensive, but the risks of attack are greater. Technological security measures are popular in countries where people are expensive to hire and the risks are lower.

Balancing risks and trade-offs are the point of our five-step process. In Step 2, we determine the risks. In Steps 3 and 4, we look for security solutions that mitigate the risks. In Step 5, we evaluate the trade-offs. Then we try to balance the pros and cons: Is the added security worth the trade-offs? This calculation is risk management, and it tells us what countermeasures are reasonable and what countermeasures are not.

Everyone manages risks differently. It's not just a matter of human imperfection, our inability to correctly assess risk. It also involves the different perspectives and opinions each of us brings to the world around us. Even if we both have the same knowledge and expertise, what might seem like adequate security to me might be inadequate to you because we have different tolerances for risk. People make value judgments in assessing risk, and there are legitimate differences in their judgments. Because of this fact, security is subjective and will be different for different people, as each one determines his own risk and evaluates the trade-offs for different countermeasures.

I once spoke with someone who is old enough to remember when a front-door lock was first installed in her house. She recalled what an imposition the lock was. She didn't like having to remember to take her key each time she went out, and she found it a burden, on returning home, to have to fumble around to find it and insert it in the lock, especially in the dark. All this fuss, just to get into her own home! Crime was becoming a problem in the town where she lived, so she understood why her parents had installed the lock, but she didn't want to give up the convenience. In a time when we've all become security-conscious, and some have become security-obsessed, an attitude like hers may seem quaint. Even so, I know people who still leave their doors unlocked; to them, the security trade-off isn't worth it.

It's important to think about not just the huge variety of security choices open to us, but the huge variety of personal responses to those choices. Security decisions are personal and highly subjective. Maybe a trade-off that's totally unacceptable to you isn't at all a problem for me. (You wouldn't consider living without theft insurance, whereas I couldn't be bothered.) Maybe countermeasures that I find onerous are perfectly okay with you. (I don't want my ID checked every time I enter a government building; you feel more secure because there's a guard questioning everyone.) Some people are willing to give up privacy and control to live in a gated community, where a guard (and, increasingly, a video system) takes note of each vehicle entering and exiting. Presumably, the people who live in such communities make a conscious decision to do so, in order to increase their personal security. On the other hand, there are those who most emphatically don't want their comings and goings monitored. For them, a gated community is anathema. There's not likely to be much in the way of a fruitful discussion of trade-offs between someone absolutely committed to planned community living and someone muttering about creeping encroachments on privacy. But the difference of opinion between the two (like the differences between those facing gun-control questions or workplace surveillance options) is just that—a valid difference of opinion.

When someone says that the risks of nuclear power (for example) are unacceptable, what he's really saying is that the effects of a nuclear disaster, no matter how remote, are so unacceptable to him as to make the trade-off intolerable. This is why arguments like "but the odds of a core meltdown are a zillion to one" or "we've engineered this reactor to have all sorts of fail-safe features" have no effect. It doesn't matter how unlikely the threat is; the risk is unacceptable because the consequences are unacceptable.

A similar debate surrounds genetic engineering of plants and animals. Proponents are quick to explain the various safety and security measures in place and how unlikely it is that bad things can happen. Opponents counter with disaster scenarios that involve genetically engineered species leaking out of the laboratories and wiping out other species, possibly even ours. They're saying that despite the security in place that makes the risk so remote, the risk is simply not worth taking. (Or an alternative argument: that the industry and the government don't understand the risks because they overestimate the efficacy of their security measures, or in any event can't be trusted to properly take them into account.)

For some people in some situations, the level of security is beside the point. The only reasonable defense is not to have the offending object in the first place. The ultimate example of this was the speculation that the Brookhaven National Lab's proposed ion collider could literally destroy the universe.

Sometimes perceptions of unacceptable risk are based on morality. People are unwilling to risk certain things, regardless of the possible benefits. We may be unwilling to risk the lives of our children, regardless of any rational analysis to the contrary. The societal version of this is "rights"; most people accept that things like torture, slavery, and genocide are unacceptable in civilized society.

For some, the risks of some attacks are unacceptable, as well: for example, a repetition of 9/11. Some people are willing to bear any cost to ensure that a similar terrorist attack never occurs again. For others, the security risks of visiting certain countries, flying on airplanes, or enraging certain individuals are unacceptable. Taken to the extreme, these fears turn into phobias. It's important to understand that these are personal, and largely emotional, reactions to risk. The risks can be wildly unlikely, but they are important nonetheless because people act on their perceptions.

Even seemingly absolute risk calculations may turn out to be situational. How far can you push the activist who is fearful of a runaway genetic modification? What would the activist say, for example, if the stakes were different—if a billion people would starve to death without genetically modified foods? Then the risks might be acceptable after all. (This calculation has repeatedly occurred in Africa in recent years, with different countries making different decisions. The best compromise seems to be accepting genetically modified foodstuffs that can't replicate: imported flour but not imported grain. But some famine-stricken countries still reject genetically modified flour.) In the months after 9/11, perfectly reasonable people opined that torture was acceptable in some circumstances. Think about the trade-offs made by the people who established the Manhattan Project: Having fission bombs available to the world might be risky, but it was felt that the bombs were less risky than having the Nazis in control of Europe.

These calculations are not easy. There is always an imprecision, and sometimes the uncertainty factor is very large. Even with sufficient information, the calculation involves determining what is worth defending against and what isn't, or making a judgment about what losses are acceptable. It sometimes even involves understanding that a decision will result in some deaths, but that the alternatives are unreasonable. What is the risk of another 9/11-like terrorist attack? What is the risk that Al Qaeda will launch a different, but equally deadly, terrorist attack? What is the risk that other terrorist organizations will launch a series of copycat attacks? As difficult as these questions are, it is impossible to intelligently discuss the efficacy of antiterrorism security without at least some estimates of the answers. So people make estimates, or guess, or use their intuition.

In the months after 9/11, a story circulated on the Internet about an armed air marshal who was allowed on board a plane with his firearm, but had to surrender his nail clippers. In January 2002, a US Airways pilot at the Philadelphia airport was hauled off to jail in handcuffs after he asked a screener, "Why are you worried about tweezers when I could crash the plane?" The pilot's remark was certainly ill-advised, but the reason so many laugh at the thought of tweezers being confiscated at airports is because we know that there is no increased risk in allowing them on board.

Most of us have a natural intuition about risk. We know that it's riskier to cut through a deserted alley at night than it is to walk down a well-lighted thoroughfare. We know that it's riskier to eat from a grill cart parked on the street than it is to dine in a fine restaurant. (Or is it? A restaurant has a reputation to maintain, and is likely to be more careful than a grill cart that disappears at the end of the day. On the other hand, at a cart I can watch my meat being cooked.) We have our own internal beliefs about the risks of trusting strangers to differing degrees, participating in extreme sports, engaging in unprotected sex, and undergoing elective surgery. Our beliefs are based on lessons we've learned since childhood, and by the time we're adults, we've experienced enough of the real world to know—more or less—what to expect. High places can be dangerous. Tigers attack. Knives are sharp.

A built-in intuition about risk—engendered by the need to survive long enough to reproduce—is a fundamental aspect of being alive. Every living creature, from bacteria on up, has to deal with risk. Human societies have always had security needs; they are as natural as our needs for food, clothing, and shelter. My stepdaughter's first word was "hot." An intuition about risk is a survival skill that has served our species well over the millennia.

But saying that we all have these intuitions doesn't mean, by any stretch, that they are accurate. In fact, our perceived risks rarely match the actual risks. People often underestimate the risks of some things and overestimate the risks of others. Perceived risks can be wildly divergent from actual risks compiled statistically. Consider these examples:

- People exaggerate spectacular but rare risks and downplay common risks. They worry more about earthquakes than they do about slipping on the bathroom floor, even though the latter kills far more people than the former. Similarly, terrorism causes far more anxiety than common street crime, even though the latter claims many more lives. Many people believe that their children are at risk of being given poisoned candy by strangers at Halloween, even though there has been no documented case of this ever happening.
- People have trouble estimating risks for anything not exactly like their normal situation. Americans worry more about the risk of mugging in a foreign city, no matter how much safer it might be than where they live back home. Europeans routinely perceive

the U.S. as being full of guns. Men regularly underestimate how risky a situation might be for an unaccompanied woman. The risks of computer crime are generally believed to be greater than they are, because computers are relatively new and the risks are unfamiliar. Middle-class Americans can be particularly naïve and complacent; their lives are incredibly secure most of the time, so their instincts about the risks of many situations have been dulled.

• Personified risks are perceived to be greater than anonymous risks. Joseph Stalin said, "A single death is a tragedy, a million deaths is a statistic." He was right; large numbers have a way of blending into each other. The final death toll from 9/11 was less than half of the initial estimates, but that didn't make people feel less at risk. People gloss over statistics of automobile deaths, but when the press writes page after page about nine people trapped in a mine—complete with human-interest stories about their lives and families—suddenly everyone starts paying attention to the dangers with which miners have contended for centuries. Osama bin Laden represents the face of Al Qaeda, and has served as the personification of the terrorist threat. Even if he were dead, it would serve the interests of some politicians to keep him "alive" for his effect on public opinion.

• People underestimate risks they willingly take and overestimate risks in situations they can't control. When people voluntarily take a risk, they tend to underestimate it. When they have no choice but to take the risk, they tend to overestimate it. Terrorists are scary because they attack arbitrarily, and from nowhere. Commercial airplanes are perceived as riskier than automobiles, because the controls are in someone else's hands—even though they're much safer per passenger mile. Similarly, people overestimate even more those risks that they can't control but think they, or someone, should. People worry about airplane crashes not because we can't stop them, but because we think as a society we should be capable of stopping them (even if that is not really the case). While we can't really prevent criminals like the two snipers who terrorized the Washington, DC, area in the fall of 2002 from killing, most people think we should be able to.

• Last, people overestimate risks that are being talked about and remain an object of public scrutiny. News, by definition, is about anomalies. Endless numbers of automobile crashes hardly make news like one airplane crash does. The West Nile virus outbreak in 2002 killed very few people, but it worried many more because it was in the news day after day. AIDS kills about 3 million people per year worldwide—about three times as many people each day as died in the terrorist attacks of 9/11. If a lunatic goes back to the office after being fired and kills his boss and two co-workers, it's national news for days. If the same lunatic shoots his ex-wife and two kids instead, it's local news ... maybe not even the lead story.

In America, automobiles cause 40,000 deaths every year; that's the equivalent of a full 727 crashing every day and a half—225 total in a year. As a society, we effectively say that the risk of dying in a car crash is worth the benefits of driving around town. But if those same 40,000 people died each year in fiery 727 crashes instead of automobile accidents, you can be sure there would be significant changes in the air passenger systems. (I don't mean to harp on automobile deaths, but riding in a car is the riskiest discretionary activity the majority of Americans regularly undertake.) Similarly, studies have shown that both drivers and passengers in SUVs are more likely to die in accidents than those in compact cars, yet one of the major selling points of SUVs is that the owner feels safer in one.

This example illustrates the problem: People make security decisions based on perceived risks instead of actual risks, and that can result in bad decisions.

. . .

Technology Creates Security Imbalances

When dynamite was invented in 1886, European anarchists were elated. Finally they had a technology that they believed would let them dismantle the state ... literally. It didn't work out that way, of course; European anarchists found that blowing up a building wasn't as easy as it seemed. But shaped charges and modern explosives brought down the U.S. Marine and French military barracks in Beirut in October 1983, killing 301.

Throughout history, technological innovations have altered the balance between attacker and defender. At the height of its empire, Rome's armies never lost a set battle, precisely because their equipment and training were far better than their opponents'. The stirrup revolutionized warfare by allowing people to fight effectively on horseback, culminating in the formidable medieval mounted knight. Then the crossbow came along and demonstrated that mounted knights were vulnerable after all. During the American Civil War, the Union Army used the railroad and telegraph to create a major security imbalance; their transport and communications infrastructure was unlike anything the Confederacy could muster, and unlike anything that had come before.

In the eighteenth century, Tokugawa Japan, Manchu China, and Ottoman Turkey all maintained large armies that were effective for internal security; but having eschewed modern technology, they were highly vulnerable to European powers. The British army used the machine gun to create another security imbalance at the Battle of Ulundi in 1879, killing Zulus at a previously unimaginable rate. In World War I, radio gave those who used it intelligently an enormous advantage over those who didn't. In World War II, the unbalancing technologies were radar, cryptography, and the atomic bomb.

When NORAD's Cheyenne Mountain complex was built in the Colorado Rockies in 1960, it was designed to protect its occupants from nuclear fallout that would poison the air for only a few weeks. Provide a few weeks' sustenance, and Cheyenne Mountain workers would be safe. Later the cobalt bomb was invented, which could poison the air for many years. That changed the balance: Use a cobalt bomb against the mountain, and a few weeks' air supply means nothing.

What has been true on the battlefield is true in day-to-day life, as well. One hundred years ago, you and a friend could walk into an empty field, see no one was nearby, and then have a conversation, confident that it was totally private. Today, technology has made that impossible.

It's important to realize that advanced tools—a technological advantage—do not always trump sheer numbers. Mass production is its own technological advantage. In the city of Mogadishu alone there are a million assault rifles (and only 1.3 million people); they're for sale in the local markets and are surprisingly cheap. And they are relatively easy to use. En masse, these small arms can do just as much damage as a biological attack, or a dozen Apache helicopters. Given all these guns, or a million machetes in Rwanda—mass-produced weapons of mass destruction—even a disorganized mob, if it is large enough, can create devastation.

Nonetheless, smart attackers look for ways to make their attack as effective as possible, and technology can give them more leverage and more ambition. Physically robbing a bank can net you a bagful of money, but breaking into the bank's computer and remotely redirecting bank transfers can be much more lucrative and harder to detect. Stealing someone's credit card can be profitable; stealing her entire identity and getting credit cards issued in her name can be more profitable; and stealing an entire database of credit card numbers and associated personal information can be the most profitable of all.

The advance of technology is why we now worry about weapons of mass destruction. For the first time in history, a single attacker may be able to use technology to kill millions of people. Much has been written about these weapons—biological and nuclear—and how real the threats are. Chemical weapons can kill people horribly, though on a smaller scale. But the point is that if a certain weapon is too cumbersome, or too expensive, or requires too much skill for attackers to use today, it's almost certain that this will be less true a year from now. Technology will continue to alter the balance between attacker and defender, at an ever-increasing pace. And technology will generally favor the attacker, with the defender playing catch-up.

This is not an immediately obvious point, because technological advances can benefit defenders, as well. Attackers know what they're doing and can use technology to their benefit. Defenders must deploy countermeasures to protect assets from a variety of attacks, based on their perception of the risks and trade-offs. Because defenders don't necessarily know what they're defending against, it is more expensive for them to use technological advances at the precise points where they're needed. And the fact that defenders have to deal with many attackers magnifies the problem from the defenders' point of view. Defenders benefit from technology all the time, just not as efficiently and effectively as attackers do.

Another important reason technology generally favors attackers is that the march of technology brings with it an increase in complexity. Throughout history, we've seen more and more complex systems. Think about buildings, or steam engines, or government bureaucracies, or just about anything else: Newer systems are more complex than older ones. Computerization has caused this trend to rocket skyward. In 1970, someone could tune up his VW bug with a set of hand tools and a few parts bought at the store down the street. Today he can't even begin the process without a diagnostic computer and some very complex (and expensive) equipment.

Computers are more complex than anything else we commonly use, and they are slowly being embedded into virtually every aspect of our lives. There are hundreds of millions of lines of instructions per computer, hundreds of millions of computers networked together into the Internet, and billions of pieces of data flowing back and forth.

Systems often look simple because their complexity remains hidden. A modern faucet may seem simpler than having to bring water up from the well in buckets, but that's only because there's a complex indoor plumbing system within your house's walls, a mass-production system for precision faucets, and a citywide water distribution system to hook up to. You might only do a few simple things on your computer, but there's still an unimaginable amount of complexity going on inside. When I say that systems are getting more complex, I am talking about systems as a whole, not just the part the average person sees.

As a consumer, I welcome complexity. I have more options and capabilities. I can walk to my bank to withdraw my money, or I can use my ATM card in a cash machine almost anywhere in the world. I can telephone my bank and have them print and mail a check to a third party, or even mail me a money order. Or I can do these things on the Internet. Most people continually demand more features and more flexibility, and more complexity evolves to deliver those results. The consequence is a bounty of features and conveniences not even imagined a decade or two ago.

But of course there's a hitch. As a security professional, I think complexity is terrifying. It leads to more and more subtle vulnerabilities. It leads to catastrophic failures, which are both harder to test for beforehand and harder to diagnose afterward. This has been true since the beginning of technology and is likely to be true for the foreseeable future. As systems continue to get more complex, they will continue to get less secure. This fundamental tension between ever more complex systems and security is at the core of many of today's security problems. Complexity is the worst enemy of security. Computer pioneer Niklaus Wirth once said: "Increasingly, people seem to misinterpret complexity as sophistication, which is baffling—the incomprehensible should cause suspicion rather than admiration."

Complex systems have even more security problems when they are *nonsequential* and *tightly coupled*. Nonsequential systems are ones in which the components don't affect each other in an orderly fashion. Sequential systems are like assembly lines: Each step is affected by the previous step and then in turn affects the next step. Even Rube Goldberg–type machines, as complicated as they are, are sequential; each step follows the previous step. Nonsequential systems are messier, with steps affecting each other in weirdly complex ways. There are all sorts of feedback loops, with machines obeying various do-this-if-that-happens instructions that have been programmed into them. Nonsequential systems are harder to secure because events in one place can ripple through the system and affect things in an entirely different place. Consider the air passenger transport system—a dream, and sometimes a nightmare, of nonsequential design. Each plane affects every other plane in the vicinity, and each vicinity affects still more remote locations. Weather problems in one city can affect take-offs and landings on the other side of the globe. A mechanical failure in Hong Kong, or a wildcat strike in Rome, can affect passengers waiting to board in Los Angeles or New York.

Tightly coupled systems are those in which a change in one component rapidly sets off changes in others. This is problematic because system effects occur quickly and sometimes unpredictably. Tightly coupled systems may or may not be sequential. Think of a row of dominos. Pushing one will topple all the others. The toppling is sequential, but tight coupling means that problems spread faster and that it's harder to contain them.

An AT&T phone crash in 1990 illustrates the point. A single switch in one of AT&T's centers failed. When it came back online, it sent out signals to nearby switches causing them to fail. When they all came back on, they sent out the same signals, and soon 60,000 people were without service. AT&T had a backup system, of course, but it had the same flaw and failed in exactly the same way. The "attacker" was three lines of code in a module that had been recently upgraded. The exacerbating factor was that the test environment was not the same as the real network. Similarly, a computer worm unleashed in Korea in January 2003 affected 911 service in Seattle.

A business, on the other hand, is complex but reasonably sequential and not tightly coupled. Most people work for a single boss. If someone calls in sick, or quits, the business doesn't fall apart. If there's a problem in one area of the business, the rest of the business can still go on. A business can tolerate all sorts of failures before it completely collapses. But it's still not perfect; a failure at the warehouse or the order entry desk could kill the entire business.

By changing the functionality of the systems being defended, new technologies can also create new security threats. That is, attacks not possible against the older, simpler systems are suddenly possible against the newer, more complex systems.

In 1996, Willis Robinson worked at a Taco Bell in Maryland. He figured out how to reprogram the computerized cash register so that it would record a $2.99 item internally as a one-cent item while visibly showing the transaction as the proper amount, allowing him to pocket $2.98 each time. He stole $3,600 before he was caught. The store manager assumed a hardware or software error and never suspected theft—remember what happens when attacks are rare—and Robinson got caught only because he bragged. This kind of attack would not have been possible with a manual or even an old electric cash register. An attacker could under-ring a sale and then pocket the difference, but that was a visible act and it had to be repeated every single time $2.98 was stolen. With the computer helping him, Robinson was able to steal all the money at the end of the day no matter who was ringing up purchases.

Before we had the ability to bank by phone, there was no way for an attacker to steal money out of an account remotely. No system was in place for him to attack. Now such a system exists. Before companies started selling on the Internet, it was impossible for an attacker to manipulate the prices on Web pages and convince the software to sell the merchandise at a cheaper price. Now that kind of attack takes place regularly. Manual systems are not vulnerable to attacks against the power system; electrical systems are. Pencil-and-paper systems aren't vulnerable to system crashes; computerized systems are.

Technological systems also require *helpers*: mediators and interfaces between people and the system they interact with. Telephone keypads, aircraft instrument panels, automatic language translation software, and handheld GPS receivers are all helpers. They add even more complexity to a system, and they introduce a level of abstraction that can obscure they system's security properties from the people using it.

Imagine you're sitting on an isolated park bench talking to a friend. That's a relatively simple system to secure against eavesdropping. If you're talking on the telephone, it's a much harder security problem. Not only is there a complex technological system between you and your friend, but there is also a helper device—a telephone—mediating your interaction with both the telephone network and your friend. In this case, you have to secure not only your environment and your friend's environment, but two telephones and the vast and unseen telephone system, as well.

Securing a theater against someone sneaking in and watching the movie is a much simpler problem than securing a pay-per-view television program. Making paper money hard to counterfeit is much easier than securing a credit card commerce system or an electronic funds transfer system. As helpers mediate more and more interactions, the complexity of the systems supporting those interactions increases dramatically. And so do the avenues of attack. It's like the difference between a manual missile launcher and a remote satellite missile launcher. In the former case, the only way to attack the system is to attack the people doing it. In the latter case, you could also attack the satellite system.

In addition, helpers obscure the security-related properties of a system from the users and make it even harder for them to assess risk accurately. It's easy to understand security systems surrounding cash sitting in a wallet. In contrast, the security systems required to support a check-writing system and, even more so, a credit card or a debit card system are much less obvious. By hiding the details, helpers make a system easier to operate at a basic level, but much harder to understand at the more detailed level necessary to evaluate how its security works.

This problem is larger than security. Designers put a good deal of effort, for example, into coming up with familiar metaphors for technological systems. We're all familiar with the trash can icon on computer desktops, e-mail symbols that show sealed envelopes flying out of your out box when you press the "send now" key, and online photo albums. New computer users are regularly surprised when they throw a file away into the trash can and it later turns out to be still in existence on their hard drives, or when they learn that an e-mail message is really much more like a postcard than it is like a sealed letter. Science fiction author Neal Stephenson calls this phenomenon "metaphor shear"—the technological complexities or unpalatable facts that we hide or smooth over in the interests of functionality or easy understanding.

Technological advances bring with them standardization, which also adds to security vulnerabilities, because they make it possible for attackers to carry out *class breaks*: attacks that can break every instance of some feature in a security system.

For decades, phone companies have been fighting against class breaks. In the 1970s, for example, some people discovered that they could emulate a telephone operator's console with a 2600-Hz tone, enabling them to make telephone calls for free, from any telephone.

In the mid-1980s, someone discovered that a Mexican one-peso coin (then worth about half a cent) was indistinguishable from the $1.50 token used for toll machines on the Triborough Bridge and Tunnel Authority's bridges in New York. As word spread, use of the Mexican coins became more common, eventually peaking at 7,300 coins in one month. Around the same time, another clever attacker discovered that Connecticut Turnpike tokens ($0.175 each—they were sold in packages of ten) would work in New York City subway turnstiles ($0.90); the same company made both tokens. (Coin sensors to detect the precise composition of the metal as well as the size and weight, an effective countermeasure, are not cost-effective for such a low-value application. They are more likely to be found on high-value applications like slot machines, where an attacker can win thousands of dollars.) Class breaks can be even worse in the computer world, in which everyone uses the same operating systems, the same network protocols, and the same applications. Attacks are regularly discovered that affect a significant percentage of computers on the Internet.

Picking the lock to a garage requires the attacker to do some work. He has to pick the lock manually. Skill acquired picking other garage locks will certainly make the job easier, but each garage he attacks is still a new job. With the invention of automatic garage-door openers, another type of attack became possible. A smart attacker can now build a single garage-opening device that allows him to effortlessly open all the garages in the neighborhood. (Basically, the device cycles through every possible code until the garage door responds.)

Class breaks mean that you can be vulnerable simply because your systems are the same as everyone else's. And once attackers discover a class break, they'll exploit it again and again until the manufacturer fixes the problem (or until technology advances in favor of the defender again).

Automation also exacerbates security vulnerabilities in technological systems. A mainstay of technology, automation has transformed manufacturing (from automobiles to zippers), services, marketing, and just about every aspect of our lives. Automation is also a friend to attackers. Once they figure out a class break, they often can attack a greater number of systems by automating the attack. Automation makes individual attackers, once they've perfected a break, much more dangerous.

Manually counterfeiting nickels is an attack hardly worth worrying about, but add computers to the mix and things are different. Computers excel at dull and repetitive tasks. If our counterfeiters could "mint" a million electronic nickels while they slept, they would wake up $50,000 richer. Suddenly the task would have a lot more appeal. There are incidents from the 1970s of computer criminals stealing fractions of pennies from every interest-bearing account in a bank. Automation makes attacks like this both feasible and serious.

Similarly, if you had a great scam to pick someone's pocket, but it worked only once every 100,000 tries, you'd starve before you successfully robbed anyone. But if the scam works across the Internet, you can set your computer to look for the 1-in-100,000 chance. You'll probably find a couple dozen every day. This is why e-mail spam works.

During World War II, the British regularly read German military communications encrypted with the Enigma machine. Polish mathematicians first discovered flaws in the cryptography, but basically the system was broken because the British were able to automate their attack in ways the German cryptographers never conceived. The British built more than 500 specialized computers—the first computers on the planet—and ran them round-the-clock, every day, breaking message after message until the war ended.

Technology, especially computer technology, makes attacks with a marginal rate of return and a marginal probability of success profitable. Before automation, it was possible for defenders to ignore small threats. With enough technology on the attackers' side, defenders often find that the small stuff quickly gets too big to ignore.

Automation also allows class breaks to propagate quickly because less expertise is required. The first attacker is the smart one; everyone else can blindly follow his instructions. Take cable TV fraud as an example. None of the cable TV companies would care much if someone built a cable receiver in his basement and illicitly watched cable television. Building that device requires time, skill, and some money. Few people could do it. Even if someone built a few and sold them, it wouldn't have great impact.

But what if that person figured out a class break against cable television? And what if the class break required someone to press some buttons on a cable box in a certain sequence to get free cable TV? If that person published those instructions on the Internet, it could increase the number of nonpaying customers by millions and significantly affect the company's profitability.

In the 1970s, the Shah of Iran bought some intaglio currency printing presses and, with the help of the U.S. government, installed them in the Tehran mint. When Ayatollah Khomeini seized power, he realized that it was more profitable to mint $100 bills than Iranian rials. The U.S. Treasury Department calls those bills "supernotes" or "superbills," and they're almost indistinguishable from genuine notes. It got so bad that many European banks stopped accepting U.S. $100 bills. This is one of the principal reasons the U.S. redesigned its currency in the 1990s; the new anticounterfeiting measures employed on those bills are beyond the capabilities of the presses in Iran. But even though there was no way to prevent the supernotes from entering circulation, the damage was limited. The Iranian presses could print only so much money a year, which put a limit on the amount of counterfeit money that could be put into circulation. As damaging as the attack was, it did not affect monetary stability.

Now imagine a world of electronic money, money that moves through computer networks, into electronic wallets and smart cards, and into merchant terminals. A class break against a system like that would be devastating. Instead of someone working in his basement forging individual bills, or even Iranians working in factories forging stacks of bills, a forger could write a computer program that produced counterfeit electronic money and publish it on the Internet. By morning, it could be in the hands of 1,000 first-time counterfeiters; another 100,000 could have it in a week. The U.S. currency system could well collapse as everyone realized that most of the electronic money in circulation was counterfeit. This attack could do unlimited damage to the monetary system.

This kind of thing happens on computer networks all the time; it's how most of the major Internet attacks you read about happen. Smart hackers figure out how to break into the networks, and they write and promulgate automated tools. Then anyone can download the ability to break into computer networks. These users of automated tools are known as script kiddies: clueless teenagers who don't understand how the attacks work but can use them anyway. Only the first attacker needs skill; everyone else can just use the software.

Encapsulated and commoditized expertise expands the capabilities of attackers. Take a class break, automate it, and propagate the break for free, and you've got a recipe for a security disaster. And as our security systems migrate to computerized technology, these sorts of attacks become more feasible. Until recently, we were able to think about security in terms of average attackers. More and more today, we have to make sure that the most skilled attackers on the planet cannot break the system, because otherwise they can write an automatic tool that executes the class break and then distribute that tool to anyone who can use it anywhere in the world.

Another effect of modern technology is to extend reach. Two hundred years ago, the only way to affect things at a distance was with a letter, and only as quickly as the physical letter could be moved. Warfare was profoundly different: A ruler would give his general orders to do battle in a foreign land, and months would go by before he heard word of how his armies did. He couldn't telephone his general for an update.

Today he can. Not only can he instantly learn how his armies are doing, but he can also approve individual bombing targets from the comfort of his office. Airplanes, long-range missiles, remote drones, satellite reconnaissance: These give a country the ability to effect such weighty acts at a distance. Soon we'll have remote-controlled tanks and artillery pieces. Already some aspects of war are becoming an extension of a video game; pilots sitting at an American base can steer UAVs (Unmanned Aerial Vehicles) in remote war zones using a computer screen and a joystick.

Technology facilitates action at a distance, and this fact changes the nature of attacks. Criminals can commit fraud with their telephones. Hackers can break into companies using networked computers. And eventually—not today, for the most part—cyberterrorists will be able to unleash undreamed-of havoc from the safety of their home countries or anywhere else.

Action at a distance makes systems harder to secure. If you're trying to secure your house against burglary, you only have to worry about the group of burglars for whom driving to your house is worth the trouble. If you live in Brussels, it doesn't matter to you how skilled the house burglars are in Buenos Aires. They are not going to get on an airplane to rob your house. But if you run a computer network in Brussels, Argentine attackers can target your computer just as easily as they can target any other computer in the world. Suddenly the list of potential attackers has grown enormously.

The notion of action at a distance also affects prosecution, because much of our criminal justice system depends on attackers getting close to victims and attacking. If attackers are safe within their own country but attack you in your country, it's harder to have them arrested. Different countries have different laws. The student who wrote the ILOVEYOU computer virus in 2000 lived in the Philippines. He did an enormous amount of damage to networks around the world, but there was no Philippine law he could be accused of breaking. One of the primary motivations for the U.S. invading Afghanistan was that the ruling Taliban was not willing to arrest Al Qaeda members.

The international nature of the Internet will continue to be a persistent problem. Differences in laws among various countries can even lead to a high-tech form of jurisdiction shopping: An organized crime syndicate with enough money to launch a large-scale attack against a financial system would do well to find a country with poor computer crime laws, easily bribable police officers, and no extradition treaties. Even in the U.S., confusion abounds. Internet financial fraud, for example, might be investigated by the FBI, the Secret Service, the Justice Department, the Securities and Exchange Commission, the Federal Trade Commission, or—if it's international—the Customs Service. This jurisdictional confusion makes it easier for criminals to slip through the seams.

Data aggregation is another characteristic of technological systems that makes them vulnerable. The concept is old, but computers and networks allow data aggregation at an unprecedented level. Computers, by their nature, generate data in ways that manual and mechanical systems do not. In supermarkets, computerized cash registers attached to bar code product scanners generate a list of every item that every customer purchases, along with the date and time he purchased it. Early mechanical cash registers generated no persistent data at all; later models generated a simple register tape. Online bookstores generate not only a list of everything every customer buys, but everything every customer looks at and how long the customer looks at it. Modern digital cell phones can generate a record of every place the phone is while it is turned on, which presumably is also a record of where the user is, day and night.

This data generation goes hand in hand with new breakthroughs in data storage and processing; today data is easily collected, correlated, used, and abused. If you wanted to know if I used the word "unguent" in a paper copy of this book, you would have to read it cover to cover. If you had an electronic copy of this book, you could simply search for the word. This makes an enormous difference to security. It's one thing to have medical records on paper in some doctor's office, police records in a police station, and credit reports in some bank's desk; it's another thing entirely to have them in electronic form. Paper data, even if it is public, is hard to search and correlate. Computerized data can be searched easily. Networked data can be searched remotely and then collated, cross-referenced, and correlated with other databases. Automated accessibility is one of the emergent properties of the cheapness of data storage combined with the pervasiveness of networks.

Under some circumstances, using this data is illegal. People have been prosecuted for peeking at confidential police or tax files. Many countries have strong privacy laws protecting personal information. Under other circumstances, it's called data mining and is entirely legal. For example, the big credit database companies have mounds of data about nearly everyone in the U.S. This data is legally collected, collated, and sold. Credit card companies have a mind-boggling amount of information about individuals' spending habits: where they shop, where they eat, what kind of vacations they take. Grocery stores give out frequent-shopper cards, allowing them to collect data about the food-buying proclivities of individual shoppers. Other companies specialize in taking this private data and correlating it with data from other public databases.

And the data can be stolen, often en masse. Databases of thousands (in some cases millions) of credit card numbers are often stolen simultaneously. Criminals break into so-called secure databases and steal sensitive information all the time. The mere act of collecting and storing the data puts the data at risk.

Banks, airlines, catalog companies, and medical insurers are all saving personal information. Many Web sites collect and sell personal data. The costs to collect and store the data are so low that many companies just say, "Why not?" These diverse data archives are moving onto the public networks, where they're not only being bought and sold, but are vulnerable to all sorts of computer attackers. The system of data collection and use is now inexorably linked to the system of computer networks, and the vulnerabilities of the latter system now affect the security of the former.

The common theme here is leverage. Technology gives attackers leverage because they can do more in an attack. Class breaks give attackers leverage because they can exploit one vulnerability to attack every system within a class. Automation gives attackers leverage because they can exploit vulnerabilities millions of times. Technique propagation gives attackers leverage because now they can try more attacks, including ones they can't even understand. Action at a distance and aggregation also give attackers leverage because now there are many more potential targets.

Attackers can exploit the commonality of systems among different people and organizations. They can use automation to make unprofitable attacks profitable. They can parlay a minor vulnerability into a major disaster.

Leverage is why many people believe today's world is more dangerous than ever. Thomas Friedman calls this new form of attacker the "superempowered angry young man." A lone lunatic probably caused the anthrax attacks in October 2001. His attack wouldn't have been possible twenty years ago and presumably would have been more devastating if he'd waited another twenty years for biowarfare technology to improve. In military terminology, a leveraged attack by a small group is called an "asymmetric threat." Yale economist Martin Shubik has said that an important way to think about different periods in history is to chart the number of people ten determined men could kill before being stopped. His claim is that the curve didn't vary throughout much of history, but it's risen steeply in the last few decades. Leverage is one of the scariest aspects of modern technology because we can no longer count on the same constraints to limit the effectiveness of attackers.

Thinking about how to take technologies out of the hands of malicious attackers is a real conundrum precisely because many of the most effective tools in the hands of a terrorist are not, on the face of it, weapons at all. The 9/11 terrorists used airplanes as weapons, and credit cards to purchase their tickets. They trained in flight schools, communicated using e-mail and telephone, ate in restaurants, and drove cars. Broad bans on any of these technologies might have made the terrorists' mission more difficult, but it would change the lives of the rest of us in profound ways.

Technology doesn't just affect attackers and attacks, it affects all aspects of society and our lives. It affects countermeasures. Therefore, when we make a technology generally available, we are trading off all of the good uses of that technology against the few bad uses, and we are risking unintended consequences. The beneficial uses of cell phones far outweigh the harmful uses. Criminals can use anonymous e-mail to communicate, but ban it and you lose all the social benefits of anonymity. It's the same with almost every technology: Automobiles, cryptography, computers, and fertilizer are all good for society, but we have to accept that they will occasionally be misused.

This is not to say that some technologies shouldn't be hard to obtain. There are weapons we try to keep out of most people's hands, including assault rifles, grenades, antitank munitions, fighter jets, and nuclear bombs. Some chemicals and biological samples can't be purchased from companies without proper documentation. This is because we have decided that the bad uses of these technologies outweigh the good.

But since most of technology is dual-use, it gives advantages to the defenders, as well. Moonshiners soup up their cars, so the revenuers soup theirs up even more (although these days it's more likely drug smugglers with speedboats in the Caribbean). The bad guys use encrypted radios to hide; the good guys use encrypted radios to coordinate finding them.

And sometimes the imbalances from a new technology naturally favor the defenders. Designed and implemented properly, technological defenses can be cheaper, safer, more reliable, and more consistent—everything we've come to expect from technology. And technology can give the defenders leverage, too. Technology is often a swinging pendulum: in favor of attackers as new attack technologies are developed, then in favor of defenders as effective countermeasures are developed, then back in favor of attackers as even newer attack technologies are developed, and so on.

Technology can also make some attacks less effective. Paper filing systems are vulnerable to theft and fire. An electronic filing system may be vulnerable to all sorts of new computer and network attacks, but the files can be duplicated and backed up a thousand miles away in the wink of an eye, so burning down the office building won't destroy them.

Still, fast-moving technological advances generally favor attackers, leaving defenders to play catch-up. Often the reasons have nothing to do with attackers, but instead with the relentless march of technology. In the 1980s, Motorola produced a secure cell phone for the NSA. Cell phone technology continued to march forward, phones became smaller and smaller, and after a couple of years the NSA cell phone was behind the times and therefore unpopular. After we spent years making sure that telephone banking was secure enough, we had to deal with Internet banking. By the time we get that right, another form of banking will have emerged. We're forever fixing the security problems of these new technologies.

The physicist C. P. Snow once said that technology "... is a queer thing. It brings you great gifts with one hand, and it stabs you in the back with the other." He's certainly right in terms of security.

31   "Questioning Surveillance and Security"

Torin Monahan

Surveillance is nothing new. People have been conducting surveillance for thousands of years. But new technologies allow people to gather new kinds of information, in new places, on a grander scale. With these new abilities come new ethical questions. Techniques like fingerprint analysis, DNA fingerprinting, wiretapping, e-mail surveillance, x-ray machines, thermal imaging, closed-circuit television (CCTV) cameras, facial recognition computers, satellites, manned and unmanned aircraft, and even Google Earth are all used to gather information and identify risky individuals and groups. As with any technology, each of these technologies privileges certain values and tends to inhibit others. They also tend to empower certain groups and disenfranchise other groups. In this chapter, Torin Monahan explores the effects that systems of surveillance can have. He does not deny the need for surveillance, but he does believe that those who tout their importance should be able to demonstrate that they live up to their claims. He fears—like Schneier—that our infatuation with having the most advanced technology will not only lead to unfortunate side effects, but perhaps distract us from the best methods of achieving our original goals. Monahan encourages us to think carefully through the technological choices we make to ensure that we are privileging the values we want to privilege and empowering those who should be empowered.

Unfortunately, security and liberty form a zero-sum equation. The inevitable trade-off: To increase security is to decrease liberty and vice versa.
Walter Cronkite, journalist

Now we all know that in times of war and certainly in this post-9/11 world, one of the most difficult questions we face is how to balance security and liberty.
Charles E. Schumer, U.S. senator

Since the 9/11 terrorist attacks, the government is charged with protecting the rights of the individual as well as ensuring our collective safety. The antiterrorist policies the government institutes will, by necessity, be more invasive.
Lynn M. Kuzma, political scientist

Why are questions about surveillance and security always framed in terms of trade-offs? Regardless of the forum, from popular media broadcasts to political speeches to academic publications, trade-offs are taken as the starting point for any discussion.

From *Surveillance and Security: Technological Politics and Power in Everyday Life* (New York: Routledge, 2006), pp. 1–23. Reprinted with permission.

Some of the most common expressions of trade-offs are security versus liberty, security versus privacy, security versus freedom, and security versus cost. But, seemingly, once the issues are presented in these terms, the only thing left to decide is whether the public is willing to make the necessary sacrifices to bring about greater national security. Absent are discussions about the politics behind surveillance and security systems, what one means by "security," what (or who) gets left out of the conversation, and the veracity of such assumptions about trade-offs to begin with. Occasionally, more astute critics will ask about the efficacy of surveillance systems in bringing about greater national security. The question is usually along the lines of "Do they work?"—meaning, are surveillance systems efficacious at preventing crime or terrorism? Although important, this type of question is really just an extension of the logic of trade-offs proffered in the opening quotes, because the implication is that if systems are not sufficiently effective, then they are not worth the sacrifice or investment.

[T]hese are the wrong questions because they obscure the real changes underway and issues at play with the incorporation of surveillance technologies into public life. The questions, in other words, function as a rhetorical smoke screen, hiding deeper motivations and logics behind surveillance and security. Some of the obvious issues not discussed when talking about trade-offs are how surveillance contributes to spatial segregation and social inequality, how private high-tech industries are benefiting from the public revenue generated for these systems, and what the ramifications are of quantifying "security" (e.g., by the number of video cameras) for political purposes.

This chapter...aims to dispel some of the smoke concealing deeper issues about surveillance and security. It starts, for the sake of fairness, by taking the wrong questions seriously, with a specific focus on the question of how efficacious surveillance systems are at bringing about greater security. Next, it proposes and discusses some of the questions that I see as being the right ones: why do we believe in trade-offs, what social relations are produced by surveillance systems, and how can surveillance be used to increase security without sacrificing civil liberties, if at all? In raising alternative questions of this sort, my goal is not to provide definitive answers but instead to open up the field of inquiry and to move beyond the fog surrounding current debates over these critically important topics.

Taking the Wrong Questions Seriously

On February 12, 1993, two ten-year-old schoolboys kidnapped and murdered two-year-old Jamie Bulger in Merseyside, United Kingdom. Closed-circuit television (CCTV) footage showed Bulger being led by the hand out of a shopping center unbeknownst to his distracted mother. The boys proceeded to take him on a two-and-a-half mile walk, periodically beating him and taunting him along the way. When confronted by several concerned bystanders, the boys claimed that Jamie was their younger brother and that they were looking out for him, and no one intervened. When they reached a secluded railway line, the boys threw paint in Jamie's face and then beat him with stones, bricks, and an iron bar. Finally, he was laid across the railroad tracks with stones stacked on his head and was later run over by a train (Wikipedia 2004). The assailants could not be identified in the grainy video footage from the shopping center, but friends later turned them in. Nevertheless, the media played the tape countless times to a shocked public, and this had the effect of galvanizing tremendous support for public video surveillance in the United Kingdom (Rosen 2001).

Now, more than ten years after the Jamie Bulger killing, Great Britain boasts the most extensive system of public surveillance in the world, with more than four million cameras throughout the United Kingdom (Rice-Oxley 2004) and more than half a million in London alone (Norris 2004).[1] With the equivalent of one camera for every fourteen people, it is estimated that the average person in a large city like London is filmed three hundred times a day (Coaffee 2004). Yet in spite of this proliferation of video surveillance, surprisingly little evaluative research has been conducted on the effectiveness of surveillance in preventing crime, and the independent research that has been done is largely inconclusive.

Two of the most cited studies about surveillance efficacy were carried out in Airdrie and Glasgow, Scotland, in the mid-1990s. The Airdrie research compared total recorded crimes from two years before and two years after 1992—the year when twelve open street CCTV cameras were installed. The research found a 21 percent drop in recorded crimes in the area, so surveillance was determined to be a "success" (Short and Ditton 1995). Nonetheless, the report raises some doubts because it did not explicitly make mention of social factors such as population changes and unemployment rates in the area, which criminologists consider to be crucially important variables in explaining crime rates (Reiman 2000; LaFree 1998; Collins and Weatherburn 1995). The issue of geographical displacement of crime from one area to another is also problematic in this study, even though the authors claim otherwise:

[Adjacent] areas recorded slight increases in total crimes and offenses in the 2 years following the installation of CCTV. This increase is almost entirely accounted for by the growth in crimes relating to the possession or supply of drugs and to offences committed whilst on bail. Displacement would be suggested if these crimes declined in the CCTV area. However this was not the case. (Short and Ditton 1995: 3)

The interpretation here is that even though crimes did increase in surrounding areas, these were "natural" occurrences and therefore should not be attributed to displacement. In other words, drug offenses or offenses perpetrated while on bail do not count as crimes unless they are occurring (or declining) in CCTV areas. Because these crimes do not seem to fit the researchers' model of displacement, they are discounted.[2] Still, this can be considered a qualified success for surveillance.

The Glasgow research compared recorded crime offenses from two years before and one year after the installation of thirty-two open street CCTV cameras in 1994. In addition to looking at crime occurrences, this study also measured public perceptions of the system and observed camera monitoring by security personnel in a control room.

The findings with regard to efficacy were a wash. As the report states, "The researchers suggest that the cameras were relatively successful, with some reductions in certain crime categories. Overall, however, the reductions in crime are no more significant than those in the control areas outwith [beyond] the camera locations" (Ditton et al. 1999: 1). Thus, the report continues, "CCTV cameras could not be said to have had a significant impact overall in reducing recorded crimes and offences" (Ditton et al. 1999: 2). The explanation provided for this lack of success is that people were generally unaware of the cameras, and without awareness there is no deterrence.

More recent research does nothing to clear up this muddy water about video surveillance efficacy. The *Christian Science Monitor* reports that after ten years of CCTV projects in the United Kingdom at a publicly funded cost of £250 million ($460 million)[3] that

research has yet to support the case for CCTV. A government review 18 months ago [in 2002] found that security cameras were effective in tackling vehicle crime but had limited effect on other crimes. Improved streetlighting recorded better results. (Rice-Oxley 2004: 1–2)

In a government review, which was mandated by the Home Office (the U.K. department in charge of public security) to see what general conclusions could be drawn from existing research, only twenty-four studies were found to be methodologically sound, and the overall outcome was that "CCTV appears to have no effect on violent crimes, a significant effect on vehicle crimes and it is most effective when used in car parks" (Armitage 2002: 5).

On the whole, what these studies from the United Kingdom indicate is that as gruesome as the Jamie Bulger murder was, it would not have been prevented with a more comprehensive system of video surveillance. Indeed, most crimes—violent or otherwise—are not prevented by surveillance. One bright spot within the evaluation literature on video surveillance is that it does appear to enable apprehending and convicting criminals after the fact (Gill 2004). But if the criterion for a worthwhile trade-off (of civil liberties, of privacy, of cost, etc.) is *prevention* of crime, then one must respond negatively to the question "Is it worth it?"

Oddly enough, given the astronomical crime rates in the United States, relatively speaking, one is hard pressed to find *any* independent evaluations of video surveillance in that country. There are several reasons for this. First, unlike many CCTV schemes in the United Kingdom, video surveillance in the United States is largely implemented in an ad hoc way by private companies rather than through public funds or with public oversight. This makes it difficult to even locate where the operational cameras are, let alone evaluate their effectiveness in some controlled way.[4] Second, the most obvious governmental agency for evaluating surveillance—the federal Office of Technology Assessment—was dissolved in 1995 because, as some say, they too often produced reports that suggested politically unattractive regulation of private industries (Coates 1995).[5]

Third, in the United States, publicly funded video surveillance is most often used for generating revenue from traffic violations, such as running red lights, or it is trained on the urban poor on streets, on public transit, or in schools (Nieto, Johnston-Dodds, and Simmons 2002; Monahan, 2006b). Because of the stigma attached to poor minorities in the United States and the public's perception of surveillance systems as crime deterrents, it is highly unlikely that the general public would demand evaluation and oversight of surveillance, especially when those "public" systems are seldom focused on the more affluent.[6] Finally, for reasons that are explored in the next section, evaluations of technological systems, generally speaking, are simply not funded. Thus, of the more than 200 U.S. police agencies that employ CCTV systems, 96 percent conduct *no evaluation* of their effectiveness (Nieto, Johnston-Dodds, and Simmons 2002: 13).

One of the most well-known studies of video surveillance efficacy in the United States was conducted in low-income public housing in the late 1970s (Musheno, Levine, and Palumbo 1978). The researchers found that the use of video surveillance in New York City's public housing did not reduce crime or fear of it, even though CCTV's implementation came at great public cost of an estimated $10,000 per apartment (in three public buildings). The reasons for this "failure," the authors explain, stemmed from a conceptual deficiency as much as from technical limitations. The design strategy in public housing was predicated on the concept of "defensible space" (O. Newman 1972), implying that the agents of crime existed outside of the immediate community and that close collaboration between community members and police officers would keep deviants out. In fact, crime emerged from within the community, poor relations between residents and police prevented community members from contacting the police, vandals routinely disabled the surveillance equipment, and residents chose not to watch the video feeds, which were routed through their television sets.

There is more recent evidence to suggest that criminals are appropriating video surveillance systems that were originally intended to thwart them.[7] In the Frederick Douglas Towers, a public housing complex for seniors in Buffalo, New York, drug dealers established a crack cocaine operation using existing CCTV systems to monitor customers and keep a lookout for police. According to one law enforcement official, "The dealers were using all the security features of the senior apartments at Douglas to their advantage ... to screen who was coming up to the apartment and buzzing people inside the building" (Herbeck 2004). In another case in Virginia, four teenagers were "arrested on charges of operating a largescale, well-organized crime ring that used surveillance, two-way radios, lookouts and disguises to stage at least 17 commercial burglaries over a 14-month period" (Branigin 2003). As an added twist to this story, the teenagers established their base of operations within a private, fortified, gated community with its own police force (Aquia Harbour 2004). When surveillance technologies originally intended to prevent crime are employed to facilitate crime or protect criminals, it lends a whole different meaning to the question of "Do they work?"

On the subject of traffic violations, cities with red-light surveillance programs do report a significant reduction in red-light runners at those intersections. A Washington,

D.C., program reported a 63 percent decrease in red-light runners; Oxnard, California, reported a 42 percent decrease; and Fairfax, Virginia, reported a 40 percent decrease (Nieto, Johnston-Dodds, and Simmons 2002: 20). So, at least for this type of traffic crime, there has been demonstrated effectiveness. This conclusion is somewhat complicated, however, by the potential for increased rear-end collisions when people brake abruptly to avoid fines (Nieto, Johnston-Dodds, and Simmons 2002: 21).[8]

The history of eschewing publicly funded surveillance and security systems in the United States is shifting rapidly in the wake of the 9/11 attacks. Instead of being conceived of as deterrents to ordinary crimes, these systems are now being embraced by policy makers as counterterrorism and intelligence-gathering tools (Lyon 2003a). Perhaps the hottest area of development, along these lines, is in biometrics, meaning the range of technologies designed to measure and classify unique human attributes. Biometrics can include fingerprinting systems, face-recognition technologies, hand-geometry scanning, iris and/or retinal scans, odor identification, thermal face print scans, voice recognition, and so on (Woodward, Orlans, and Higgins 2003). These technologies are varied and complex and present many sociotechnical obstacles for "successful" use (contingent on the social context, the goals of the system designers and users, the interoperability of systems, etc.). The professional biometrics community, for instance, actively debates the appropriateness of some systems versus others (e.g., whether identifiers should be stored in a general database or within portable documents), and they frequently criticize each other for trying to push proprietary biometric "solutions" from which individual companies stand to benefit enormously should their technologies become industry standards.[9] In this respect, knowledge of these technologies is carefully regulated by a professional group, much like with the construction of "facts" in other scientific fields (Latour 1987; D. Hess 1997; M. Fortun and Bernstein 1998). The primary policy goal in the United States is to integrate unique biometric markers into identification documents, such as passports or national ID cards, and then harmonize these identity tokens with massive databases designed to screen for potential terrorists or to monitor the movements and activities of people more broadly. It is worthwhile noting that U.S. security agencies and industries were already moving toward the widespread application of biometric and other surveillance systems prior to 9/11. The attacks, however, provided the impetus for rapidly deploying the systems with as little public scrutiny or debate as possible (Lyon 2003b; Winner 2004).

But do biometrics work for the purpose of locating and stopping terrorists? According to the U.S. General Accounting Office,[10] although "the desired benefit is the prevention of the entry of travelers who are inadmissible to the United States" (Kingsbury 2003: 6), or "keeping the bad guys out" in President George W. Bush's parlance, the challenges to the success of biometric systems are manifold. Obstacles include labor increases, travel delays, tourism reduction, inadequate training, grandfathering arrangements, reciprocal requirements from other countries, exemptions, false IDs, "significant" costs, and circumvention of border systems by more than 350,000 illegal entries a year (U.S. Citizenship and Immigration Services 2002). In addition, more technical obstacles include managing a massive database of up to 240 million records and maintaining accurate "watch lists" for suspected terrorists.

A recent report by Privacy International is forceful in its denunciation of biometrics and national identity cards. The report argues that because no evidence exists that these systems can or do prevent terrorism, any link between these systems and antiterrorism is merely rhetorical:

> Of the 25 countries that have been most adversely affected by terrorism since 1986, eighty per cent have national identity cards, one third of which incorporate biometrics. This research was unable to uncover any instance where the presence of an identity card system in those countries was seen as a significant deterrent to terrorist activity. Almost two thirds of known terrorists operate under their true identity ... It is possible that the existence of a high integrity identity card would provide a measure of improved legitimacy for these people. (Privacy International 2004: 2)

Thus, not only might biometric systems fail to perform their intended functions, they might have the opposite effect of deflecting inquiry away from terrorists who possess *valid* high-tech biometric IDs. This point should give policy makers pause, because all of the 9/11 attackers entered the United States legally with the requisite visas (Seghetti 2002). Finally, even with completely operational biometric and national ID systems in place, there are numerous ways to circumvent them, for instance, by pretending to be an "outlier" (or a person unable to provide accurate biometric data), acquiring a false identity, escaping watch lists (by providing false information or by virtue of being a "new recruit"), or spoofing identity (for instance, by using custom-made contact lenses to fool iris scanners) (Privacy International 2004: 7–8). Regardless of the cost or complexity of implementing and harmonizing biometric systems across countries, it is clear that they can never be foolproof, and it is questionable whether they would even diminish threats.

This section has sought to take seriously some of the questions about surveillance and security, as they are typically mobilized. Although the technologies discussed are clearly varied, complex, and contextually dependent, the purpose has been to probe the common underlying assumption of effectiveness that undergirds their deployment. *Efficacy operates, in a sense, as a prerequisite for any determination of whether trade-offs are worth it.* Concerning crime, evaluative studies of video surveillance indicate some success with car burglaries or traffic-related crimes but little or no success with the prevention of other crimes. The general inadequacy of surveillance for stopping violent crime has been acknowledged for some time and is usually attributed to the spontaneous nature of these crimes, which are often called "crimes of passion." One unanticipated consequence of CCTV, then, is that it may provide people with a false sense of security whereby they expose themselves to increased risks. With regard to terrorism, new biometric systems appear even more ill conceived: the technical and social difficulties are seemingly insurmountable, borders are porous (if incredibly dangerous for illegal immigrants), and costs are significant. Most important, when terrorists can and have entered countries like the United States and United Kingdom legally (or when they are already legal citizens or residents), then complex systems of documentation may do little to prevent legal entry in the future.

If we are to take the question "Do they work?" on its own terms, we are led to other questions: Why are there so few evaluative studies? And why are more independent evaluative studies not funded? One possible answer is that most people do not really want to know if surveillance and security systems work; people are afraid to hear that they might not work or that they are as (or more) vulnerable with them as without them. Although this may be true, it is perhaps too individualistic a response, which neglects the political and institutional forces at work. Another answer... is that *surveillance and security are important components of emerging neoliberal sensibilities and structures.* Contracts for surveillance systems are enormously lucrative for private industries, the likes of which influence local and national security policies. There are also overtly political reasons for the lack of evaluation studies. For example, in January 2004, the U.S. Department of Homeland Security disbanded an independent task force charged with evaluating security systems at U.S. points of entry. This move baffled some lawmakers, because the task force had "a lengthy research agenda, dedicated staff and budget to carry its work through 2004" (Strohm 2004). It seems that the fatal move of this group was to recommend an independent evaluation of the "U.S. Visitor and Immigrant Status Indicator Technology [US-VISIT] program, a biometric entry–exit system for the nation's borders" (Strohm 2004). By dissolving the task force, the Department of Homeland Security was able to postpone any conversation of US-VISIT's inadequacies and thereby avoid the need to justify the agency's (and the administration's) commitment to a flawed system.

Another related explanation for (inter)national commitment to systems with no demonstrable efficacy at preventing crime or terrorism could be strong cultural desires for retaliatory criminal justice, for catching and punishing criminals after the fact. Even if violent crimes like the murder of Jamie Bulger cannot be prevented, surveillance technologies nourish retributive impulses in societies by supporting judicial mechanisms of payback. Thus, punitive tendencies gain strength when the public, the media, politicians, and academics continue to ask questions that presume the effectiveness of technologies for meeting intended purposes but ignore unintended social changes. Surveillance and security systems may, of course, serve a largely symbolic function. If publics perceive enhanced safety, then this may ensure social order and renew faith in policy makers. Unfortunately, such widespread awareness of and subjection to invasive surveillance may actually increase public fears and aggravate existing social and economic vulnerabilities....

The belief in trade-offs is contingent on efficacy, so questions about efficacy can potentially undermine the dominant political discourse about what we are willing to give up to achieve security. This, in turn, would require a more nuanced political debate about security. Efficacy questions can also challenge widespread faith in technological progress by implying that real answers to threats of crime or terrorism will involve complex social arrangements that defy quick technological fixes. However, as the next section takes up, even if the answer was "Yes, they do work for their intended purposes," questions about efficacy and trade-offs are dangerously reductive to begin with.[11]

Asking the Right Questions

The main problems with questions about trade-offs or efficacy are that root causes for crime or terrorism are not engaged and that deeper social changes brought about by surveillance and security systems are left uninterrogated. One need not embrace technological determinism—or the simplistic belief that technology drives social change of its own accord without any human agency or intervention—to recognize the profound effects that security regimes have on social life. Surveillance and security systems are simultaneously social and technical, and in some ways this is not a new phenomenon: even before the automation of surveillance, modern bureaucracies and architectures functioned as pervasive technical systems of social control (Weber 2000; Foucault 1977). Technologies are neither separate from society nor are they neutral tools that can be applied discretely to social problems (e.g., crime or terrorism). Instead, technologies are thoroughly social inventions to begin with and are part of the social problems they are intended to correct (Winner 1977). As sociotechnical systems, then, surveillance and security are intimately intertwined with institutions, ideologies, and a long history of social inequality (Lyon 2001; Gandy 1993). From this standpoint, one can begin to ask the kinds of questions worth asking and answering—questions about power.

Why Do We Believe in Trade-offs?

A simple answer to the question of why we believe in trade-offs is that, generally speaking, most people—academics included—think badly about technology. Popular opinion perceives technologies as somehow separate from society; they are neutral, efficient, accurate, and discrete tools used to achieve rational and intentional ends. When technologies fail, people blame "human error" or insufficiently evolved social institutions. And when technologies create more problems, sometimes disastrous ones, they are labeled as "side effects" or "unintended consequences" rather than addressed as problems inherent in the design of technologies themselves (Winner 1986).

Take the following argument as an example of how narrow conceptions of surveillance technologies promulgate the logic of trade-offs. In *The Costs of Privacy*, Steven Nock (1993) claims that surveillance arises out of necessity in modern societies, as a way to simulate traditional monitoring by people and to regulate social norms in a society now based on anonymity. Nock writes,

As traditional methods of family supervision decline, institutional methods of surveillance arise that serve the same social control functions ... New methods of information-gathering and dissemination by employers, creditors, and governments that strike many as worrisome, are not necessarily violations of privacy ... Almost all [of these developments] depend on *voluntary self-disclosure* (the completion of credit, insurance or drivers license, or employment forms, for example) ... It is certainly legitimate to be concerned about the elaboration of computerized methods of monitoring and tracking people. The use of those techniques, however, is governed by widespread standards of propriety and personal autonomy. (Nock 1993: 4, 13–14; italics added)

In Nock's formulation, surveillance technologies simply automate social control functions that existed previously, without any other meaningful changes in social relations. Moreover, as rational actors, each of us has evaluated the options and voluntarily chosen to participate in new surveillance regimes, seemingly without any coercion or without any sanctions if we had (somehow) chosen to opt out instead.

This view of surveillance technologies lends itself to a discussion of trade-offs because it implies that individuals have total control and intentionality with technology use. It perceives all people as equal rational actors, without any power asymmetries, and intimates that social relations or spaces cannot be altered unintentionally. Technological fixes, from this perspective, are natural social progressions, but—at the same time—technologies somehow operate outside of society, as tools that can be applied to social problems (Weinberg 2003). All that is left to do is for societies to collectively weigh the options and choose intelligently.

What is left out of this view of surveillance? Mainly, all the ways that technological systems produce social relations or have the capacity for such production.[12] The pure view of technology articulated by Nock ignores—is bound to ignore—ways that technologies operate not only as tools but as creators of social worlds.[13] For instance, much like architecture, surveillance "programs" spaces for particular, acceptable activities so that nonsanctioned uses of space are discouraged by the environment. So, schools are for learning, malls are for shopping, streets are for driving, and so on. Provided that one adheres to the official program of a space, he or she will encounter little resistance, but should one try to appropriate a space for other uses, such as socializing, sleeping, or protesting, surveillance systems will be employed to discipline those activities. Thus, surveillance on college campuses is intended to protect property and provide public safety, but security personnel freely admit that they also monitor and record public protests and rallies, just to keep people in line (Brat 2004).

Surveillance technologies clearly alter social behavior and are intended to do so, usually as planned deterrents to deviant behavior *but not always with the outcomes intended*. They act as forms of social engineering that legislate norms for acceptable and unacceptable behaviors and actions, and they accomplish this task by individualizing people. As Jason Patton (2000) explains, when people cannot adjust their behavior to the reactions they perceive in others (i.e., physically removed observers), the social context becomes an ambiguous one where everyone is presumed to be individually deviant until proved otherwise. The result is a "panoptic" effect on social behavior (Foucault 1977), meaning that people tend to police themselves and refrain from any actions that might verify their presumed status as deviants in the eyes of unseen others. Rather than surveillance indicating a rationalized and distributed imposition on individual privacy,[14] however, surveillance is often applied selectively and with varying intensities according to one's social address (Phillips and Curry 2003); as such, surveillance can—and does—structure unequal power relations in societies (Cameron 2004; Van der Ploeg 2005; Kupchik and Monahan 2006).

Hille Koskela (2000), writing about video surveillance in Finland, adds to these observations a strong feminist critique. She finds that public surveillance does not deter violent crime against women, but the use of cameras does tend to objectify women, sterilize actions, and thereby masculinize space. The emphasis on visual surveillance is completely gendered, with women more often than not subjected to the disembodied gaze of men who operate the cameras that are concentrated in public spheres frequented by women (e.g., shopping malls, public transportation). Furthermore, even while under the presumably paternalistic eye of security cameras, any nonvisual harassment of women remains undocumented and uncorrected—from the official viewpoint, then, verbal abuse or threats never happen. The masculinization of space, which makes women the objects of surveillance, may be completely unintentional but is nevertheless a real production of social relations brought about by surveillance.

We can believe in trade-offs so long as we pretend that the only affective powers technologies have on social spaces, relations, or meanings are rationally chosen and intended. Thus, surveillance advocates can say, "A camera is just like having another officer on the beat" (Conde 2004: 1) or "There is no theoretical difference between surveillance through a camera lens and a naked eye" (Conde 2004: 2). And these conclusions are believable to the extent that any unintended social effects of the kinds described previously are discounted as side effects and to the extent that data are analyzed from afar without delving into the messy materialities of how surveillance systems work. Whereas side effects are seen as *unintended* consequences of surveillance systems, trade-offs are presented as *anticipated* undesirable outcomes, such as the loss of privacy or civil liberties. Contrary to this position, ethnographic studies of the coordination of CCTV security forces and the police in the United Kingdom reveal labor intensification rather than reduction for police personnel who must now respond to additional disturbances witnessed by camera operators (Goold 2004). Another compelling study finds antagonism caused by competing forms of expertise, such that CCTV operators tell the police to mind their own business, try to take credit for arrests, and sometimes come to blows—quite literally—fighting over jurisdiction (McCahill and Norris 2003). These observations reveal one dimension of how surveillance systems are thoroughly social and could never be just like having more police on the street.

Thinking badly about technology is only one answer for why people believe in trade-offs between what are seen as two goods, such as security and liberty. A perhaps more deep-rooted reason has to do with Western systems of logic predicated on dualities: good–bad, black–white, friend–enemy, and so on. This ingrained way of looking at the world explains the rhetorical power of statements such as President Bush's "Either you're for us, or you're against us" (G. W. Bush 2001), and it also explains the social value attributed to clarity and rationality. It is unfortunate that dualistic thinking also instills a profound intolerance for ambiguity and for the necessary messiness that characterizes social worlds (Derrida 1988). Social perceptions of technology are certainly not immune to dualistic logics, which are usually articulated as being "for" technological progress or being "anti-technology," with no middle ground in between. But there are many ways to measure progress (e.g., social, economic, environmental, emotional) and many possibilities for the design and incorporation of surveillance technologies into social spaces and public institutions.

What Social Relations Do Surveillance and Security Systems Produce?
The question of what social relations are produced through the incorporation of surveillance into daily life directs inquiry toward a rich set of data, far less constrained than questions about trade-offs or efficacy. A different way of phrasing the question might be, "What effects do surveillance and security systems have on power, inequality, or democracy?" This question is intended to be not an argument for causality or determinism but, instead, following Foucault's lead, a recognition of the capacity of power to manifest in quotidian institutional operations that simultaneously generate and sustain social relations apart from any property of control that might be possessed by individuals (Foucault 1977, 1980). Clearly, surveillance is part of larger trends toward sociospatial segregation in modern societies (Caldeira 2000; Low 2003), but the social relations produced by these technologies may be difficult to spot when looking at high-tech systems (such as biometrics or video surveillance) alone. Instead, *by attending to the embedding of surveillance technologies into existing institutional systems and social practices, power relations are much easier to detect.*

Consider the following superb example of asking some of the right questions about everyday surveillance. Virginia Eubanks (2004) writes about a small urban city in upstate New York where welfare and food stamp recipients have had their lives dramatically altered by the introduction of "electronic benefit transfer" (EBT) systems. Mandated of all states by the Welfare Reform Act of 1996, these systems signify an effort to crack down on food stamp fraud and, ostensibly, to reduce the stigma attached to using food stamps in public places. The EBT tracking, as a form of electronic surveillance, is intended to increase efficiency and reduce fraud, but at what social and financial cost?

Whereas current holders of EBT cards, who are more often than not women, were previously able to walk to local grocery stores to purchase food as they and their families needed it, they now must endure the added expense and inconvenience of hiring a cab or taking a bus some three miles to the nearest large-chain supermarket that accepts the magnetic-strip EBT cards. The local markets cannot afford, or choose not to implement, the systems necessary to accept the welfare cards as a method of payment. Even if the cardholders did elect to walk the additional distance, the main street that one must use to get to the large supermarket doubles as a state highway, at times without sidewalks, making the trip virtually impossible by foot, especially in winter months. This situation is certainly an impediment to "normal" living or economic assimilation, and the burdens of this card system are unduly shouldered by the poor.

EBT systems can be seen as important precursors to biometric national IDs, where the technologies are tested on the most vulnerable members of society first (Gilliom 2001). These systems can integrate biometric identifiers, as has been proposed by the General Accounting Office (1995), and they have the potential to track the movements and spending habits of individuals. Meanwhile, as public agencies and private companies slowly work out flaws in the system, they are draining much needed resources from the poor. For instance, the cards also double as mechanisms for receiving welfare benefits other than food stamps, and people are charged fees for requesting "cash back" at stores or withdrawing cash from ATMs. A *New York Times* article reports that a mother allotted $448 a month for her family to live on pays up to $2.35 for each transaction and that in 1999 the total number of fees charged to the poor per month was around $275,000 (Barstow 1999). A 2001 audit of the New York EBT system placed the surcharges at up to $700,151 per month (Feig 2001: 13). Moreover, few ATM machines accept the cards, cards often do not work across state lines, and—unlike ATM cards—no protections are offered if the cards are stolen and used by others.[15]

The EBT system serves as a case study of the complex deployment of surveillance technologies in everyday life. The question remains, What social relations are produced by it? Reinforced sociospatial segregation of and increased burden on the poor are two clear outcomes. This is seen with the ghettoizing of the poor in upstate New York: they must now endure added inconvenience and cost to purchase food from grocery stores in more affluent areas and then return to their economically segregated downtown apartments. This example also reveals one more dimension to the radically asymmetrical monitoring and tracking of the poor in the United States, whether in public schools, public transportation, public housing, or places of commerce. Finally, *this example draws attention to the vast profits that private companies stand to accrue at public expense*. As an example, with the privatization of the food stamp program, Citicorp Services, Inc., has been awarded lucrative contracts with 34 states, as well as with Guam and the Virgin Islands (Stegman, Lobenhofer, and Quinterno 2003: 14). And although the outsourcing of public services by states makes it difficult to determine total public costs, Citicorp's contract with California alone is for $250 million over seven years (Bartholow and Garcia 2002), with the potential for up to $450 million (*San Francisco Bay Guardian* 2001).

Pursuing the question of "What social relations are produced?" into the arena of privatized surveillance and security systems reveals a pattern of increased dependency and disempowerment of the poor, coupled with the state's relinquishment of its responsibility to meet the basic needs of citizens. A New York audit of Citicorp and Continental Card Services concluded that

neither contractor produced all of the contract deliverables or regularly met performance standards. As a result, the EBT system is not meeting client expectations, is not providing the level of service to its users that was anticipated, and may be resulting in clients needlessly incurring surcharge fees to access their benefits. (Feig 2001: 4)

Although purportedly saving money for the public, privatization leaves little recourse to the poor when the system imposes serious difficulties or fails. Furthermore, once states have awarded contracts, costly and protracted legal action is their only alternative if they wish to correct problems. This example illustrates the destructiveness of neoliberal ideologies as they are hardwired into institutions and technological systems. The dual outcome of such arrangements is increased profitability for private companies and increased surveillance and marginalization of the poor (Duggan 2003; Comaroff and Comaroff 2000; Giroux 2004).

This is but one example, taken in detail to show how different surveillance and security regimes could be analyzed from a perspective of social change rather than from one of trade-offs or efficacy. Inquiry into border control and biometrics would likely yield similar findings. For example, the U.S. Department of Homeland Security has awarded a 10-year contract of up to $10 billion to the private company Accenture for biometric systems at U.S. ports of entry (Lichtblau and Markoff 2004). Meanwhile, the increased militarization of the border in California and Texas has produced a funnel effect with immigrants crossing in the most dangerous parts of the desert in Arizona and dying at record rates (Cornelius 2001). The social relations produced are those of empowerment for private industries, disempowerment, dependency, and danger for poor or marginalized groups, and inflexibility for the nation-state to provide both police security and human security for the people within—and outside—its borders. Indeed, security in terms of providing for the well-being of people (i.e., "human security" or "population security") has recently been fused with and largely eclipsed by national security apparatuses and logics (Collier, Lakoff, and Rabinow 2004). Thus, "natural" disasters like those caused by Hurricane Katrina serve both as symbols of this lack of institutional "preparedness" and, strangely enough, as rationales for further neoliberal undermining of social and environmental support mechanisms.

How Can Surveillance Be Used to Increase Security without Sacrificing Civil Liberties?

If the important questions about surveillance and security revolve around the production of social relations, as I have claimed, and if trade-offs are attractive, in part, because technologies are seen as somehow divorced from society, then the challenge lies in how to govern surveillance technologies well—with an awareness of their social embeddedness and an eye toward their social ramifications. It may be that most public surveillance systems are misguided and inappropriate to begin with. Clearly, mechanisms for evaluating and contesting such systems need to be developed. Nonetheless, civil libertarians, academics, and progressively minded citizens have been able to make precious few inroads in this direction given the current political climate of "the war on terror." Democratizing surveillance practices—in addition to strategic opposition—may be a second, complementary strategy for intelligent technology design and use.

The question of how to govern surveillance technologies well does not imply seeking a *balance* between security and liberty, because this scale metaphor connotes the same either–or logic of trade-offs: an increase on one side necessarily diminishes the other. Rather, it means asking questions about how surveillance can be used to increase security without sacrificing liberties, if at all, and perhaps even to augment liberties. Jeffrey Rosen (2004) writes, as a telling example of a technical solution to this problem, about two different kinds of body screening technologies for passengers at airports. The first displays naked bodies in anatomically correct detail, including any hidden objects that people may be carrying; the second "extracts the images of concealed objects and projects them onto a sexless mannequin" (Rosen 2004: 4). Both systems, which Rosen refers to as "the naked machine" and "the blob machine," respectively, provide the same degree of security, but the blob machine is less invasive by design. This example demonstrates that there are social and technical choices to be made when it comes to surveillance and security, should we take the time to inquire.

The comparison between the "naked" and the "blob" machines is intended to illustrate both the contingency of technological systems and the need for alternatives. It may be the case that neither machine is desirable or sufficiently democratic, for even the blob machine objectifies, scrutinizes, and individualizes people while shifting power to those doing the monitoring. If democratic or liberty-safeguarding designs are not readily available, then perhaps societies should insist on them before proceeding further. Most of the time, there will not be easy answers to the question of how to ensure national security without sacrificing liberties, but until this is seen as a question worth asking, it is likely that surveillance and security systems will continue to disproportionately impose upon and discriminate against women and poor, ethnic minorities.

A starting point would be to make surveillance systems more transparent and democratic. For most people, especially in the United States, surveillance is inherently ambiguous. It is unclear where the cameras (or other information-gathering devices) are, who owns the equipment, who is watching, what the policies are for collecting and disposing of data, to what use data will be put, and what rights people have. In the United Kingdom, under the Data Protection Act of 2000, there are strict rules governing data collection and retention,[16] including the disclosure of surveillance monitoring through signage (e.g., signs telling people when they are under surveillance), but even so, it is estimated that 73 percent of CCTV cameras in London alone are in noncompliance with these rules (McCahill and Norris 2002: 21).[17] The United States is far behind in even establishing basic disclosure policies and does not appear to be interested in catching up. Transparency would mean dissolving some of the many layers of ambiguity around surveillance and recognizing that just because data can be collected and saved indefinitely does not mean that they should be or that collecting and saving data is productive for maintaining and protecting civil society. Indeed, social forgetfulness is a core value in American society, tied to its frontier history (seen in idioms such as "a clean slate," "a fresh start," "forgive and forget"), so data collection, retention, and disposal policies should be critical elements in the governance of surveillance systems (Blanchette and Johnson 2002).

It stands to reason that the best way to increase transparency is to increase public participation in the governance of surveillance. From a policy perspective, this could be done by conducting surveys or interviews about the social effects of surveillance systems (not just about public approval) and using that data to inform public policy. It could be done by requiring a public vote on all surveillance systems and policies, just like for other infrastructure-related projects, but with choices that extend beyond "yes" or "no" to provide a range of options concerning the policies for such systems. Informational pamphlets on ballot initiatives could be distributed wherein one could find evaluations of existing systems elsewhere, discussions of the pros and cons, and so forth. Or, in a much stronger vein, incentives could be provided to enroll citizens of all walks of life into the policy-making process, including participation on subcommittees, citizen review panels, and oversight committees (Sclove 1995).

Some might argue that democratic transparency and participation may work well for local contexts and for relatively mundane purposes but not for national security, where secrecy is somehow mandated. I disagree with this objection. Greater transparency is needed on the level of national security so that individuals know their rights, security agents are held accountable, and contracts with private security industries are kept in check. Given recent revelations that President Bush authorized the U.S. National Security Agency to spy on citizens illegally, the pressing need for transparency and accountability to preserve civil liberties could not be more apparent. Moreover, the call for secrecy with national security neglects (rather than cultivates) public expertise—effectively forcing the public into passive identity roles instead of those of active, democratic agents. Although the U.S. Department of Homeland Security's efforts to enroll citizens into surveillance operations are obviously misguided and problematic, especially for their authoritarian approach to "participation," members of the public are often acutely aware of security vulnerabilities but simply do not communicate them for fear of becoming targets of increased suspicion or legal retaliation. Public involvement may, in fact, help to limit violations of civil liberties, detect fraud, correct security vulnerabilities, and decrease the need for extensive surveillance systems.

Public involvement in data monitoring presents another venue for increasing transparency through participation. In combination with neighborhood-watch initiatives, the public could assist with monitoring cameras, as has been tried with reported success in public housing communities in Boston (Nieto 1997), or could get involved with "copwatch" organizations, which, if sensitive to community needs, could help protect vulnerable members of society (see Huey, Walby, and Doyle 2006). Unlike the case described earlier, where community members did not watch surveillance feeds on their television sets (Musheno, Levine, and Palumbo 1978), far better results could likely be produced by designating responsibility to specific community members (or to volunteers) in on-site control rooms or on the streets. The difference revolves around the "valence" (C. G. Bush 1997) of sociotechnical systems: watching television is a passive and removed social experience, but being directly responsible for community safety is a uniquely active experience. At the very least, security personnel doing the monitoring can remain proximate to communities, visible to and approachable by people within communities rather than located in remote "surveillance farms" far away both physically and socially from the people they observe. Naturally, an informed public debate about the merits of public surveillance should precede any community-watching scenario. Part of this should include asking questions of how to provide adequate oversight of surveillance practices, identifying—in advance—specific criteria for "successful" surveillance interventions, and specifying when and under what conditions the systems will be disabled. Absent such discussions, this recommendation could easily fold into a "snitch" or "tattling" culture, where community members spy on each other and contribute to a society of widespread suspicion, discrimination, and social control.

Unfortunately, efforts at achieving transparency and democracy are not only absent from the current surveillance landscape but being pushed further beyond the horizon, making them harder to imagine, let alone attain, with every passing moment. As the example of the EBT system for welfare recipients demonstrates, the privatization of surveillance, security, and public services delegates technical decisions to companies with profit imperatives rather than social equality agendas. The same could be said of private security forces in malls, gated communities, business improvement districts, war zones, and disaster areas. And the same could be said of vast urban surveillance systems outsourced to private companies by cities or implemented by the private sector without any public oversight or jurisdiction. Finally, the policy aftershocks of 9/11— namely, the USA PATRIOT and Homeland Security Acts—have made public surveillance at once more secretive and pervasive, so the public sector does not exactly provide a model worth emulating in this regard.

Increasing transparency and democratic participation in the governance of surveillance systems are not guaranteed mechanisms for achieving national security or human security or for preserving civil liberties, of course, but they are surely steps in the right direction. The approach advocated here, then, takes the social embeddedness and anticipated ramifications of technologies as a departure point and is therefore predisposed to notice social inequalities earlier in the process and better equipped to mitigate them (Woodhouse and Nieusma 2001; Guston and Sarewitz 2002). The key is seeing surveillance systems as political entities with the capacity to produce social relations—whether intended or not—and then asking how they can be employed to achieve democratic outcomes. From this perspective, "good" surveillance systems would be those that corrected power asymmetries and increased human security in societies. One example might be the website Scorecard.org, which collects and disseminates information about toxic releases in local neighborhoods, assigns blame for environmental contamination (when possible), and provides action items for people to get involved in monitoring industries and cleaning up their communities (K. Fortun 2004). Surveillance systems are more likely to meet the goal of power correction if they are designed for "structural flexibility" (Monahan 2005a), meaning that they are democratic, participatory, localized, and open to alteration.

Conclusion

This chapter has set out to destabilize the framing of surveillance and security in terms of trade-offs. Although conversations about trade-offs—between security and liberty, for example—may serve a strategic purpose of drawing attention to matters of importance and values worth preserving, these debates artificially constrain inquiry by offering little room to talk about deeper social changes underway with the incorporation of surveillance technologies into everyday life. These changes include the ongoing privatization of public spaces and services; increased social and spatial segregation along class, race, and gender lines; and disproportionate burdens and risks placed on marginalized groups in society. Moreover, questions about trade-offs or balances or efficacy are all predicated on an uninterrogated assumption that taking national security seriously must perforce threaten liberty or other social goods. It is worth probing the veracity of such assumptions and the reasons why they are so attractive.

I began by taking questions about trade-offs on their own terms, specifically evaluating the efficacy of surveillance systems in preventing crime or terrorism. It turns out that there are very few independent evaluative studies and that they are inconclusive at best. There is evidence to suggest that surveillance systems may deter vehicular and traffic crimes but that they do not deter violent crimes at all. In the domain of national security, there is no evidence to suggest efficacy, in spite of the great financial costs, institutional labor, and public inconvenience. In fact, surveillance and biometric systems may provide a false sense of security, thereby increasing vulnerability across the board. The absence of studies and debates about efficacy could mean that most people—or at least most policy makers and industry contractors—do not really want to know if surveillance and security systems work.

Even if surveillance and security systems were highly effective, I assert that questions about trade-offs are still misguided. Better questions worth asking include the following: Why do we believe in trade-offs? What social relations are produced by surveillance systems? How can surveillance be used to increase security without sacrificing civil liberties? Tentative answers might be that relations of inequality are produced, that technologies are not seen as the social and political agents that they are, and that transparent policies and democratic governance of surveillance would help amend the situation. My purpose has been not to present these alternative questions as the only ones worth asking or to answer them definitively but instead to open up the conversation, moving beyond trade-offs to a fuller consideration of the role of surveillance in society.

Acknowledgments

Special thanks to Gary T. Marx and Michael Musheno for their generous comments on this chapter.

Notes

1. This is not meant to imply direct causality between the Bulger killing and the rise of CCTV systems in the United Kingdom, because certainly other factors such as fear of ter-

rorists contribute to this trend. That said, immediately following the Bulger murder, "John Major's Conservative government decided to devote more than three-quarters of its crime-prevention budget to encourage local authorities to install CCTV" (Rosen 2001).

2. Other scholars have criticized the Airdrie study for similar reasons: that crime did rise in peripheral areas and even increased in the district by 20 percent (S. Graham 1998; Dawson 1994; Davies 1995).

3. Other reports calculating public and private expenditures on CCTV put the figure at anywhere from £225 million to £450 million being spent *per year* in the United Kingdom (Nieto, Johnston-Dodds, and Simmons 2002: 9).

4. In fact, a few activist countersurveillance groups have emerged to respond to this lack of knowledge and oversight with cameras monitoring public spaces (Monahan 2006a; Institute for Applied Autonomy 2004; New York Surveillance Camera Players 2002).

5. It is more likely that the Office of Technology Assessment (OTA) produced balanced reports about the complexity of technologies and that policy makers were frustrated that these reports could not translate into simple or clear-cut policy recommendations (Bimber 1996; Sarewitz 1996).

6. Of course the affluent are filmed regularly in places of commerce, like shopping malls or banks, but these are almost exclusively privately owned surveillance systems deployed on private property, not public systems monitoring public space. A similar observation could be made of the monitoring of the affluent in private gated communities.

7. I would categorize these appropriations of surveillance systems as instances of *countersurveillance*: intentional, tactical uses or disruptions of surveillance technologies to correct institutional power asymmetries (Monahan 2006a). Like other appropriations of technology (Eglash et al. 2004), countersurveillance reveals the underdetermination of technology and destabilizes deterministic views of technological progress. Gary T. Marx (2003) calls such acts of resistance to dominant uses of surveillance "a tack in the shoe," exploiting ironic vulnerabilities in larger projects of total public surveillance.

8. Potential conflicts of interest also exist when cities and private companies profit handsomely from the operation of these red-light systems. As a California report relates, "In San Diego, a judge dismissed nearly 300 tickets in a class-action lawsuit, ruling that the evidence was unreliable because the system is privately run and the company is paid through a percentage of the fines" (Nieto, Johnston-Dodds, and Simmons 2002: 21).

9. For an example of one such professional community, see http://biometrics.propagation.net/forums/.

10. In 2004, the U.S. General Accounting Office was officially renamed the "Government Accountability Office." The legislation that enacted this change was the "GAO Human Capital Reform Act," which was signed into law by President Bush on July 7, 2004. Among other things, this legislation "will allow the agency [the Government Accountability Office] to break its link to the federal employee pay system and adopt compensation practices that are more closely tied to job performance and other factors" (Barr 2004). This means increased instability for government workers and signals the gradual elimination of unionized

labor in the federal government. Of course, the symbolism of the name change is crucial: it signals the embracing of neoliberal ideologies, new managerial practices, and disciplinary organizational structures. Elsewhere, I have called these trends *fragmented centralization*, indicating the simultaneous centralization of decision-making authority and decentralization of accountability for (and instability brought about by) those decisions (Monahan 2005a, 2005b).

11. Some technology critics may instead seek to question the purposes served by surveillance—or the stated intended goals of these technologies in specific contexts. This line of inquiry would be a fine starting point if surveillance policies were transparent and rationales were clear. For almost all deployments of surveillance on the public (whether by state agents or by industry agents), this is not the case. There is no enlightened, objective perspective one could achieve to parse policy goals, technologies, and social contexts. Questions of power are more complicated than that, and policy motives are often obscure, influenced by multiple ideological and professional interests.

12. A recognition of the contingent design of all technologies is also often absent from these formulations. This perspective is known as the "social construction of technology" (e.g., Bijker, Hughes, and Pinch 1987; Bijker and Law 1992) and is one way to track the complex design processes that lead to the systems that we often take for granted. Rather than being outside of society and impinging on it in some deterministic way, technologies and social practices exist in dynamic and mutually shaping relationships.

13. Staples (2000) offers a compelling case for the many ways that new forms of electronic, "postmodern" surveillance are radically different from previous, "modern" ones. Mainly, contemporary surveillance is systematic and impersonal, targets bodies more than people, is locally integrated into everyday practices, and scrutinizes and profiles everyone as potentially "deviant," in advance of any evidence or informed suspicion to that effect. Haggerty and Ericson (2000) similarly theorize the distributed, decentralized power and politics of contemporary surveillance regimes. The potential of electronic surveillance for monitoring everyone equally, however, should not imply the removal of asymmetrical power relations, discrimination, or profiling; if anything, these particularistic inequalities are perpetuated, extended, and simultaneously masked by the rhetoric of universalistic (read "objective") surveillance and security (Curry 2004).

14. Privacy is, of course, an ambiguous and hyperindividualized concept that does not account very well for encroachments on social spaces and practices absent targeted individual scrutiny, usually in "private" domains. One way to overcome the limitations of privacy as a conceptual category is to expand it beyond legal definitions to include multiple forms of information generation, access, and expression in modern societies (DeCew 1997; Phillips forthcoming). Another approach is to focus on trust relations, which hold communities and cultures together—manifested either in contestations of social power or in voluntary disclosures for the sake of intimacy or social cohesion (Bourdieu 1977; de Certeau 1984; de Certeau, Giard, and Mayol 1998).

15. In another example, in August 2001, a computer glitch incorrectly registered close to six thousand EBT transactions, double-charging many people (Shesgreen and Hollinshed 2001).

16. The data protection guidelines issued by the Organization for Economic Cooperation and Development provide a related template for regulating surveillance technologies; however, such guidelines were crafted with the primary aim of facilitating trade, not protecting privacy, so their use may be limited for thinking about the power relations engendered by new technologies (Clarke 1989).

17. Goold (2004) also cautions that police officers may require additional oversight to ensure that they do not interfere with control room operators or tamper with surveillance data—two practices that were identified in a study he carried out in the United Kingdom.

References

Aquia Harbour. 2004. Security, http://www.aquiaharbour.org/pages/security.asp (cited May 16, 2004).

Armitage, R. 2002. To CCTV or not to CCTV? London: Nacro, http://www.epic.org/privacy/surveillance/spotlight/0505/nacro02.pdf.

Barr, S. 2004. GAO gets a new name, permission to launch new compensation system. *Washington Post*, July 12, B02.

Barstow, D. 1999. A.T.M. Cards Fail to Live Up to Promises to Poor. *New York Times*, August 16, 1999, http://www.nytimes.com/library/politics/081699ny-welfare-atm.html.

Bartholow, J., and Garcia, D. 2002. An Advocate's Guide to Electronic Benefit Transfer (EBT) in California: Consumers Union, http://www.consumersunion.org/finance/ebt/ebt-rpt1.htm.

Bijker, W. E., and Law, J. (eds.) 1992. *Shaping Technology/Building Society: Studies in Sociotechnical Change*. Cambridge, MA: MIT Press.

Bijker, W. E., Hughes, T., and Pinch, T. (eds.) 1987. *The Social Construction of Technological Systems: New Directions in the Sociology and History of Technology*. Cambridge, MA: MIT Press.

Bimber, B. 1996. *The Politics of Expertise in Congress: the Rise and Fall of the Office of Technology Assessment*. Albany: State University of New York Press.

Blanchette, J. F., and Johnson, D. G. 2002. Data retention and the panoptic society: The social benefits of forgetfulness. *The Information Society* 18:33–45.

Bourdieu, P. 1997. *Outline of a Theory of Practice*. Trans. R. Nice. Cambridge, MA: Cambridge University Press.

Branigin, W. 2003. Stafford teens charged in burglary ring. *Washington Post*, July 14, 2003.

Brat, I. 2004. Always under surveillance. *The State Press (Arizona State University)*, April 28, 2004, 1, 8–9.

Bush, C. G. 1997. Women and the assessment of technology. In *Technology and the Future*, ed. A. H. Teich. New York: St. Martins Press.

Bush, G. W. 2001. President unveils back to work plan. Washington, DC: White House, http://www.whitehouse.gov/news/releases/2001/10/20011004-8.html.

Caldeira, T. P. R. 2000. *City of Walls: Crime, Segregation, and Citizenship in Sao Paulo*. Berkeley: University of California Press.

Cameron, H. 2004. CCTV and (In)dividuation. *Surveillance and Society* 2(2–3):136–144, http://www.surveillance-and-society.org/articles2(2)/individuation.pdf.

Clarke, R. 1989. The OECD data protection guidelines: A template for evaluating information privacy law and proposals for information privacy law, Canberra: Xamax Consultancy, http://www.anu.edu/au/people/Roger.Clarke/DV/PaperOECD.html (accessed December 2005).

Coaffe, J. 2004. Rings of steel, rings of concrete, and rings of confidence: Designing out terrorism in Central London pre and post September 11th. *International Journal of Urban and Regional Research* 28(1):201–211.

Coates, V. 1995. On the demise of OTA. *TA-Datenbank-Nachrichten* 4(4), http://www.itas.fzk.de/deu/TADN/TADN1295/inst.htm#inst1.

Collier, S. J., Lakoff, A., and Rabinow, P. 2004. Biosecurity: Towards an anthropology of the contemporary. *Anthropology Today* 20(5):3–7.

Collins, M. F., and Weatherburn, D. 1995. Unemployment and the dynamics of offender populations. *Journal of Quantitative Criminology* 11(3):231–245.

Comaroff, J., and Comaroff, J. L. 2000. Millennial capitalism: First thoughts on a second coming. *Public Culture* 12(2):291–343.

Conde, C. 2004. *The Long Eye of the Law: Closed Circuit Television, Crime Prevention and Civil Liberties*. St. Leonards, Australia: Centre for Independent Studies, http://www.cis.org.au/IssueAnalysis/ia48/IA48.pdf.

Cornelius, W. A. 2001. Death at the border: Efficacy and unintended consequences of US immigration control policy. *Population and Development Review* 27(4):661–85.

Curry, M. 2004. The profilers question and the treacherous traveler: Narratives of belonging in commercial aviation. *Surveillance and Society* 1(4):475–499.

Davies, S. 1995. Welcome home big brother. *Wired*, May 1995, 58–62.

Dawson, T. 1994. Framing the villain. *New Statesman and Society*, January 28, 1994.

De Certau, M. 1984. *The Practice of Everyday Life*. Trans. S. Rendall. Berkeley: University of California Press.

De Certau, M., Giard, L., and Mayol, P. 1998. *The Practice of Everyday Life*. Trans. T. J. Tomasik, ed. L. Giard. New revised and augmented ed., Vol. 2: Living and Cooking. Minneapolis: University of Minnesota Press.

DeCew, J. W. 1997. *In Pursuit of Privacy: Law, Ethics, and the Rise of Technology*. Ithaca, NY: Cornell University Press.

Derrida, J. 1988. *Limited Inc*. Trans. S. Weber and J. Mehlman. Evanston, IL: Northwestern University Press.

Ditton, J., Short, E., Phillips, S., Norris, C., and Armstrong, G. 1999. *The Effect of Closed Circuit Television Cameras on Recorded Crime Rates and Public Concern about Crime in Glasgow*. Edinburgh: Scottish Office Central Research Unit, http://www.scotcrim.u-net.com/researchc2.htm.

Duggan, L. 2003. *The Twilight of Equality? Neoliberalism, Cultural Politics, and the Attack on Democracy*. Boston: Beacon Press.

Eglash, R., Croissant, J. L., Di Chiro, G., and Fouché, R. 2004. *Appropriating Technology: Vernacular Science and Social Power*. Minneapolis: University of Minnesota Press.

Eubanks, V. 2004. Popular technology: Citizenship and inequality in the information economy. Doctoral diss., Science and Technology Studies, Rensselaer Polytechnic Institute, Troy, NY.

Eubanks, V. 2006. Technologies of citizenship: Surveillance and political learning in the welfare system. In *Surveillance and Security: Technological Politics and Power in Everyday Life*, ed. T. Monahan. New York: Routledge.

Feig, B. E. 2001. *Audit of Office of Temporary and Disability Assistance Electronic Benefit Transfer System*. New York: Office of the State Comptroller, http://www.osc.state.ny.us/audits/allaudits/093001/99s51.pdf.

Fortun, K. 2004. Environmental information systems as appropriate technology. *Design Issues* 20(3):54–65.

Fortun, M., and Bernstein, H. J. 1998. *Muddling Through: Pursuing Science and Truths in the 21st Century*. Washington, DC: Counterpoint.

Foucault, M. 1977. *Discipline and Punish: The Birth of the Prison*. New York: Vintage Books, Random House.

Foucault, M. 1980. *Power/Knowledge: Selected Interviews and Other Writings*, 1972–1977. Brighton, Sussex: Harvester Press.

Gandy, O. H. 1993. *The Panoptic Sort: A Political Economy of Personal Information*. Boulder, CO: Westview.

General Accounting Office. 1995. *Electronic Benefits Transfer: Use of Biometrics to Deter Fraud in the Nationwide EBT Program*. Washington, DC: General Accounting Office, http://www.gao.gov/archive/1995/os95020.pdf.

Gill, M. 2004. Offenders' views and CCTV. Paper read at CCTV and Social Control, January 8–9, Sheffield, United Kingdom.

Gilliom, J. 2001. *Overseers of the Poor: Surveillance, Resistance, and the Limits of Privacy*. Chicago: University of Chicago Press.

Giroux, H. A. 2004. *The Terror of Neoliberalism: Authoritarianism and the Eclipse of Democracy*. Boulder, CO: Paradigm.

Goold, B. J. 2004. *CCTV and Policing: Public Area Surveillance and Police Practices in Britain*. Oxford and New York: Oxford University Press.

Graham, S. 1998. Towards the fifth utility? On the extension and normalization of public CCTV. In *CCTV Surveillance and Social Control*, ed. C. Norris and G. Armstrong. Avebury: Gower.

Guston, D. H., and Sarewitz, D. R. 2002. Real-time technology assessment. *Technology in Society* 24:93–109.

Haggerty, K. D., and Ericson, R. V. 2000. The surveillant assemblage. *British Journal of Sociology* 51(4):605–622.

Herbeck, D. 2004. Raids target drug dealing in housing projects. *Buffalo News*, February 26, 2004.

Hess, D. 1997. *Science Studies: An Advanced Introduction*. New York: New York University Press.

Huey, L., Walby, K., and Doyle, A. 2006. Cop watching in downtown Eastside: Exploring the use of (counter)surveillance as a tool of resistance. In *Surveillance and Security: Technological Politics and Power in Everyday Life*, ed. T. Monahan. New York: Routledge.

Institute for Applied Autonomy. 2004. http://www.appliedautonomy.com/.

Kinsbury, N. 2003. *Border Security: Challenges in Implementing Border Technology*. Washington, DC: General Accounting Office, http://www.gao.gov/new.items/d03546t.pdf.

Koskela, H. 2000. "The gaze without eyes": Video-surveillance and the changing nature of urban space. *Progress in Human Geography* 24(2):243–265.

Kupchick, A., and Monahan, T. (2006). The New American School: Preparation for post-industrial discipline. *British Journal of Sociology and Education* 27(5):617–631.

Lackoff, A. 2006. Techniques of preparedness. In *Surveillance and Security: Technological Politics and Power in Everyday Life*, ed. T. Monahan. New York: Routledge.

LaFree, G. 1998. *Losing Legitimacy: Street Crime and the Decline of Social Institutions in America*. Boulder, CO: Westview.

Latour, B. 1987. *Science in Action: How to Follow Scientists and Engineers through Society*. Cambridge, MA: Harvard University Press.

Lichtblau, E., and Markoff, J. 2004. Accenture is awarded U.S. contract for borders. *New York Times*, June 2, 2004.

Low, S. M. 2003. *Behind the Gates: Security and the New American Dream*. New York: Routledge.

Lyon, D. 2001. *Surveillance and Society: Monitoring Everyday Life*. Buckingham, UK: Open University.

Lyon, D. 2003a. *Surveillance after September 11*. Malden, MA: Polity Press.

Lyon, D. 2003b. Technology vs. "Terrorism": Circuits of City Surveillance since September 11. *International Journal of Urban and Regional Research* 27(3):666–678.

Marx, G. T. 2003. A tack in the shoe: Neutralizing and resisting the new surveillance. *Journal of Social Issues* 59(2):369–390.

Marx, G. T. 2006. Soft surveillance: The growth of mandatory volunteerism in collecting personal information—"Hey Buddy Can You Spare a DNA?" In *Surveillance and Security: Technological Politics and Power in Everyday Life*, ed. T. Monahan. New York: Routledge.

McCahill, M., and Norris, C. 2002. *CCTV in London*. Hull, UK: UrbanEye Porject, http://www.urbaneye.net/results/ue_wp6.pdf.

McCahill, M., and Norris, C. 2003. *CCTV Systems in London: Their Structures and Practices*. Hull, UK: Urban Eye Project, http://www.urbaneye.net/results/ue_wp10.pdf.

Monahan, T. 2005a. *Globalization, Technological Change, and Public Education*. New York: Routledge.

Monahan, T. 2005b. The School System as a Post-Fordist Organization: Fragmented Centralization and the Emergence of IT Specialists. *Critical Sociology* 3(4):583–615.

Monahan, T. 2006a. Counter-surveillance as Political Intervention? *Social Semiotics* 16(4):515–534.

Monahan, T. 2006b. The surveillance curriculum: Risk management and social control in the Neoliberal School. In *Surveillance and Security: Technological Politics and Power in Everyday Life*, ed. T. Monahan. New York: Routledge.

Musheno, M. C., Levine, J. P., and Palumbo, D. J. 1978. Television surveillance and crime prevention: Evaluating an attempt to create defensible space in public housing. *Social Science Quarterly* 58(4):647–656.

Newman, O. 1972. *Defensible Space*. New York: Macmillan.

New York Surveillance Camera Players. 2002. How to make maps of camera locations, http://www.notbored.org/map-making.html.

Nieto, M. 1997. *Pubic Video Surveillance: Is It an Effective Crime Prevention Tool?* Sacramento: California Research Bureau, http://www.library.ca.gov/CRB/97/05/.

Nieto, M., Johnston-Dodds, J., and Wear Simmons, C. 2002. *Public and Private Applications of Video Surveillance and Biometric Technologies*. Sacramento, CA: California Research Bureau, http://www.library.ca.gov/crb/02/06/02-006.pdf.

Nock, S. L. 1993. *The Costs of Privacy: Surveillance and Reputation in America*. New York: Aldine De Gruyter.

Norris, C. 2004. Introductory Remarks. Paper read at CCTV and Social Control, Sheffield, United Kingdom, January 8–9.

Patton, J. 2000. Protecting privacy in public? Surveillance technologies and the value of public places. *Ethics and Information Technology* 2:181–187.

Phillips, D. J. Forthcoming. *Knowing Glances: Identity, Visibility, and Power in Information Environments*. Cambridge, MA: MIT Press.

Phillips, D. J., and Curry, M. 2003. Privacy and the phonetic urge: Geodemographics and the changing spatiality of local practice. In *Surveillance as Social Sorting: Privacy, Risk, and Digital Discrimination*, ed. D. Lyon. London: Routledge.

Privacy International. 2004. *Mistaken Identity: Exploring the Relationship between National Identity Cards and the Prevention of Terrorism*. London, http://www.privacyinternational.org/issues/idcard/uk/id-terrorism.pdf.

Reiman, J. 2000. *The Rich Get Richer and the Poor Get Prison: Ideology, Class, and Criminal Justice*. 7th ed. Boston: Pearson.

Rice-Oxley, M. 2004. Big Brother in Britain: Does more surveillance work? *Christian Science Monitor*, February 6, 2004. http://www.csmonitor.com/2004/0206/p07s02-woeu.htm.

Rosen, J. 2001. A cautionary tale for a new age of surveillance. *New York Times Magazine*, October 7, 2001, http://www.schizophonia.com/archives/cctv.htm.

Rosen, J. 2004. *The Naked Crowd: Reclaiming Security and Freedom in an Anxious Age*. 1st ed. New York: Random House.

San Francisco Bay Guardian. 2001. Food stamps and ATMs. June 6, 2001, http://www.sfbg.com/News/35/36/36edwelf.html.

Sarewitz, D. 1996. *Frontiers of Illusion: Science, Technology, and the Politics of Progress*. Philadelphia, PA: Temple University Press.

Sclove, R. E. 1995. *Democracy and Technology*. New York: Guilford.

Seghetti, L. M. 2002. *Immigration and Naturalization Service: Restructuring Proposals in the 107th Congress*. Washington, DC: Congressional Research Service, http://212.111.49.124/news-archive/crs/10094.pdf.

Shesgreen, D., and Hollinshed, D. 2001. Computer glitch leaves food stamp recipients in lurch. *St. Louis Post-Dispatch*, August 18, 2001.

Short, E., and Ditton, J. 1995. *Does Closed Circuit Television Prevent Crime? An Evaluation of the Use of CCTV Surveillance Cameras in Airdrie Town Centre*. Edinburgh: Scottish Office Central Research Unit, http://www.scotland.gov.uk/cru/resfinds/crf08-00.htm.

Staples, W. 2000. *Everyday Surveillance: Vigilance and Visibility in Postmodern Life*. Lanham, MD: Rowman and Littlefield.

Stegman, M. A., Lobenhofer, J. S., and Quinterno, J. 2003. *The State of Electronic Benefit Transfer (EBT)*. Chapel Hill: Center for Community Capitalism, University of North Carolina at Chapel Hill, http://www.ccc.unc.edu/documents/cc_ebt.pdf.

Strohm, C. 2004. Lawmaker questions demise of government technology task force, *GovExec.com*, August 20, 2004, http://www.govexec.com/dailyfed/0804/082004c1.htm.

U.S. Citizenship and Immigration Services. 2002. *Yearbook of Immigration Statistics*. Washington, DC: Office of Immigration Statistics, http://uscis.gov/graphics/shared/aboutus/statistics/Illegal2002.pdf.

Van der Ploeg, I. 2005. *The Machine-Readable Body: Essays on Biometrics and the Informatization of the Body*. Maastricht: Shaker.

Van der Ploeg, I. 2006. Borderline identities: The enrollment of bodies in technological reconstruction of borders. In *Surveillance and Security: Technological Politics and Power in Everyday Life*, ed. T. Monahan. New York: Routledge.

Weber, M. 2000. *The Protestant Ethic and the Spirit of Capitalism*. New York: Routledge.

Weinburg, A. 2003. Can technology replace social engineering? In *Teghnology and the Future*, ed. A. H. Teich, 9th ed. Belmont, CA: Thomson/Wadsworth. (Orig. pub 1966.)

Wikipedia. 2004. James Bulger murder case, http://en.wikipedia.org/wiki/James_Bulger_murder_case (cited May 20, 2004).

Winner, L. 1977. *Autonomous Technology: Technics-Out-of-Control as a Theme in Political Thought*. Cambridge, MA: MIT Press.

Winner, L. 1986. *The Whale and the Reactor: A Search for Limits in an Age of High Technology*. Chicago: University of Chicago Press.

Winner, L. 2004. Trust and terror: The vulnerability of complex socio-technical systems. *Science as Culture* 13(2):155–172.

Winner, L. 2006. Technology studies for terrorists: A short course. In *Surveillance and Security: Technological Politics and Power in Everyday Life*, ed. T. Monahan. New York: Routledge.

Woodhouse, E. J., and Nieusma, D. 2001. Democratic expertise: Integrating knowledge, power, and participation. In *Knowledge, Power and Participation in Environmental Policy Analysis*, ed. M. Hisschemöller, R. Hoppe, W. N. Dunn, and J. R. Ravetz (pp. 73–96). New Brunswick, NJ: Transaction.

Woodward, J. D., Orlans, M., and Higgins, P. T. 2003. *Biometrics*. New York: McGraw-Hill/Osborne.

32   Energy, Society, and Environment: Technology for a Sustainable Future

David Elliott

Decisions about technology made over the course of the past several centuries have had enormous effects on the availability of energy resources, patterns of use, and patterns of living. The energy systems on which we now depend have powerful impacts on the environment. In recent years, these effects on the environment have come sharply into focus. With an eye to the future and a broader appreciation of the implications of past decisions about energy, a good deal of concern is now being expressed about how to address the problems of American (as well as global) dependence on particular kinds of energy. One of the goals that is most commonly put forward is that we must achieve sustainability. Sustainability has been defined in numerous ways, but the basic idea is to create a system—of people, institutions, and the environment—that can be maintained indefinitely. This chapter by David Elliott points to some of the challenges that will have to be overcome and offers some tools that may help us make progress towards the goal of sustainability. Specifically he argues that we must take into account the different stakeholders and decision makers that help form sociotechnical energy systems. He encourages us to think about the goals and desires of producers, consumers, and shareholders, as well as the governments that help regulate and control the activities of these groups. All groups have an economic stake in the strategies that are adopted, but they also have a stake in the state of the environment. How we negotiate these conflicting interests will make a huge difference in whether we achieve sustainability and what type of sustainability that will be.

People and the Planet

Human beings have developed a capacity to create and use tools—or what is now called technology. Technology provides the means for modifying the natural environment for human purposes—providing basic requirements like shelter, food, warmth, as well as communications, transport and a range of consumer products and services. All these activities have some impact on the environment. The sheer scale of human technological activity puts an increasing stress on the natural environment to the extent that it cannot absorb our wastes, while our profligate lifestyles lead us to increasingly exploit the planet's limited resources.

Energy resources are an obvious example of limited resources whose use can have major impacts. Figure 32.1 shows the gigantic leap in energy use since the industrial revolution. Certainly energy use is now central to most human activities and many of our environmental problems could be described in terms of our energy-getting and

From *Energy, Society and Environment: Technology for a Sustainable Future* (New York: Routledge, 2002), pp. 3–16. Reprinted with permission.

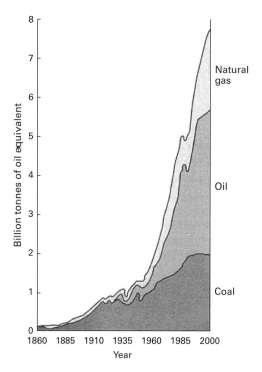

Figure 32.1
Growth in total fossil fuel consumption worldwide since the industrial revolution (in billion tonnes of oil equivalent). Source: *Physics Review* 2, no. 5 (May 1993), based on data from D. A. Lashof and D. A. Tirpak, eds., *Policy Options for Stabilizing Global Climate*, U.S. Environmental Protection Agency. Draft Report to Congress, Washington, D.C., 1989. Reproduced by permission of Philip Allan Updates. Updated from 1985 to 2000 using data from "vital statistics," Worldwatch Institute, 2002.

energy-using technologies. The most obvious environmental impacts are the physical impacts of mining for coal and drilling for oil and gas, and distributing the resultant fuels to the point of use. However, increasingly it is the use of these fuels that presents the major problems. Burning these fuels in power stations to generate electricity, or in homes to provide heat, or in car engines to provide transport, generates a range of harmful gases and other wastes, and also, inevitably, generates carbon dioxide, a gas which is thought to play a key role in the greenhouse "global warming" effect. We will be discussing global warming and climate change in detail later, but certainly there seems likely to be a link between the continuing, seemingly inexorable, rise in global carbon dioxide emissions, shown in figure 32.2, the gradual resultant rise of carbon dioxide levels in the atmosphere, shown in figure 32.3, and the continuing rise in planetary average surface temperature, shown in figure 32.4.

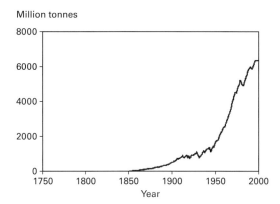

Figure 32.2
Global carbon emissions from fossil fuel combustion (in millions of tonnes of carbon). Source: Worldwatch Institute, "State of the world," © 2002, http://www.worldwatch.org.

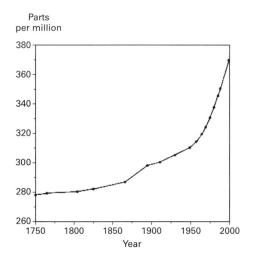

Figure 32.3
Concentrations of carbon dioxide in the atmosphere (in parts per million). Source: Worldwatch Institute, "Slowing global warming: A worldwide strategy," Worldwatch Paper 91, © 1989, updated to 2000 from data in "Reading the weathervane," Worldwatch Paper 160, 2002, http://www.worldwatch.org.

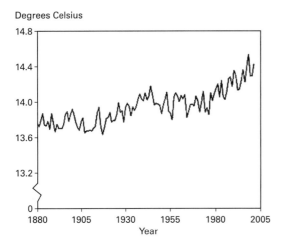

Figure 32.4
Global average temperature at the earth's surface (in degrees Celsius). Source: Worldwatch Institute, "Reading the weathervane," Worldwatch Paper 160, © 2002, http://www.worldwatch.org.

If this trend continues, the world climate could be significantly changed, leading, for example, to the melting of the ice caps, serious floods, droughts and storms, all of which could have major impacts on the ecosystem and on human life. Some of the predictions are certainly very worrying. For example, according to the UK Meteorological Office Hadley Centre "failure to act now on Climate Change could mean the Amazon rainforest is devastated; large sections of the global community go short of food and water; many heavily populated low-lying coastal areas flooded and deadly insect-borne diseases such as malaria spread across the world." However, it is not just a matter of warming, or only a problem for developing countries or tropical areas. There is the possibility that cold water flowing south due to the melting of the polar ice cap could disturb the Gulf Stream which could result in average temperatures in the UK falling by around 10°C.

Of course there remain many uncertainties over the nature and likely future rate of climate change. Nevertheless, in simple terms it seems obvious that something must change if significant amounts of carbon dioxide are released into the atmosphere. This gas was absorbed from the primeval carbon dioxide-rich atmosphere and was trapped in underground strata in the form of fossilised plant and animal life. We are now releasing it by extracting and burning fossil fuels. It took millennia to lay down these deposits, but a large proportion of these reserves may well be used up, releasing trapped carbon back into the atmosphere, within a few centuries.

In addition to major global impacts like this, there are a host of other environmental problems associated with energy extraction, production and use—acid emissions from the sulphur content of fossil fuels being just one. Air quality has become an urgent issue in many countries, given the links it has to health. The release of radioactive materials from the various stages of the nuclear fuel cycle represents an equally worrying problem, even assuming there are no accidents. Accidents can happen in all industries—and they present a further range of environmental problems, the most familiar being in relation to oil spills.

Ever since mankind started burning wood, air pollution has been a problem, but it became worse as the population increased and was more concentrated in cities, and as industrialisation, based on the use of fossil fuels, expanded. Concern over environmental pollution became particularly strong in the 1960s and 1970s, in part following some spectacular oil spills from tankers, including the *Torrey Canyon* off Cornwall in 1967 and the *Amoco Cadiz* off Brittany in 1979. However, the main concern amongst the emerging environmental movement was over longer term strategic problems: in the mid-1970s, following a series of long-range energy resource predictions (and, as we shall see later, the experience of the 1974 oil crisis), it was felt that some key energy resources might run out, or at least become scarce and expensive in the near future (Meadows *et al.* 1972).

Certainly, a substantial part of the world's fossil fuel reserves have been burnt off and a substantial part of the world's uranium reserves have been used, although there are disagreements about precisely what the reserves are, and from the 1980s resource scarcity was seen as a less urgent problem. The more important strategic question nowadays is whether what resources are left can be used safely.

Sustainability

The basic issue is one of environmental sustainability: can the planet's ecosystem survive the ever increasing levels of human technological and economic activity? The planetary ecosystem consists of a complex, dynamic, but also sometimes fragile, network of interactions, some of which can be disrupted or even irreversibly damaged by human activities.

In recent years, following the report on *Our Common Future* by the Brundtland Commission on Environment and Development in 1987, and the UN Conference on Environment and Development held in Rio de Janeiro in 1992, the term *sustainable development* has come into widespread use to reflect these concerns. The Brundtland Commission defined it in human terms, as "development that meets the needs of the present generation without compromising the ability of future generations to meet their own needs" (Brundtland 1987: 43).

In this formulation, the emphasis is mainly on material levels of resource use and on pollution, but the term "needs" is also wider, and might be thought of as also reflecting concerns about lifestyle and quality of life, as well perhaps as global inequalities and redistribution issues.

Some radical critics of current patterns of energy and resource use go beyond just the issue of environmental impacts, resource scarcity and ecosystem disruption. For some, it is not just a matter of pollution or global warming but also a matter of how human beings live. For them, as well as being environmentally unsustainable, modern industrial technology underpins an unwholesome and unethical approach to life. Technology, at least in the service of modern industrial society, leads, they say, not to social progress but to social divisions, conflicts and alienation, and underpins a rapacious, consumerist society in which materialism dominates. Some therefore call for sustainable alternatives to consumerist society (Trainer 1995). Other critics go even further and, from a radical political perspective, challenge the whole industrial project, and the basic concept of "development," which they see as, in practice, reinforcing inequalities, marginalising the poor, exploiting the weak and disadvantaging minorities, destroying whole cultures and species, and irreversibly disrupting the ecosystem, all in the name of economic growth for a few.

Even leaving extreme views like this aside, from a number of perspectives, the interaction between technology, the environment and society would seem to be a troubled one. Clearly, it is not possible, in just one book, to try to explore, much less resolve, all these issues. Nevertheless, it may be possible to get a feel for some of the key issues and factors involved.

A Model of Interactions

What follows is a very simplified model of the conflicting interests that exist in society, which may help our discussion, in that they may influence interaction between humanity and the environment. Put very simply, there would seem to be three main human "domains" that interact on this planet with each other and with the rest of the natural environment. First, there are the *producers*—those engaged in using technology to make things or provide services. Second, there are the *consumers*—those who use the products or services. Third, there are those who own, control, and make money from the process of production and consumption, chiefly these days, *shareowners*. In addition you might add a fourth, meta-group, that is *governments*, nationally and supra-nationally, who, to some extent, "hold the ring," i.e. they seek to control the activities of the other human groups, for example by developing rules, regulations and legislation.

Obviously, these are not exclusive groups: in reality people have multiple roles. Producers also consume, even if all consumers do not produce. Producers and consumers may also share in the economic benefits of the production-consumption process, e.g. as shareholders. Moreover, there will be differences and conflicts within each of these groups: not all the producers or consumers or shareholders will necessarily have the same vested interests. However, in general terms, the conflicts between and amongst the three groups or roles will probably be larger than those within each group.

In the past, there has always been a conflict between producers and "owners"—labour versus capital if you like. While the interests of "capitalists" (i.e. owners or shareholders) are to get more work for less money, battles have been fought by trade unionists to squeeze out more pay. A similar but less politically charged conflict also exists between consumers and capitalists/shareowners. Consumers want good cheap products and services and capitalist/shareowners want profits and dividends. Over the years governments have intervened to control some aspects of both these interactions—for example, to limit the health and safety risks faced by workers and to ensure that certain quality standards are maintained in terms of consumer products and services. In parallel consumers themselves have organised to protect their interests.

In some circumstances the interests of consumers and producers may also clash: consumers want good, safe, cheap products and producers want reasonable pay and job security. The "capitalists," i.e. the owners and their managerial representatives, tend to have the advantage in most situations: they can set the terms of the conflicts. Thus they may argue that pay rises will, in a competitive consumer market, lead to price increases, reduced economic performance and therefore to job losses, and in general, they can set the terms of employment and of trade. However, they can be constrained by effective trade unions or by market trends created collectively by consumers.

The Environmental Part of Interaction

So much for the human side of the model. The other element is the natural environment: the source of resources from which producers can make goods for consumers and profits for capitalists. The natural environment has no way of responding actively to the human actors, unless you subscribe to the simplified version of the Gaia hypothesis, as originally developed by James Lovelock, which suggests that the planet as a whole has an organic ability to act to protect itself; the various elements of the ecosystem act together to ensure overall ecosystem survival (Lovelock 1979). Nevertheless, even if the natural environment is passive, it represents a constraint on human activities.

Describing this situation more than a century ago, the German philosopher Karl Marx argued that there could in principle be a conflict between the human actors in the system and the natural environment, but that the constraints on resource availability, and the environmental limits on getting access to resources, were far off. He and his followers therefore devoted themselves to the other more immediate conflicts—between the human actors. Nevertheless, it was recognised that at some point, as human economic systems expanded, they would come into serious conflict with the environment.

You could say that this point has now been reached. The rate of economic growth and technological development has brought industrial society to its environmental limits. Some radical critics argue that such is the desire for continued profit by capitalists that, in the face of effective trade unions on the one hand, and tight consumer markets on the other, there has been a tendency to increase the rate of exploitation of nature, thus heightening the environmental crisis (O'Connor 1991).

Certainly, mankind has always exploited nature, just as capitalists have exploited producers and consumers, and this process does seem to have increased. However, just as the latter two human groups have fought back, so now the planet is beginning, as it were, to retaliate, by throwing up major environmental problems. Now, as already suggested, apart from putting constraints on some human activities, the natural environment cannot fight back very actively or positively: it requires the help of human actors, environmental pressure groups and governments to protect and promote what they see as its interests.

The Model Reviewed

The model is now complete: there are the three conflicting human groups (producers, consumers and investors/shareholders) locked into economic conflict; governments active nationally and globally to varying extents; and the natural environment. The environment is mainly dependent for protection on the interventions of people and governments, but perhaps the environment is also able to constrain human activities by, as it were, imposing costs on human activities if these disturb key natural processes, and, in the extreme, making human life on earth unviable.

The issues facing those involved with trying to diffuse this complex situation are many and varied, and, by focusing on economic conflicts, our "interest" model only partly reflects the political, cultural and ideological complexity of society. In reality people occupy a range of roles: they are not just consumers, producers or shareholders. Moreover, while our model may reflect some of the economic conflicts between various groups within any one country, it does not reflect the conflicts amongst nations or groups of nations. For example, there are the massive imbalances in wealth and resources amongst the various peoples around the world, and conflicts over ideas about how these imbalances and the inequalities should be dealt with. Moreover, the global nature of industrialisation means that there are global-level problems and conflicts, not least since pollution is no respecter of national sovereignty and national boundaries. These problems seem likely to be further heightened by the rapid expansion of global markets and global corporate economic power, in effect strengthening the power of the "owner/shareholder" group in the model and increasingly excluding some groups from economic and political involvement and influence. Indeed, it could be argued that we should add this excluded and disenfranchised group to our model—an "underclass" of the dispossessed poor who, around the world, operate on the margins of society, and often experience the worst of the environmental problems. The world's environmental problems cannot be addressed without considering these wider issues, and our model only partly reflects them.

However, despite these failings, the model does at least provide a framework for discussing some of the key human-environment conflicts. The model abstracts the human element for purpose of analysis, but it should be clear that human beings are not in reality separate from the natural environment. Albert Einstein once said the environment was "everything except me," but in this book the term "environment" will mean the entire planetary ecosystem and all that exists in it, including human beings. This definition, with humans as part of the environment, is important, in that too often the environment is seen as something outside of humanity—just as a context for human action. The fact that human beings are part of nature of course makes the problem of solving environmental problems even harder: can the part understand the whole? Will mankind's much vaunted intellectual capacity enable it to rise to the challenge? Or will nature impose its constraints on mankind?

Negotiating Conflicts of Interest

...It seems clear that, assuming that human beings can act usefully to protect the environment, there will be a need to find some way in which the conflicting interests of the four main "domains" in the model outlined above can be balanced. The battles among the three human elements can no longer be allowed to dominate the political and planetary scene: the fourth element, the environment, must also be considered. Indeed some of the purists amongst the environmental movement would go further: rejecting the idea that minor "pale green" adjustments and accommodations will suffice, the "deep greens" would argue, adopting a fundamentalist view, that the interests of the natural world should dominate all others, even to the extent of seriously limiting human activities. This view goes well beyond the idea that human beings have an ethical or moral responsibility for environmental stewardship, a view which is seen as patriarchal or paternalistic. "Deep ecology" writers like Devall argue that human beings should stop thinking of themselves as being the centre of creation and instead adopt an "eco-centric" viewpoint (Devall 1988).

Some of the more pessimistic "deep greens" seem to believe that the planet will never be safe until the impact of human beings on the environment has been returned to the level it was before human civilisation, or at least industrialisation, got going on a significant scale. At times it almost seems as if the ultimate "deep green" or "deep ecology" prognosis is that the planet would only be safe without a human presence, and that if mankind does not mend its ways, Gaia will arrange just that.

Social Equity

Even leaving aside such ultra-pessimistic views, on the assumption that humankind can respond in time and effectively (a rather big assumption of course), some "doom" scenarios still have some force. For example, it seems possible that some responses to environmental problems could involve major social dislocations. Those in control might feel it necessary, in order to protect their own interests, to impose socially inequitable solutions, seriously disadvantaging some specific human group or groups.

If this is to be avoided, then some way must be found to combine environmental sustainability and social equity. There will be a need for some sort of accommodation or balance that will not only ensure a more sustainable relationship between humanity and the rest of the ecosystem, but will also attempt to reduce rather than increase social and economic inequalities.

The implication would seem to be that no one human group should be expected to meet all the costs of environmental protection: the costs must be shared. For example, let us assume that consumers want greener products. Industrialists will reply that they can be made available but they will cost more. Similarly workers may press their employers for safer and cleaner production technologies for their own sake but also for the sake of the communities in which they live. They are likely to be told by industrial managers that this would add cost and could lead to lower wages or even job losses. What is missing in this formulation is the interests of shareholders. After all, you could argue that everyone should carry the burden: consumers, producers and shareholders.

In reality of course, shifting to greener products and cleaner production processes may not in the end cost more, or at least it may not be a bad move, in the longer term, commercially. As pressures for a "clean up" grow, for example through government legislation on environmental protection or consumer pressure for greener products, companies that take the initiative will have a competitive advantage compared to those that do not. Of course, if left just to market competition, that means there will be some commercial losers and overall there may well be short-term costs and dislocations in terms of employment and profits. That is precisely why a proper negotiation process is so important—to reach some sort of agreement on the distribution of costs and benefits amongst all the human stakeholders, in the wider context of overall environmental protection.

Clearly, it will not be easy to establish this sort of negotiation process, even if we stay within the confines of our model and apply it just to one country. Quite apart from the powerful vested interests of the competing groups, regulations take time to have an impact, and there are usually ways of avoiding pressures for change, at least in the short term. And once we step outside of a single country focus, the opportunities for evasion become even greater. For example, companies can move to countries where regulatory pressures are weak. So to be effective, an attempt has to be made to extend the framework of negotiation and regulation to all levels, local, national and global.

This process of negotiation of interests, at whatever level, will depend to some extent on actions taken by governments, for example by setting new environmental standards and regulations, although, equally, all the human actors in the system can also play a part, by including environmental considerations in their otherwise partisan negotiations. Despite the difficulties, there have already been attempts to do this, for example in relation to technological development choices, at all levels around the world.

We [now] turn to exploring some of the environmental problems the world faces in more detail and look at some of the potential technological solutions and their limitations. For, as will become apparent, purely technical solutions may not suffice if the aim is to develop a genuinely sustainable and equitable future: social and economic changes and adjustments may also be necessary.

The Growth of Environmental Concern

The social and political dimensions of the problem of devising a sustainable future may become clearer if we look back to the beginnings of the contemporary environmental debate. In the late 1960s and early 1970s there was a perhaps unique concurrence of ideas from a number of social and political movements.

The early 1960s had seen the beginnings of environmental concern, symbolised by the publication in 1962 of Rachel Carson's *Silent Spring* which, amongst other things, warned of the ecological dangers of pesticides like DDT.

Subsequently there was a growth in environmental concern amongst young people, many of whom formed part of a counter-culture, which flowered briefly in the late 1960s and early 1970s. The young people involved were often from relatively affluent backgrounds, but they challenged the ideas of the conventional consumerist and materialist society in which they had grown up. Some were not content with simply objecting to the way things were done at present, but also wanted to create alternatives: alternative lifestyles and alternative technologies to support them. There were self-help experiments in rural retreats with windmills, solar collectors and so on, with their decentralist, communitarian philosophy being underpinned by books like Fritz Schumacher's seminal *Small Is Beautiful* (Schumacher 1973), which argued for the use of smaller, human-scale technology supporting and reflecting a more humane and caring society.

In parallel, the late 1960s and early 1970s saw a rise in radical politics, reflected most visibly by the student "protest" movements around the world. The "new left," which emerged as one of the many strands in this movement, challenged the political dogma of the traditional left. The latter held that capitalism, with its single-minded concern for the economic interests of those who owned and controlled technology, denied society as a whole the full benefits of technology; but once freed from capitalist constraints the same technology could be used to meet human needs more effectively. The classic interpretation of this view in practice had been fifty years earlier, when, following the Russian Revolution, Lenin adopted Western production technology and Western technology generally, since he argued that it was the best available at that historical stage. The theoretical point was that the existing technology could simply be redirected to meet new ends. Eighty years on, the shortcomings of this view have become clear: industrial development during the Soviet period seems to have replicated many of the worst examples of environmentally destructive Western technology.

The "new left" in the late 1960s and early 1970s to some extent foresaw this problem. They argued that technical means and political ends inevitably interacted: old means could not be used to attain the new ends and a new set of technologies was required. Like many environmentalists and the "alternativists" in the counter-culture, they were arguing for an alternative technology.

Alternative Technology

This line of argument had been put forward by a British writer, David Dickson, in the seminal paperback *Alternative Technology and the Politics of Technical Change* (1974). However, Dickson, along with many members of the counter-culture, also felt that a simple switch of technology would not be sufficient: technology and society interacted, so there was a need for an alternative society as a base for the alternative technology. As Dickson put it, "A genuine alternative technology can only be developed—at least on any significant scale—within the framework of an alternative society" (Dickson 1974: 13). Thus the "soft" and "hard" paths to a sustainable future outlined by influential US energy activist Amory Lovins, could not just be defined by "soft" and "hard" technology (for example, solar and nuclear respectively), but also required different social and economic arrangements (Lovins 1977).

Indeed, some, following Dickson, felt that society determined technology, so you would need social change first, although a less deterministic, two-way interaction, was usually seen as a more realistic model, implying linked changes in both society and technology (Elliott and Elliott 1976).

With this in mind, rather than a comprehensive confrontational approach, some adopted a more strategic approach to social and technological change, operating in pathfinder mode, initially on the fringe. For example, Peter Harper, an English enthusiast for what he called "radical technology," claimed that: "premature attempts to create alternative social, economic and technical organisation for production can contribute in a significant way to the achievement of political conditions that will finally allow them to be fully implemented" (Harper 1974: 36).

Others again warned that any attempts to introduce radical alternatives would be co-opted by commercial interests, and would be shorn of their values and their radical political edge. The emphasis could thus be on selling "technical fixes," that is just the hardware, and not on implementing the social changes with which alternative technology was meant to be associated. Thus, for example, solar collectors could become just conventional consumer products rather than harbingers of social transformation.

Of course, some would say this did not matter. Certainly, an interest in "alternatives" does not always necessarily imply or require a collective commitment to "progressive" social policies of the sort espoused by liberal environmentalists or radicals on the left of the political spectrum. Some people have sought simply to be free of social norms and state control on an individual basis. For some, alternative technologies simply provide a way of escaping from society. In the USA this libertarian response has sometimes shaded into the militantly independent "survivalist" ideology, often involving extreme right-wing views.

These sorts of developments highlight the weakness in the simple proposition that alternative technologies are automatically linked to progressive social change.

The Current "Green" Debate

These debates are relevant to our contemporary situation in that the issues are now much clearer. Some alternative technology has been co-opted, some green products have just become luxury items for the well heeled, and some technical fixes are being offered as solutions to our environmental problems, while at the same time some radicals in the contemporary "green" movement are still arguing that only a radical transformation of society will be sufficient. More optimistically, some argue that alternative technology has simply moved out of a counter-cultural ghetto, beyond a marginal niche market, and into the mainstream (Smith 2002).

Inevitably, there is a wide range of views on these issues in the "green" movement. This is hardly surprising since the green movement, which emerged in the 1980s, has a multitude of strands. It is much more than just the members of green political parties or activist groups, even though some of the latter are now quite large: tens of thousands of people belong to organisations like Friends of the Earth and Greenpeace. It might also be thought to cover anyone who has some concern for the environment, as reflected, for example, in their support for wildlife protection or in their consumer behaviour.

All these levels of involvement can have an impact, although equally there can also be tactical and strategic divergences and disagreements. Certainly, the growth of consumer awareness has led to pressure for environmentally friendly products and, in turn, for environmentally sound production processes: greener products and cleaner production technologies, which have fewer impacts and use less energy. There is considerable activity in this field at present, but equally there are those who would ask whether this is enough to achieve real sustainability. For example, it has been argued (by Chris Ryan, a leading Australian exponent of ecodesign) that if the various global environmental problems are to be properly addressed, pollution levels and global energy and material resource use must be cut by around 95 per cent, but that this may not be possible just by "technical fixes." There may also be a need for social change, for example, in qualitative patterns and quantitative levels of consumption (Ryan 1994).

For radical "greens" the real issue is this one: can and should growth in material and energy consumption be continued, stimulated by ever growing expectations concerning living standards? Is there not a need for more radical changes—in society as well as technology? Increasingly it is argued that there is a need for a more radical transformation of technology and also possibly of society—an alternative set of technologies better matched to environmental protection, linked to an alternative set of social and cultural perspectives and structures.

. . .

References

Brundtland, G. H. 1987. *Our Common Future*, Commission on Environment and Development, Oxford University Press, Oxford.

Devall, B. 1988. *Simple in Means, Rich in Ends: Practicing Deep Ecology*, Peregrine Smith Books, Salt Lake City, UT.

Dickson, D. 1974. *Alternative Technology: The Politics of Technical Change*, Fontana, London.

Elliott, D., and Elliott, R. 1976. *The Control of Technology*, Wykeham, London.

Harper, P. 1974. "What's left of AT," *Undercurrents*, 6:35–8.

Lovelock, J. 1979. *Gaia: A New Look at Life on Earth*, Oxford University Press, Oxford. (See also Lovelock, J. 1988. *The Ages of Gaia*, Oxford University Press, Oxford.)

Lovins, A. 1977. *Soft Energy Paths: Towards a Durable Peace*, Penguin, Harmondsworth.

Meadows, D., Meadows, D., and Randers, J. 1972. *Limits to Growth*. Earth Island, London.

O'Connor, J. 1991. Capitalism, Nature, Socialism, (CNS) Conference Papers, CES/CNS Pamphlet 1, Centre for Ecological Socialism, Santa Cruz.

Ryan, C. 1994. "The practicalities of ecodesign," in Harrison, M. (ed.), *Ecodesign in the Telecommunications Industry*, RSA Environmental Workshop, 3–4 March 1994, RSA, London.

Schumacher, F. 1973. *Small Is Beautiful*. (First published 1973; many editions subsequently, e.g., Hutchinson, London, 1980).

Smith, A. 2002. "Transforming technological regimes for sustainable development: a role for appropriate technology niches?" Electronic Working Paper 86, Environment & Energy Program, SPRU, University of Sussex, Brighton, http://www.sussex.ac.uk/spru/.

Trainer, T. 1995. *The Conserver Society*, Zed Press, London.

33 Introduction to *Environmental Justice: Creating Equality, Reclaiming Democracy*

Kristin S. Shrader-Frechette

The environment is affected by a range of factors in addition to energy production and consumption; as awareness of these effects has increased, a wide range of groups have been formed to protect the environment. In certain respects, the environmental movement has been quite successful; it has been embraced by engineers and scientists, has led to the creation of new professional fields, and has even been integrated into education from kindergarten through graduate school. But as Elliott mentions in the previous chapter, the causes and effects are quite broad and, hence, require a comprehensive response. Increasingly a group of scholars, NGOs, and politicians have become concerned with the social equity issues that can be seen in environmental degradation; in other words, issues of environmental justice have come into focus. Scholars in this area argue that the human consequences of damage done to the environment have not been evenly distributed. Environmental impacts have had markedly greater negative effects on specific groups of already disadvantaged people. In this chapter, Shrader-Frechette gives an overview of the field of environmental justice. She offers numerous examples of how those without political power (because of race, class, or poverty) have had to suffer further injustices because those with power diverted the negative environmental effects of development to their "backyards." The environmental justice movement is based on the idea that the world is interconnected; we cannot simply measure the environmental effects of a given project, but must also take into account the structure and arrangement of society to ensure that the benefits and burdens are fairly distributed.

World War III has already begun, according to environmental activist Dave Foreman. In this struggle of humans against the earth, he says "there are no sidelines, there are no civilians."[1] Founder of Earth First!, Foreman and his followers have been fighting this world war by performing acts of "monkeywrenching," or "ecotage" (ecological sabotage, the destruction of machines or property that are used to destroy the natural world). Monkeywrenching includes acts such as pulling up survey stakes, destroying tap lines, putting sand in the crankcases of bulldozers, cutting down billboards, and spiking trees so they cannot be logged. Foreman claims such acts of ecological sabotage are part of a proud American tradition of civil disobedience, like helping slaves escape through the Underground Railroad or dumping English tea into Boston Harbor. Rather than slaves or colonists, monkeywrenchers say they are not protecting humans, but earth itself.

From *Environmental Justice: Creating Equality, Reclaiming Democracy* (New York: Oxford University Press, 2002), pp. 3–18. © Oxford University Press, Inc. Reprinted with permission.

As Foreman's remarks suggest, environmentalists have tended to focus on protecting the earth rather than the humans who inhabit it. This book argues not only for protection of the planet but also for public-interest advocacy on behalf of people victimized by environmental injustice. Environmental injustice occurs whenever some individual or group bears disproportionate environmental risks, like those of hazardous waste dumps, or has unequal access to environmental goods, like clean air, or has less opportunity to participate in environmental decision-making. In every nation of the world, poor people and minorities face greater environmental risks, have less access to environmental goods, and have less ability to control the environmental insults imposed on them.

This chapter begins the task of diagnosing, analyzing, and resolving problems of environmental injustice (EJ). It focuses on six key questions: (1) Why have so many environmentalists called for protection of the environment, even as they remained misanthropic and ignored the plight of humans? (2) How did environmentalists come to recognize problems of environmental justice? (3) What are the characteristics of environmental injustice? (4) What are some key examples of environmental injustice, both in developed and in developing nations? (5) Why do some people deny EJ problems, and how defensible are their denials? (6) Why do critics of the EJ movement tend to reject various solutions to EJ problems, and are their rejections reasonable?

Environmentalism and Biocentrism

To understand why people have ignored environmental injustices for so long, it might be helpful to examine the attitudes and priorities of various environmentalists, like Dave Foreman. Foreman's priorities were called into question several years ago after an accident at the Louisiana-Pacific sawmill in Cloverdale, California. On May 8, 1987, a band saw struck an 11-inch spike embedded in a redwood log. The saw shattered, and pieces of blade flew across the room. A large section hit workman George Alexander, 23. It broke his jaw and knocked out several teeth. Foreman called the California accident "tragic"; nonetheless, the attitudes and writings of many environmentalists seem to encourage disrespect for humans even as they call for a greater respect for nature and the earth. Such writings often are exclusively nature centered (biocentric) rather than also human centered (anthropocentric).[2]

In "Animal Liberation: A Triangular Affair," J. Baird Callicott claims that "the extent of misanthropy in modern environmentalism ... may be taken as a measure of the degree to which it is biocentric." And most environmentalists have heard Edward Abbey's famous remark that he would rather shoot a human than a snake. Garrett Hardin even went so far as to recommend that people injured in wilderness areas not be rescued; he worried that rescue attempts would damage pristine wildlife. Even Paul Taylor, in *Respect for Nature*, writes that "in the contemporary world the extinction of the species *Homo sapiens* would be beneficial to the Earth's Community of Life as a whole." In *Eco-Warriors*, Rik Scarce advocates extermination of humanity as "an environmental cure-all."[3]

Gene Hargrove believes that several factors explain the misanthropy of many environmentalists. One reason is that the early U.S. environmentalists, like Teddy Roosevelt, were the most educated and powerful people in the country. Their environmentalism frequently consisted of bird-watching or expensive ecotourism, not addressing areas of greatest pollution where poor people live. Another reason is that there was no significant conflict between environmentalists and the government until the 1950s, when the Sierra Club had a falling out with the U.S. Forest Service over logging policy.[4] Prior to this time, environmentalists were aligned with powerful commercial and government interests, not with poor people. A third reason for traditional environmentalists' emphasis on protection for nature, rather than humans, is that many environmental ethicists have claimed that problems of planetary degradation can be blamed on anthropocentrism, or human-centered values. Callicott's remark, just quoted, is a good example. Rejecting anthropocentric ethics, many environmental philosophers have called for biocentric norms. They have argued for evaluating human actions on the basis of how well they promote ecological, not human, welfare.

Often this biocentrism or ecocentrism is coupled with an appeal to holism, to valuing nature as a whole, rather than valuing its individual species or parts, like humans. Because biocentrists focus on the good of the whole (ecosystems, habitats, and so on), philosophers like Tom Regan have charged them with "environmental fascism." Regan and others believe an ethics of maximizing biotic or ecological welfare could lead to violating human rights in order to serve environmental welfare. Indeed, the misanthropic words of Callicott, Hardin, and Taylor, already quoted, give some credence to the charge of environmental fascism.[5]

Contrary to environmental fascists and misanthropic biocentrists, this book [*Environmental Justice*] argues that protection for people and the planet go hand in hand. Recognizing the importance of environmental justice, the book points out that poor and minorities are the most frequent victims of all societal risks, including environmental degradation. To help reclaim the democratic birthright of people everywhere, these chapters suggest methodological and procedural reforms in the way society evaluates and distributes environmental risks. They also argue for correcting unequal opportunities to participate in environmental decision-making. Finally, the book explains why everyone ought to assume responsibility for the actions of those who pollute, develop, and threaten either the land or the most vulnerable people on it.[6]

From Environmentalism to Environmental Justice

Early in the twentieth century many environmentalists were aligned with governmental and industrial interests. The environmental movement of that era conjured up images of backpackers and bird-watchers, Boy Scouts and nature lovers. The images were of white upper- or middle-class people concerned with conserving a pristine wilderness or an important sanctuary. The environmental movement often focused on action to protect threatened forests, rivers, and nonhuman species, not humans. Even in the academic community, environmental scholarship and particularly environmental ethics traditionally have focused on esoteric topics such as whether to give "rights" to trees and rocks and whether nature has intrinsic or inherent value.[7] Have they been playing the violin while Rome burned?

Two decades ago, while wealthy environmentalists focused on leisure activities and environmental scholars wrote about ivory-tower topics, the grassroots environmental movement began to notice society's most vulnerable groups. They recognized that poor and minorities have been especially damaged by societal threats such as environmental pollution, runaway development, and resource depletion. This grassroots movement saw farmworker communities victimized by pesticides, Native American tribes devastated by radioactive waste, African-American ghettos beset with urban pollutants, Latino settlements plagued by hazardous waste incinerators, and Appalachian towns controlled by absentee-owned coal companies.[8] They saw minority communities forced to trade unemployment for environmental pollution, to exchange a shrinking local tax base for toxic dumps, to trade no bread for a bloody half loaf. Such tradeoffs arose in communities more worried about starvation, unemployment, and violent crime than about health threats from industrial pollution. As Professor Bob Bullard, U.S. sociologist and EJ advocate, notes, this situation has changed. Most minority communities are no longer willing to make such no-win exchanges. They realize they constitute the path of least resistance for polluters and developers, and they have begun to take action. In fact, Bullard says that 80 percent of minority-community resistance groups began as environmental organizations. The tactics of such groups have been demonstrations, marches, hearings, public workshops, research, and lawsuits.[9]

Many people believe that traditional environmental activists, as opposed to EJ advocates, have different goals and backgrounds because often they come from different worlds. This book suggests, however, that the two movements are merely different sides of the same coin. What affects the welfare of the planet affects us all. And once polluters and developers learn that their costs of doing business must be borne by everyone and not shifted to the poor and the powerless, "greening" the ghetto may be the first step in "greening" the entire society.

Understanding Environmental Injustice

The grassroots, minority-led movement for political equality, self-determination, and EJ has sprung up mainly in the urban centers of America. Led largely by women of color, this movement combines many of the philosophies and goals of civil rights and environmental activism. But what is the environmental justice movement? It is the attempt to equalize the burdens of pollution, noxious development, and resource depletion. Environmental justice requires both a more equitable distribution of environmental goods and bads and greater public participation in evaluating and apportioning these goods and bads. Evidence indicates that minorities (e.g., African Americans, Appalachians, Pacific Islanders, Hispanics, and Native Americans) who are disadvantaged in terms of education, income, and occupation not only bear a disproportionate share of environmental risk and death but also have less power to protect themselves.[10] Even children represent a minority victimized by environmental injustice. They are more sensitive to all forms of environmental pollution, and frequently schools have been built atop closed hazardous waste sites.[11] Studies consistently show that socioeconomically deprived groups are more likely than affluent whites to live near polluting facilities, eat contaminated fish, and be employed at risky occupations. Research also confirms that they are less able to prevent and to remedy such inequities.[12] Because minorities are statistically more likely to be economically disadvantaged, some scholars assert that "environmental racism" or "environmental injustice" is the central cause of these disparities. Other social scientists have found that race is an independent factor, not reducible to socioeconomic status, in predicting the distribution of air pollution, contaminated fish consumption, municipal landfills and incinerators, toxic waste dumps, and lead poisoning in children.[13] Members of communities facing such threats typically are too poor to "vote with their feet" and move elsewhere.

Often the sources of environmental injustice are the corporations and governments who site questionable facilities among those least able to be informed about, or to stop, them. Zoning boards, influenced by politically and economically powerful developers and their friends, also have helped create much environmental injustice. If the arguments of this book are correct, however, we the people ultimately are responsible for environmental injustice. We have allowed corporate and government abuses to disenfranchise the weakest among us.

To understand environmental injustice, consider a typical situation that began several decades ago in Texarkana, Texas. Patsy Ruth Oliver, a former resident of Carver Terrace, a polluted African-American suburb of Texarkana, began to notice dark patches of "gunk" seeping up through withered lawns, around puddles, and into the cracked centers of streets. The suburb also had an unusual cluster of medical problems. Their cause finally emerged in 1979, one year after residents of Love Canal, New York, discovered leaking barrels of dioxin beneath their homes. When Congress ordered the largest chemical firms in the United States to identify their hazardous waste sites, the Koppers Company of Pittsburgh identified Carver Terrace as one of its problem areas. For over 50 years, Koppers had used creosote (a known carcinogen) to coat railroad ties. In 1961, when it closed its Carver-Terrace operation, it bulldozed over most of its facilities, including the creosote tanks. Not realizing the dangers left by Koppers, poor families eagerly bought plots in the new Carver Terrace. When Koppers finally admitted the risks at the site, the Environmental Protection Agency (EPA) brought in scientists in full protective gear. They declared the Carver Terrace soil contaminated, but the scientists did not bother to interview the residents. Instead they claimed that the area posed "no immediate health threat" to citizens. Oliver and her neighbors were enraged. They formed the Carver Terrace Community Action Group and soon discovered that the EPA had failed to notify them of two other EPA studies that concluded the site posed immediate health hazards. Oliver argued that the government should "buy out" her community, just as it did for Love Canal. She also concluded that racism was the only reason her neighborhood was treated differently from Love Canal. "I have a master's degree in Jim Crow," she said. Eventually Oliver forced the government to purchase the homes in Carver Terrace, although the buyout destroyed the African-American community there. In 1984, Texas officials asked the U.S. EPA to place Carver Terrace on the Superfund list, the $1.3 billion trust that Congress established in 1980 to clean up toxic waste dumps.[14]

Bob Bullard says that the Patsy Olivers of the world are typical of the EJ movement. Struggling to protect their families and homes, they are not traditional activists. They are just trying to survive. On December 17, 1993, the day demolition of homes began in Carver Terrace, Patsy Oliver died of a heart attack.

Environmental Injustice at Home and Abroad
Inspired by the example of Patsy Oliver, many EJ activists also trace their beginnings to 1982 when North Carolina decided to build a polychlorinated biphenyl (PCB) disposal site in Shocco Township in Warren County. The township is 75 percent African American, and the average per capita income of the county is 97th (of 100 counties) in North Carolina. The U.S. EPA allowed state officials to place the waste only 7 feet above the water table instead of the normal 50 feet required for PCBs. Outraged by this discrimination, 16,000 residents (mostly African Americans and Native Americans) organized marches and protests. Officials arrested more than five hundred local residents. They lost their battle, the state opened the dump, and PCBs have been leaching into the soil. Their actions, however, helped begin the EJ movement.[15]

As in the North Carolina PCB case, African-American communities have been among those hardest hit by environmental injustice. Often the government is the culprit, as in West Dallas, Texas, where, in 1954, the Dallas Housing Authority built a large public housing project—3,500 units—immediately adjacent to a lead smelter. During its peak operations in the 1960s, each year the smelter released 269 tons of lead into the air. West Dallas children had blood lead levels that were 36 percent higher than those in children in control areas. Such exposures are significant because even small amounts of lead can impair learning, interfere with red blood cell production, and damage the liver and brain. Despite repeated studies showing the public-housing children were in danger from the smelter, officials did nothing. For 20 years local and federal officials ignored citizens of West Dallas who asked merely that the city and state enforce existing lead-emission standards. Finally, in 1983 the city and state sued the smelter for violations of city, state, and federal lead-emissions standards. Within two years, the smelter agreed to clean up lead-contaminated soil, to screen children and pregnant women for lead poisoning, and to provide $45 million in compensation to several generations, including hundreds of children exposed to the lead.[16]

Perhaps the most notorious example of environmental injustice against African Americans has occurred in the "Cancer Alley" region of Louisiana. An 85-mile stretch of the Mississippi River between Baton Rouge and New Orleans, Cancer Alley produces one-quarter of the nation's petrochemicals. More than 125 companies there produce fertilizers, paints, plastics, and gasoline. Each year more than a billion pounds of toxic chemicals are emitted in the alley. An advisory committee to the U.S. Civil Rights Commission concluded that African-American communities have been disproportionately impacted by Cancer Alley for at least two reasons. One is that the system of state and local permitting for Louisiana hazardous facilities is unfair. The other reason is that citizens living in Cancer Alley have low socioeconomic status and limited political influence.[17]

Besides African Americans, indigenous peoples repeatedly have been victims of environmental injustice. Among Native Americans, some of the most serious abuses have occurred in connection with uranium mining in the West. Churchrock, New Mexico, in Navajo Nation, the territory of the largest Native-American tribe, is a case in point. Churchrock is the site of the longest continuous uranium mining in Navajo Nation, from 1954 until 1986. Navajo tribal governments leased mining rights to companies such as Kerr-McGee, but they did not obtain either the consent of Navajo families or any information as to the consequences of company activities. Because rainfall at Churchrock is about only 7 inches per year, mining companies withdrew as much as 5,000 gallons of water per minute from the Morrison aquifer to support construction and operation of the mines. Once this groundwater was contaminated with uranium, the companies released it into the Rio Puerco, the main water source for the Navajos. As a result, companies like Kerr-McGee not only significantly reduced the groundwater from which many families drew well water but also contaminated the only main surface water supply. For years, the two main companies, Kerr-McGee and United Nuclear Corporation, argued that the Federal Water Pollution Control Act did not apply to them. They said their activities took place on Native-American land that is not subject to any environmental protections. It was not until 1980 that the courts forced the companies to comply with U.S. clean water regulations.[18]

Among Latinos, one of the most common forms of environmental injustice is that faced by farmworkers exposed to pesticides. In 1972, the United States banned many chlorinated hydrocarbon pesticides such as DDT, aldrin, dieldrin, and chlordane, in part because they were so long-lived and remained on fruits and vegetables when they were consumed by the public. Instead farmers began using the much shorter-lived but much more toxic pesticides known as "organophosphates." The pesticides pose less threat to consumers because they are less persistent, but they are a greater threat to farmworkers. A large proportion of farmworkers are Mexican Americans, often illegal aliens who work for less-than-minimum wage and typically under difficult or illegal working conditions. Given such circumstances, the workers are in no position to complain about pesticide exposure. Moreover, what pesticide laws exist typically are not enforced, so farmworkers have little protection.[19]

People in developing nations usually face similar or worse environmental threats. In the case of pesticides, for example, after the United States banned many chlorinated hydrocarbons, U.S. and multinational chemical companies merely began shipping them abroad. Currently about one-third of the pesticides manufactured in the United States are not allowed to be used in the United States and are exported, mostly to developing nations. According to the World Health Organization, the chemicals contribute to approximately 40,000 pesticide-related deaths annually in the developing world.[20] The case of Gammalin 20 is fairly typical. A highly toxic relative of DDT known as "lindane," Gammalin 20 has been banned in the United States for about 30 years. After it was imported into Ghana for use as a pesticide, the local fishermen along the shores of Lake Volta found it had another use as well. When they dumped it into the water, many dead fish floated to the top of the water, and the fishermen could easily collect them, sell them, and feed them to their families. Soon the fish population began dropping off at the rate of about 10 percent per year, and the Ghana villagers began experiencing the classic symptoms of nausea, vomiting, convulsions, circulatory disorders, and liver damage. The people did not connect their ailments to the chemical they dumped into the lake, and their problems continued until a Ghanaian nongovernmental organization explained what had happened.[21]

The 1984 chemical spill in Bhopal, India, also illustrated that people in developing nations receive far less protection from environmental threats than do citizens in the developed world. When a toxic gas, MIC, leaked from a Union Carbide pesticide plant in Bhopal, the accident killed nearly 4,000 people and permanently disabled another 50,000. The company later settled, with survivors and the disabled, for only several thousand dollars per person. After Bhopal, the predominantly African-American community of Institute, West Virginia, became the center of a violent conflict. West Virginia's Kanawha Valley, "the chemical capitol of the world," is the site of the only Union Carbide facility in the United States that manufactures MIC. On the one side, Union Carbide workers fought for their jobs. On the other side, local residents said they fought for their lives. Both the company and the EPA stonewalled citizens' demands for investigation of their health complaints and the chemical odors that saturated the valley's air. Citizens claimed that the EPA attempted to show there was no public health threat by continually revising its risk-assessment methods[22] so as to obtain the answers Union Carbide wanted.

Apart from the lax standards that U.S. and multinational corporations employ in their plants in poor areas, including developing nations like India, groups in the industrialized world also often intentionally dump toxic wastes in the Third World. Each year companies and local governments offer nations in the Caribbean and in West Africa hundreds of dollars for every 55-gallon barrel of toxic waste that can be dumped legally. For example, in 1988, the city of Philadelphia hired a Norwegian company, Bulkhandlung, to transport 15,000 tons of toxic incinerator ash to the African nation of Guinea. After plant and animal life died at the waste site, the African government ordered Bulkhandlung to remove the ash and return it to Philadelphia. The Africans appealed to the 1989 Basel Convention on the Control of Transboundary Movements of Hazardous Wastes and Their Disposal, ratified by more than one hundred nations, including the United States. According to the convention, companies wishing to ship hazardous waste must notify the receiving country. In fact, exporters must receive written permission from the importing nation. Because the Basel Convention allows any country to refuse permission, it has helped address waste-related EJ problems. Nevertheless, corruption and lack of information often keep the citizens of waste-receiving countries from knowing what their leaders have accepted in exchange for payment. Thus it is questionable whether people in many developing nations actually give free informed consent to imports of hazardous waste that may threaten them.[23]

A chief economist from the World Bank recently created a massive controversy when he wrote an internal memo explaining the economic rationale for such waste transfers. The memo was leaked to the press in 1991. It said: "Just between you and me, shouldn't the World Bank be encouraging MORE migration of the dirty industries to the LDCs [less-developed countries]?" The memo further enraged ethicists and environmentalists by offering three reasons that developing nations were a good place to dump toxics: their citizens already had a lower life expectancy; such countries were relatively "under-polluted"; and impairing the health of the people with the lowest wages made the "greatest economic sense."[24]

Over the last two decades, many studies have documented the fact that polluters, both at home and abroad, appear to be following the advice of the World Bank economist. In 1983, Bob Bullard showed that, from the late 1920s to the late 1970s, Houston placed all of its city-owned landfills in largely African-American neighborhoods. Although they comprised 28 percent of the city's population, African-American communities received 15 of 17 landfills and 6 of 8 incinerators. Bullard pointed out that such dumping has magnified the myriad social ills—crime, unemployment, poverty, drugs—that already plague inner-city areas.[25] Journalists also have shown that the dirtiest zip code in California, a one-square-mile section of Los Angeles County, is filled with waste dumps, smokestacks, and wastewater pipes from polluting industries. In one zip code, where 18 companies discharge five times as much pollution as they emit in the next-worst zip code, the population is 59 percent African-American and 38 percent Latino.[26]

In 1984, Cerell Associates, a private consulting firm hired by the California Waste Management Board, issued a report titled "Political Difficulties Facing Waste-to-Energy Conversion Plant Siting." The report concluded that all socioeconomic groups resist the siting of hazardous facilities in their neighborhoods and adopt positions of NIMBY ("Not in My Back Yard"). Nevertheless, the study showed that because lower-income groups have fewer resources to fight corporate and government siting decisions, they usually lose.[27] Further confirming the Cerell findings, in 1986 the Center for Third World Organizing in Oakland, California, issued the report, "Toxics and Minority Communities." It showed that 2 million tons of radioactive uranium tailings, left from uranium mining, had been dumped on Native-American lands. As a result, the study argued, cancers of the reproductive organs among Navajo teenagers had climbed to 17 times the national average. Later, in April 1987 the United Church of Christ Commission for Racial Justice released a widely quoted report that documented environmental racism throughout the United States.[28] Ben Chavis, the executive director of the National Association for the Advancement of Colored People (NAACP), organized a study that later showed 60 percent of African Americans live in communities endangered by hazardous waste landfills. The report revealed that the largest U.S. hazardous waste landfill, which receives toxins from 45 states, is in Emelle, Alabama;

Emelle is 79 percent African American. The study also demonstrated that the greatest concentration of hazardous waste sites in the United States is in the predominately minority South Side of Chicago. Typically minority communities have agreed to take the sites in exchange for jobs and other benefits that have never become a reality. A more recent report, published in 1992 in the *National Law Journal*, concluded that government agencies do not guarantee equal political power and equal participation to all groups victimized by environmental injustice. In fact, the study showed that government agencies treat polluters based in minority areas less severely than those in largely white communities. The same report showed that toxic cleanup programs, under the federal Superfund law, take longer and are less thorough in minority neighborhoods.[29]

A 1992 EPA report likewise found significant evidence that low-income, nonwhite communities are disproportionately exposed to lead, air pollution, hazardous waste facilities, contaminated fish, and pesticides. When the report recommended greater attention to environmental injustice,[30] the EPA established the Office of Environmental Equity (OEE). Also in 1992 the General Accounting Office (GAO) began an ongoing study to examine the EPA's activities relating to EJ.[31] The Clinton administration likewise emphasized environmental justice when it selected a prominent leader of the EJ movement, Bob Bullard, to serve on the Clinton-Gore transition team.[32] On February 11, 1994, Clinton signed an executive order that directed each federal agency to develop an EJ strategy for "identifying and addressing ... disproportionately high and adverse human health or environmental effects of its programs, policies, and activities on minority and low-income populations."[33]

Bullard says that Clinton's actions are not enough. He claims the United States and other nations need an EJ equivalent of the 1964 Civil Rights Act and the 1968 Fair Housing Act. Every year since 1994, Congress has been debating bills designed to guarantee environmental justice. Because none has ever passed, current efforts to promote EJ rest on three bases: Clinton's executive order, the environmental justice division of the EPA, and the 1969 National Environmental Policy Act (NEPA).[34] Before leaving office in January 2001, President Clinton set the budget of the EJ branch of EPA at roughly the same amounts for 2001 and for 2002 as it was for the year 2000. President Bush is expected to cut both the overall EPA budget and the environmental justice program of the EPA.

Why have local, national, and international media not helped more to promote EJ? One reason is that small-town leaders like Patsy Oliver are typically unknown women. Both sexism and racism combine to silence them in the press. Another reason is that the Patsy Olivers of the world typically do not want media attention and public glory. They want results: health and safety for their families and communities. A third reason is that even the EPA has been slow to acknowledge environmental justice. Only in 1990, in its report "Environmental Equity: Reducing Risks for All Communities," did it finally admit that minority communities have borne more than their "fair share" of environmental pollution.[35] Policymakers bear some of the blame for the failure to confront environmental racism. They typically use quantitative risk assessment and benefit-cost analysis in ways that are not sensitive to justice issues. Both methods incorporate aggregation methods that often hide inequitable impacts. Those using both methods also usually try to trace the causes of specific problems to particular hazardous substances.[36] However, EJ proponents say that scientists should assess the total risks that a given community faces because many health threats are a combination of several factors. They also argue that often no one addresses the cumulative and synergistic public health and environmental burdens that minority communities often bear.

Apart from deficiencies in media attention, science, and law, another reason that society has been slow to confront issues of environmental injustice is the backwardness of environmental organizations. Groups like the Sierra Club sometimes mirror the biases of the larger society. Organizing at a time when discrimination was the norm, early Sierra Club leaders did not link social justice to the conservation cause. In fact, in 1959 the Sierra Club vetoed an explicit antidiscrimination policy and said membership already was open to everyone. And in 1971 members voted against addressing conservation issues related to the poor and minorities. Even today, many environmentalists view alliances with the disenfranchised as "too political." Nevertheless, in Los Angeles, Virginia, and Florida, many Sierra Club groups have taken up EJ issues on behalf of Latinos, Native Americans, and African Americans.[37]

Denial of Environmental Injustice Charges

In response to repeated calls for EJ, critics typically make two responses, one based on denying environmental injustice and another based on excusing it. The "denial" retort is that although EJ is desirable, because flaws in existing research make it almost impossible to identify particular instances of environmental injustice, most supposed cases can be challenged. The "excuse" response is to admit that there are instances of environmental injustice but to claim that the benefits of avoiding them do not outweigh the costs of correcting them. Proponents of the first, or "denial," argument often say that although poor and minority communities appear to be victimized, much of the evidence for their discrimination is "largely anecdotal." Attacking Bob Bullard's early study of environmental racism in Houston, they note that the lawsuit based on it, *Bean v. Southwestern Waste Management Corp.*, was unsuccessful. They also claim that authors often assume rather than prove that actual risks near hazardous facilities are higher than elsewhere.[38]

While it is wrong to assume that risks always are higher near dangerous facilities, critics of EJ research ignore the fact that, all things being equal, public health risks probably are higher near noxious facilities, and research is needed to determine their level. Proponents of the denial argument also ignore the fact that such sites lower nearby property values.[39]

Many proponents of the "denial" argument specifically attack a widely discussed General Accounting Office (GAO) analysis that alleges environmental racism. This 1983 report examined community demographics near commercial waste treatment, storage, and disposal facilities. After assessing data from four noxious facilities in EPA Region IV (the Southeast), the GAO researchers found that the populations in three of the four areas surrounding the problematic sites were predominantly African American, even though they were only a minority in the state's population. Objecting to the GAO study, critics argue that it is ambiguous with respect to how one ought to characterize a community as minority. Christopher Boerner and Thomas Lambert, for example, claim that defining a minority community as one in which the percentage of minority residents exceeds the percentage in the entire population may be problematic. According to this definition, they note that Staten Island, New York, home of the nation's largest landfill, is a minority community even though more than 80 percent of its residents are white.[40] One problem with the preceding Boerner-Lambert criticism, however, is that it confuses the neighborhood near the landfill with all of Staten Island. Just because Staten Island is only 20 percent nonwhite does not mean that the area immediately around the landfill is only 20 percent nonwhite. Because most residents within several miles of the landfill are African American, Boerner's and Lambert's attempted criticism is questionable.

Critics of EJ research use the "denial" argument to make other allegations. They claim many EJ studies err in ignoring population density when they characterize a community as "minority." They say the real issue is the total number of people affected by some noxious facility, not just the percentage of nonwhites around it.[41] While the total number of people affected is important, this criticism begs the question of the importance of distributive justice. It arguably is worse for some people to be discriminated against, as subsequent chapters show, than for everyone to be treated the same and exposed to similar threats. Such discrimination is worse because it entails threats both to life and to equal treatment, whereas the same treatment of different groups may jeopardize only rights to life and not also rights to equal treatment.

Critics of the EJ movement also employ the "denial" argument to challenge the 1987 report of the Commission for Racial Justice (CRJ) of the United Church of Christ. Correlating percentages of nonwhites, within zip codes, with numbers of waste plants, the CRJ analysis showed that the percentage of nonwhites in zip codes with one facility was twice that in zip codes having no such plant. For zip codes with more than one waste facility, the percentage of nonwhites was three times that in zip codes with no such plant. The CRJ also revealed that race was statistically more significant than either mean household income or mean value of owner-occupied housing as a determinant of where noxious facilities were located.[42]

In response to the CRJ findings, proponents of the "denial" argument allege that environmental injustice often disappears once one stops aggregating data from large areas such as zip codes. They say that how one defines the relevant geographic area determines whether or not there is environmental injustice.[43] Such criticisms, of course, are reasonable. One often can gerrymander geographic regions so as to exhibit or to cover up some spatially related effect. Nevertheless, the criticism is beside the point. If the area closest to a noxious facility tends to have a population of nonwhites rather than whites, then regardless of what zip codes (or any other systems of aggregation) reflect, there is likely to be environmental racism. Moreover, if even large aggregates appear to reveal evidence of environmental injustice, the appropriate response is to determine whether the apparent disparate impact is real. The appropriate response is not to say that there are ways of aggregating the data so that the injustice "disappears," because the real question is the defensibility of such methods of aggregation. And this question should be analyzed on a case-by-case basis. It would be surprising if there were never any real environmental injustice, and if poor or powerless people never were subject to more noxious facilities than wealthier ones.[44]

Utilitarian Excuses for Environmental Injustice

Using the "excuse" response, critics of the EJ movement do not deny environmental injustice. Instead they give two arguments to put the alleged injustice into perspective. They argue that (1) on balance, victims of alleged environmental insults may benefit from living near noxious facilities. They say victims might suffer worse from higher unemployment and housing costs if they did not live near dangerous sites. Likewise they charge that (2) the mere correlation of hazardous sites and the presence of poor or minority communities does not prove that racism or injustice actually caused the siting there. They say that African Americans, for example, may have moved to risky or undesirable areas because housing was cheaper or because of some other factor.[45] Both of these "excuse" arguments are questionable. Complaint (1) ignores the fact that, apart from the ultimate balance of costs and benefits (such as more employment) near a risky facility, the evidence of what residents want is clear. Poor people and minorities usually do not want most of the dangerous or undesirable sites to be located near them. And nearby residents have the right to control the risks that others impose on them. Critics of the EJ movement who use this "excuse" response seem to forget principles of equal human rights and instead to use utilitarian grounds to attempt to defend injustice. Such a defense is obviously flawed because all people, especially innocent potential victims, have rights to exercise their preferences regarding what threatens their welfare—particularly when others profit from the threats.

"Excuse" argument (2), that the correlation between race and risky facilities does not prove discrimination, is correct. Nevertheless, it is misleading. The issue is not whether people, corporations, or governments deliberately discriminate against poor people or minorities in siting decisions and therefore cause them to live in polluted areas. Even if minorities moved to an area after it was polluted, the issue is whether some citizens ought to have less than equal opportunity to breathe clean air, drink clean water, and be protected from environmental toxins. If they do have less than equal opportunity, even though no one may have deliberately discriminated against them, the situation may need to be remedied, at least in part because people have rights to equal treatment. Moreover, racism or injustice need not be deliberate. Many people behave in racist or sexist ways even when they have no idea of their prejudices. Their ignorance of their own faults may limit their guilt, but it provides no evidence of the absence of those faults. Absence of evidence for deliberate discrimination is not the same as evidence of the absence of deliberate discrimination. Admittedly, in the landmark case of *Washington, May of Washington, D.C., et al. v. Davis et al.*, the court set a stringent standard of proof for damage awards in cases of environmental harm.[46] The standard is stringent because the court ruled that a plaintiff seeking damages must prove that harmful actions taken by an individual or group were intended to cause the plaintiff harm and not merely that the harm occurred as an unexpected by-product of the action. Just because such a standard of proof is required before a defendant must pay legal damages, however, does not mean that environmental injustice occurs only when the same standard of proof is met. Rather, the legal standard is stricter (1) because defendants must be presumed innocent until proved guilty, (2) because courts must be conservative in meting out punishment, and (3) because courts must be cautious in making damage awards. Although the "discriminatory intent" ruling in the *Washington* case damages some civil rights and environmental justice cases, because it is almost impossible to prove the subjective motivations of a decision-maker, it applies only to legal rulings. The limits of truth or moral responsibility are not the same as the limits of what can be proved in a court of law as a basis for a damage award. Lack of legal proof for deliberate discrimination does not entail the absence of environmental injustice. Besides, even if citizens, corporations, and governments do not deliberately discriminate, they nevertheless may be responsible for the institutional structures that indirectly cause disparate impacts on poor or minority groups. At least in democracies, citizens typically have the governments they deserve and create. And if so, then citizens have duties to monitor and to correct government policies, especially those allowing discrimination against poor and minorities.

Many critics of the EJ movement use the "excuse" argument in a third way. They claim that alleged solutions to environmental injustice are even worse than the original injustice. They tend to focus on three such solutions: (1) eliminating all social costs (like pollution) of industrial processes; (2) allocating these costs evenly throughout the population; or (3) compensating the individuals who bear more of these costs.[47] With respect to the first solution to environmental injustice, critics of the EJ movement say that it would cause greater harm to society than does environmental injustice, and they probably are right, insofar as it is impossible to eliminate all pollution. In the case of pesticides, for example, critics claim (correctly) that because some pollution is inevitable, the "costs to society" of completely eliminating these chemicals are far higher than those of environmental injustice.[48] Nevertheless, proponents of the "excuse" argument beg a crucial question. Costs to whom? Costs to poor and minority communities might not be greater if society reduced or eliminated pollution near them. Moreover, in the specific case of pesticides, experts have argued that most of these chemicals are not essential to society and agriculture but instead are used to make foods look more appetizing. The same experts argue that biological forms of pest control are safer alternatives than chemicals.[49] The most basic problem, however, with this first solution to environmental injustice—eliminating all pollution—is that it is not realistic. It is a straw-man solution, one easy to reject because it is so extreme. A more realistic solution would be to reduce pollution to levels as low as practical. But critics of EJ do not consider this less extreme option.

What about a second solution to EJ problems, distributing pollution equally? Critics of the EJ movement also reject this alternative on the grounds that not siting noxious facilities in poor neighborhoods would have undesirable consequences, such as reducing the tax base and employment in areas needing them most.[50] This criticism, however, ignores the fact, that residents of poor neighborhoods typically do not feel deprived of economic benefits when someone protects them from dangerous facilities. And if not, then rejecting this second solution to EJ problems errs because it ignores the authentic consent and the well-confirmed opinions of those who have been most victimized by environmental injustice. To argue that communities desire health threats in exchange for economic benefits presupposes that the communities have given free informed consent to the noxious facilities. But proponents of the "excuse" argument typically have not established this presupposition. The argument also assumes that there is no right to a liveable environment. Probably EJ advocates would argue that all people do have such rights and that they ought not be traded for money, especially if what is traded is the health and safety of innocent victims such as children.[51]

Critics of the EJ movement also reject a third solution to EJ problems, compensating individuals who are disproportionately impacted by pollution from which society benefits. They reject this compensation solution on the grounds that paying the poor to take health risks amounts to bribery or coercion. To avoid bribery or coercion, they claim that society should compensate only nonpoor or nonminorities, those who can freely consent to the risks. But if only they are paid, proponents of the "excuse" argument say the payment schemes ultimately would raise the level of unemployment and poverty.[52] Are they correct? No: this third objection is flawed in that it ignores the fact that if compensation is owed, then some is better than none. It also begs the question of whether compensation, as such, would increase poverty and unemployment. After all, there are ways to increase employment and reduce poverty, independent of compensating people for accepting noxious facilities. The criticism likewise errs because it presupposes that society has no responsibility to help correct unemployment and poverty, independent of its solutions to EJ problems. Moreover, it is desirable to consider the option of compensation in part because it forces society to ask whether the pollution costs associated with a proposed facility may be so high as to make it undesirable in any location.[53] It forces society to ask whether polluters genuinely are able to pay the full market costs of their actions. A key benefit of compensation schemes thus is that they force polluters to internalize the social costs of pollution and not to try to save money by dumping their burdens on the unwilling, the vulnerable, and the poor. In this regard, one model of compensating host communities for noxious facilities may be the 1982 Wisconsin program for landfill negotiation/arbitration.[54] One compensation model that appears not to have worked is the one created by the U.S. Department of Energy (DOE) for the proposed Yucca Mountain radioactive waste facility. This model failed, in part, because the DOE did not secure free informed consent from potential victims, did not disclose the complete risks to them, and severely limited all liability for the site. The conclusion to draw from cases like Yucca Mountain is not that compensation for environmental injustice is unworkable but that not all compensation schemes are just and reasonable.[55]

. . .

Notes

1. Dave Foreman, *Confessions of an Eco-Warrior* (New York: Harmony Books, 1991), pp. viii–ix, hereafter cited as: Foreman, *CEW*.

2. Foreman, *CEW*, pp. ix, 118–119, 149, 165. See also Dave Foreman, "Ecotage Updated," *Mother Jones* 15, 7 (November–December 1990): 49, 76, 80–81.

3. Callicott, Scarce, and others are quoted in Gene Hargrove, foreword to *Faces of Environmental Racism*, edited by Laura Westra and Peter Wenz (Lanham, MD: Rowman and Littlefield, 1995), pp. ix–xiii; hereafter cited as: Hargrove, Foreword, in Westra and Wenz, *FER*. See also David E. Newton, *Environmental Justice* (Oxford, England: ABC-CLIO, 1996); hereafter cited as Newton, *EJ*.

4. Hargrove, Foreword, pp. ix–xi.

5. Hargrove, Foreword, in Westra and Wenz, *FER*, pp. x–xiii.

6. See Wendell Berry, *A Continuous Harmony* (New York: Harcourt, Brace, Jovanovich, 1972), p. 79; see also Avner De-Shalit, *The Environment in Theory and in Practice* (Oxford: Oxford University Press, 2000).

7. J. Baird Callicott, *In Defense of the Land Ethic* (Albany: State University Press of New York, 1989); Holmes Rolston, *Environmental Ethics* (Philadelphia: Temple University Press, 1988); Laura Westra, *Our Environmental Proposal for Ethics* (Lanham, MD: Rowman and Littlefield, 1994); Byron G. Norton, "Environmental Ethics and the Rights of Non-humans," *Environmental Ethics* 4 (1982): 17–36.

8. See Kristin Shrader-Frechette and Lynton K. Caldwell, *Policy for Land* (Savage, MD: Rowman and Littlefield, 1993); see also Kristin Shrader-Frechette, *Environmental Justice: Creating Equality, Reclaiming Democracy* (New York: Oxford University Press, 2002), chapter 3.

9. Robert Bullard, *Confronting Environmental Racism: Voices from the Grassroots* (Boston: South End Press, 1996); hereafter cited as: *Racism*. See P. Cotton, "Pollution and Poverty Overlap Becomes Issue, Administration Promises Action," *Journal of the American Medical Association* 271, 13 (April 6, 1994): 967–969; Mary E. Northridge and Peggy M. Shepard, "Environmental Racism and Public Health," *American Journal of Public Health* 87 (May 1997): 730–732.

10. R. D. Bullard, *Dumping in Dixie: Race, Class, and Environmental Quality* (Boulder, CO: Westview Press, 1990); hereafter cited as: Bullard, *Dumping*; Bullard, *Racism*; R. D. Bullard, ed., *Unequal Protection: Environmental Justice and Communities of Color* (San Francisco: Sierra Club Books, 1994); hereafter cited as: Bullard, *Unequal Protection*; US Environmental Protection Agency (US EPA), *Environmental Equity: Reducing Risks for All Communities*, EPA-230-R-92-008 (Washington, DC, 1992); hereafter cited as US EPA, *Equity*; United Church of Christ (UCC) Commission for Racial Justice, *Toxic Wastes and Race in the United States: A National Report on the Racial and Socioeconomic Characteristics of Communities with Hazardous Waste Sites* (New York: UCC, 1987); hereafter cited as: UCC, *Toxic Wastes*; David N. Pellow, "Environmental Inequity Formation," *American Behavioral Scientist* 43, no. 4 (January 2000): 581–602.

11. Stacy A. Teicher, "Schools atop Dumps: Environmental Racism?" *Christian Science Monitor* 91, no. 238 (November 4, 1999): 3.

12. Regarding living near polluting facilities see J. Gould, *Quality of Life in American Neighborhoods: Levels of Affluence, Toxic Waste, and Cancer Mortality in Residential Zip Code Areas* (Boulder, CO: Westview Press, 1986); UCC, *Toxic Wastes*; Bullard, *Dumping*; B. A. Goldman, *The Truth about Where You Live: An Atlas for Action on Toxins and Mortality* (New York Times Books, 1991); B. Bryant and P. Mohai, eds., *Race and the Incidence of Environmental Hazards: A Time for Discourse* (Boulder, CO: Westview Press, 1992): R. L. Calderon et al., "Health Risks from Contaminated Water: Do Class and Race Matter?" *Toxicology and Industrial Health* 9, no. 5 (1993): 879–900; J. Tom Boer, Manuel Kastor, and James L. Sadd, "Is There Environmental Racism? The Demographics of Hazardous Waste in Los Angeles County," *Social Science Quarterly* 78, no. 4 (December 1997): 793–810; Evan J. Ringquist, "Equity and the Distribution of Environmental Risk: The Case of TRI Facilities," *Social Science Quarterly* 78, no. 4 (December 1997): 811–829. See also Timothy Maher, "Environmental Oppression," *Journal of Black Studies* 28, no. 3 (January 1998):357–368.

Regarding employment at risky occupations, see, for example, chapter 7, and Karen Messing, Jeanne Stellmar, and Carmen Sirianni, eds., *One-Eyed Science: Occupational Health and Women Workers* (Philadelphia: Temple University Press, 1998); William Burgess, *Recognition of Health Hazards in Industry* (New York: Wiley, 1995); Mara Klein, *Public Health and Industrial Contaminants* (Philadelphia: Pennsylvania Department of Environmental Resources, 1994); Colin Soskolne, *Ethical, Social, and Legal Issues Surrounding Studies of Susceptible Populations and Individuals* (Washington, DC: National Institute of Environmental Health Sciences, 1997); Robert F. Herrick, *Exposure Assessment for Occupational Risks* (Boca Raton, FL: Lewis, 2000).

13. R. D. Bullard, "Environmental Racism in America?" *Environmental Protection* 206 (1991): 25–26. See R. D. Bullard, "Anatomy of Environmental Racism and the Environmental Justice Movement," in Bullard, *Racism*, p. 21.

14. Quoted by Ruth Rosen, "Who Gets Polluted? The Movement for Environmental Justice," in *Taking Sides: Clashing Views on Controversial Environmental Issues*, edited by Theodore D. Goldfarb (Guilford, CT: Dushkin/McGraw-Hill, 1997), pp. 67–68; hereafter cited as: Rosen, "Who Gets Polluted?" See also Daniel Faber, *The Struggle for Ecological Democracy* (New York: Guilford 1998).

15. Newton, *EJ*, pp. 1–2.

16. Ibid., pp. 6–7.

17. Ibid., pp. 9–11.

18. Ibid., pp. 7–9.

19. Ibid., pp. 11–12.

20. See Shrader-Frechette, *Environmental Justice*, chapter 8, note 2, for references.

21. Newton, *EJ*, pp. 13–14.

22. Rosen, "Who Gets Polluted?" pp. 64–65; see Tracy Baxter, "Environmental Justice for All," *Sierra* 82, no. 2 (March–April 1997): 101–104.

23. H. Shue, "Exporting Hazards," in *Boundaries*, edited by P. Brown and H. Shue (Totowa, NJ: Rowman and Littlefield, 1981), p. 107. For the Burkhandlung case, see Newton, *EJ*, pp. 47–48; for the text of the Basel Convention, see Newton, *EJ*, pp. 131–134. See also Jan Marie Fritz, "Searching for Environmental Justice: National Stories, Global Possibilities," *Social Justice* 26, no. 3 (fall 1999): 174–190; Yozo Yokota, "International Justice and the Global Environment," *Journal of International Affairs* 54, no. 2 (spring 1999): 583–599; and Francis O. Adeola, "Cross-National Environmental Injustice and Human Rights Issues," *American Behavioral Scientist* 43, no. 4 (January 2000): 686–707.

24. Rosen, "Who Gets Polluted?" p. 66; see Meena Singh, "Environmental Security," *Social Justice* 23, no. 4 (winter 1996): 125–134.

25. Bullard, *Dumping*.

26. Rosen, "Who Gets Polluted?" pp. 65–66; see Raquel Pinderhughes, "The Impact of Race on Environmental Quality," *Sociological Perspectives* 39, no. 2 (summer 1996): 231–249; hereafter cited as: Pinderhughes, "Impact."

27. Rosen, "Who Gets Polluted?" p. 66.

28. UCC, *Toxic Wastes*.

29. *National Law Journal* 15, no. 3, special issue, "Unequal Protection: The Racial Divide in Environmental Law" (September 21, 1992); Rosen, "Who Gets Polluted?" p. 66.

30. EPA, *Equity*.

31. See Sexton, "Environmental Justice," 688–692.

32. D. Ferris, "A Call for Justice and Equal Environmental Protection," in Bullard, *Unequal Protection*, pp. 298–320.

33. Executive Order No. 12898, Sec. 1–101. For the full text of the order, see *Environment* 36, no. 4 (May 1994): 16–19.

34. Thomas M. Parris, "Spinning the Web of Environmental Justice," *Environment* 39, no. 4 (May 1997): 44–46; Rosen, "Who Gets Polluted?" pp. 69–70. The 1967 NEPA requires assessors to take account of distributive inequities in environmental impacts. For the text of NEPA, see Newton, *EJ*, pp. 92–97.

35. Pinderhughes, "Impact," pp. 231–249.

36. See Kristin Shrader-Frechette, *Risk and Rationality* (Berkeley: University of California Press, 1991); Kristin Shrader-Frechette, *Risk Analysis and Scientific Method* (Boston: Kluwer, 1985).

37. Baxter, "Environmental Justice."

38. Christopher Boerner and Thomas Lambert, "Environmental Injustice: Industrial and Waste Facilities Must Consider the Human Factor," in Goldfarb, *Taking Sides*, pp. 73–75; hereafter cited as: Boerner and Lambert, "Environmental Injustice." The same objection to charges of environmental injustice also is made by Vicki Been, "Locally Undesirable Land Uses in Minority Neighborhoods: Disproportionate Siting or Market Dynamics?" *Yale Law*

Journal 103 (1994): 1383–1422, and by Henry Payne, "Environmental Injustice," *Reason* 29, no. 4 (August–September 1997): 53–56. See also David Friedman, "The 'Environmental Racism' Hoax," *American Enterprise* 9, no. 6 (November–December 1994): 75.

39. See Shrader-Frechette, *Environmental Justice*, chapter 2.

40. Boerner and Lambert, "Environmental Injustice," p. 74; see also the third section of chapter 2 and Ralph M. Perhac, Jr., "Environmental Justice: The Issue of Disproportionality," *Environmental Ethics* 21, no. 1 (1999): 81–92.

41. Boerner and Lambert, "Environmental Injustice," p. 75.

42. See Baxter, "Environmental Justice"; Parris, "Spinning the Web," pp. 44–46. For a similar criticism of claims of environmental justice, see James Hamilton, "Politics and Social Costs," *Fand Journal of Economics* (spring 1993): 101–125.

43. Boerner and Lambert, "Environmental Injustice," p. 75. The same argument is also made by Douglas Anderson et al., "Hazardous Waste Facilities: 'Environmental Equity' Issues in Metropolitan Areas," *Evaluation Review* 18, no. 2 (1994): 123–140.

44. See Pinderhughes, "Impact," pp. 231–249.

45. Deb Starkey, "Environmental Justice," *State Legislature* 20, no. 3 (March 1994): 27–31; Boerner and Lambert, "Environmental Injustice," p. 75. Payne, "Environmental Injustice," also makes this objection, as do Been, "Locally Undesirable" (note 38), and John S. Baker, "Dissent," in Louisiana Advisory Committee to the U.S. Commission on Civil Rights, *The Battle for Environmental Justice in Louisiana* (Kansas City: U.S. Commission on Civil Rights, 1993).

46. For discussion of the *Washington* case, see Newton, *EJ*, pp. 44–45, 142–144.

47. Boerner and Lambert, "Environmental Injustice," p. 76. For criticisms of additional solutions, see Been, "Locally Undesirable."

48. Boerner and Lambert, "Environmental Injustice," pp. 76–77.

49. Kristin Shrader-Frechette, *Environmental Ethics* (Pacific Grove, CA: Boxwood Press, 1991), pp. 270–324; David Pimentel et al., "Assessment of Environmental and Economic Impacts of Pesticide Use," in *Technology and Values*, edited by Kristin Shrader-Frechette and Laura Westra (New York: Rowman and Littlefield, 1997), pp. 375–414.

50. See Starkey, "Environmental Justice," pp. 27–31; Boerner and Lambert, "Environmental Injustice," p. 79.

51. See Shrader-Frechette, *Environmental Justice*, chapter 8.

52. Boerner and Lambert, "Environmental Injustice," pp. 80–81.

53. Ibid., p. 81.

54. Ibid., pp. 81–82.

55. Kristin Shrader-Frechette, *Burying Uncertainty* (Berkeley, CA: University of California Press, 1993), pp. 15–23, 96–98, 204–207.

34 "Icarus 2.0: A Historian's Perspective on Human Biological Enhancement"

Michael Bess

Some of the major triumphs of twentieth century technologies have been the amazing advances in health and medicine. New understandings about diet, the development of innovative medicines, advances in food production and distribution, prenatal care, and the invention of diagnostic devices like X-ray and MRI machines have generated profound changes in the health and welfare of people around the world. As noted in part I in the Fukuyama piece, the evidence of these successes can be seen in many ways, but perhaps most simply in the way that life spans have dramatically increased. In 1900 the average lifespan of an American was 47 years. In the year 2000 that average had risen to 77 years. But as Michael Bess argues in this article, we must not let these great advances cloud our vision and cause us to think that all technological change will effortlessly generate social goods. There have been heated debates about a number of medical technologies and research programs including genetic screening, stem cell research, human cloning, the focus of pharmaceutical companies on profitable drugs for Westerners rather than the needs of people in developing countries, and—the subject of this article—human enhancements. These are not simply disputes between people who like technology and people who dislike technology. They are debates about what we should take as our priorities. In order to create a future in which we want to live, we must reflect on our values, consider the possible implications of our actions, and take active steps to realize these values. Bess warns us, however, that this will not be easy. Not only will it be difficult to convince others to change, it may be difficult for us as individuals to resist the lure of certain technologies, abilities, and enhancements. But Bess insists that we must put forth our best effort. He offers hope that even if we can't dictate the path of technological and social change, we can continue discussion, increase awareness of the issues and implications, and guide the dialogue to higher ground.

Some of the most important watersheds in human history have been associated with new applications of technology in everyday life: the shift from stone to metal tools, the transition from hunting and gathering to settled agriculture, the substitution of steam power for human and animal energy. Today we are in the early stages of an epochal shift that will prove as momentous as those other great transformations. This time around, however, the new techniques and technologies are not being applied to reinventing our tools, our methods of food production, our means of manufacturing. Rather, it is we ourselves who are being refashioned. We are applying our ingenuity to the challenge of redesigning our own physical and mental capabilities. Technologies

From *Technology and Culture* 49, no. 1 (2008): 114–126. Reprinted with permission.

of human enhancement are developing, ever more rapidly, along three major fronts: pharmaceuticals, prosthetics/informatics, and genetics.[1] Though advances in each of these three domains are generally distinct from those in the other two, their collective impact on human bodies and minds has already begun to manifest itself, raising profound questions about what it means to be human. Over the coming decades, these technologies will reach into our lives with increasing force. It is likely that they will shake the ethical and social foundations on which contemporary civilization rests.[2]

One fascinating feature of this phenomenon is how much it all sounds like science fiction. The bionic woman, the clone armies, the intelligent robot, the genetic mutant superhero: these images all form part of contemporary culture. And yet, this link with science fiction is potentially misleading. Precisely because we associate human enhancement with the often bizarre futuristic worlds of novels and movies, we tend to dismiss the evidence steadily accumulating around us. Technologies of human enhancement are incrementally becoming a reality in today's society, but we don't connect the dots. Each new breakthrough in genetics, robotics, prosthetics, neuroscience, nanotechnology, psychopharmacology, brain-machine interfaces, and similar fields is seen as an isolated, remarkable event occurring in an otherwise unaltered landscape. What we miss, with this fragmentary perspective, is the importance of all these developments, taken together.

The technological watersheds of the past came about gradually, building over centuries. People and social systems had time to adapt. Over time they developed new values, new norms and habits, to accommodate the transformed material conditions. This time around, however, the radical innovations are coming upon us suddenly, in a matter of decades. Contemporary society is unprepared for the dramatic and destabilizing changes it is about to experience, down this road on which it is already advancing at an accelerating pace.[3]

Let me begin with two brief stories.[4] They are, in a sense, Promethean parables, tales of the human aspiration to rise above earthly limits. But they are also anti-Promethean, in that both begin with tragedy and end on a cautiously hopeful note.

In 1997, a fifty-three-year-old man named Johnny Ray had a massive stroke while talking on the telephone. When he woke up several weeks later, he found himself in a condition so awful that most of us would have a hard time imagining it. It is called "locked-in" syndrome: you are still you, but you have lost all motor control over your body. You can hear and understand what people say around you, but you cannot respond. You have thoughts and feelings but cannot express them. You cannot scream in frustration or despair; you can only lie there. The only way Johnny Ray could communicate was by blinking his eyelids.

In March 1998 two neurologists at Emory University and Georgia Tech inserted a wireless implant into the motor cortex of Ray's brain. The implant transmitted electrical impulses from Ray's neurons to a nearby computer, which interpreted the patterns of brain activity and translated them into cursor movements on a video display. After several weeks of training, Ray was able to think "up" and thereby will the cursor to move upward onscreen. After several more months, he was able to manipulate the cursor with sufficient dexterity to type messages. By that point, the brain-computer interface had become so natural to him that using it seemed almost effortless. When the doctors asked him what it felt like to move the cursor, he spelled out, n-o-t-h-i-n-g. Johnny Ray had escaped from his terrible isolation and returned to the rich world of language.

My second story is about a girl named Ashanti DeSilva, born in 1985 with the genetic disorder known as "bubble boy disease." Her body lacked the gene required for making the protein adenosine deaminase, or ADA. Without it, her immune system was drastically impaired: just about any virus or bacteria she encountered threatened her life. She lived in total isolation at home, kept alive by injections of synthetic ADA. Her parents knew that the effectiveness of the injections would diminish over time, and that their daughter would eventually die of her disease. The only other alternative, a bone-marrow transplant, was impossible, because no compatible donor could be found.

Out of desperation, Ashanti's parents turned to what was at that time cutting-edge experimental medicine. In 1990, a team of doctors at the National Institutes of Health Clinical Center in Maryland extracted blood cells from her veins, then used a hollowed-out virus vector to insert working copies of the ADA gene into those blood cells. They were, in effect, repairing the deficient gene that had caused her disease. When the modified blood cells were injected back into Ashanti's body, the results were dramatic. Within six months her immune system became sufficiently active to allow her to go safely out of the house. Within two years she was enrolled in school and began for the first time to experience a normal childhood. Ashanti DeSilva is alive and healthy today, though she still requires periodic renewal of the gene therapy to boost her immune response. Hers is the first case of successful gene therapy on humans.

The stories of Johnny Ray and Ashanti DeSilva have two striking features in common. First, it is remarkable how far the science and technology have come in the short time since these pioneering feats took place. Johnny Ray's brain implant possessed only a single electrode for linking up with his nervous system. A mere eight years later, an owl monkey at Duke University was equipped with a similar implant containing seven hundred electrodes and a high-bandwidth interface. This far more powerful device allowed the monkey—staring at a video screen, arms dangling motionless at its side—to control the movements of a robot arm in another room by thought alone. The monkey played games using its new arm, and appeared to have seamlessly incorporated this machine appendage into the functioning of its own body. Human trials on this technology are already in the works; more than a dozen universities and private companies are currently in a race to push this line of research still further. Meanwhile, the progress in genetic technologies has been even more dramatic. In 1997 Dolly the sheep became the first successful clone of a mammal; in 2003 the Human Genome Project produced the first complete map of human genetic material; in 2004 there were 987 gene-therapy trials under way around the world.

A second common feature in these two stories is less obvious, but equally important. The same pathbreaking techniques that render healing possible usually also render enhancement possible. If I can place a brain implant in Johnny Ray to let him out of his "locked-in" world, I can also use a similar device, down the road, to let healthy people manipulate robotic arms by thought alone. If I can insert new genetic instructions into Ashanti's blood cells, making them produce ADA, then I can also use a similar procedure, down the road, to make other human cells produce other proteins of my own choosing. The technologies for repairing a malfunctioning human body are inseparable from the technologies that allow us to push human capabilities to ever higher levels. Where we can heal, we can also tweak, boost, reconfigure, redesign.[5]

The implications have not been lost on scientists and technology developers. Large numbers of them are busily at work today, in universities, government labs, and private companies, extending the biotechnologies of healing ever further into the domain of enhancement. They are, in effect, working to build a better human.

Three Major Areas of Enhancement

People are using pharmaceuticals in increasingly sophisticated and powerful ways to reshape their bodies and minds. I need not belabor the highly publicized rise of chemicals such as steroids, which enhance physical traits like speed, strength, and endurance, and have caused major upheavals in the world of competitive athletics. But the realms of human cognition, learning, and emotion are being shaken up in equally profound ways. Behavioral traits such as restlessness and short attention span, formerly viewed as problems of character and will power, are being medicalized, redefined as illnesses treatable with potent drugs like Ritalin. Conditions such as depression, which used to be approached through endless hours on the psychiatrist's couch, are increasingly being handled through the administration of an ever growing array of neurotransmitters, hormones, and other mood-altering chemicals.

In the process, our society's sense of what constitutes normal ability and basic mental well-being is being destabilized. As Carl Elliott describes it in his 2003 book *Better Than Well*, we are engaged today in a sort of chemical arms race, seeking to push our own physical and mental abilities to ever higher levels. When college students discovered, for example, that certain attention-deficit hyperactivity disorder (ADHD) drugs like Ritalin also enhance the cognitive performance of purportedly normal individuals, the outcome was thoroughly predictable. A black market rapidly developed among healthy students, many of whom reported that the drug helped them think more clearly, concentrate better, and remember new information more accurately than before. The motivation to enhance was strong, given the competitive nature of our educational system and broader society. Moreover, the line between healing and enhancing proved extremely difficult, in practice, to draw.

A second important area of human enhancement lies in the field of neuroscience and its intersection with the technologies of prosthetics, robotics, informatics, and artificial intelligence. As the story of Johnny Ray makes clear, the boundaries between human body and information-processing machine are beginning to blur. Ray became a kind of human-machine hybrid, in the sense that key aspects of his individuality—his ability to communicate in language with other people—came to be linked to the functioning of machine components that he had incorporated into his being. For now, such deliberate blurring of boundaries occurs only in animal experiments or extreme cases like Ray's. But over the coming decades it is likely that such human-machine hybrids will proliferate. We will have the ability to link directly into the human nervous system or sensorium with an increasingly broad array of electromechanical and informatic devices. Within thirty or forty years, some of the blind will see again, some of the deaf will hear again, some of the paralyzed will walk again.

Prosthetic technologies that already exist today are bringing such "futuristic" capabilities closer and closer to reality. In 2002, for example, the brain researcher William Dobelle created a media sensation by partially restoring sight to a totally blind patient. Dobelle implanted electrodes in the man's visual cortex and linked them through a portable computer to a tiny video camera mounted on the man's glasses. The result was grainy, blurred vision—but vision nonetheless. Dr. Dobelle's blind patients could see well enough to drive a car around a parking lot (slowly!) and carry out simple everyday tasks. Equally remarkable advances are taking place with cochlear and brain implants to restore hearing, and with prosthetic devices and neurosurgery to restore motility to paralyzed patients.

Here again, technologies of healing will be inseparable from technologies of enhancing. If I can put a functional artificial eye into a blind patient, then it is but a short step, technologically speaking, to add extra features to the implanted device, such as a telescopic lens or an infrared sensor. The result would be a formerly blind person who not only can see normally, but who can also zoom in clearly on very distant objects and see extremely well at night. She would see, in Carl Elliott's apt formulation, better than well. It would be remarkable, under such circumstances, if some people with normal vision did not hanker to have their own optical sensorium similarly tweaked, as long as the technology was safe and affordable.

These kinds of developments, not surprisingly, have elicited great interest—and significant funding—from the military. In the United States, the Defense Advanced Research Projects Agency (DARPA) envisions a battlefield of the not-so-distant future in which enhanced humans and potent machines are deeply interwoven at all levels. Imagine a soldier who can sprint at top speed for five miles with a heavy backpack, yet not get tired, because his blood has been modified to carry oxygen more efficiently. Imagine a soldier who can stay awake and alert for seventy-two hours because his nervous system has been augmented accordingly. Imagine a pilot who controls his aircraft directly through a brain-machine interface. Imagine a flexible, semi-intelligent armature, worn like an exoskeleton, that allows a soldier to lift 250 pounds effortlessly. Even if *you* are not imagining such things, DARPA is, and it is supplying considerable amounts of money to advance both the basic science and the practical technology for such capabilities.

Not all this research will bear fruit, of course. But that should not obscure the broader point. We are gaining an ever more sophisticated understanding of how the human brain works, how the nervous system and sensory organs function. We are building ever more powerful robotic and informatic devices. And, most significantly, we are getting better and better at linking these two realms, human and machine, and teaching them to work as one. Over the next few decades, these functional hybrids will become more and more a part of our lives.

Direct intervention at the level of the human genome is potentially the most powerful form of enhancement, because it can modify not just a single individual in the here and now, but entire lineages of humans down through the generations. No one knows for sure today how great a role genes play in making us who we are, how each of us is shaped by inherited genetic predispositions, and to what extent our personalities and capabilities are the result of nongenetic factors in our upbringing and life experience. But we do know a lot more than we did a mere ten or twenty years ago. Moreover, with the decoding of the entire human genome in 2003, we possess powerful tools for learning more quickly. Breakthroughs in genetics come almost every month.

Three basic principles undergird genetic intervention. First, some diseases are caused by malfunctions in a mere one or two genes. Fixing the gene removes the disease. Second, some intangible human traits, such as intelligence or shyness, are probably linked to complex systems of genes rather than to isolated genes. To adopt a musical metaphor, they depend not on single notes but on chords or even symphonies. Third, by altering individual components in certain systems of genes, we can directly affect complex and intangible traits in predictable ways.

In 1999, for example, a Princeton biologist, Joe Tsien, modified a single gene in laboratory mice that controls production of a chemical known as nerve growth factor (NGF). To his astonishment, the NGF-enhanced mice performed up to five times better than normal mice in tests of memory, learning, and intelligence. Other biologists, such as Eric Kandel and Tim Tully, have tinkered with a different gene, responsible for the production of a chemical that strengthens brain synapses. Through manipulation of a single gene, they have significantly boosted the learning abilities of mice, fruit flies, and sea slugs.

Let me be clear: this does not mean that genetic enhancement of human intelligence is just around the corner. But it does point to a conclusion that should get our full attention: genetic enhancement of basic human traits is no longer a topic of fantasy. The pieces of the scientific and technological puzzle are coming together, in real developments happening today. Neuroscience and psychology are telling us more and more about the electrochemical basis of how brains function and produce specific states of mind. Genetics is telling us more and more about how particular genes regulate the production of certain chemicals. Our technological ability to modify individual genes is growing rapidly.

Taken together, these elements form a recipe for powerful genetic interventions to redesign human bodies and minds. The time frame, depending on which expert you consult, is probably a matter of three to five decades. Within the lifetime of today's college seniors, our society is going to face some very tough choices about whether to use, and how to use, these extraordinary genetic powers.

Let us suppose, for the sake of argument, that a majority of U.S. citizens were to decide today that human enhancement is a bad road, and that we should refuse to go down it.[6] Could we stop this process? What would that entail?

The answer is sobering. Bringing the enterprise of human enhancement to a halt would require a vast, draconian system of surveillance and regulation. Precisely because the technologies of healing and the technologies of enhancement are intrinsically connected, a ban on enhancement would prove ineffectual unless it severely curtailed research in such areas as computers and informatics, genetics, robotics, neuroscience, nanotechnology, cognitive psychology, pharmaceuticals, and many fields of contemporary medicine. Moreover, it would require that this highly restrictive system be imposed with equal rigor in all the world's nations at the same time, to prevent the research and innovation from simply migrating overseas to the least-regulated regions of the planet. The chances of such a coordinated global relinquishment are small indeed.

Nevertheless, to admit that we cannot ban enhancement technologies outright is not to say that we are powerless to exert any control at all over the situation. I believe the history of the environmental movement offers a useful model in this regard. Here, too, humankind faced a similarly daunting challenge: reorienting the totality of our economic system by profoundly changing not only products, laws, and industrial practices but also consumer habits and mentalities. A mere forty-five years ago, in the early 1960s, almost no one even knew there *was* such a thing as "the environment," much less a serious set of problems associated with it. In the decades since, humankind has become aware of the crisis and mobilized to shift economies and habits toward ecological sustainability. We remain far indeed from reaching that goal, but it would be foolish to deny the substantial progress that has been made in less than half a century.[7]

The green movement is particularly instructive because it has made a significant impact despite vehement disagreement about the nature of the ecological crisis and the best ways to deal with it. There was never any point at which a straightforward consensus developed in the population at large; rather, the terms of debate gradually became clearer and more concrete. Can a factory run cleanly and still turn a profit? Can we find economically viable alternatives to fossil-fuel energy? Animated by pragmatic questions like these, public opinion slowly shifted, incrementally incorporating ideas that had once seemed marginal or downright radical. Ultimately, the overall trajectory of social practices was successfully deflected down a new course. Humanity saw the problem and partially changed direction.

There is hope in this green story. It offers reasonable grounds for optimism that humankind may be able, in a similar fashion, to exert some measure of control over the immense social and economic forces involved in human enhancement. We are not helpless before these technological changes; we can have some say in how they transform our lives. In particular, I see four main issues that these accelerating developments will compel us to confront.

The Challenges We Face

Human enhancement is going to be very hard to resist, once you and I personally are offered it. It not only taps into our instinct for self-preservation, but also draws upon our concern for those we love. Most parents go to extreme lengths to give their children the greatest possible chance of leading healthy, educated, fulfilling lives. If biotechnology safely increases that chance, or appears to do so, how many will be able to resist? This pressure will ratchet up even further as parents see their children competing with others whose capabilities have been augmented in various ways. We have here, in short, a classic case of a slippery slope.

Biotechnological modifications are likely to come in discrete, incremental packages, each offering a slight improvement in some aspect of our bodies or minds, along a steadily increasing gradient of potency and sophistication, as the science and technology advance each year. Over decades these increments will add up to significant qualitative changes in our physical and mental makeup, but at any given moment they will seem like small, sensible extensions of capabilities we already possess. The net result will be a social context in which the very meaning of the word "normal" is constantly shifting. What was normal last year becomes slightly subpar this year; what was normal ten years ago is completely obsolete today. Once enhancement technologies become widespread, people will have to accept a continual, unending process of upgrades and boosts, simply to keep up with the ever-shifting baseline of normal human performance.

A second challenge of the enhanced future will stem from the sheer outlandishness of the traits and capabilities that many citizens of that era will be able to choose for themselves. Popular science fiction has not been much help in this regard. From *Star Trek* to *Star Wars*, we see a lot of strange critters running around: intelligent robots, not-so-intelligent robots, bizarre species from galaxies far, far away. But these aliens exist alongside perfectly ordinary-looking human beings. For the most part, the only humans who are profoundly modified are the evil ones, like Darth Vader in *Star Wars* or the Borg in *Star Trek*.

This is a telling point: it indicates that we are psychologically unprepared for what is actually far more likely to happen. Over the coming century, some of us—perhaps many of us—will be increasingly merging with our machines, while at the same time modifying our own biology in ever deeper ways. By the year 2050, our society is likely to include a wide variety of truly hybrid beings, part genetically modified human, part machine. To be sure, some individuals, and some entire family lines, will no doubt follow a conservative path, rejecting major modifications. But others will push their enhancement possibilities to the limit. No one can foresee what those more aggressively modified people will look or act like. But it is probable that, from today's standpoint, they will be deeply unsettling to behold, both when they are at rest and when they do the things they do. Many of their behaviors will lie well beyond the range of current human capabilities.

A skeptic might argue that humans have been enhancing both their abilities and their appearances for centuries, if not millennia, and that most societies have adapted rapidly and seamlessly to such innovation. How long did it take, for example, for people to accept the wearing of eyeglasses as perfectly normal, or to consider an airplane flight across the Atlantic routine? Might not the myriad enhancements of the mid-twenty-first century find a similarly swift and easy embrace? What many would consider a freak today might seem utterly mundane tomorrow.

This is a valid point, but it should be qualified in two important respects. The enhancements of the mid-twenty-first century will be far more potent than anything witnessed thus far in human history. They will affect the qualities we deem most centrally and deeply human. Personality, emotions, cognitive ability, talents, memory, perception, physical sensation, the boundaries between one person and another—all these will be subject to deliberate manipulation. It is hard to see how such an unrelenting succession of profound changes would not produce a disorientation—a continually destabilized identity—among the citizenry of the coming era.

It is also worth noting how much of today's economic activity and technology are oriented toward fashion performance, entertainment, embellishment, or sexual behavior. What will be the equivalent of cosmetics, body piercings, tattoos, Botox, and Viagra for the people of 2050? What will take the place of recreational drugs like marijuana or Ecstasy? What will the people of that era do to their bodies and minds for the simple purpose of signaling their individuality, or experiencing a new form of erotic pleasure? The mind boggles. We should not underestimate the sheer strangeness of the future that awaits us, just a few decades down the road.

A third key issue confronting us in that future will be the socially disruptive potential of enhancement technologies. They hold out the possibility of liberating humans from many of their constraints and afflictions, but also of dividing humankind more profoundly than at any time in recorded history. It is not at all clear whether a population of highly enhanced humans can coexist peacefully alongside a population of unmodified humans. There will be plenty of opportunities for prejudice, resentment, and dehumanizing stereotypes—going both ways. Within the population of the enhanced, moreover, we are also likely to see ever growing levels of heterogeneity. The people of that era will not only look far more different from each other than we do today, they will also possess a much wider variety of physical and mental capabilities, arrayed in all manner of combinations. Diversity, in such a context, will be based on varying biologies, dissimilar machine components, sharply contrasting abilities. Can our culture absorb such a riotous level of heterogeneity? The historical track record in this regard is not very promising.

Finally, what happens to the moral ideals of equality and human dignity in such a world? By the concept of "human dignity" we usually mean a quality of intrinsic and absolute value that all humans possess in the same measure—whether a tiny baby, a genius, or a mentally handicapped person. It is the quality that leads us, for example, to consider murder an equally serious crime regardless of the victim's personal characteristics.

The technologies of enhancement threaten human dignity precisely because they tempt us to think of a person as an entity that can be "improved." To take this step is to break down human personhood into a series of quantifiable traits—resistance to disease, intelligence, and so forth—that are subject to augmentation or alteration. The danger in doing this lies in reducing individuals to the status of products, artifacts to be modified and reshaped according to our own preferences, like any other commodity. In this act, inevitably, we risk losing touch with the quality of intrinsic value that all humans share equally, no matter what their traits may be. In this sense, the well-intentioned effort to enhance a person can result in treating them as a mere *thing*.

The eugenics movements of the late nineteenth and early twentieth centuries showed us where such dehumanizing lines of thought can lead. One place they did lead was to Auschwitz. A central moral challenge of the coming decades will be to prevent the technologies of enhancement from eroding the foundations of equality and human dignity on which our political and social systems rest.

None of us has a crystal ball, of course. We cannot know with any precision what shape the civilization of enhancement will take. Nevertheless, the broad outlines of what is coming are already clear enough: enhancement of human bodies and minds will become a defining feature of our society, whether we as individuals approve of it or not. Even if this future takes twice as long as anticipated to arrive—say, eighty years instead of forty—we are still speaking here of one of the great disjunctions in human history.

To some, putting the matter this way may smack of technological determinism. Ultimately, they would say, it is humans who make technology, and therefore it is up to humans to choose whether or not to go down this road. This is undoubtedly true. I have argued, however, that the very nature of these technologies will make many people *want* to have them, and that the collective impact of those desires, expressed as social and economic forces, will exert tremendous pressure to bring these technologies into being. Short of a cataclysmic war or ecological collapse, and short of a radical transformation in the basic profile of human aspirations and needs, the overall "parallelogram of forces" (to use Engels's artful formulation) will propel industrial civilization down this path.

It will be up to us as citizens and as consumers, of course, to decide just how, and at what pace, and in what configurations and distributions, these enhancements enter our lives. That is most certainly not predetermined. Our responsibility as citizens, therefore, requires that we start preparing ourselves as best we can today to make those decisions.

This will demand a great deal from us as individuals. It will require that we educate ourselves about the underlying science and monitor closely the ongoing developments in the many areas of innovation I have described. Most importantly, it will require that we clarify in our own minds the basic social, political, and moral values we wish to defend during this period of swift technological change.

It will also demand a great deal from us as a society. Our government will need to address basic issues of safety, devising effective ways to regulate new enhancement technologies without stifling scientific innovation in the process. This will not be an easy balance to strike.[8] Equally important will be the question of fairness: ensuring that opportunities for enhancement do not become the exclusive prerogative of a select few. If we fail in this, and the rich gain preferential access to the most potent enhancements, we will witness a further widening of the already cruel gap that separates people into haves and have-nots. This time around, however, that gap will not merely be expressed outwardly in social status and power: it will be written in biology itself.

Finally, we will need to create a civic culture that can deal constructively with ever-deepening diversity among the citizenry. In the end, this will probably require nothing less than a new ethics of personhood—an expanded conception of human dignity, a more generous understanding of the word "us." I will need to be able to stand before you, acknowledge how radically different you are from me—in looks, perceptions, abilities—and still feel that, underneath it all, we are members of a common family of beings.

Safety, fairness, social solidarity: these are not new moral imperatives in the history of human society. But the advent of enhancement technologies casts ancient social and political challenges in a particularly urgent light. We face a situation akin to the one lamented by atomic scientists like Albert Einstein and Leo Szilard in the late 1940s, as they contemplated the predicament of a nuclear-armed humanity. The invention of these radical new weapons, as these scientists saw it, confronted humankind with a basic choice: either to find a way, once and for all, to resolve international conflicts through peaceful means, or to face eventual annihilation in a final world war.

Enhancement technologies may well force a similar moral reckoning over the coming decades. Either human beings will learn how to reconfigure their societies along more equitable and civically inclusive lines, or the dehumanizing tendencies, identity tensions, and centrifugal forces unleashed by these technologies will risk tearing their societies apart. As with nuclear weapons, these devices confront humankind with the fateful disparity that Einstein repeatedly underscored in the last years of his life: the gap between human power and human wisdom, between our extraordinary technological mastery and our still primitive capacity for just coexistence. The innovations have gone on accelerating in the years since Einstein's death, and today we encounter a paradox that might have astonished even him: it is not just our weaponry that threatens us, but our technologies of healing as well. Our inventions have reached a degree of such potency that, turned back upon humans themselves, even the most seemingly benign of them risk turning our world inside-out.

Notes

This essay was supported in part by the National Human Genome Research Institute's Program on Ethical, Legal, and Social Implications (grant no. RO3HG003298-01A1), the

College of Arts and Science at Vanderbilt University, and the endowment of the Chancellor's Chair in History at Vanderbilt.

1. The scholarly and popular literature on human biological enhancement is voluminous and growing rapidly. For a bibliography (presently comprising some 260 titles and counting), see my project website at http://www.vanderbilt.edu/historydept/michaelbess/Currentbookprojects (accessed 14 November 2007).

2. Some authors argue that the biotechnology revolution will be rivaled in its transformative social impact by concurrent revolutions in nanotechnology, informatics, and robotics. They maintain that, rather than think of the mid-twenty-first century as being defined primarily by biotechnology, we should conceive of it as a period of conjoined and intertwined innovations: the nano-bio-info-cogno (NBIC) era. This makes good sense at one level, because it is undoubtedly true that human enhancement technologies will be partly a result of, and deeply imbricated with, radical new capabilities in these other areas. But I believe the defining feature of the era will still be the technologies of human enhancement. It is one thing to alter the nature of the objects and devices with which we surround ourselves, and quite another to alter fundamentally our own bodies and minds. In one case we are reshaping our tools; in the other we are reshaping ourselves. Of the two, it is the latter change that cuts deeper qualitatively, and that will, I think, come to be seen as the more important transformation. See Mihail C. Roco and William Sims Bainbridge, eds., *Converging Technologies for Improving Human Performance: Nanotechnology, Biotechnology, Information Technology, and Cognitive Science* (Springer: Dordrecht, 2003); and Ray Kurzweil, *The Singularity Is Near: When Humans Transcend Biology* (Viking Adult: New York, 2005).

3. The discussion that follows focuses on the United States, but the gist of my argument applies to most other industrialized nations as well—despite the significant differences in culture, institutions, and traditions of science and technology policy that apply in individual countries. The importance of cross-national variation should not be underestimated, of course. For example, the average Western European citizen today faces a quite different array of health policy options than the average American, and over the coming decades this disparity will probably translate into similarly divergent possibilities for human enhancement. Nevertheless, such dissimilarities should not obscure the underlying fundamentals that all the industrialized nations share in common. A banking crisis or an oil spill might well produce different responses in France than in the United States, but the basic features of such a crisis—its causes and extent, and the range of possible solutions—will be determined by the many underlying social, technological, and economic factors that clearly characterize both nations. In a similar way, the basic tectonics of the coming civilization of human enhancement will possess fundamental common features across national boundaries, despite important differences from country to country.

4. For the illustrative anecdotes in this essay, I draw extensively from the excellent book by Ramez Naam, *More Than Human: Embracing the Promise of Biological Enhancement* (Broadway: New York, 2005).

5. This point is forcefully made in Joel Garreau, *Radical Evolution: The Promise and Peril of Enhancing Our Minds, Our Bodies—and What It Means to Be Human* (Doubleday: New York, 2004); and in James Hughes, *Citizen Cyborg: Why Democratic Societies Must Respond to the Redesigned Human of the Future* (Basic Books: Cambridge, Mass., 2004). Hughes is one of the

leading figures in the World Transhumanist Association, an international body devoted to promoting human enhancement; see http://www.transhumanism.org/index.php/WTA/index (accessed October 2007).

6. Three thoughtful advocates of this position (albeit from sharply differing ideological backgrounds) are: Bill McKibben, *Enough: Staying Human in an Engineered Age* (Times Books: New York, 2003); Francis Fukuyama, *Our Posthuman Future: Consequences of the Biotechnology Revolution* (Farrar, Straus, and Giroux: New York, 2002); and Leon Kass, *Life, Liberty, and the Defense of Dignity: The Challenge for Bioethics* (Encounter Books: San Francisco, 2002).

7. See Michael Bess, *The Light-Green Society: Ecology and Technological Modernity in France, 1960–2000* (University of Chicago Press: Chicago, 2003).

8. Both McKibben and Fukuyama propose concrete political and legislative measures that they believe could be taken today to bring the forces of enhancement under greater control. It remains unclear, however, whether the measures they advocate would suffice to rein in the broader social and economic processes propelling enhancement forward.

Index

Abbey, Edward, 580
Abortion, 58–59, 195–196, 243
Académie des Sciences, 169
Acheson-Lilienthal report, 82
Actor network theory, 151, 433
Advanced cell technology, 39
Aeolipile, 99
Ageism, 43–44, 46
AIDS, 524
Air bags, 94–95
Air Line Pilots Association (ALPA), 268
Akrich, Madeleine, 157, 434
Alexander, George, 580
All India People's Science Network, 504
Al Qaeda, 522, 524, 533
Alternating current, 144
Alvarez, Luis, 84
Alzheimer's disease, 44–45
American Academy of Arts & Sciences, 89
American College of Obstetricians, 246
American Physical Society, 327
Amish
 buggies, 306–307
 and electricity, 306, 310, 312, 314
 entrepreneurs, 312
 kitchens, 313
 milk regulations, 309–311
 Ordnung, 301–304, 308, 310–311, 313–314
 rumspringa, 309
 telephones, 313
Amniocentesis, 196, 243
Anomie, 102
Anthrax attacks, 534
Anthropocentrism, 580–581
Anthropometrics, 266
Anthropomorphism, 159–160, 163
Antibiotics, 81
Antiglobalization movement, 446, 453, 455

Anti-misting kerosene (AMK), 408
AOL, 190
Apgar test, 245
Aquaculture, 239
Archimedes, 155
Arc lights, 260
Argonne National Laboratories, 148
Aristotle, 84
Armstrong, Louis, 277
Arthur, IL, 311
Artificial insemination, 241–242
Artificial photosynthesis systems, 67
Asimov, Isaac, 74
Aspen Institute, 89
Aspin, Les, 269
Assemblers, 78–80
ATLAS particle detector, 357–359
Atom bomb, 81–82, 220
Attali, Jacques, 87–88
AT&T phone crash (1990), 528
Autarky, 147
Authoritarianism, 217–218
Automat, 236
Automated Robotic Crew Helper (ARCH), 229–230
Automation, 530–531, 534
Automobile crashes, 524–525
Automobiles, as system of production and use, 149
Autonomy of technology, 93
Ayatollah Khomeini, 531
Ayres, Russell W., 223

Babbage, Charles, 100
Backlighting, 261
Badische Anilin und Soda Fabrik (BASF), 146–147
Bakelite, 109
Baran, Paul, 11
Bardeen, John, 327

Barton, Amy, 215
Baruch, Bernard, 82
Baruch Plan, 82
Basel Convention on the Control of Transboundary Movements of Hazardous Wastes, 586
Basic research, 280–281, 287, 289
Bayh, Birch, 335
Bayh, Evan, 335
Bayh-Dole Act (1980), 338
Bel Geddes, Norman, 342
Berkeley Unix, 75
Berliner key, 172, 174
Bethe, Hans, 87
Bharat Gyan Vigyan Samithi (BGVS), 504, 508–510
Bhopal chemical spill (India), 586
Bicycle
 facile, 128
 high-wheeler, 126–127
 Lawson's bicyclette, 114, 121, 127
 "Penny farthing," 114
 Rudge Ordinary, 126
 safety, 121, 125
 tricycles, 118
Bijker, Wiebe, 111, 142
Bill of Rights (US), 185
bin Laden, Osama, 524
Binnig, Gerd, 327
Biocentrism, 580–581
Biological Weapons Convention (BWC) (1972), 86
Biometrics, 542–544, 549–550
Biotechnology, 490
Black box, 108
Blok, Vlaams, 42
Blow-by (of O-rings), 381–382, 384–385, 387–388
Body screening technologies, 551
Boeing, 148, 268
Boerner, Christopher, 590
Boisjoly, Roger, 370, 378, 381–382, 384, 386
Bond, Christopher, 336
Bönig, Jürgen, 111
Boorstin, Daniel, 210
Borg (from *Star Trek*), 72–73, 80, 446, 606
Born, Georgina, 432, 434

Bosch, Carl, 146
Boulton, Matthew, 100
Boyle, James, 183
Brand, Stewart, 446
Brattain, Walter, 327
Bread, 239
Brookhaven National Laboratories, 148, 522
Brundtland Commission on Environment & Development, 569
Bubble boy disease, 601
Buchanan, Jack, 380
Buckyballs, 333, 339
Bulger, Jamie (murder), 538–540, 544
Bulkhandlung, 586
Bullard, Bob, 582, 584, 587–589
Bulletin of the Atomic Scientists, 83
Bureaucracies, 102, 232
Bush, George W., 238, 338
Bush, Vannevar, 331
Butler, Samuel, 170

Caesarean section, 246
California NanoSystems Institute, 323
California Rural Legal Assistance, 215
California Waste Management Board, 587
Callicott, J. Baird, 580–581
Callon, Michel, 110, 143
Cancer Alley, Louisiana, 584–585
Canton, James, 68
Caro, Robert A., 212
Carpal tunnel syndrome, 397
Carson, Rachel, 575
Castells, Manuel, 460–461
Castro, Fidel, 43
Catastrophic failures, 527
Celluloid, 109
Center for Biological & Environmental Nanotechnology, 341
Center for Third World Organizing, 587
Central Electricity Generating Board (CEGB), 407–409, 414–415, 420
Centre d'Histoire des Sciences et des Techniques, 155, 162
Cerell Associates, 587
CERN (European Organization for Nuclear Research), 356, 361, 364–365
CFCs, xii

Chandler, Alfred D., 220–221
Chaos theory, 76
Chavis, Ben, 587
Chemical Weapons Convention (CWC) (1993), 86
Chicken factories, 240
Childbirth, 244–245
Chlorinated hydrocarbons, 585
Chorionic villus sampling (CVS), 243
Churchill, Winston, 85
Citicorp, 549
Civil disobedience, 579
Clarke, Arthur C., 84
Class breaks, 530–532, 534
Clinton, William J., 323, 335, 588
Clonaid, 242
Cloning, 51, 86
Closed-circuit television (CCTV), 538–541, 543, 547, 551
Closure, 127, 129
Cobalt bomb, 525
Collins, Harry, 111
Colvin, Vicki L., 341
Colwell, Rita, 335–336
Combined Insurance, 233
Commission for Racial Justice (CRJ) of the United Church of Christ, 590
Commons, 185
Complex systems, 526–527
Comprehensive Test Ban Treaty, 86
Computer calls, 234
Constant, Edward W., 129
Constitutionalism, 181, 183–185
Continental Card Services, 549
Convivia, 449–450
Core set, 112
Cotton spinning mills, 217
Counter culture, 575
Counterfeiting, 531
Credit card fraud, 526, 534
Crick, Francis, 327
Crossbow, 525
Cryptography, 531
Customer service, 234, 463
Cybernetics, 183
Cyberspace, 182–184, 190
Cyberterrorism, 532

Dabur Research Foundation, 476
Dalai Lama, 87–88
Dallas Housing Authority, 584
Darwin, Charles, 142
Data mining, 534
Davis, Gray, 337
Death, 247–248
Defense Advanced Research Projects Agency (DARPA), 603
Defense Advisory Committee on Women in the Service (DACOWITS), 271
De Lee, Joseph, 244–245
Delphic, 427
Dennet, Daniel, 53, 76
Design work, 366
DeSilva, Ashanti, 601
Deskilling, 157
Determinism
  social, 143–210
  soft, 102
  technological, 93–94, 97–98, 102–105, 107, 141–142, 144, 145, 148, 211, 275, 340, 441, 461, 545, 608
Deutsch, John, 272
Devall, Bill, 573
Diagnostic related groups (DRG), 231
Dickson, David, 575–576
Diesel engines, 312
Disruptive technology, 488
Dobelle, William, 603
Dolly the sheep, 39, 51, 242, 601
Doomsday Clock, 83
Door closer, 155–158
Dorn, Edwin, 271
Dosi, Giovanni, 109–110
Drexler, K. Eric, 78, 84, 328–329, 332–333, 335–336, 340–342
Dreyfuss, Henry, 342
Dunlop, John Boyd, 129
Dworkin, Ronald, 58
Dynamite, 525
Dyson, Freeman, 82–83
Dyson, George, 76

Earth First!, 579
Easterbrook, Gregg, 77
Eberstadt, Nicholas, 40–41

E-commerce, 67
Ectogenes, 53, 58
Ectogenesis, 51, 53, 59
Eigler, Donald, 328, 335
Einstein, Albert, 572, 609
Electric Bond & Share Company (EBASCO), 143–145
Electronic benefit transfer (EBT) systems, 548–549, 553
Electronic money, 531
Electron microscopy, 327, 333
Elliott, Carl, 602–603
Ellul, Jacques, 142
E-mail, 428, 531
Embryos, 59, 195
Emory University, 600
Empirical Programme of Relativism (EPOR), 111–112, 121, 123, 127, 130
Encryption, 67
Energy crops, 9–10
Engels, Friedrich, 217–219, 224, 608
Engineering science, 146
Enigma machine, 531
Environment, 572–573
Environmental injustice (EJ), 580, 584, 588, 591–593
Environmental movement, 573, 576–577, 581–582, 605
Environmental Protection Agency (EPA), 588
Episiotomy, 245
Ergonomics, 265
ETC Group, 491
Eugenics movement, 608
European Economic Community (EEC), 390
Euthanasia, 248
Eyeglasses, 607

Factor, Max, 261
Faraday, Michael, 412
Fast food restaurants, 228–230, 235–236
Federal Aviation Authority (FAA), 415
Female labor in the developing world, 470
Feminization
 and technology, 199
 labor, 462, 470
 of management, 465–467
 of service work, 463
Feschner's law, 170
Fetus, rights of, 58–59
Feynman, Richard, 78, 327, 330, 333, 335, 340, 370, 387
Film stock, 260
Finagle's law, 71
Ford, Henry, 147, 233
Fordism, 144
Foreman, Dave, 579–580
Foresight Institute, 79, 328
Foucault, Michel, 156, 390, 396, 548
Fourastié, Jean, 40
Fox, Robert, 153, 155
Franco, Francisco, 43
Frankenstein, 171
Freedom Party, 42
Free software movement, 454, 456
Freling, Bob, 7
Friedland, William, 215
Friedman, Emanuel, 245
Friedman, Thomas, 41, 534
Friends of the Earth, 577

Gaia hypothesis, 571
Galison, Peter, 365
Gammalin, 20, 586
Gelernter, David, 71
Gell-Mann, Murray, 76
Gendarme couché, 166
Gender, 196–198, 200–203, 242, 403
Gene therapy, 601
General Accounting Office (GAO), 589
General Electric, 143, 148
General Motors, 148
Genetically modified foods, 77–78, 522
Genetically modified organisms, 490
Genetic defects, 243
Genetic engineering, 10, 70, 73, 77–78, 84–85, 341, 521
Genetics, nanotechnology, and robotics (GNR), 73, 80, 85–87
Georgia Tech, 600
Geron Corporation, 39
Gerontology, 38, 40
Gibson, William, 183

Gilder, George, 69
Gilfillan, S. Colum, 102
Gingrich, Newton L., 332
Gleick, James, 282
Glick, Isaac, 306
Globalization, 285, 287, 445, 454–455, 471
Global warming, 566, 568, 570
Glover, Jonathan, 53
Gooding, David, 412
Gore, Albert, 329, 331
Granny pregnancies, 242
Gray goo problem, 79–80, 336
Greenpeace, 414–415, 420, 577
Groom (door), 155–157
Groves, General Leslie, 148
Guarante, Leonard, 39
Guatemala City, 520

Haber-Bosch, 146
Hackers (computer), 186, 465
Haider, Jörg, 42
Haldane, JBS, 53, 57
Halle aux Cuirs, La, 153–156, 158, 161–162
Hammer, Claus, 60
Ham radio, 74
Handley, Charles, 260
Hardin, Garrett, 580
Hardy, George, 371, 385–387
Hargrove, Gene, 581
Harper, Peter, 576
Harrison, Michelle, 244–245
Haseltine, William, 45
Hasslacher, Brosl, 75–76, 79
Have Spacesuit Will Travel, 74
Hayes, Denis, 210
Hayflick limit, 38–39
Heart transplants, 57
Heinlein, Robert, 74
Helper devices, 529
Helsinki Declaration, 55, 57
Hennion, Antoine, 170
Herbert, Frank, 72
Hero of Alexandria, 99–100
Hierarchical organization of work, 101, 221
High technology, 284, 452
Hilbert's Grand Challenges in Mathematics, 481

Hillis, Danny, 72, 76–77
Hinges (door), 154, 156, 158
Hiroshima, Japan, xi, 82, 87
History of technology, 108–109, 112
HIV/AIDS, 489
Holding Company Act of 1935, 145
Homeland Security, Department of, 544, 550, 552
Home pregnancy tests, 242, 244
Hôtel-Dieu de Beaune, 163
Household appliances, 469–470
Hu, Evelyn, 331
Hubble Space Telescope, 370
Human cloning, 77
Human engineering, 266
Human factors, 265, 267
Human genome project, 10, 244, 340, 601
Humphry, Derek, 248
Hurricane Katrina, 550
Husserl, Edmund, 211
Huxley, Aldous, 1
Hydraulic piston, 158
Hydroburst test, 374–375
Hydrogen bombs, 83

ICI (Imperial Chemical Industries), 408, 416
IG Farben, 146–147
"il buon vivere," 457
ILOVEYOU computer virus, 533
Imutran, 57
Industrial Revolution, 75
Information age, 67, 282, 459–460
Information technology (IT), 461–463
Innovation, 107, 276, 279, 287
Innovation, linear models of, 108, 319
Institute for Global Futures, 68
Intaglio currency printing presses, 531
Intel Corporation, 447–448
Intelligent machines, 76
Interagency Working Group on Nanoscience, Engineering & Technology (IWGN), 63, 65, 67, 332
Interchangeable parts, 103
International Network for Participatory Irrigation Management, 507
International Water Management Institute, 507

Internet, 11, 181
Interpretative flexibility, 107, 111, 121, 123, 125, 324
Intracytoplasmic sperm injection, 241
Intuition, 523
Invasive pneumococcal disease, 291
Invention, simultaneous, 99
In vitro fertilization (IVF), 51–54, 241–242
I, Robot, 74

James, William, 102
Japanese Technology Evaluation Center, 331
Jasanoff, Sheila, 298
Java, 75–76, 79, 87
Jini, 75, 79, 87
Johnston, Ron, 109–110
Joint Primary Aircraft Training System (JPATS), 267, 271–272
Jones, David E. H., 330
Joy, Bill, 336, 341, 448
Joy, K. J., 506, 508

Kaczynski, Theodore, 71–72, 449
Kalahari bushmen, 103
Kalil, Thomas, 333
Kandel, Eric, 604
Katsh, Ethan, 191
Kauffman, Stuart, 76, 80
Kennedy, John F., 369
Kerr-McGee, 585
Kevorkian, Jack, 46, 248
Kilminster, Joe, 387
Kindem, Gorham, 262
KinderCare, 230
Kirkwood, Tom, 39
Klieg eyes, 260
Knolls Atomic Power Laboratory, 148
Knowledge economy, 459, 461
Knowledge-enabled mass destruction (KMD), 73
Koppers Company, 583
Koppleman, Lee, 212
Koskela, Hille, 547
Kowinski, William Severini, 237
Kraybill, Donald, 306

Kroc, Ray, 229, 233
Kropotkin, Peter, 210
Kurzweil, Ray, 69, 71–73, 76–77, 83, 87–88, 448–449
Kuzma, Lynn M., 537

Lambert, Thomas, 590
Lane, Neal, 64, 332–333, 335–336, 339
Last mile, problem of, 11
Latour, Bruno, 143, 126, 135
Lawrence Livermore Laboratory, 148
Layton, Edwin T., Jr., 108
Lazonick, William, 111
Lécaille, Pascal, 357
Lega Lombarda, 42
Lenin, V. I., 142, 575
Leslie, John, 83
Leverage, 534–535
Libertarianism, 182, 187
Life expectancies, 37, 40
Lift-irrigation, 506–507
Lilienthal, David, 210
Locked-in syndrome, 600
Lockheed, 148
Long Island Railroad, 212
Long Now Foundation, 72
Los Alamos, 86, 148
Love Canal, New York, 583–584
Lovelock, James, 571
Lovins, Amory, 77, 78, 576
Lovins, Hunter, 77, 78
Luddites, 71, 75, 78
Lund, Robert, 386
Lux Research, 493
Lycurgus Cup, 325

Macbeth, Bill, 386
Machine gun, 525
Macintosh computer, 157
Mackay, Hugh, 426
Maddox, Tom, 184
Magnetic resonance imaging (MRI), 281
Maharashtra Krishna Valley Development Corporation (MKVDC), 502, 505–510
MAJC (Microprocessor Architecture for Java Computing), 76
Makeup, 260

Management & Administrative Computing (MAC), 427
Manhattan Project, 85, 87, 148, 522
Marshall, Sir Walter, 407–410, 413–415, 420
Marshall Center, Huntsville, AL (NASA), 374–376, 379–380
Martin, Paul, 481
Marx, Karl, 98, 101, 102, 142, 183, 218, 571
Masculinity, 198, 465, 468
Mason, Jerry, 386
Mass consumerism, 456
Mass production, 526
Maternal serum alpha-fetoprotein (MSAFP) testing, 243
Maxwell, James Clerk, 154
Maxwell's demon, 154
McAuliffe, Christa, 370–371
McCormick, Cyrus, 213
McCurdy, Howard, 326
McDonaldization, 227, 248
McDonnell Douglas, 148, 268
McGrath, Jim, 51
Megamachines, 143
Mehta, Michael, 490
Methanol, 147
Michael, Mike, 425–426
Michelangelo, 74
Microsoft, 184
Midwifery, 244
Mikulski, Barbara, 336–337
Military Standard 1472, 266
Mill, John Stuart, 186–187
Ministry of International Trade & Industry (Japan), 330
MIT (Massachusetts Institute of Technology), 148
Mitchell, William, 184, 190
Molecular electronics, 76, 79
Monkeywrenching, 579
Moore, Gordon, 447
Moore's Law, 76, 340, 447, 449
Moravec, Hans, 71–72, 76–77, 83, 448
Morris, William, 210
Morton Thiokol, 370–371, 373–379, 382–388
Moses, Robert, 212

Motorola, 535
Mulkay, Michael, 110, 125
Mulloy, Larry, 384–386
Mumford, Lewis, 143, 209, 217
Muon chambers, 358, 360–364
Muscle Shoals Dam, 147
Mussolini, Benito, 453

Nagasaki, Japan, xi, 82
National Front, 42
Nanotechnology, 64, 67–68, 70, 73, 78–79, 84, 323–330, 332–336, 338, 340–342, 486–487, 490–491
National Air & Space Administration (NASA), 148
National Institutes of Health (NIH), 481
National Institutes of Health (NIH) Clinical Center, 601
National Institute of Standards & Technology, 331
National Nanofabrication Facility, 331
National Nanotechnology Initiative (NNI), 325–326, 332–336, 338, 339–340, 342
National Science & Technology Council (NSTC), 65, 332
National Security Agency (NSA), 536, 552
National socialism, 81, 147
National Union of Iron Molders, 213
Navajo Nation, 585
Neofascism, 453
Neohumanism, 454
Nerve growth factor (NGF), 604
Networks in place, 426, 430–431
Network Society, 460–461
New left, 575
Nicquevert, Bertrand, 357, 364
Nietzsche, Frederik, 84
NIMBY (not in my back yard), 587
Nitrogen fixation, 146
Noble, David, 110, 216
Nock, Steven, 545
Nonhumans, 155–157
Nonlinear systems, 76
Nonsequential systems, 527
Norris, George, 147
Novartis International AG, 57
Nozick, Robert, 53

Nuclear, biological & chemical (NBC), 73, 80, 85, 87
Nuclear fuel flask, 409
Nuclear power, 222, 341, 521
Nye, David, 340
Nylon, 281

Office of Environmental Equity (OEE) (US), 588
Office of Management & Budget (OMB), 65
Office of Science & Technology Policy (OSTP), 65
Office of Technology Assessment, 540
Oliver, Patsy Ruth, 583–584, 588
Open Source software, 186, 454, 456
Oppenheimer, J. Robert, 81–83
Oracle, 427
Organophosphates, 585
O-rings, 369–384, 387–388
Orwell, George, 1
Ove Arup, 414
Ozanne, Robert, 213

Paclitaxel, 476
Panacea Biotec, 475
Parfit, Derek, 52
Parfit teletransporter, 52
Participatory resource mapping (PRM), 504–505, 509
Pasteur, Louis, 169
Patents, 287, 489–490, 493
Patil, R. K., 509
Peachey Church, 306
Pesticides, 585, 592
Petrini, Carlo, 449
Pfaffenberger, Bryan, 423
Pharmaceutical industry and the developing world, 489–490, 492
Phenol, 109
Photography, 257–259, 263
Pinch, Trevor, 142
Plato, 218–219, 221, 224
Plutarch, 155
Plutonium recycling, 223
Pneumatic tools, 312
Pollution, 569–570

Polychlorinated biphenyl (PCB), 584
Popular Mechanics, 67
Porter. *See* Door closer
Presidential Council of Advisors on Science & Technology (PCAST), 334
Prisoners, 234–235
Privacy International, 543
Progress, 275–277, 280, 282–283, 304
Project-affected peoples (PAPs), 502–503, 507
Pugwash conference, 89

Quality, 463
Quantum computers, 67
Quantum dots, 476, 489
Quasimodo, Salvatore, 454

Radical technology, 576
Ray, Johnny, 600–603
Ray, Leon, 375, 377–378
Reading machine for the blind, 69
Reed, Mark, 76
Regan, Tom, 581
Relevant social groups, 116, 118–119, 121
Research & development (R&D), 276, 285
Revolution of 1848 (France), 213
Rhetorical closure, 128
Rhodes, Richard, 82
Ricochet, 11
Riehl, Bill, 385
Risk, security, 519, 523–524
Risk management, 519–520
Ritalin, 602
Road bumper, 166
Robinson, Willis, 528
Robots, 69, 77, 84, 87, 229, 248
Rockwell International, 148
Roco, Mihail C., 330, 332–334, 336–337, 339–341
Rogers, William, 369
Rohrer, Heinrich, 327
Roosevelt, Teddy, 581
Rosen, Jeffrey, 551
Royal Institution (UK), 412
Rwanda, 526
Ryan, Chris, 577

Sagan, Carl, 81
Salamanca-Buentello, Fabio, 488–492
Satellites and the internet, 8
Scalpel, 245–246
Scarce, Rik, 581
School of Mines (École des Mines, Paris), 153, 158
Schumacher, E. F., 575
Schumer, Charles E., 537
Schweizer, Erhard, 328
Science shops, 499
Scientific management, 232
Scorecard.org, 553
Script kiddies, 532
Scripts, 425–426
Sea farms, 239–240
Searle, John, 69, 88
Seat belts, 151–152, 168
Self-acting mule, 111
Shapin, Steven, 412
Shapiro, Stuart, 434
Shockley, William, 327
Shramik Mukti Dal (SMD), 503–504, 506, 508
Shubik, Martin, 534
Sierra Club, 581, 589
Siegel, Richard, 331
Simons, Menno, 301
Singer, Peter, 59, 240–241
SIR2, 39
Slow cities movement, 450–454, 456
Slow food movement, 449–450, 452–453
Smalley, Richard E., 64, 333–335, 339–340
Smart bombs, 249
Smith, Nick, 333
Smoking, 188–189
Snow, C. P., 536
Social childbirth, 244
Social construction, 93–94, 107, 110–112, 123, 127, 130, 142, 144–145, 148
Social technology, 492
Society for Promoting Participative Ecosystem Management (SOPPECOM), 504–506, 508–510
Sociologism, 162
Sociology of technology, 109, 111, 130

Sociotechnical systems, xiii–xiv, 94, 144, 205
Software reliability, 89
Solar Electric Light Fund (SELF), 7–9
Solar energy, 7–9, 11, 219
Solid Rocket Boosters (SRB), 371, 373, 379–380
Solow, Robert M., 276
Solter, Davor, 51
South African Nanotechnology Initiative, 476
Soviet Union, 101
Space exploration, 341
SPARC, 76
Stabilization, 121, 127
Stage magic, 413
Stalin, Joseph, 524
Stanford University, 148
Stanley Exhibition of Cycles, 118, 123, 125
Star Trek, 74, 446, 606
Star Wars, 606
Stefik, Mark, 184
Stem cells, 39, 52, 57
Stephenson, Neal, 529
Stone, Irving, 74
Stone, W. Clement, 233
Stormer, Horst, 64
Strategic Defense Initiative (SDI), 84, 148
Suchman, Lucy, 427
Sung, Kim Il, 43
Sun Microsystems, 75, 336, 448
Superconductivity, 340
Supernotes, 531
Sustainability, 569
Sylvan Learning Center, 230
Synthetic gasoline, 147
Szilard, Leo, 609

Taylor, Frederick Winslow, 232–233
Taylor, Paul, 580
Technicolor, 262
Technological fixes, 576
Technological imperative, 445–446, 448–449
Technological momentum, 141–142, 145–146, 148–149

Technological systems, 144, 148–149
Technologies as forms of life, 216
Technologism, 162
Technology, 142, 227, 565
Technology and identity, 300
Technology transfer, 290
Telecosm conference, 69
Teledisc system of satellite communication, 11
Teledyn, 148
Telephone banking, 528, 536
Telephone call centers, 234–235
Teleworking, 468–469
Teller, Edward, 82
Telomerase, 39
Telomeres, 38–39
Tennessee Valley Authority, 147
Thalidomide, xii
Thinking Machines Corporation, 72
Thompson, Arnie, 378, 386
Thomson, Judith Jarvis, 53
Thoreau, Henry David, 85, 87
Tightly coupled systems, 528
Tire, pneumatic, 123, 125, 129
Tissue engineering, 57
Tomato harvester, 214–216
Tong, Rosemarie, 53
Toxic wastes, 586–587
Trade-offs, 537, 545, 547, 553–554
Transistors, lithographically drawn, 76
Transparency, 551–553
Trinity Test, 81–82, 85
TRW, 148
Tsien, Joe, 604
Tully, Tim, 604
Turing test, 166
Turkle, Sherry, 465

UAVs (Unmanned Aerial Vehicles), 532
Uchangi dam (India), 502, 504–506, 508
Ultrasound, 243
Unabomber (see Kaczynski, Theodore)
Union Carbide, 586
United Airlines, 234
United Church of Christ Commission for Racial Justice, 587
Universities Funding Council (UFC), 427

University of California, 214–215
UN Millennium Development Goals (MDGs), 480–481, 486, 490
USDA, 78
US Forest Service, 581
Utopia, 87–88, 342

Vaccines, 290–291
Valentine, Joseph, 261
Van Oost, Ellen, 111
Vatican Television, 238
Vaughan, Diane, 369, 371, 386
Vi (text editor), 75
Viagra, 241
Video surveillance, 540–541, 547
Villette, La, 153, 155, 162, 166–167, 170, 172
Vinge, Vernor, 184
von Braun, Wernher, 374
von Neumann probes, 83

Waddington's "chreod," 162
Walker, Joseph, 261
Wantagh Parkway, 212
Washing machines, 389, 391
Watson, James, 327
Watt, James, 100
Watson-Crick base-pairing, 80
Weapons of mass destruction (WMD), 73, 86–87, 526
Wear, Larry, 385
Weber, Max, 149
Webster, Juliet, 462
Wells, Deane, 59
Westinghouse, 390
West Nile virus, 524
WGBH Educational Foundation, 262
White, L. Meadows, 118
White, Lynn, 142
White Plague, 78
Wildson, James, 493
Wilson Dam, 147
Winner, Langdon, 158, 320
Winston, Brian, 261
Wirth, Niklaus, 527
Wittgenstein, Ludwig, 375
Wolfram, Stephen, 75

Women cyclists, 118
Woolgar, Steve, 434
World Bank, 507, 587
World Health Organization (WHO), 585
World Technology Evaluation Center (WTEC), 331
World Trade Organization (WTO), 489
Wurn technique, 241

Xenotransplantation, 51, 53–55, 57, 60

Young, Michael, 53
Yucca Mountain, 593

Zanussi, 390
Zapatistas, 455

Inside Technology
edited by Wiebe E. Bijker, W. Bernard Carlson, and Trevor Pinch

Janet Abbate, *Inventing the Internet*

Atsushi Akera; *Calculating a Natural World: Scientists, Engineers and Computers during the Rise of US Cold War Research*

Charles Bazerman, *The Languages of Edison's Light*

Marc Berg, *Rationalizing Medical Work: Decision-Support Techniques and Medical Practices*

Wiebe E. Bijker, *Of Bicycles, Bakelites, and Bulbs: Toward a Theory of Sociotechnical Change*

Wiebe E. Bijker and John Law, editors, *Shaping Technology/Building Society: Studies in Sociotechnical Change*

Stuart S. Blume, *Insight and Industry: On the Dynamics of Technological Change in Medicine*

Pablo J. Boczkowski, *Digitizing the News: Innovation in Online Newspapers*

Geoffrey C. Bowker, *Memory Practices in the Sciences*

Geoffrey C. Bowker, *Science on the Run: Information Management and Industrial Geophysics at Schlumberger, 1920–1940*

Geoffrey C. Bowker and Susan Leigh Star, *Sorting Things Out: Classification and Its Consequences*

Louis L. Bucciarelli, *Designing Engineers*

H. M. Collins, *Artificial Experts: Social Knowledge and Intelligent Machines*

Paul N. Edwards, *The Closed World: Computers and the Politics of Discourse in Cold War America*

Herbert Gottweis, *Governing Molecules: The Discursive Politics of Genetic Engineering in Europe and the United States*

Joshua M. Greenberg, *From Betamax to Blockbuster: Video Stores and the Invention of Movies on Video*

Kristen Haring, *Ham Radio's Technical Culture*

Gabrielle Hecht, *The Radiance of France: Nuclear Power and National Identity after World War II*

Kathryn Henderson, *On Line and On Paper: Visual Representations, Visual Culture, and Computer Graphics in Design Engineering*

Christopher R. Henke, *Cultivating Science, Harvesting Power: Science and Industrial Agriculture in California*

Christine Hine, *Systematics as Cyberscience: Computers, Change, and Continuity in Science*

Anique Hommels, *Unbuilding Cities: Obduracy in Urban Sociotechnical Change*

David Kaiser, editor, *Pedagogy and the Practice of Science: Historical and Contemporary Perspectives*

Peter Keating and Alberto Cambrosio, *Biomedical Platforms: Reproducing the Normal and the Pathological in Late-Twentieth-Century Medicine*

Eda Kranakis, *Constructing a Bridge: An Exploration of Engineering Culture, Design, and Research in Nineteenth-Century France and America*

Christophe Lécuyer, *Making Silicon Valley: Innovation and the Growth of High Tech, 1930–1970*

Pamela E. Mack, *Viewing the Earth: The Social Construction of the Landsat Satellite System*

Donald MacKenzie, *Inventing Accuracy: A Historical Sociology of Nuclear Missile Guidance*

Donald MacKenzie, *Knowing Machines: Essays on Technical Change*

Donald MacKenzie, *Mechanizing Proof: Computing, Risk, and Trust*

Donald MacKenzie, *An Engine, Not a Camera: How Financial Models Shape Markets*

Maggie Mort, *Building the Trident Network: A Study of the Enrollment of People, Knowledge, and Machines*

Peter D. Norton, *Fighting Traffic: The Dawn of the Motor Age in the American City*

Helga Nowotny, *Insatiable Curiosity: Innovation in a Fragile Future*

Nelly Oudshoorn and Trevor Pinch, editors, *How Users Matter: The Co-Construction of Users and Technology*

Shobita Parthasarathy, *Building Genetic Medicine: Breast Cancer, Technology, and the Comparative Politics of Health Care*

Paul Rosen, *Framing Production: Technology, Culture, and Change in the British Bicycle Industry*

Susanne K. Schmidt and Raymund Werle, *Coordinating Technology: Studies in the International Standardization of Telecommunications*

Wesley Shrum, Joel Genuth, and Ivan Chompalov, *Structures of Scientific Collaboration*

Charis Thompson, *Making Parents: The Ontological Choreography of Reproductive Technology*

Dominique Vinck, editor, *Everyday Engineering: An Ethnography of Design and Innovation*